Current Topics in Microbiology and Immunology

Volume 372

For further volumes:
http://www.springer.com/series/82

Larry J. Anderson · Barney S. Graham
Editors

Challenges and Opportunities for Respiratory Syncytial Virus Vaccines

Responsible series editor: Richard W. Compans

Springer

Editors
Larry J. Anderson
Department of Pediatrics and Children's Healthcare of Atlanta
Emory University
Atlanta, GA
USA

Barney S. Graham
Vaccine Research Center
National Institute of Allergy and Infectious Diseases
National Institutes of Health
Bethesda, MD
USA

ISSN 0070-217X ISSN 2196-9965 (electronic)
ISBN 978-3-642-38918-4 ISBN 978-3-642-38919-1 (eBook)
DOI 10.1007/978-3-642-38919-1
Springer Heidelberg New York Dordrecht London

Library of Congress Control Number: 2013955565

Printed on acid-free paper

Springer is part of Springer Science+Business Media (www.springer.com)

Preface

Respiratory syncytial virus (RSV) was first isolated from chimpanzees in 1955, and shortly thereafter from young children, and recognized as an important cause of serious lower respiratory tract disease in infants and young children, i.e., pneumonia and bronchiolitis. Shortly after its discovery efforts to develop a vaccine began. It is now over 50 years since its discovery and no safe and effective vaccine is yet available. The fact that the peak of disease occurs in young infants despite the presence of maternally acquired antibody and that humans have multiple disease-associated infections througout life, foretell the challenge in inducing protective immunity with a vaccine. The first RSV vaccine, formalin-inactivated tissue culture grown virus (FI-RSV) formulated with alum, was ineffective, associated with enhanced respiratory disease, and raised concerns that other vaccines might also predispose to FI-RSV vaccine enhanced respiratory disease. Older children did not experience FI-RSV vaccine enhanced respiratory disease suggesting that prior infection establishes a safe immune response pattern. Therefore, development of RSV vaccines for the RSV naïve infant has focused on live-attenuated RSV or other live viruses expressing RSV genes, while other types of vaccines are being developed for older children and adults. Subsequent to the FI-RSV vaccine trial, a number of live-attenuated RSV strains, live virus and other gene expression systems, and protein subunit vaccines in different platforms have been developed and tested in animals and a few have also being studied in humans. No vaccine has yet shown sufficient promise to move toward licensure. However, the prospects for many candidate vaccines, as well as different vaccine platforms, will remain unknown until clinical trials in the selected target population for the vaccine are performed. Though the lack of success to date highlights the biological difficulties in developing an RSV vaccine, the efficacy of immune prophylaxis suggests a safe and effective vaccine is achievable. The availability of ever more powerful tools to study the immune response and pathogenesis of disease and ability to construct a wide variety of vaccines using different vaccine platforms suggest that an RSV vaccine should be within reach. Caren Hall who made so many contributions to our understanding of RSV wisely noted in a poem one of the ongoing challenges to achieving an RSV vaccine (Anderson and Heilman 1995).

Immunity and RSV
What is this thing we call immunity?
Does it exist for ills from RSV?
Perhaps for mice within their splenic soul,
But in the babe, the old, what is its role?
Is there a pattern of response we can discern?
Or is it like night skies that change with each earth's turn,
With season, angle viewed, and light years passed?
Has each from different, fluid molds been cast?
Thus, is it only solitary stars we see.
And not a constellation called immunity?

Caroline Breese Hall

We feel this book brings together in one place what we know about RSV and helps to organize the constellation of facts about the virus and host factors that can guide successful development of an RSV vaccine. We feel that a better understanding of the of clinical and epidemiologic features of infection, functional and structural features of the virus, pathogenesis of the associated disease, and the immune response it induces provide the underpinnings for success.

Finally, we mourn the passing of Caren Hall. She has been the queen of RSV and authored most of the seminal studies on the clinical and epidemiologic features of infection and transmission of the virus. Equally important, she has been

Caroline Breese Hall (center) teaching and mentoring

a wonderful friend to so many in the field and an example to all of an outstanding scientist with a kind and nurturing spirit. In recognition of what she accomplished and the person she was, we dedicate this book to her. We refer the reader to the obituaries published in the Journal of Pediatric Infectious Disease Society in May 2013 Plotkin et al. (2013); Caserta and Long (2013); Englund et al. (2013). These three tributes provide a glimpse into why we felt so honored to have known and worked with her.

Larry J. Anderson
Barney S. Graham

References

Anderson LJ, Heilman CA (1995) Protective and disease-enhancing immune responses to respiratory syncytial virus. J Infect Dis 171:1–7
Caserta M, Long C (2013) Caroline Breese Hall Obituary. J Ped Inf Dis Soc 2:2–3
Englund J, Karron R, Dennehy P, Kimberlin D (2013) Caroline Breese Hall Obituary. J Ped Inf Dis Soc 2:3–4
Plotkin SA, Peter G, Katz (2013) Tribute to Caroline Hall. J Ped Inf Dis Soc 2:1–2

Contents

Part I
Virology, Pathogenesis, and Clinical and Epidemiological Features of Disease

Respiratory Syncytial Virus: Virology, Reverse Genetics, and Pathogenesis of Disease

Peter L. Collins, Rachel Fearns and Barney S. Graham

Abstract Human respiratory syncytial virus (RSV) is an enveloped, nonsegmented negative-strand RNA virus of family *Paramyxoviridae*. RSV is the most complex member of the family in terms of the number of genes and proteins. It is also relatively divergent and distinct from the prototype members of the family. In the past 30 years, we have seen a tremendous increase in our understanding of the molecular biology of RSV based on a succession of advances involving molecular cloning, reverse genetics, and detailed studies of protein function and structure. Much remains to be learned. RSV disease is complex and variable, and the host and viral factors that determine tropism and disease are poorly understood. RSV is notable for a historic vaccine failure in the 1960s involving a formalin-inactivated vaccine that primed for enhanced disease in RSV naïve recipients. Live vaccine candidates have been shown to be free of this complication. However, development of subunit or other protein-based vaccines for pediatric use is hampered by the possibility of enhanced disease and the difficulty of reliably demonstrating its absence in preclinical studies.

P. L. Collins (✉)
Laboratory of Infectious Diseases, National Institute of Allergy and Infectious Diseases, National Institutes of Health, Building 50, Room 6517, 50 South Dr MSC 8007, Bethesda, MD 20892-8007, USA
e-mail: pcollins@niaid.nih.gov

R. Fearns
Department of Microbiology, Boston University School of Medicine, Boston, MA 02118, USA

B. S. Graham
Vaccine Research Center, National Institute of Allergy and Infectious Diseases, National Institutes of Health, Bethesda, MD 20892, USA

L. J. Anderson and B. S. Graham (eds.), *Challenges and Opportunities for Respiratory Syncytial Virus Vaccines*, Current Topics in Microbiology and Immunology 372,
DOI: 10.1007/978-3-642-38919-1_1, © Springer-Verlag Berlin Heidelberg 2013

Contents

1 Introduction

Respiratory syncytial virus (RSV) was first isolated in 1955, but its biochemical and molecular characterization remained rudimentary for many years due to its relatively inefficient growth in cell culture, pleomorphic and cell-associated nature, and physical instability. Detailed characterization began in 1981 with the molecular cloning and sequencing of RSV RNA. The resulting molecular tools have greatly increased our understanding of RSV and revolutionized research towards treatment and prevention. For example, candidate live-attenuated RSV vaccine viruses designed by reverse genetics are presently in clinical studies (see chapters by H.Y. Chu and J.A. Englund, and by R.A. Karron et al., this volume). Vectored vaccines and recombinantly expressed viral antigen vaccines also are under development (see chapters by T.G. Morrison and E.E. Walsh, and by R.J. Loomis and P.R. Johnson, this volume). The murine monoclonal antibody that was the basis for palivizumab, used clinically for passive immunoprophylaxis in high-risk infants and children, was produced using recombinantly expressed antigen.

2 Classification

RSV is the type species of Genus *Pneumovirus*, Subfamily *Pneumovirinae*, Family *Paramyxoviridae*, Order *Mononegavirales*. Human RSV exists as two antigenic subgroups, A and B, that exhibit genome-wide sequence divergence (Table 1). The other members of this genus are bovine RSV (BRSV), ovine RSV(ORSV), and pneumonia virus of mice (PVM) (see Table 1 for amino acid sequence

Table 1 Amino acid sequence identity between the proteins of RSV subgroup A (RSV-A) and the indicated members of subfamily *Pneumovirinae*[a]

Viruses compared		Amino acid sequence identity for the indicated protein (%)										
		NS1	NS2	N	P	M	SH	G	F	M2-1	M2-2	L
RSV-A versus	RSV-B	87	92	96	91	91	76	53	89	92	72	93
	BRSV	69	84	93	81	89	38	30	81	80	42	84
	PVM	16	20	60	33	42	23	12	43	43	10	53
	HMPV-A	–[b]	–[b]	42	35	38	23	15	33	36	17	45
	AMPV-A	–[b]	–[b]	41	32	38	19	16	35	37	12	43

[a] Viruses are listed in order of decreasing relatedness to HRSV-A. RSV-B is RSV subgroup B; HMPV-A and AMPV-A are subgroup A of human and avaian human metapneumovirus. Viruses in *Paramyxovirinae* are not shown for comparison because the percent identity is <20 % at the level of the entire protein

[b] This virus does not have this gene

Table 2 Notable features of RSV

Replication and budding in vitro are inefficient, infectivity is unstable, particles grown in vitro are mostly large filaments
RSV encodes additional proteins that are either unique to the genus (NS1 and NS2) or found only in a subset of viruses in *Paramyxoviridae* (SH, M2-1, and M2-2)
Two genes, NS1 and NS2, are dedicated to expressing proteins that interfere with the host type I interferon system, among other functions
Overlapping ORFs in the M2 mRNA encode factors that confer transcription processivity (M2-1) or shift viral RNA synthesis from transcription to RNA replication (M2-2)
The M2 and L genes overlap and L mRNA is expressed by a backtracking mechanism
The small hydrophobic (SH) protein forms a pentameric ion channel, but its function is unclear
The F protein precursor is activated by cleavage at two furin recognition sites
The F protein activates TLR-4 signaling pathways, but this is inhibited by the G protein
Viral attachment appears to involve both the F and G proteins, but F fuses independently of G
The G protein is heavily glycosylated, nonglobular, and highly variable
The G protein bears a CX3C fractalkine-like motif that may modify the cellular immune response
The G protein is expressed in membrane-bound and secreted forms; the latter interferes with antibody-mediated neutralization, and interacts directly with antigen presenting cells to modify their function

relationships). More pneumoviruses remain to be identified: recent wide-ranging fieldwork provided sequence evidence of RSV-like viruses in African bats (Drexler et al. 2012). Subfamily *Pneumovirinae* contains a second genus, *Metapneumovirus*, which consists of human and avian metapneumoviruses (HMPV and AMPV). The other subfamily of Family *Paramyxoviridae*, *Paramyxovirinae*, includes animal and human parainfluenza viruses (PIVs), mumps, measles, Nipah and Hendra viruses, and numerous other viruses whose number continues to expand (Drexler et al. 2012). Some of the notable features of RSV are summarized in Table 2.

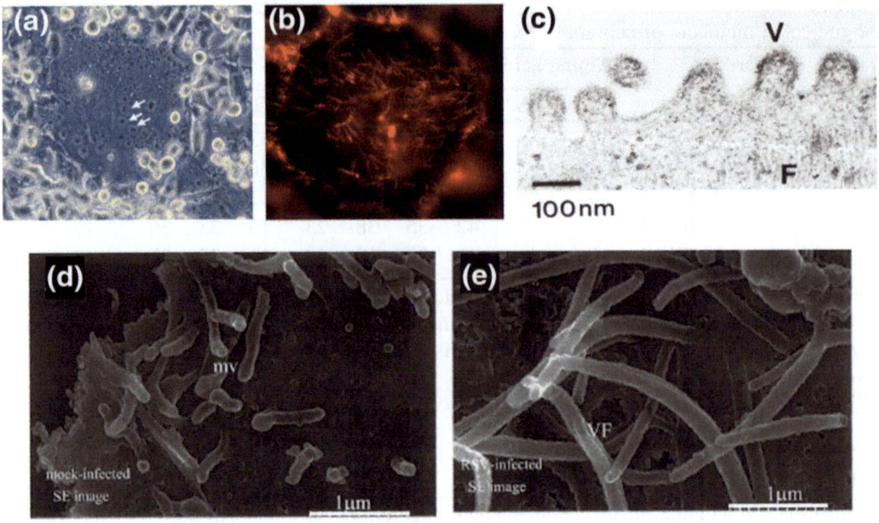

Fig. 1 Photomicrographs (**a** and **b**) and electron micrographs (**c**–**e**) of RSV-infected cells and associated viral structures. **a** is a photomicrograph of a syncytium in an RSV-infected cell monolayer (several nuclei are indicated with *arrows*; courtesy of Dr. Alexander Bukreyev). **b** is a fluorescence photomicrograph of a syncytium in an RSV-infected cell monolayer (not the same one as in **a**) stained with an antibody specific to the F protein, showing filamentous viral projections (courtesy of Dr. Ursula J. Buchholz). **c** is an electron micrograph of negatively stained budding RSV virions: V indicates a budding virion and F indicates filamentous cytoplasmic structures that likely are nucleocapsids (courtesy of Dr. Robert M. Chanock) (Kalica et al. 1973). **d** and **e** are field emission scanning electron micrographs of the surface of uninfected (**d**) and RSV-infected (**e**) cells, illustrating viral filamentous structures (VF in **e**) that are thought to form at sites of virus budding and may yield filamentous particles; also shown are microvilli (mv in **d**) that are found in uninfected cells (courtesy of Dr. Richard Sugrue) (Jeffree et al. 2003)

3 Virion

The RSV virion consists of a nucleocapsid packaged in a lipid envelope derived from the host cell plasma membrane (Figs. 1, 2). Virions produced in cell culture consist of spherical particles of 100–350 nm in diameter and long filaments that usually predominate and are 60–200 nm in diameter and up to 10 μm in length (Jeffree et al. 2003) (Fig. 1). In vitro, 95 % of progeny virus remains associated with the cell surface as particles that seemingly have failed to fully bud. In preparing virus stocks, infected cells typically are subjected to freeze-thawing, sonication, or vortexing to release attached virus, although this reduces infectivity and increases cellular contamination. RSV readily loses infectivity during handling and freeze-thawing due to particle instability and aggregation, although this can be partly overcome by excipients such as sucrose (Ausar et al. 2007). There is indirect evidence that the surface glycoproteins, especially F, are factors in the instability (Sastre et al. 2007; Rawling et al. 2011). The long filamentous shape of the particle likely also confers fragility.

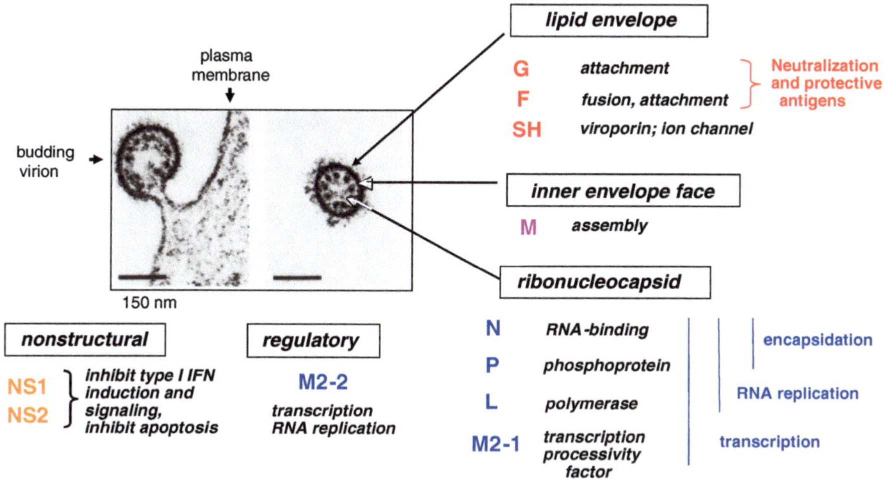

Fig. 2 RSV proteins and their functions and location in the virion, shown in reference strained electron micrographs of negatively stained budding (**a**) and free (**b**) virions (courtesy of Dr. Robert M. Chanock)

The RSV envelope contains three viral transmembrane surface glycoproteins: the large glycoprotein G, the fusion protein F, and the small hydrophobic SH protein (Fig. 2). The nonglycosylated matrix M protein is present on the inner face of the envelope. The viral glycoproteins form separate homo-oligomers that appear as short (11–16 nm) surface spikes. RSV lacks neuraminidase or hemagglutinin activity and the F is known to be heavily sialylated, presumably because of the lack of a neuraminidase. There are four nucleocapsid/polymerase proteins: the nucleoprotein N, the phosphoprotein P, the transcription processivity factor M2-1, and the large polymerase subunit L (Fig. 2).

4 RNAs

The RSV genome (Fig. 3) is a single-stranded nonsegmented negative-sense RNA of 15,191–15,226 nt for six sequenced strains (subgroup A strain A2, 15,222 nt, GenBank accession number M74568, is the reference strain). RNA replication involves a complementary copy of the genome called the antigenome (Fig. 4). The genome and antigenome lack 5′ caps or 3′ polyA tails. The first 24–26 nt at the 3′ ends of the genome and antigenome have 88 % sequence identity (Fig. 3b), representing conserved promoter elements that will be described later. The genome and antigenome are bound separately for their entire length by the N protein to form stable nucleocapsids. These are the templates for RNA synthesis and remain intact throughout the replicative cycle and in the virion. In addition, encapsidation

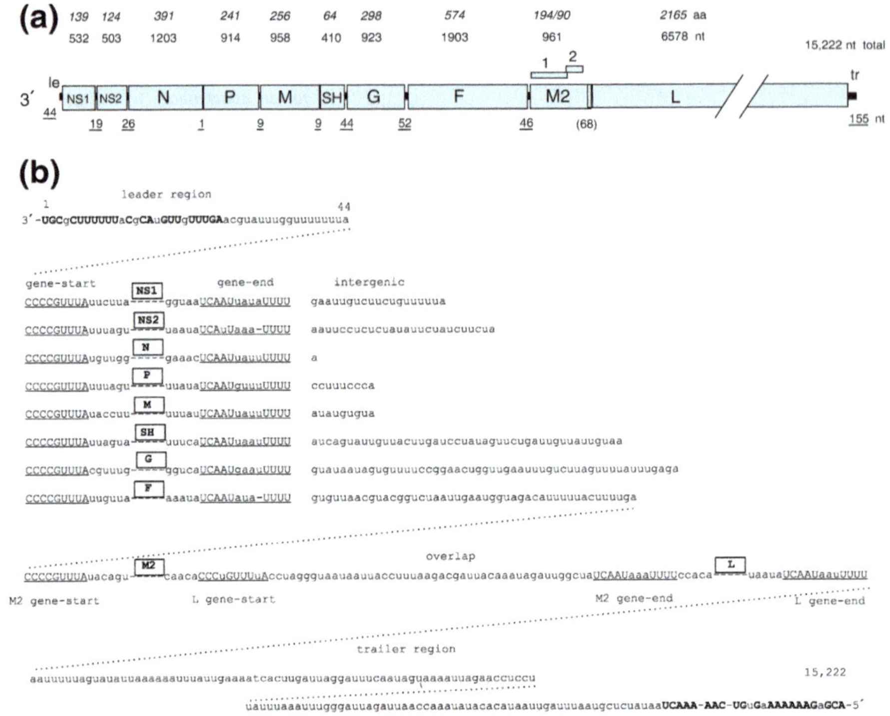

Fig. 3 Diagram of the 3′ to 5′ negative-sense RSV genome (approximately to scale, strain A2), and sequences of the leader region, gene junctions, overlap, and trailer region. **a** Genome diagram: each *box* represents a gene encoding a separate mRNA. The *first row* of numbers over the diagram indicates the nucleotide lengths of the genes, and the second, *upper row* of numbers (*italicized*) indicates the amino acid lengths of the primary, unmodified proteins. The overlapping ORFs of the M2 mRNA are illustrated over the M2 gene. The numbers under the diagram indicate the lengths of the leader, intergenic, and trailer regions (*underlined*) and the gene overlap (*parentheses*). **b** Sequences of the leader region, gene junctions, overlap, and trailer region (3′ to 5′, negative-sense). This shows the leader region, followed by the NS1, NS2, N, P, M, SH, G, and F genes and their intergenic regions, followed by the overlapping M2 and L genes and the trailer region. The main body of each gene is deleted and is represented by a *box* with the gene name. Nucleotide assignments that are conserved between the 3′ ends of the genome and the antigenome (represented here as the reverse complement in the trailer region) are in *bold capitals*. The gene-start and gene-end signals of the gene are *underlined*, and conserved asignments are in *capitals*

protects the RNA from degradation and shields it from recognition by host cell pattern recognition receptors that initiate innate immune responses.

The genome has 10 genes in the order 3′ NS1-NS2-N-P-M-SH-G-F-M2-L (Fig. 3). Each gene encodes a corresponding mRNA (Fig. 4) (Collins et al. 1986). The mRNAs have methylated 5′ caps and 3′ polyA tails. Each mRNA encodes a single major protein except for M2, which has two separate ORFs that overlap slightly and encode the M2-1 and M2-2 proteins. The downstream M2-2 ORF is

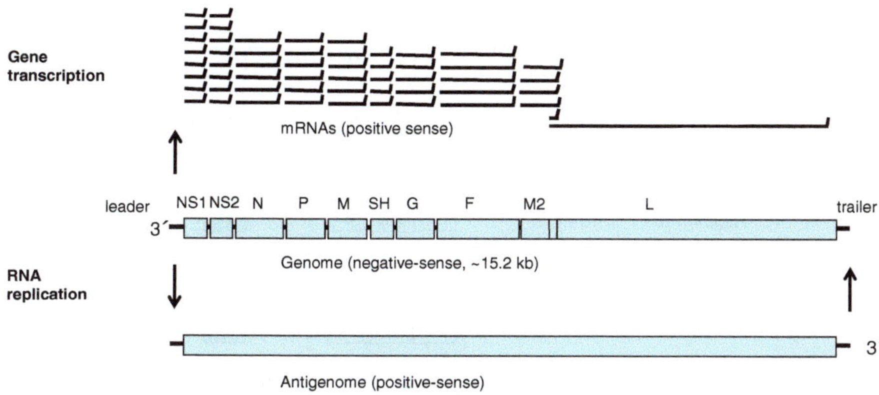

Fig. 4 Overview of RSV transcription and RNA replication. The polymerase enters the negative-sense genome at its 3′ end executes transcription to yield positive-sense subgenomic mRNAs (in a polar gradient) or executes the first step in RNA replication to yield full-length positive-sense antigenome. The polymerase enters the antigenome at its 3′ end and executes the second step of RNA replication to yield full-length progeny genomes. Note that the L gene yields two polyadenylated mRNAs: a very short species due to termination in the gene overlap, and full-length L mRNA

accessed by ribosomes that exit the M2-1 ORF and reinitiate, a process that is influenced by upstream structure in the M2 mRNA (Gould and Easton 2007).

The 3′ end of the genome consists of a 44-nt extragenic leader region that precedes the NS1 gene. The 5′ end of the genome consists of a 155-nt extragenic trailer region that follows the L gene (Fig. 3). Each gene begins with a highly conserved 9-nt gene-start (GS) signal and terminates with a moderately conserved 12–14-nt gene-end (GE) signal that ends with 4–7 U residues (genome-sense) that encode the polyA tail by polymerase stuttering (Fig. 3b). The first nine genes are separated by intergenic regions that vary in length from 1 to 58 nt for the strains sequenced to date. These lack any conserved motifs, are poorly conserved between strains, and appear to be unimportant spacers, except that at some gene junctions the first nucleotide of the intergenic region is important for mRNA termination (Bukreyev et al. 2000; Harmon and Wertz 2002). A tolerance for intergenic variability is illustrated by the finding that incrementally increasing the length of an intergenic region in recombinant RSV up to 160 nt had little effect on gene expression or viral replication in vitro; however, this was moderately attenuating in mice, indicating that excessive length is restrictive (Bukreyev et al. 2000). The last two genes, M2 and L, overlap by 68 nt: specifically, the L GS signal is located 68 nt upstream of the end of the M2 gene (Collins et al. 1987) (Fig. 3b). The same overlap occurs in BRSV, and gene overlaps occur for some genes in some members of *Rhabdoviridae* and *Filoviridae*.

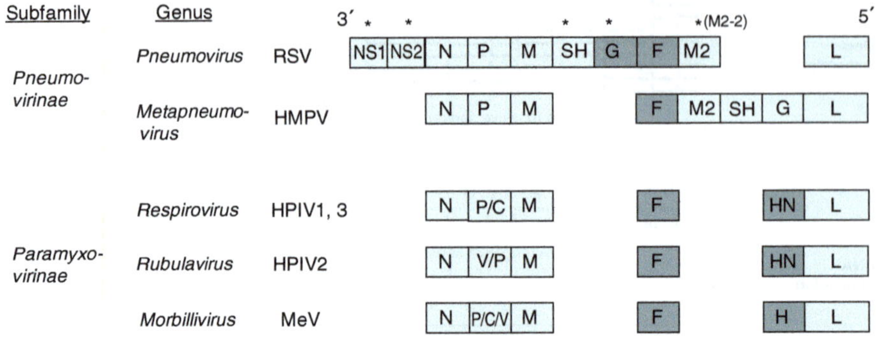

Fig. 5 Comparison of the genes and gene order of RSV with those of selected members of *Paramyxoviridae*: HMPV, human parainfluenza virus (HPIV) serotypes 1, 2, and 3, and measles virus (MeV). The genes are shown in their 3′ to 5′ order in genomic RNA, which is the direction of transcription. Genes are not to scale, and orthologous genes are aligned vertically as much as possible (the only genes that could not be appropriately aligned are SH and G of RSV and SH of HMPV), with gaps introduced to maximize the alignments. Genes encoding major protective antigens are in *dark shading*. *Asterisks* indicate proteins that can be deleted from RSV without loss of replication, although this may be reduced. Proteins that have no direct ortholog in RSV include: *C* small accessory protein, *V* cysteine-rich accessory protein, *HN* hemagglutinin-neuraminidase glycoprotein, *H* hemagglutinin glycoprotein. The *Henipavirus* and *Avulavirus* genera and a number of unclassified viruses within *Paramyxovirinae* are not represented

5 Proteins

RSV encodes 11 separate proteins, and thus is more complex than most members of *Paramyxovirinae*, which typically have 6–7 mRNAs encoding 7–9 separate proteins (Fig. 5). The N, P, M, F, and L proteins of RSV have clear orthologs throughout *Paramyxoviridae*, and their relative genome order is conserved (Fig. 5). Amino acid sequence relatedness between RSV and *Paramyxovirinae* is low and is evident primarily for the F and L proteins and segments in the *C*-terminal region of N. The NS1, NS2, M2-1, and M2-2 proteins of RSV have no counterparts in *Paramyxovirinae*, and an SH protein (which is found in all members of *Pneumovirinae*) is present only in a few members of *Paramyxovirinae*.

The RSV F and G glycoproteins are the only viral neutralization antigens and are the major protective antigens. These proteins are reviewed elsewhere (see chapter by J.S. McLellan et al., this volume) and will only be briefly described here. The 574-amino acid F protein directs viral penetration and syncytium formation, like a typical F protein of *Paramyxoviridae*. The RSV F protein has general structural similarity to the F proteins of *Paramyxovirinae*, and similarly is synthesized as an inactive F0 precursor that is activated by cellular endoprotease to yield two disulfide-linked subunits, NH2-F2–F1-COOH. However, RSV F has two, rather than one, cleavage sites (Gonzalez-Reyes et al. 2001): one site (KKRKRR↓F-137) corresponds to that found in other paramyxoviruses, and the second site (RARR↓E-110) is located 27 amino acids upstream. RSV F0 is readily cleaved intracellularly by furin-like protease and is not a limiting factor for viral infectivity and tropism. The RSV F protein also binds to TLR-4, initiating signal transduction and innate immune responses (Haynes et al. 2001).

The 298-amino acid RSV large glycoprotein G, involved in attachment, appears to be unrelated to the *Paramyxovirinae* HN, H, or G attachment proteins. RSV G has a membrane anchor near its *N*-terminus, with the *C*-terminal two-thirds of the molecule being external. G also is produced as a secreted form—estimated to account for 80 % of released G protein (Hendricks et al. 1988)—that lacks the membrane anchor due to translational initiation at the second AUG in the ORF followed by proteolytic trimming. The ectodomain of G consists of two large divergent domains flanking a short central conserved segment. The divergent domains have a high frequency of amino acid differences among RSV strains; in addition, variants of G have been noted in nature containing partial intra-gene duplications (Eshaghi et al. 2012), small frame shifts, and *C*-terminal extensions. The central conserved region has a cysteine noose stabilized by two disulfide bonds, and this includes a CX3C motif. Surprisingly, this conserved region can be deleted with little effect on replication in vitro or in mice (Teng and Collins 2002). The large divergent domains have a high content of serine, threonine, and proline residues, as well as a high content of N-linked and, especially, O-linked sugars, and these large domains are thought to have extended, unfolded structures. These features also are characteristic of mucins, suggesting possible mucin mimickry by G, although the significance of this is unknown. The sugar side chains increase the estimated Mr of G from 32,000 for the polypeptide backbone to 80,000–90,000, and possibly 180,000 (Kwilas et al. 2009). The amino acid divergence and the presence of a sheath of host-specified sugars are thought to reduce immune recognition. Surprisingly, given its involvement in attachment, RSV lacking the G gene replicates in some cell lines as efficiently as wt RSV. A live-attenuated RSV vaccine candidate lacking most of the SH and G genes due to spontaneous deletions during passage in vitro appeared to be competent for replication in children, although it was highly restricted (Karron et al. 1997a). RSV isolates have been found, from infants exposed to or infected with human immunodeficiency virus, with deletions spanning most of the G ectodomain, indicating that loss of most of G can occur in nature in some situations (Venter et al. 2011). Thus, G is a malleable, variable protein that is absolutely not essential for replication.

In addition to its role in attachment, G helps RSV evade host immunity. The region of G containing the CX3C motif noted above has been reported to mimic the CX3C chemokine fractalkine, with the effect of reducing the influx of immune cells into the lungs of RSV-infected mice (Tripp et al. 2001). The secreted form of G has been shown to interfere with antibody-mediated neutralization, acting as an antigen decoy as well as impeding cell-mediated neutralization of RSV by Fc receptor-bearing immune cells (Bukreyev et al. 2008). G has been speculated to mimic the receptor for tumor necrosis factor alpha (TNF-α), with the possible effect of inhibiting the antiviral effects of that cytokine (Langedijk et al. 1998). G can interact with DC-SIGN on human dendritic cells (DCs) and alter signaling pathways associated with antigen presentation (Johnson et al. 2012). Also, the central conserved domain of the G protein has been shown to inhibit the activation of several TLRs including TLR-4, thus countering the effect of the F protein (Polack et al. 2005).

The 64-amino acid SH protein is a transmembrane protein that is anchored near the *N*-terminus, with the *C*-terminus oriented extracellularly. Most of the SH protein is unglycosylated (Mr $\sim$ 7,500), but SH also accumulates in a variety of forms from Mr 4,500 to up to 60,000 or more due to differences including N-linked sugar, polylactosaminoglycan, and translational initiation at the second methionine codon. This array of isoforms is conserved but their significance is unknown. SH forms pentameric pore-like structures that confer cation-selective channel-like activity (Carter et al. 2010; Gan et al. 2012), although the significance of this for RSV is not clear. Thus, the SH protein appears to be a viroporin, a class of small viral proteins that can modify membrane permeability and can affect budding and apoptosis. SH was reported to reduce apoptosis, but the effect was small (Fuentes et al. 2007). SH also appeared to inhibit signaling from TNF-α, an antiviral cytokine (Fuentes et al. 2007). Recombinant RSV lacking SH can replicate somewhat more efficiently in vitro than its wt parent—presumably due to its smaller genome size and smaller number of genes—and was slightly attenuated in mice and chimpanzees (Whitehead et al. 1999).

The 256-amino acid M protein plays key roles in virion morphogenesis. Early in infection, M is detected in the nucleus and may be responsible for the modest inhibition of host transcription during RSV infection, whereas later M is found associated with cytoplasmic viral inclusion bodies—thought to be the site of viral RNA synthesis—and the plasma membrane—the site of virion formation (Ghildyal et al. 2006). M appears to silence viral RNA synthesis by nucleocapsids, presumably in preparation for their packaging into virions (Ghildyal et al. 2006), and appears to be required for the transport of nucleocapsids from viral inclusion bodies to the plasma membrane (Mitra et al. 2012). M is not required to initiate the formation of viral filaments (thought to be the precursor to infectious virus), but in the absence of M the filaments remain stunted and immature (Mitra et al. 2012). Crystallography revealed an M protein monomer that is organized into compact *N*-terminal and *C*-terminal domains joined by a short linker (Money et al. 2009). The monomer surface contains a large positively charged area that extends across the two domains and may mediate association with nucleocapsids and the negatively charged plasma membrane (Money et al. 2009).

The 391-amino acid N protein binds tightly to the genome and antigenome to form helical nucleocapsids, creating the templates for RNA synthesis. N protein expressed in bacteria bound to host RNA to form decamer rings that resembled one turn of the helical nucleocapsid (Tawar et al. 2009). Determination of the atomic structure of the N-RNA rings indicated that each N monomer consists of *N*-terminal and *C*-terminal domains separated by a hinge. Each N monomer was associated with seven nt of RNA, with the RNA groove at the hinge. Adjacent monomers were oriented in the same direction and loosely connected, providing flexibility that would allow polymerase access without disassembling the helix.

Of the seven nt associated with each N monomer, nt 2–4 are oriented into the groove while the other four nt face outward. Passage of the polymerase may induce a transient hinge movement that makes the three buried nucleotides flip out to be accessible (Tawar et al. 2009). The N protein also has a role in antagonizing host innate immunity: N binds to the dsRNA-regulated protein kinase PKR and prevents it from phosphorylating eIF-2a and inhibiting protein synthesis (Groskreutz et al. 2010).

The 241-amino acid P protein is an essential polymerase co-factor. It also acts as an adapter that binds to the N, M2-1, and L proteins to mediate interactions in the nucleocapsid/polymerase complex. In addition, P binds to free N protein monomers and delivers them to nascent genomes/antigenomes, thus preventing N from self-aggregating or binding to nonviral RNA (Castagne et al. 2004). The expression of N and P alone are sufficient to form viral inclusion bodies, which are large, dense cytoplasmic structures that are thought to be the sites of viral RNA synthesis. P exists as a homotetramer formed through a multimerization domain in the middle of the molecule, which is flanked by intrinsically disordered domains (Castagne et al. 2004; Llorente et al. 2006). The *C*-terminal region of P was shown to interact with the nucleocapsid by binding to a pocket on the surface of the N protein that includes discontinuous residues from positions 46–151 brought together in the folded structure (Galloux et al. 2012); other P-N binding sites may also exist, perhaps depending on conformation. P may contribute to conformational changes that help the polymerase access the RNA template (Castagne et al. 2004) and appears to be necessary for promoter clearance and chain elongation by the viral polymerase (Dupuy et al. 1999). It also appears to have a role in dissociating the M protein from the nucleocapsid during uncoating to initiate infection (Asenjo et al. 2008). P is the major phosphorylated RSV protein and contains phosphate at more than 10–12 sites, with different sites exhibiting differing rates of turnover due to interplay between cellular kinases and phosphatases (Asenjo et al. 2005). Many of the activities of P described above appear to be affected by dynamic phosphorylation/dephosphorylation at a subset of these sites, apparently involving a small percentage of the total phosphate content (Asenjo et al. 2006, 2008). Experiments in which P phosphorylation was reduced by mutational ablation or the use of inhibitor against cellular kinase supported the idea that phosphorylation is important for RSV replication, but that much of the low-turnover phosphate is not essential (Lu et al. 2002).

The 2,165-amino acid RSV L protein is very similar in length to its *Paramyxovirinae* counterparts and shares low but unambiguous sequence relatedness along nearly its entire length. Specific segments are conserved within and beyond *Mononegavirales* that are thought to represent catalytic domains involved in polymerization. Analysis of RSV mutants has provided preliminary identification of functional regions in L, including the polymerization domain (Fix et al. 2011), a putative nucleotide-binding site involved in capping (Liuzzi et al. 2005) as well as residues that affect the efficiency of recognition of GE signals (Cartee et al. 2003).

The 194-amino acid M2-1 protein is an essential transcription processivity factor (Fearns and Collins 1999b; Collins et al. 1996, 1999). M2-1 accumulates in phosphorylated and nonphosphorylated forms and forms a homotetramer via an oligomerization domain at residues 32–63 (Tran et al. 2009; Cartee and Wertz 2001). The M2-1 protein binds RNA: the specificity of this interaction remains somewhat unclear, but M2-1 may preferentially bind RSV mRNAs (Cartee and Wertz 2001; Blondot et al. 2012). M2-1 also interacts with the P protein: binding to RNA or P involves partially overlapping domains in the center of the molecule (Blondot et al. 2012). M2-1 can be found in viral inclusion bodies and its presence there depends on interaction with P (Blondot et al. 2012). Interactions with RNA or the P protein are essential for the ability of M2-1 to support RNA synthesis and are competitive (Tran et al. 2009; Blondot et al. 2012), suggesting that P delivers M2-1 to the nucleocapsid and is then displaced. M2-1 also binds to the M protein and mediates its transport to inclusion bodies and interaction with nucleocapsids (Li et al. 2008). M2-1 contains a CCCH zinc finger motif near its *N*-terminus (residues 7–25) that is essential for its activity in viral RNA synthesis (Hardy and Wertz 2000). This unusual CCCH motif is also found in tandem in a family of cellular zinc finger proteins that includes tristetraprolin (TTP), which binds to AU-rich elements present in a number of host response mRNAs including cytokine mRNAs and affects their stability. Like TTP, M2-1 was recently shown to associate with cellular stress granules, which are involved in translational regulation under stress conditions, but the significance of this possible similarity is unclear (Fricke et al. 2013). M2-1 is unique to *Pneumovirinae*, although it shares structural homology with the VP30 transcriptional activator of *Filoviridae* (Blondot et al. 2012).

The M2-2 protein (88 or 90 amino acids, depending on the start site; Chang et al. 2005) is expressed at a low level in infected cells, and its status as a virion component is not known. Deletion of M2-2 from recombinant RSV results in a virus that exhibits delayed and reduced RNA replication and increased "runaway" transcription; this contrasts with wt RSV, for which transcription appears to be downregulated later in infection in favor of RNA replication (Bermingham and Collins 1999). These results suggest that M2-2 plays a role in regulating RNA synthesis; specifically, as the level of M2-2 increases during the time course of infection, it reduces transcription and promotes RNA. Consistent with a direct effect on RNA synthesis, over-expression of M2-2 inhibited RNA synthesis by mini-replicons and inhibited replication of complete RSV (Collins et al. 1996; Cheng et al. 2005). In addition, in experiments designed to produce virus-like particles, expression of M2-2 increased the efficiency of packing: this might reflect its effects on RNA synthesis or might be an unrelated activity (Teng and Collins 1998). Replication of ΔM2-2 RSV in vitro is delayed but reaches titers comparable to wt RSV, whereas in mice and chimpanzees the virus was restricted approximately 500- to 1000-fold compared to wt RSV (Bermingham and Collins 1999; Teng et al. 2000).

The NS1 and NS2 proteins (139 and 124 amino acids, respectively) are thought to be nonstructural. While the two proteins can function separately, they appear to form complexes and may have synergistic effects, but this is poorly understood (Spann et al. 2005; Swedan et al. 2011). NS1 and NS2 interfere with innate immune responses including interferon induction and signaling (Spann et al. 2005;

Swedan et al. 2011). They also inhibit apoptosis, thereby prolonging the life of the cell and increasing viral yield (Bitko et al. 2007). In a mini-replicon system, co-expression of NS1—and, to a lesser extent, NS2—inhibited transcription and RNA replication, affecting both the genomic and antigenomic promoters (Atreya et al. 1998). These effects remain to be further investigated, but they suggest that NS1 and possibly NS2 might downregulate and restrain viral RNA synthesis. This may be comparable to effects shown for the C and V proteins of some members of *Paramyxovirinae* that, by downregulating viral RNA synthesis, avoids the accumulation of viral dsRNA that otherwise activates innate immunity. Recombinant RSV lacking the NS1 and/or NS2 genes have increased sensitivity to interferon, cause increased apoptosis, and replicate with reduced efficiency in cultured cells and experimental animals, with the effect of deleting NS1 being greater (Teng et al. 2000; Whitehead et al. 1999) (see chapter by S. Barik, this volume).

6 Transcription and RNA Replication

For transcription, the polymerase enters at the 3′ end of the genome and copies the genes into their corresponding mRNAs (Fig. 4) by a sequential stop–start process guided by the GS and GE signals (Kuo et al. 1996). Synthesis of each mRNA initiates opposite the first nucleotide of the GS signal. Surprisingly, the initiating nucleotide can be selected by the polymerase independent of the template (Kuo et al. 1997). Thus, the GS signal triggers the start of mRNA synthesis; in addition, its complementary sequence present in the nascent transcript is thought to act as a signal for capping and cap methylation by the polymerase, based on analogy with other nonsegmented negative-strand RNA viruses (Wang et al. 2007). Capping and/or methylation appears to be essential for mRNA elongation: when RSV capping was blocked by a specific inhibitor, transcription produced uncapped abortive RNAs of ~45 to 50 nt (Liuzzi et al. 2005). Polymerase that is engaged in mRNA synthesis is unresponsive to encountering an additional GS signal, but encountering a GE signal triggers polyadenylation/termination of the mRNA and makes the traversing polymerase responsive to a GS signal (Kuo et al. 1996; Fearns and Collins 1999a). The triggering of polyadenylation/termination at the various GE signals is not completely efficient, and the polymerase occasionally continues synthesis through the next gene. This produces various readthrough mRNAs that account for approximately 10 % of total mRNA (Collins and Wertz 1983). RSV transcription has a polar gradient in which gene transcription decreases along the gene order (Fig. 4). This is typical for *Mononegavirales* and occurs because some of the transcribing polymerases disengage and exit the genome at the various gene junctions.

Studies with mini-replicons showed that N, P, and L are the viral proteins necessary for transcription, but under these conditions transcription terminates prematurely and nonspecifically within several hundred nt, and genes that are further downstream are not significantly transcribed (Collins et al. 1995, 1996,

1999; Fearns and Collins 1999b). Fully processive transcription requires in addition the M2-1 protein, which can be present in relatively low relative molar amounts (Fearns and Collins 1999b; Collins et al. 1996). In mini-replicon experiments, M2-1 also decreased the efficiency of termination at the GE transcription signals—possibly a reflection of the same processivity activity—resulting in increased production of readthrough mRNAs (Hardy and Wertz 1998). These activities raise the possibility that M2-1 might affect the relative levels of expression of the various viral genes, such as by promoting sequential transcription and reducing the transcriptional gradient, but this has not been observed in infected cells (Fearns and Collins 1999b).

One-way sequential transcription does not provide for initiation at the L GS signal, since it is located upstream of the M2 GE signal, as noted (Fig. 3a). Studies with mini-replicons showed that, upon completing transcription of the M2-1 gene, the polymerase backtracks by retrograde scanning to initiate at the L GS signal (Fearns and Collins 1999a). Apart from the M2 GE and L GS signals, the overlap region did not appear to contain any other *cis*-acting elements needed for this activity. Furthermore, the polymerase was found to scan in both directions. This led to the realization that scanning may occur at each gene junction and may be the mechanism by which the next GS signal is located, and could explain the tolerance for intergenic variability noted above. The presence of the M2 GE signal within the L gene (due to the overlap) causes 90 % of newly initiated L gene transcripts to add polyA and terminate at the signal, producing a 68-nt polyadenylated RNA that does not appear to encode a protein and is not known to have any further significance (Collins et al. 1987). The synthesis of full-length L mRNA depends on the "error" of polymerase readthrough at the M2 GE signal (Collins et al. 1987). Whether the level of M2-1 protein affects this process remains to be evaluated. Although 90 % of L gene transcripts terminate at the M2 GE signal, there is not a steep drop in the transcriptional gradient at the L gene (Fearns and Collins 1999a; Kwilas et al. 2010). This indicates that backtracking is very active. The gene overlap appears to be an accidental arrangement that may provide no advantage but can be tolerated due to the scanning function of the polymerase.

The relative level of expression of the various RSV genes is determined mostly by the polar gradient of transcription. Thus, the most abundant mRNAs are for the NS1 and NS2 proteins that antagonize host responses. The expression of the L mRNA, the last gene in the order, is further reduced by an unidentified effect that appears to be post-transcriptional (Kwilas et al. 2010), possibly mRNA stability. Differences in the efficiency of polyadenylation/termination by the various GE signals due to natural sequence variation may affect the relative levels of gene expression by changing the amount of transcriptional readthrough (Harmon and Wertz 2002). On the one hand, increased readthrough spares the polymerase from disengagement at the gene junctions, providing more polymerase to downstream genes. On the other hand, ORFs that are at internal positions in readthrough transcripts are not efficiently translated, reducing the synthesis of proteins from genes downstream of inefficient GE signals. Thus, the overall effect is complex, and the impact of GE variation is not clear.

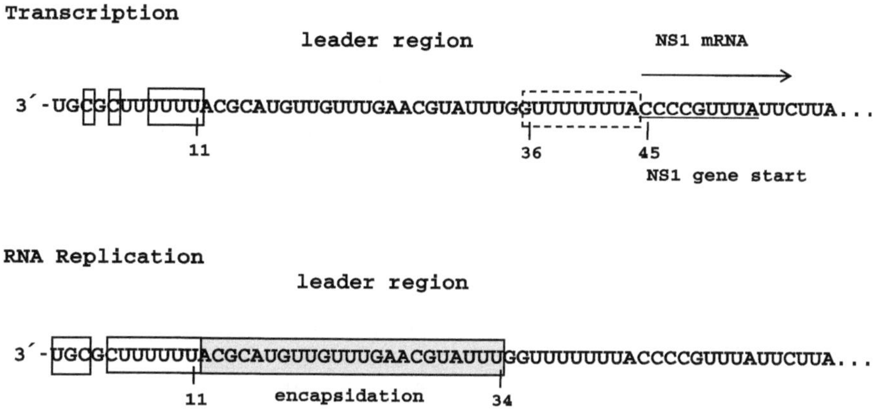

Fig. 6 Nucleotide positions in the leader region of genomic RNA that are important for transcription (*top*) and RNA replication (*bottom*). Important residues present in positions 1–11 are indicated with *open boxes*; note that those that are important for transcription are a subset of those important for RNA replication. A region that increases the efficiency of transcription is indicated with a *dashed box*. The GS signal of the first gene, necessary for transcriptional initiation but not involved in RNA replication, is *underlined* in the diagram for transcription. A region that contains an apparent encapsidation signal necessary to produce full-length replication products is indicated with a *shaded box*. Sequences are in negative-sense

RSV RNA replication initiates opposite the first nucleotide at the 3′ end of the genome. As with transcription, the initiating nucleotide can be selected independent of the template (Noton et al. 2010; Noton and Fearns 2011). The polymerase ignores the GS and GE signals and produces a full-length positive-sense replicative intermediate, the antigenome (Fig. 4). A fraction of antigenomes are modified at their 3′ termini by addition of 1–3 nt, which are not copied into the genome RNA. The significance of the addition is not known, but it might represent a mechanism of promoter regulation (Noton et al. 2012). At least 10-fold more genome is produced than antigenome, both because the antigenome promoter is more efficient than that of the genome, and because synthesis from the genome is divided between RNA replication and transcription (Fearns et al. 2000; Hanley et al. 2010). Whereas efficient RNA replication by members of subfamily *Paramyxovirinae* requires that the genome nucleotide length be an even multiple of six ("rule of six"), RSV has no comparable length requirement. Studies with minireplicons showed that N, P, and L are the viral proteins that are necessary and sufficient to direct RNA replication, and that RNA replication is unaffected by M2-1 (Collins et al. 1996; Grosfeld et al. 1995).

The *cis*-acting sequence elements at the 3′ end of the genome that are involved in initiating transcription and RNA replication were analyzed using mini-replicons (Fig. 6). Transcription was found to require two sequence elements: the 3′-terminal 11 nt of the genome, and the presence of a nearby, downstream GS signal. In addition, the efficiency of transcription was increased by the presence of a

U-rich sequence found at the end (nt 36–44) of the leader region (Fig. 6) (McGivern et al. 2005). Furthermore, the efficiency of transcription was affected by the length of the nucleotide chain between the 3′-terminal 11 nt and the GS signal: a length similar to that in wt RSV was the most efficient (McGivern et al. 2005; Fearns et al. 2000). For RNA replication, the 3′-terminal 11 nt of the genome was sufficient for initiation, but the resulting transcripts terminated prematurely and were unencapsidated: the synthesis of full-length encapsidated RNA depended on the additional presence of nt 16–34, implying the presence of an encapsidation signal within the first 34 nt of the genome (Fig. 6) (Cowton and Fearns 2005; McGivern et al. 2005). Thus, these studies identified the 3′-terminal 11 nt of the genome as a promoter element necessary for both transcription and RNA replication. Saturation mutagenesis showed that transcription was particularly dependent on positions 3, 5, 8, 9, 10, and 11. RNA replication depended on these same positions, as well as positions 1, 2, 6, and 7 (Fig. 6). The assignment at position four differentially affected transcription versus RNA replication: the wt assignment of 4G (negative-sense) favored transcription over RNA replication, whereas the substitution of 4C or 4U had the opposite effect (Fearns et al. 2002).

The *cis*-acting elements at the 3′ end of the antigenome (i.e., containing the promoter involved in producing progeny genomes) are less well characterized. As noted, there is considerable sequence identity between the 3′-terminal 24–26 nt of the genome and antigenome, reflecting conserved promoter elements (Fig. 3b). In particular, the first 11 nt of the genome and antigenome differ only at position 4: the assignment in the genome (4G) is optimal for transcription, whereas that in the antigenome (4U) is optimal for RNA replication (Fearns et al. 2002). Mutation analysis of the antigenomic promoter showed that single nucleotide substitutions at positions 1–7 of this promoter have similar effects on RNA replication as the corresponding mutations in the genome, although they do not behave identically (Peeples and Collins 2000; Noton and Fearns 2011). Mini-replicon studies showed that the first 36 nt of the antigenome were sufficient for initiation and synthesis of full-length genomes; however, production was increased by including nt 37–155 (Fearns et al. 2000). In infectious recombinant virus, nt 37–155 could be deleted with only minimal restriction of viral replication in vitro and in mice although, consistent with the mini-replicon studies, the synthesis of genomes was reduced (Hanley et al. 2010). In the same study, a mutant virus containing the complement of the 44-nt Le in place of the trailer region (i.e., in which the antigenomic promoter was replaced with the genomic promoter), replicated with reduced efficiency over multiple cycles and exhibited increased accumulation of stress granules (Hanley et al. 2010). One plausible interpretation of these data are that the 5′ terminal 36 nt of the wt trailer RNA inhibits stress granule formation, which would otherwise reduce RSV replication, although another study has indicated that stress granules might benefit RSV replication (Lindquist et al. 2010).

The involvement of the 3′- terminal 11 nt of the genome in both transcription and RNA replication indicates that the first step in either process involves recognition of this segment by the polymerase. RNA replication initiates opposite the first nucleotide at the 3′ end of the genome, as noted. For transcription, early events remain unclear. One possibility is that synthesis also initiates opposite the first position and produces a short transcript of the leader region. The polymerase may release this leader RNA—perhaps as abortive synthesis due to its uncapped nature—and reinitiate at the first GS signal. In an alternate model for transcription, the polymerase does not synthesize a leader RNA and instead scans without synthesis along the leader region to locate the first GS signal, similar to the proposed scanning at intergenic regions. These two models are discussed in detail by Cowton et al. (2006). It is not known whether a single form of polymerase complex is partitioned between transcription and RNA replication, or whether there are functionally distinct transcriptase and replicase complexes (which seems more likely). A common hypothesis for this type of virus is that there is a balance between transcription and RNA replication, and that the availability of soluble N protein promotes RNA replication at the expense of transcription, but this was not observed in studies with RSV mini-replicons (Fearns et al. 1997). The detailed mechanism of RNA synthesis, and how proteins such as P, M2-1, and M2-2 interact with the polymerase and nucleocapsid template, remains largely unknown. Cellular proteins likely are involved in RSV transcription and RNA replication. Soluble actin is important for RSV transcription, and its activity is augmented by profilin. Actin appears to function by binding directly to the viral N-RNA template and recruiting profilin to the complex (Harpen et al. 2009). Also, heat shock protein 70 associates with the RSV polymerase complex and there is evidence it augments RNA synthesis activity (Brown et al. 2005). A cell-free system for RSV RNA synthesis, in which polymerase activity is reconstituted with recombinant proteins and a synthetic template, has recently been developed and will provide the means for detailed characterization (Noton et al. 2012).

7 Summary of *cis*-Acting RNA Sequences

As detailed above, essential cis-acting signals include much of the 44-nt leader region, the 10 GS and GE signals, and the last ~36 nt of the trailer region (which encodes the 3′ end of the antigenome). These segments contain a total of ~300 nt, accounting for 2 % of the genome. It is likely that further fine-mapping will find that the essential positions in these segments comprise substantially fewer than 300 nt. As noted, there may be additional sequence in the M2 mRNA that promotes translation of the M2-2 ORF, and the efficiency of virus replication in vivo was increased slightly by including the complete trailer regions. There is presently no evidence of additional essential *cis*-acting signals.

8 Viral Replicative Cycle

RSV attachment and entry are mediated by the G and F glycoproteins, with no apparent contribution by SH, and are reviewed elsewhere (see chapter by J.S. McLellan et al., this volume).Entry occurs by fusion of the viral envelope with the cell plasma membrane. There is also evidence of entry by clathrin-mediated endocytosis, although endosomal acidification was not required and thus this pathway involves the same fusion mechanism as at the plasma membrane (Kolokoltsov et al. 2007; Srinivasakumar et al. 1991). Genome transcription and replication occur in the cytoplasm and the virus can grow in enucleated cells and in the presence of actinomycin D, indicating a lack of essential nuclear involvement.

By extrapolation from prototype members of *Mononegavirales*, incoming nucleocapsids engage in transcription by preformed polymerases (primary transcription). The availability of newly synthesized soluble viral N and P proteins promotes elongation of RNA replication products, leading to the production of full-length encapsidated antigenomes and genomes. Progeny genomes engage in transcription (secondary transcription) and RNA replication. Transcription and RNA replication occur concurrently. RSV mRNAs and proteins can be detected intracellularly at 4–6 h after infection and reach a peak accumulation by 15–20 h. At this time, transcription may be downregulated in favor of RNA replication and the production of genomes needed for packaging, and this appears to be mediated by the M2-2 protein (Bermingham and Collins 1999). However, there is no evidence of a change in the relative molar amounts of the various viral mRNAs during the infection time course (Fearns and Collins 1999b). The release of progeny virus begins by 10–12 h post infection, reaches a peak after 24 h, and continues until the cells deteriorate by 30–48 h.

As already noted, RSV-infected cells develop large cytoplasmic inclusion bodies that become evident by 12 h post-infection (Lindquist et al. 2011; Lifland et al. 2012). These have been shown to contain the viral N, P, M2-1, and L proteins, as well as viral RNA. As noted, the inclusion bodies are thought to be sites of RNA synthesis. More recently, the viral inclusion bodies have been shown to sequester key cellular signaling components and thereby inhibit cellular responses to infection. These include Mda5 and MAVS involved in interferon induction (Lifland et al. 2012) as well as p38 mitogen-activated protein kinase and O-linked N-acetylglucosamine tranferase involved in stress responses and stress granule formation (Fricke et al. 2013). Infected cells develop filamentous surface projections that bear viral glycoproteins and may give rise to viral filaments (Fig. 1). Infected cell lines develop syncytia (Fig. 1) that are a major viral cytopathic effect and lead to the destruction of the monolayer, but that are much less evident in differentiated, polarized epithelium in vitro and in vivo (Johnson et al. 2007; Zhang et al. 2002).

RSV assembly and budding occur at the plasma membrane. In polarized cells, this occurs at the apical surface (Roberts et al. 1995; Zhang et al. 2002). These regions contain localized virus-modified lipid rafts involving all three viral surface proteins and the M protein (McDonald et al. 2004; Jeffree et al. 2003; McCurdy and Graham 2003; Henderson et al. 2002; Yeo et al. 2009). The minimum viral protein requirements for the formation of virus-like particles capable of delivering the viral genome to target cells are the F, M, N, and P proteins (Teng and Collins 1998), and expression of these proteins induced the formation of viral filaments (Utley et al. 2008). Both genome and antigenome have been detected in virions, suggesting a lack of selective packaging. Genome-containing nucleocapsids are much more abundant in the infected cell, and are correspondingly more abundant in virions. RSV appears to hijack cellular apical recycling endosomes for budding, a pathway that is distinct from that described for a number of other enveloped RNA viruses (Brock et al. 2003; Utley et al. 2008).

9 Spontaneous Mutations

RSV has a high rate of nucleotide substitution (10^{-3} to 10^{-4}), as is typical of RNA viruses. Virus passaged in vitro occasionally can acquire inserts of one (usually) or more U residues (negative sense) into the U tract at the end of GE signals or into untranslated sequence. Spontaneous deletion of the G and SH genes has been noted in vitro, and deletion of most of the G gene has been noted in vivo (Venter et al. 2011). Two different spontaneous intragenic duplications of a segment of the G gene have been documented in nature (Eshaghi et al. 2012), and in one case the resulting virus has spread worldwide and continued to evolve. The mechanism for this duplication is not known, but might involve backtracking of the polymerase following pausing at a secondary structure in the RNA. Recombination can occur when the polymerase jumps between templates during synthesis. This can create defective interfering genomes, which are typical for *Mononegavirales* but are poorly described for RSV. This also has the potential to create replication-competent mosaic genomes, although this appears to be very rare (Spann et al. 2003). Fresh clinical isolates of RSV may undergo some sort of adaptation to cell culture (Marsh et al. 2007), but this is poorly understood. In any event, passaged laboratory strains retain their virulence for chimpanzees and humans (Karron et al. 1997b; Whitehead et al. 1998). Circulating RSV appears to accumulate progressive changes in sequence and antigenicity, primarily in the G protein, in response to immune pressure, but this is a slow process occurring over decades.

10 Reverse Genetics

Reverse genetics involves the production of mini-replicons or complete virus from cDNA. Mini-replicons are versions of the genome or antigenome in which most or all of the viral genes are deleted and may be replaced by one or more convenient marker genes, such as chloramphenicol acetyl transferase or luciferase (Fig. 7a).

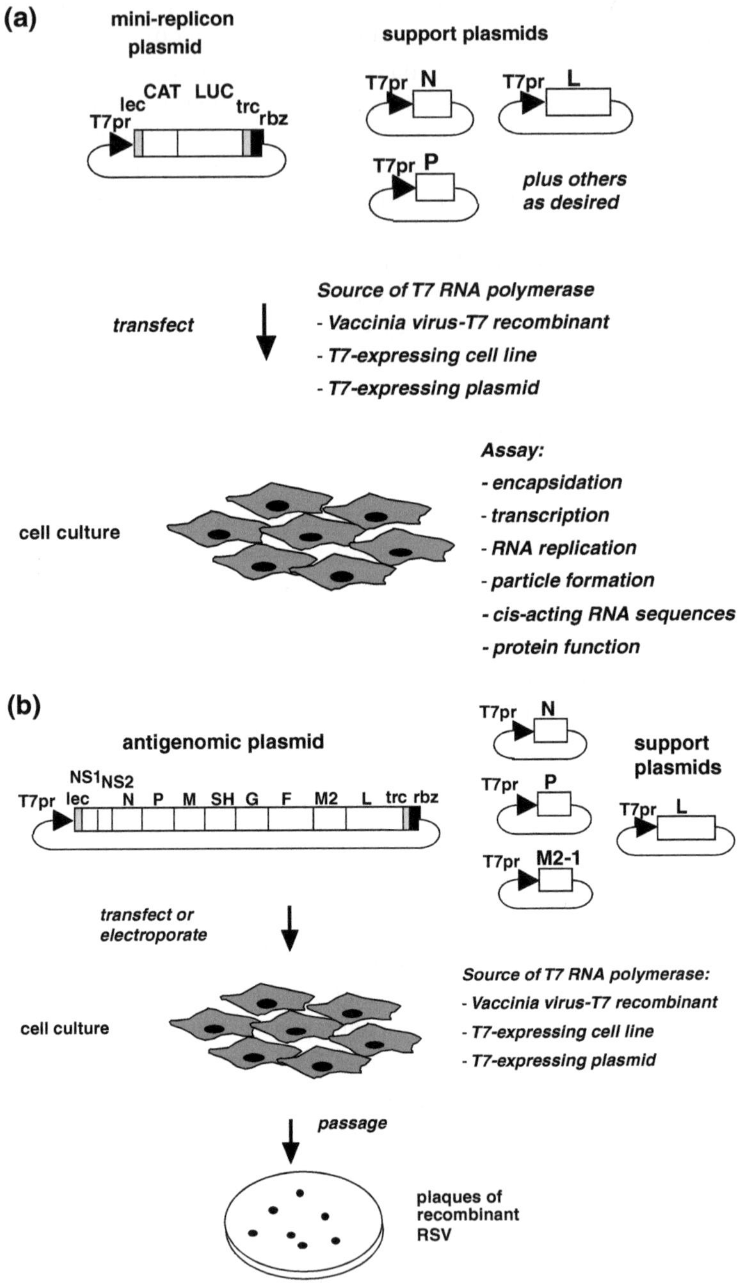
(a)
mini-replicon plasmid
support plasmids
T7pr lec CAT LUC trc rbz
T7pr N
T7pr L
T7pr P
plus others as desired
transfect
Source of T7 RNA polymerase
- Vaccinia virus-T7 recombinant
- T7-expressing cell line
- T7-expressing plasmid
cell culture
Assay:
- encapsidation
- transcription
- RNA replication
- particle formation
- cis-acting RNA sequences
- protein function
(b)
antigenomic plasmid
T7pr lec NS1 NS2 N P M SH G F M2 L trc rbz
T7pr N
T7pr P
T7pr M2-1
T7pr L
support plasmids
transfect or electroporate
cell culture
Source of T7 RNA polymerase:
- Vaccinia virus-T7 recombinant
- T7-expressing cell line
- T7-expressing plasmid
passage
plaques of recombinant RSV

◀ **Fig. 7** Reverse genetic systems. **A** Helper-dependent mini-replicon system, illustrated with a dicistronic mini-replicon (shown here as an antigenome-sense replicon) containing chloramphenicol acetyl transferase (CAT) and luciferase (LUC) marker genes under the control of RSV GS and GE signals (not shown) and flanked by the complements of the RSV leader and trailer regions (*gray boxes*, lec and trc, respectively). The mini-replicon cDNA is flanked by a T7 RNA polymerase promoter (*arrow*, T7 pr) and a self-cleaving ribozyme (*black box*, rbz). This is complemented by support plasmids encoding various RSV proteins. **b** Recovery of complete infectious virus from a full-length antigenome expressed from a transfected plasmid in the presence of support plasmids encoding the N, P, M2-1, and L proteins. The complements of the leader and trailer regions are shown (*gray boxes*, lec and trc, respectively), as are the T7 RNA polymerase promoter (*arrow*, T7pr) and ribozyme (*black box*)

The marker genes are each under the control of a set of RSV GS and GE signals. In the plasmid, the mini-replicon cDNA is flanked by a promoter for T7 RNA polymerase and a self-cleaving ribozyme sequence, which provide for synthesis of a negative- or positive-sense (depending on the orientation of the insert) mini-replicon with nearly correct ends. Typically, the mini-replicon plasmid is transfected into tissue culture cells together with so-called support plasmids that express individual viral proteins under the control of T7 RNA polymerase promoters. T7 RNA polymerase can be supplied by a vaccinia virus recombinant, by a constitutively expressing cell line, or by a eukaryotic expression plasmid. One can provide whatever combination of RSV proteins is desired, and in the desired relative molar amounts. Depending on the support proteins and *cis*-acting elements included in the minigenome, this system can execute mini-replicon encapsidation, transcription, RNA replication, and particle morphogenesis (Fearns and Collins 1999a; Collins et al. 1996; Teng and Collins 1998; McGivern et al. 2005; Noton et al. 2010). Due to the small size and relative simplicity of the mini-replicons, this approach is ideal for detailed structure-function studies of *cis*-acting RNA signals or *trans*-acting viral proteins. Since the viral proteins are supplied *in trans*, the supply is independent of the mini-replicon, and thus the system can be used to study mutations that have drastic effects on mini-replicon transcription or RNA replication and could not be recovered and studied in complete recombinant virus.

The production of complete infectous virus from cDNA provides the means to introduce predetermined changes into infectious virus. The recovery of cDNA-derived virus involves co-transfection of cultured cells with a plasmid encoding a copy of the antigenome and support plasmids expressing proteins of the nucleocapsid/polymerase complex, namely N, P, M2-1, and L (Fig. 7b) (Collins et al. 1995, 1999). The expressed RSV antigenome and support proteins assemble into nucleocapsids that launch a productive infection. The expression of the positive-sense antigenome rather than the negative-sense genome avoids hybridization with the positive-sense RNAs expressed from the support plasmids. Recovery of infectious virus is not very efficient but, once recovered, cDNA-derived virus is readily propagated in the same manner as biologically derived virus and is distinguishable only by introduced mutations.

The genomes of RSV and other *Mononegavirales* have proven to be very amenable to manipulation. As noted, the RSV *cis*-acting RNA signals are short and circumscribed, and the modular organization of the genes also faciliates engineering: since the mRNAs and proteins are expressed independently of each other (apart from the transcription gradient), they can readily be manipulated independently. The upper limit for added sequence has not been determined for RSV and other members of *Mononegavirales*, and there probably is no strict packaging limit. The efficiency of replication in vitro decreases with added sequence, but the effect can be for inserts of several hundred or thousand nts minimal. For example, insertion of the 3.2-kb β-galactosidase gene into RSV (increasing its nucleotide length by 21 %) caused only a marginal reduction in replication in vitro. Effects in vivo of increasing genome length have not been studied carefully but appear to be substantially more restrictive

Reverse genetics provided new tools for basic virologic and pathogenesis studies. As already described, mini-replicon systems have been used to identify *trans*-acting proteins and *cis*-acting RNA signals and structures necessary for genome encapsidation, transcription, polymerase scanning, RNA replication, and virion morphogenesis. At the level of complete virus, mutants have been made that lack, for example, the NS1, NS2, SH, G, secreted G, and M2-2 genes, or which lack portions of these genes, or which have mutations to genome structures such as intergenic regions. In addition, RSV has been engineered to express fluorescent proteins, providing for real-time monitoring of infection.

Reverse genetics also provides the means to make new live-attenuated RSV vaccine candidates. This involves identifying mutations that attenuate RSV, and introducing attenuating mutations in desired combinations into RSV to create specific, well-characterized candidates for pre-clinical and clinical evaluation. As a second approach, PIVs have been developed as vectors to express the RSV F and G proteins. Vaccine candidates based on this technology are described elsewhere (see chapter by R.A. Karron et al., this volume).

All steps in the generation and development of RSV for evaluation in humans must be done with cells and reagents qualified for human product use. For example, this avoids components derived directly from animals (e.g., bacterial growth medium is plant-derived, porcine trypsin is replaced by recombinant trypsin). Recovery is performed in qualified cells, such as qualified lots of African green monkey Vero cells. T7 RNA polymerase is supplied by a co-transfected plasmid rather than by a recombinant virus or expressing cell line, and transfection is replaced by electroporation (Surman et al. 2007). These steps substantially reduce the efficiency of recovery. Of course, clinical trial material is subjected to extensive safety testing prior to use. Although engineered virus is of recombinant origin, it is propagated and "behaves" like biologic virus. The transfecting plasmids are needed only for the initial recovery, and the content of residual plasmid DNA quickly becomes insignificant upon passage of the virus. Producing virus from cDNA provides a virus with a short and well-defined passage history, which is important for safety and regulatory reasons (Surman et al. 2007). It thus has a safety advantage compared to biologically derived virus, which originates in an infected human and typically has been passaged extensively in cell culture. Cloned cDNAs also provide a stable vaccine seed. The ability to alter the virus provides a means to update a vaccine, for example, to modify the level of attenuation or to replace the F and G genes with those of newer strains.

The relatively poor growth and lability of RSV render some reverse genetics applications impractical. For example, RSV likely is not a good choice for obtaining high-level expression of foreign antigens from inserted genes. RSV has limited suitability as a human vaccine vector because most humans are exposed to RSV early in life and have substantial immunity, which would limit vector replication and immunogenicity.

11 Pathogenesis

RSV pathogenesis is complex and variable. Disease can range from mild to lethal and can encompass a wide range of acute upper and lower respiratory tract disease manifestations, from mild rhinitis at one extreme to bronchiolitis and pneumonia at the other. The heterogeneity of RSV disease also can include an association between severe disease in infancy and subsequent airway hyperreactivity during childhood and perhaps beyond, giant cell pneumonia in immunocompromised patients, and infection of the institutionalized elderly that result in exacerbation of underlying conditions and excess mortality. Host factors play a major role in disease heterogeneity: these include premature birth, young age or frail old age, low serum antibody titer, underlying conditions such as chronic lung or heart disease or immunosuppression, narrow or reactive airways, and other host factors of a more subtle genetic nature that remain to be fully described. More recently, the question of heterogeneity in RSV strain virulence as another factor is being revisited. Many of these factors and unique host populations will be discussed in other chapters. Here we will briefly discuss the species and cell tropism of RSV, and the syndrome of vaccine-enhanced illness that occurred in children immunized with a formalin-inactivated RSV vaccine (FI-RSV) in the 1960s.

11.1 Viral Tropism

RSV infection and replication are largely restricted to the human. Although human RSV strains can infect other animals such as mice, cotton rats, sheep, and African green monkeys, these hosts are semi-permissive and transmission between animals or spread within a population does not occur. Conversely, the animal strains BRSV and PVM are strongly restricted in primates (Brock et al. 2012; Buchholz et al. 2000). Chimpanzees are the only animal host in which human RSV infects and replicates well enough to permit animal-to-animal transmission and to reliably produce respiratory tract disease.

In vivo, RSV is largely restricted to the superficial cells of the respiratory epithelium. RSV is recovered in abundance from nasal secretions, nasopharyngeal swabs, lung washes, and the sinuses. Using in situ hybridization and immunostaining for viral antigen, there is evidence in humans and cows naturally infected with their respective RSVs that epithelial cells from trachea, bronchi, and bronchioles are infected. Ciliated cells can clearly be shown to be infected in human airway epithelium, and it appears that basal epithelial cells are spared. Based on nearly circumferential staining of airway structures, it is possible that some nonciliated epithelial cells may also be infected, but this is not confirmed based on human pathology (Johnson et al. 2007). Both type I and type II alveolar pneumocytes are infected. Macrophages are sometimes shown to be immunostain-positive, but it is thought that this is most likely from phagocytized virus or virus proteins and not associated with replication. RSV RNA can be detected in the blood and in isolated reports from cerebrospinal fluid and the myocardium, and RSV antigens have been detected in circulating mononuclear leukocytes (Eisenhut 2006; Rohwedder et al. 1998), but these observations may not involve extrapulmonary infection. Infectious RSV is rarely recovered from an extrapulmonary location, and the few instances of recovery usually involve immunosuppressed individuals or experimental animals (Johnson et al. 1982; Eisenhut 2006).

Studies in vitro confirm a tropism for the superficial cells of the epithelium, but one that is not absolute. Infection of human airway epithelial cells in organ cultures is consistent with findings from human autopsy specimens and suggests RSV primarily infects ciliated cells (Zhang et al. 2002), although rare nonciliated cells may be infected in primary human airway epithelial cell cultures (Villenave et al. 2012) or adenoid organ cultures (Wright et al. 2005). Virus shedding occurs strictly at the apical membrane suggesting that polarized epithelium is the preferred cellular target. Paradoxically, studies with immortalized human cell in vitro have demonstrated that a wide variety of cell types can be infected. RSV can replicate in transformed cell lines derived from lung, kidney, liver, neural tissue, colon, breast, and ovarian tissues. Therefore, RSV replication is not necessarily restricted to its tissue of origin in the respiratory tract or confined to polarized epithelium, but highly differentiated cells (particularly ciliated cells that have undergone mesenchymal epithelial transition and achieved planar polarity) are much more permissive than other cells in the context of airway epithelium. Intriguingly, in vitro cultures of primary human airway cells (which should be a more authentic substrate than cell lines) were refractory to RSV infection and replication when freshly seeded but gained susceptibility upon differentiation over a number of days, suggesting that infection of primary epithelial cells involves a differentiation-specific factor (Zhang et al. 2002). Non-epithelial human primary cells also have been infected with RSV, but generally with substantially less efficiency: these include fibroblasts, bone marrow stromal cells (Rezaee et al. 2011), eosinophils (Dyer et al. 2009), and DCs (Johnson et al. 2011). Myeloid and monocyte-derived DCs are more permissive (5–15 % infection rate) than plasmacytoid DCs (<1 %), and other nonadherent cells of hematopoetic origin are infected at very low frequencies. In addition, RSV can infect and replicate with reasonable efficiency in cell lines derived from other species, including African green monkeys, bovines, and hamsters.

These general observations suggest that there are cell-specific, species-specific, and context-specific factors that favor the growth of RSV in the human respiratory epithelium. Cellular tropism potentially can be determined by the presence of receptors (and co-receptors) for viral attachment and entry, host structures that may be utilized by the virus for various steps in the replicative cycle, restriction elements that may interfere with viral transcription or replication, or other innate defense mechanisms. Species-specific sequence differences in host molecules involved in virus replication may contribute to reduced efficiency of viral infection and replication in non-native hosts, since viruses have co-evolved with the native host. Reduced ability to counter antiviral responses—in particular the type I IFN response—in a non-native host also can be an important factor. The multiple mechanisms that RSV has evolved to block the host IFN response (see chapter by S.M. Varga and T.J. Braciale, this volume) illustrates the importance of this inhibition for the virus. For example, in mice, RSV replication is typically restricted almost exclusively to the type I alveolar pneumocytes, but a combined MAVS and MyD88 knockout that blocks IFN induction allows RSV to replicate in bronchiolar epithelium (Bhoj et al. 2008).

A specific receptor interaction that explains tissue tropism for RSV has not yet been identified. Both the RSV G and F glycoproteins can interact with heparan sulfate and are heavily glycosylated. Therefore, RSV can bind to proteoglycans and C-type lectins indiscriminately, which has complicated the search for a more specific RSV receptor. This may also contribute to the ability to infect a variety of cell types. Although the G glycoprotein is known as the attachment protein because of its strong glycosaminoglycan binding properties, G is not required for infection of cells in vitro, as noted, and thus the F glycoprotein may be more likely to have an essential, specific receptor-binding activity. Nucleolin has recently been described as a functional receptor for the RSV F protein, and it clearly increases permissivity to RSV infection (Tayyari et al. 2011). However, nucleolin is ubiquitously expressed and thus does not explain the strong preference that RSV has for respiratory epithelium. It is possible that there may be unique tissue-specific post-translational modifications of nucleolin, or a tissue-specific co-receptor, but these possible tissue-specific factors remain to be identified. Other interactions have been noted between RSV and cellular proteins, such as between F and TLR4 or G and CX3CR1, although possible contributions to attachment and tropism remain to be defined. There is very little information available on cellular factors that may facilitate or restrict RSV replication and thereby affect cellular or species tropism, and this is an area that needs more attention.

11.2 Vaccine-Enhanced Illness

FI-RSV is an intramuscular, alum-adjuvanted, formalin-inactivated, whole-virus RSV vaccine that was evaluated in infants and young children in the 1960s. This vaccine was poorly protective and primed for enhanced disease upon RSV infection from subsequent natural exposure. The legacy of vaccine-enhanced illness has hovered over the field of RSV vaccine development for over 40 years, especially for the pediatric population. Much of the work on pathogenesis and animal model development for RSV has been devoted to understanding this phenomenon, because it was perceived to be the greatest barrier to having a robust pipeline of candidate vaccines for clinical evaluation. As a result, we have a better general understanding of the molecular mechanisms of RSV immunopathology and immunoregulation, and have developed better assays to measure the function and specificity of T cell and antibody responses. However, still there are uncertainties about the basis for the enhanced illness induced by FI-RSV and the ability to reliably determine preclinically whether new pediatric vaccine candidates are free of this phenomenon.

A series of clinical studies were done in the 1960s to evaluate the FI-RSV vaccine in children of different ages (Kim et al. 1969; Kapikian et al. 1969; Fulginiti et al. 1969; Chin et al. 1969). The vaccine did not protect against subsequent infection and, compared to control groups, caused a much greater frequency of severe disease, particularly in the youngest age group (<6 months of age). Of 31 infants immunized with FI-RSV, 25 required hospitalization following natural infection and two died (Kim et al. 1969). Only one of the 40 infants who received a control preparation of formalin-inactivated PIV3 required hospitalization. While the frequency of severe illness diminished as children became older (Fulginiti et al. 1969), enhanced disease was seen in children up to 37 months of age at the time of immunization (Fulginiti et al. 1969), and was common in children up to 23 months of age (Kapikian et al. 1969). The clinical manifestations appeared to be similar to those seen in the 3–4 % of infants in the general population who experience the most severe disease during RSV epidemics. This was suggestive of enhanced rather than altered disease, although this was not carefully evaluated. Lung pathology in two children who died showed peribronchiolar inflammation and fibrinous exudates composed of sloughed epithelial cells, mononuclear cells, neutrophils, and eosinophils causing obstruction of small airways. Eosinophilia among vaccine recipients was also noted in one study (Chin et al. 1969). This analysis was complicated by the occurrence of bacterial superinfection in both individuals. Although RSV-specific antibody responses measured by complement fixation were high, there was poor neutralizing and fusion-inhibiting activity (Murphy et al. 1986; Murphy and Walsh 1988). Delayed-type hypersensitivity (DTH) and significant lymphoproliferative responses were detected in peripheral blood mononuclear cells (Kim et al. 1976), similar to patients experiencing atypical measles syndrome which was a consequence of immunization with a whole, inactivated measles virus vaccine (Lennon et al. 1967). In addition, there was immunohistochemical evidence of immune complex deposition in small airways (Polack et al. 2002). Similar immune complex deposition has been noted in young adult patients with severe disease from pandemic influenza, with the presumption that cross-reactive low-avidity antibody failed to clear virus and caused complement deposition and inflammation resulting in more severe disease (Monsalvo et al. 2011).

FI-RSV vaccine-enhanced illness has been modeled in mice (Graham et al. 1993), cotton rats (Connors et al. 1992), calves (Gershwin et al. 1998), and non-human primates (Kakuk et al. 1993; De Swart et al. 2002). Many of the pathological features of the disease have been recapitulated in these models, but because they are all either semi-permissive hosts or use an alternative virus and host (i.e., BRSV), these models likely are inexact surrogates for human infants. Unfortunately, the original phenomenon in the human vaccinees occurred long ago and was not well characterized. Thus, data from animal models cannot exclude the possibility that a vaccine may cause an enhanced illness syndrome in humans. However, vaccine-enhanced RSV disease appeared to be specific to RSV-naive vaccinees, which also is the case in experimental animals. Studies in experimental animals indicate that enhanced disease is associated with protein-based RSV vaccines such as whole inactivated virus or purified proteins, but not with wild-type or attenuated replication-competent RSV. We know from clinical studies that replication-competent RSV given intranasally (Wright et al. 2007) or intramuscularly (Belshe et al. 1982) to human infants is not associated with enhanced RSV disease, nor is repeated exposure to RSV in nature associated with enhanced disease. Thus, replication-competent RSV vaccines are considered safe for testing in RSV-naïve infants, whereas protein-based vaccines are not.

Based on the diminished frequency of FI-RSV vaccine-enhanced illness in older children, vaccine studies in adults, and studies in animal models, it is thought that initial priming with live virus mitigates the effects of subsequent immunization with protein-based vaccines, and thus these vaccines are considered safe for use in older children and adults. We also know from animal models that the immunology of FI-RSV induced inflammation involves Th2-biased immune responses (Graham et al. 1993) including the production of IL-4, IL-5, and IL-13. Thus, enhanced disease likely involves altered immunologic priming, probably influenced by the mode of antigen presentation. Based on the available data from humans and animal models, two major immunological processes associated with the FI-RSV vaccine-enhanced disease can be articulated, and should be given careful consideration when deciding on the appropriate target population and regulatory requirements for new vaccine candidates. It is particularly important to consider these issues when the candidate vaccine is designed to be the first exposure to RSV antigen for RSV-naïve infants. First, RSV-specific antibody was induced with poor functional activity. This antibody was insufficient to restrict RSV replication, and may have contributed to immune complex deposition in small airways. Secondly, the FI-RSV enhanced disease was associated with a Th2-biased CD4 T-cell response characterized by cytokines related to allergic inflammation. Therefore, vaccines should be designed to optimize neutralizing activity and to avoid Th2-biased T-cell responses. There is evidence from epidemiological and genetic studies that associate immune components involved in allergic inflammation with severe RSV disease from primary infection. These data are discussed in other chapters in detail.

Acknowledgments PLC and BSG were funded by the NIAID Intramural Program. RF was funded by Boston University School of Medicine.

References

Asenjo A, Rodriguez L, Villanueva N (2005) Determination of phosphorylated residues from human respiratory syncytial virus P protein that are dynamically dephosphorylated by cellular phosphatases: a possible role for serine 54. J Gen Virol 86(Pt 4):1109–1120. doi:10.1099/vir.0.80692-0 86/4/1109 [pii]

Asenjo A, Calvo E, Villanueva N (2006) Phosphorylation of human respiratory syncytial virus P protein at threonine 108 controls its interaction with the M2-1 protein in the viral RNA polymerase complex. J Gen Virol 87(Pt 12):3637–3642. doi:10.1099/vir.0.82165-0 87/12/3637 [pii]

Asenjo A, Gonzalez-Armas JC, Villanueva N (2008) Phosphorylation of human respiratory syncytial virus P protein at serine 54 regulates viral uncoating. Virology 380(1):26–33. doi:10.1016/j.virol.2008.06.045 S0042-6822(08)00427-3 [pii]

Atreya PL, Peeples ME, Collins PL (1998) The NS1 protein of human respiratory syncytial virus is a potent inhibitor of minigenome transcription and RNA replication. J Virol 72(2):1452–1461

Ausar SF, Espina M, Brock J, Thyagarayapuran N, Repetto R, Khandke L, Middaugh CR (2007) High-throughput screening of stabilizers for respiratory syncytial virus: identification of stabilizers and their effects on the conformational thermostability of viral particles. Hum Vaccin 3(3):94–103. doi:10.1099/vir.0.82165-0 4149 [pii]

Belshe RB, Van Voris LP, Mufson MA (1982) Parenteral administration of live respiratory syncytial virus vaccine: results of a field trial. J Infect Dis 145(3):311–319

Bermingham A, Collins PL (1999) The M2-2 protein of human respiratory syncytial virus is a regulatory factor involved in the balance between RNA replication and transcription. Proc Natl Acad Sci U S A 96(20):11259–11264

Bhoj VG, Sun Q, Bhoj EJ, Somers C, Chen X, Torres JP, Mejias A, Gomez AM, Jafri H, Ramilo O, Chen ZJ (2008) MAVS and MyD88 are essential for innate immunity but not cytotoxic T lymphocyte response against respiratory syncytial virus. Proc Natl Acad Sci U S A 105(37):14046–14051. doi:10.1073/pnas.0804717105

Bitko V, Shulyayeva O, Mazumder B, Musiyenko A, Ramaswamy M, Look DC, Barik S (2007) Nonstructural proteins of respiratory syncytial virus suppress premature apoptosis by an NF-kappaB-dependent, interferon-independent mechanism and facilitate virus growth. J Virol 81(4):1786–1795. doi:10.1128/JVI.01420-06 JVI.01420-06 [pii]

Blondot ML, Dubosclard V, Fix J, Lassoued S, Aumont-Nicaise M, Bontems F, Eleouet JF, Sizun C (2012) Structure and functional analysis of the RNA- and viral phosphoprotein-binding domain of respiratory syncytial virus M2-1 protein. PLoS Pathog 8(5):e1002734. doi:10.1371/journal.ppat.1002734

Brock SC, Goldenring JR, Crowe JE Jr (2003) Apical recycling systems regulate directional budding of respiratory syncytial virus from polarized epithelial cells. Proc Natl Acad Sci U S A 100(25):15143–15148

Brock LG, Karron RA, Krempl CD, Collins PL, Buchholz UJ (2012) Evaluation of pneumonia virus of mice as a possible human pathogen. J Virol 86(10):5829–5843. doi:10.1128/JVI.00163-12

Brown G, Rixon HW, Steel J, McDonald TP, Pitt AR, Graham S, Sugrue RJ (2005) Evidence for an association between heat shock protein 70 and the respiratory syncytial virus polymerase complex within lipid-raft membranes during virus infection. Virology 338(1):69–80. doi:10.1016/j.virol.2005.05.004

Buchholz UJ, Granzow H, Schuldt K, Whitehead SS, Murphy BR, Collins PL (2000) Chimeric bovine respiratory syncytial virus with glycoprotein gene substitutions from human respiratory syncytial virus (HRSV): Effects on host range and evaluation as a live-attenuated HRSV vaccine. J Virol 74(3):1187–1199

Bukreyev A, Murphy BR, Collins PL (2000) Respiratory syncytial virus can tolerate an intergenic sequence of at least 160 nucleotides with little effect on transcription or replication in vitro and in vivo. J Virol 74(23):11017–11026

Bukreyev A, Yang L, Fricke J, Cheng L, Ward JM, Murphy BR, Collins PL (2008) The secreted form of respiratory syncytial virus G glycoprotein helps the virus evade antibody-mediated restriction of replication by acting as an antigen decoy and through effects on Fc receptor-bearing leukocytes. J Virol 82(24):12191–12204. doi:10.1128/JVI.01604-08 JVI.01604-08 [pii]

Cartee TL, Wertz GW (2001) Respiratory syncytial virus M2-1 protein requires phosphorylation for efficient function and binds viral RNA during infection. J Virol 75(24):12188–12197. doi:10.1128/JVI.75.24.12188-12197.2001

Cartee TL, Megaw AG, Oomens AG, Wertz GW (2003) Identification of a single amino acid change in the human respiratory syncytial virus L protein that affects transcriptional termination. J Virol 77(13):7352–7360

Carter SD, Dent KC, Atkins E, Foster TL, Verow M, Gorny P, Harris M, Hiscox JA, Ranson NA, Griffin S, Barr JN (2010) Direct visualization of the small hydrophobic protein of human respiratory syncytial virus reveals the structural basis for membrane permeability. FEBS Lett 584(13):2786–2790. doi:10.1016/j.febslet.2010.05.006 S0014-5793(10)00395-9 [pii]

Castagne N, Barbier A, Bernard J, Rezaei H, Huet JC, Henry C, Da Costa B, Eleouet JF (2004) Biochemical characterization of the respiratory syncytial virus P-P and P-N protein complexes and localization of the P protein oligomerization domain. J Gen Virol 85(Pt 6):1643–1653

Cheng X, Park H, Zhou H, Jin H (2005) Overexpression of the M2-2 protein of respiratory syncytial virus inhibits viral replication. J Virol 79(22):13943–13952. doi:10.1128/JVI.79.22.13943-13952.2005 79/22/13943 [pii]

Chin J, Magoffin RL, Shearer LA, Schieble JH, Lennette EH (1969) Field evaluation of a respiratory syncytial virus vaccine and a trivalent parainfluenza virus vaccine in a pediatric population. Am J Epidemiol 89(4):449–463

Collins PL, Wertz GW (1983) cDNA cloning and transcriptional mapping of nine polyadenylylated RNAs encoded by the genome of human respiratory syncytial virus. Proc Natl Acad Sci U S A 80(11):3208–3212

Collins PL, Dickens LE, Buckler-White A, Olmsted RA, Spriggs MK, Camargo E, Coelingh KV (1986) Nucleotide sequences for the gene junctions of human respiratory syncytial virus reveal distinctive features of intergenic structure and gene order. Proc Natl Acad Sci U S A 83(13):4594–4598

Collins PL, Olmsted RA, Spriggs MK, Johnson PR, Buckler-White AJ (1987) Gene overlap and site-specific attenuation of transcription of the viral polymerase L gene of human respiratory syncytial virus. Proc Natl Acad Sci U S A 84(15):5134–5138

Collins PL, Hill MG, Camargo E, Grosfeld H, Chanock RM, Murphy BR (1995) Production of infectious human respiratory syncytial virus from cloned cDNA confirms an essential role for the transcription elongation factor from the 5′ proximal open reading frame of the M2 mRNA in gene expression and provides a capability for vaccine development. Proc Natl Acad Sci USA 92:11563–11567

Collins PL, Hill MG, Cristina J, Grosfeld H (1996) Transcription elongation factor of respiratory syncytial virus, a nonsegmented negative-strand RNA virus. Proc Natl Acad Sci U S A 93(1):81–85

Collins PL, Camargo E, Hill MG (1999) Support plasmids and support proteins required for recovery of recombinant respiratory syncytial virus. Virology 259(2):251–255

Connors M, Kulkarni AB, Firestone CY, Holmes KL, Morse HCd, Sotnikov AV, Murphy BR (1992) Pulmonary histopathology induced by respiratory syncytial virus (RSV) challenge of

formalin-inactivated RSV-immunized BALB/c mice is abrogated by depletion of CD4 + T cells. J Virol 66(12):7444–7451
Cowton VM, Fearns R (2005) Evidence that the respiratory syncytial virus polymerase is recruited to nucleotides 1 to 11 at the 3′ end of the nucleocapsid and can scan to access internal signals. J Virol 79(17):11311–11322. doi:10.1128/JVI.79.17.11311-11322.2005 79/17/11311 [pii]
Cowton VM, McGivern DR, Fearns R (2006) Unravelling the complexities of respiratory syncytial virus RNA synthesis. J Gen Virol 87(Pt 7):1805–1821. doi:10.1099/vir.0.81786-0 87/7/1805 [pii]
De Swart RL, Kuiken T, Timmerman HH, Amerongen Gv G, Van Den Hoogen BG, Vos HW, Neijens HJ, Andeweg AC, Osterhaus AD (2002) Immunization of macaques with formalin-inactivated respiratory syncytial virus (RSV) induces interleukin-13-associated hypersensitivity to subsequent RSV infection. J Virol 76(22):11561–11569
Drexler JF, Corman VM, Muller MA, Maganga GD, Vallo P, Binger T, Gloza-Rausch F, Rasche A, Yordanov S, Seebens A, Oppong S, Sarkodie YA, Pongombo C, Lukashev AN, Schmidt-Chanasit J, Stocker A, Carneiro AJ, Erbar S, Maisner A, Fronhoffs F, Buettner R, Kalko EK, Kruppa T, Franke CR, Kallies R, Yandoko ER, Herrler G, Reusken C, Hassanin A, Kruger DH, Matthee S, Ulrich RG, Leroy EM, Drosten C (2012) Bats host major mammalian paramyxoviruses. Nat Commun 3:796. doi:10.1038/ncomms1796
Dupuy LC, Dobson S, Bitko V, Barik S (1999) Casein kinase 2-mediated phosphorylation of respiratory syncytial virus phosphoprotein P is essential for the transcription elongation activity of the viral polymerase; phosphorylation by casein kinase 1 occurs mainly at Ser(215) and is without effect. J Virol 73(10):8384–8392
Dyer KD, Percopo CM, Fischer ER, Gabryszewski SJ, Rosenberg HF (2009) Pneumoviruses infect eosinophils and elicit MyD88-dependent release of chemoattractant cytokines and interleukin-6. Blood 114(13):2649–2656. doi:10.1182/blood-2009-01-199497
Eisenhut M (2006) Extrapulmonary manifestations of severe respiratory syncytial virus infection—a systematic review. Crit Care 10(4):R107. doi:10.1186/cc4984 cc4984 [pii]
Eshaghi A, Duvvuri VR, Lai R, Nadarajah JT, Li A, Patel SN, Low DE, Gubbay JB (2012) Genetic variability of human respiratory syncytial virus a strains circulating in Ontario: a novel genotype with a 72 nucleotide G gene duplication. PLoS ONE 7(3):e32807. doi:10.1371/journal.pone.0032807
Fearns R, Collins PL (1999a) Model for polymerase access to the overlapped L gene of respiratory syncytial virus. J Virol 73(1):388–397
Fearns R, Collins PL (1999b) Role of the M2-1 transcription antitermination protein of respiratory syncytial virus in sequential transcription. J Virol 73(7):5852–5864
Fearns R, Peeples ME, Collins PL (1997) Increased expression of the N protein of respiratory syncytial virus stimulates minigenome replication but does not alter the balance between the synthesis of mRNA and antigenome. Virology 236(1):188–201
Fearns R, Collins PL, Peeples ME (2000) Functional analysis of the genomic and antigenomic promoters of human respiratory syncytial virus. J Virol 74(13):6006–6014
Fearns R, Peeples ME, Collins PL (2002) Mapping the transcription and replication promoters of respiratory syncytial virus. J Virol 76(4):1663–1672
Fix J, Galloux M, Blondot ML, Eleouet JF (2011) The insertion of fluorescent proteins in a variable region of respiratory syncytial virus L polymerase results in fluorescent and functional enzymes but with reduced activities. Open Virol J 5:103–108. doi:10.2174/1874357901105010103
J Fricke, LY Koo, CR Brown, PL Collins (2013) P38 and OGT sequestration into viral inclusion bodies in cells infected with human respiratory syncytial virus suppresses MK2 activities and stress granule assembly. J Virol 87(3):1333–1347
Fuentes S, Tran KC, Luthra P, Teng MN, He B (2007) Function of the respiratory syncytial virus small hydrophobic protein. J Virol 81(15):8361–8366. doi:10.1128/JVI.02717-06 JVI.02717-06 [pii]

Fulginiti VA, Eller JJ, Sieber OF, Joyner JW, Minamitani M, Meiklejohn G (1969) Respiratory virus immunization. I. A field trial of two inactivated respiratory virus vaccines; an aqueous trivalent parainfluenza virus vaccine and an alum-precipitated respiratory syncytial virus vaccine. Am J Epidemiol 89(4):435–448

Galloux M, Tarus B, Blazevic I, Fix J, Duquerroy S, Eleouet JF (2012) Characterization of a viral phosphoprotein binding site on the surface of the respiratory syncytial nucleoprotein. J Virol 86(16):8375–8387. doi:10.1128/JVI.00058-12

Gan SW, Tan E, Lin X, Yu D, Wang J, Tan GM, Vararattanavech A, Yeo CY, Soon CH, Soong TW, Pervushin K, Torres J (2012) The small hydrophobic protein of the human respiratory syncytial virus forms pentameric ion channels. J Biol Chem 287(29):24671–24689. doi:10.1074/jbc.M111.332791

Gershwin LJ, Schelegle ES, Gunther RA, Anderson ML, Woolums AR, Larochelle DR, Boyle GA, Friebertshauser KE, Singer RS (1998) A bovine model of vaccine enhanced respiratory syncytial virus pathophysiology [In Process Citation]. Vaccine 16(11–12):1225–1236

Ghildyal R, Ho A, Jans DA (2006) Central role of the respiratory syncytial virus matrix protein in infection. FEMS Microbiol Rev 30(5):692–705. doi:10.1111/j.1574-6976.2006.00025.x

Gonzalez-Reyes L, Ruiz-Arguello MB, Garcia-Barreno B, Calder L, Lopez JA, Albar JP, Skehel JJ, Wiley DC, Melero JA (2001) Cleavage of the human respiratory syncytial virus fusion protein at two distinct sites is required for activation of membrane fusion. Proc Natl Acad Sci U S A 98(17):9859–9864

Gould PS, Easton AJ (2007) Coupled translation of the second open reading frame of M2 mRNA is sequence dependent and differs significantly within the subfamily Pneumovirinae. J Virol 81(16):8488–8496. doi:10.1128/JVI.00457-07 JVI.00457-07 [pii]

Graham BS, Henderson GS, Tang YW, Lu X, Neuzil KM, Colley DG (1993) Priming immunization determines T helper cytokine mRNA expression patterns in lungs of mice challenged with respiratory syncytial virus. J Immunol 151(4):2032–2040

Grosfeld H, Hill MG, Collins PL (1995) RNA replication by respiratory syncytial virus (RSV) is directed by the N, P, and L proteins; transcription also occurs under these conditions but requires RSV superinfection for efficient synthesis of full-length mRNA. J Virol 69(9):5677–5686

Groskreutz DJ, Babor EC, Monick MM, Varga SM, Hunninghake GW (2010) Respiratory syncytial virus limits alpha subunit of eukaryotic translation initiation factor 2 (eIF2alpha) phosphorylation to maintain translation and viral replication. J Biol Chem 285(31):24023–24031. doi:10.1074/jbc.M109.077321 M109.077321 [pii]

Hanley LL, McGivern DR, Teng MN, Djang R, Collins PL, Fearns R (2010) Roles of the respiratory syncytial virus trailer region: effects of mutations on genome production and stress granule formation. Virology 406:241–252. doi:10.1016/j.virol.2010.07.006 S0042-6822(10)00446-0 [pii]

Hardy RW, Wertz GW (1998) The product of the respiratory syncytial virus M2 gene ORF1 enhances readthrough of intergenic junctions during viral transcription. J Virol 72(1):520–526

Hardy RW, Wertz GW (2000) The Cys(3)-His(1) motif of the respiratory syncytial virus M2-1 protein is essential for protein function. J Virol 74(13):5880–5885

Harmon SB, Wertz GW (2002) Transcriptional termination modulated by nucleotides outside the characterized gene end sequence of respiratory syncytial virus. Virology 300(2):304–315

Harpen M, Barik T, Musiyenko A, Barik S (2009) Mutational analysis reveals a noncontractile but interactive role of actin and profilin in viral RNA-dependent RNA synthesis. J Virol 83(21):10869–10876. doi:10.1128/JVI.01271-09

Haynes LM, Moore DD, Kurt-Jones EA, Finberg RW, Anderson LJ, Tripp RA (2001) Involvement of toll-like receptor 4 in innate immunity to respiratory syncytial virus. J Virol 75(22):10730–10737. doi:10.1128/JVI.75.22.10730-10737.2001

Henderson G, Murray J, Yeo RP (2002) Sorting of the respiratory syncytial virus matrix protein into detergent-resistant structures is dependent on cell-surface expression of the glycoproteins. Virology 300(2):244–254

Hendricks DA, McIntosh K, Patterson JL (1988) Further characterization of the soluble form of the G glycoprotein of respiratory syncytial virus. J Virol 62(7):2228–2233

Jeffree CE, Rixon HW, Brown G, Aitken J, Sugrue RJ (2003) Distribution of the attachment (G) glycoprotein and GM1 within the envelope of mature respiratory syncytial virus filaments revealed using field emission scanning electron microscopy. Virology 306(2):254–267

Johnson RA, Prince GA, Suffin SC, Horswood RL, Chanock RM (1982) Respiratory syncytial virus infection in cyclophosphamide-treated cotton rats. Infect Immun 37(1):369–373

Johnson JE, Gonzales RA, Olson SJ, Wright PF, Graham BS (2007) The histopathology of fatal untreated human respiratory syncytial virus infection. Mod Pathol 20(1):108–119. doi:10.1038/modpathol.3800725 3800725 [pii]

Johnson TR, Johnson CN, Corbett KS, Edwards GC, Graham BS (2011) Primary human mDC1, mDC2, and pDC dendritic cells are differentially infected and activated by respiratory syncytial virus. PLoS ONE 6(1):e16458. doi:10.1371/journal.pone.0016458

Johnson TR, McLellan JS, Graham BS (2012) Respiratory syncytial virus glycoprotein G interacts with DC-SIGN and L-SIGN to activate ERK1 and ERK2. J Virol 86(3):1339–1347. doi:10.1128/JVI.06096-11

Kakuk TJ, Soike K, Brideau RJ, Zaya RM, Cole SL, Zhang JY, Roberts ED, Wells PA, Wathen MW (1993) A human respiratory syncytial virus (RSV) primate model of enhanced pulmonary pathology induced with a formalin-inactivated RSV vaccine but not a recombinant FG subunit vaccine. J Infect Dis 167(3):553–561

Kalica AR, Wright PF, Hetrick FM, Chanock RM (1973) Electron microscopic studies of respiratory syncytial temperature-sensitive mutants. Arch Gesamte Virusforsch 41(3):248–258

Kapikian AZ, Mitchell RH, Chanock RM, Shvedoff RA, Stewart CE (1969) An epidemiologic study of altered clinical reactivity to respiratory syncytial (RS) virus infection in children previously vaccinated with an inactivated RS virus vaccine. Am J Epidemiol 89(4):405–421

Karron RA, Buonagurio DA, Georgiu AF, Whitehead SS, Adamus JE, Clements-Mann ML, Harris DO, Randolph VB, Udem SA, Murphy BR, Sidhu MS (1997a) Respiratory syncytial virus (RSV) SH and G proteins are not essential for viral replication in vitro: clinical evaluation and molecular characterization of a cold-passaged, attenuated RSV subgroup B mutant. Proc Natl Acad Sci U S A 94(25):13961–13966

Karron RA, Wright PF, Crowe JE Jr, Clements ML, Thompson J, Makhene M, Casey R, Murphy BR (1997b) Evaluation of two live, cold-passaged, temperature-sensitive respiratory syncytial virus (RSV) vaccines in chimpanzees, adults, infants and children. J Infect Dis 176:1428–1436

Kim HW, Canchola JG, Brandt CD, Pyles G, Chanock RM, Jensen K, Parrott RH (1969) Respiratory syncytial virus disease in infants despite prior administration of antigenic inactivated vaccine. Am J Epidemiol 89(4):422–434

Kim HW, Leikin SL, Arrobio J, Brandt CD, Chanock RM, Parrott RH (1976) Cell-mediated immunity to respiratory syncytial virus induced by inactivated vaccine or by infection. Pediatr Res 10(1):75–78

Kolokoltsov AA, Deniger D, Fleming EH, Roberts NJ Jr, Karpilow JM, Davey RA (2007) Small interfering RNA profiling reveals key role of clathrin-mediated endocytosis and early endosome formation for infection by respiratory syncytial virus. J Virol 81(14):7786–7800. doi:10.1128/JVI.02780-06 JVI.02780-06 [pii]

Kuo L, Grosfeld H, Cristina J, Hill MG, Collins PL (1996) Effect of mutations in the gene-start and gene-end sequence motifs on transcription of monocistronic and dicistronic minigenomes of respiratory syncytial virus. J Virol 70(10):6892–6901

Kuo L, Fearns R, Collins PL (1997) Analysis of the gene start and gene end signals of human respiratory syncytial virus: quasi-templated initiation at position 1 of the encoded mRNA. J Virol 71(7):4944–4953

Kwilas S, Liesman RM, Zhang L, Walsh E, Pickles RJ, Peeples ME (2009) Respiratory syncytial virus grown in Vero cells contains a truncated attachment protein that alters its infectivity and dependence on glycosaminoglycans. J Virol 83(20):10710–10718. doi:10.1128/JVI.00986-09 JVI.00986-09 [pii]

Kwilas AR, Yednak MA, Zhang L, Liesman R, Collins PL, Pickles RJ, Peeples ME (2010) Respiratory syncytial virus engineered to express CFTR corrects the bioelectric phenotype of human cystic fibrosis airway epithelium in vitro. J Virol 84:7770–7781. doi:10.1128/JVI.00346-10 JVI.00346-10 [pii]

Langedijk JP, de Groot BL, Berendsen HJ, van Oirschot JT (1998) Structural homology of the central conserved region of the attachment protein G of respiratory syncytial virus with the fourth subdomain of 55-kDa tumor necrosis factor receptor. Virology 243(2):293–302

Lennon RG, Isacson P, Rosales T, Elsea WR, Karzon DT, Winkelstein W Jr (1967) Skin tests with measles and poliomyelitis vaccines in recipients of inactivated measles virus vaccine. Delayed dermal hypersensitivity. JAMA 200(4):275–280

Li D, Jans DA, Bardin PG, Meanger J, Mills J, Ghildyal R (2008) Association of respiratory syncytial virus M protein with viral nucleocapsids is mediated by the M2-1 protein. J Virol 82(17):8863–8870. doi:10.1128/JVI.00343-08

Lifland AW, Jung J, Alonas E, Zurla C, Crowe JE Jr, Santangelo PJ (2012) Human respiratory syncytial virus nucleoprotein and inclusion bodies antagonize the innate immune response mediated by MDA5 and MAVS. J Virol 86(15):8245–8258. doi:10.1128/JVI.00215-12

Lindquist ME, Lifland AW, Utley TJ, Santangelo PJ, Crowe JE Jr (2010) Respiratory syncytial virus induces host RNA stress granules to facilitate viral replication. J Virol 84(23):12274–12284. doi:10.1128/JVI.00260-10

Lindquist ME, Mainou BA, Dermody TS, Crowe JE Jr (2011) Activation of protein kinase R is required for induction of stress granules by respiratory syncytial virus but dispensable for viral replication. Virology 413(1):103–110. doi:10.1016/j.virol.2011.02.009

Liuzzi M, Mason SW, Cartier M, Lawetz C, McCollum RS, Dansereau N, Bolger G, Lapeyre N, Gaudette Y, Lagace L, Massariol MJ, Do F, Whitehead P, Lamarre L, Scouten E, Bordeleau J, Landry S, Rancourt J, Fazal G, Simoneau B (2005) Inhibitors of respiratory syncytial virus replication target cotranscriptional mRNA guanylylation by viral RNA-dependent RNA polymerase. J Virol 79(20):13105–13115

Llorente MT, Garcia-Barreno B, Calero M, Camafeita E, Lopez JA, Longhi S, Ferron F, Varela PF, Melero JA (2006) Structural analysis of the human respiratory syncytial virus phosphoprotein: characterization of an alpha-helical domain involved in oligomerization. J Gen Virol 87(Pt 1):159–169. doi:10.1099/vir.0.81430-0 87/1/159 [pii]

Lu B, Ma CH, Brazas R, Jin H (2002) The major phosphorylation sites of the respiratory syncytial virus phosphoprotein are dispensable for virus replication in vitro. J Virol 76(21):10776–10784

Marsh R, Connor A, Gias E, Toms GL (2007) Increased susceptibility of human respiratory syncytial virus to neutralization by anti-fusion protein antibodies on adaptation to replication in cell culture. J Med Virol 79(6):829–837. doi:10.1002/jmv.20892

McCurdy LH, Graham BS (2003) Role of plasma membrane lipid microdomains in respiratory syncytial virus filament formation. J Virol 77(3):1747–1756

McDonald TP, Pitt AR, Brown G, Rixon HW, Sugrue RJ (2004) Evidence that the respiratory syncytial virus polymerase complex associates with lipid rafts in virus-infected cells: a proteomic analysis. Virology 330(1):147–157

McGivern DR, Collins PL, Fearns R (2005) Identification of internal sequences in the 3′ leader region of human respiratory syncytial virus that enhance transcription and confer replication processivity. J Virol 79(4):2449–2460

Mitra R, Baviskar P, Duncan-Decocq RR, Patel D, Oomens AG (2012) The human respiratory syncytial virus matrix protein is required for maturation of viral filaments. J Virol 86(8):4432–4443. doi:10.1128/JVI.06744-11

Money VA, McPhee HK, Mosely JA, Sanderson JM, Yeo RP (2009) Surface features of a Mononegavirales matrix protein indicate sites of membrane interaction. Proc Natl Acad Sci U S A 106(11):4441–4446. doi:10.1073/pnas.0805740106 0805740106 [pii]

Monsalvo AC, Batalle JP, Lopez MF, Krause JC, Klemenc J, Hernandez JZ, Maskin B, Bugna J, Rubinstein C, Aguilar L, Dalurzo L, Libster R, Savy V, Baumeister E, Aguilar L, Cabral G, Font J, Solari L, Weller KP, Johnson J, Echavarria M, Edwards KM, Chappell JD, Crowe JE Jr, Williams JV, Melendi GA, Polack FP (2011) Severe pandemic 2009 H1N1 influenza disease due to pathogenic immune complexes. Nat Med 17(2):195–199. doi:10.1038/nm.2262

Murphy BR, Walsh EE (1988) Formalin-inactivated respiratory syncytial virus vaccine induces antibodies to the fusion glycoprotein that are deficient in fusion-inhibiting activity. J Clin Microbiol 26(8):1595–1597

Murphy BR, Prince GA, Walsh EE, Kim HW, Parrott RH, Hemming VG, Rodriguez WJ, Chanock RM (1986) Dissociation between serum neutralizing and glycoprotein antibody responses of infants and children who received inactivated respiratory syncytial virus vaccine. J Clin Microbiol 24(2):197–202

Noton SL, Fearns R (2011) The first two nucleotides of the respiratory syncytial virus antigenome RNA replication product can be selected independently of the promoter terminus. RNA 17(10):1895–1906. doi:10.1261/rna.2813411

Noton SL, Cowton VM, Zack CR, McGivern DR, Fearns R (2010) Evidence that the polymerase of respiratory syncytial virus initiates RNA replication in a nontemplated fashion. Proc Natl Acad Sci U S A 107(22):10226–10231. doi:10.1073/pnas.0913065107 0913065107 [pii]

Noton SL, Deflube LR, Tremaglio CZ, Fearns R (2012) The respiratory syncytial virus polymerase has multiple RNA synthesis activities at the promoter. PLoS Path 8(10):e1002980

Peeples ME, Collins PL (2000) Mutations in the 5′ trailer region of a respiratory syncytial virus minigenome which limit RNA replication to one step. J Virol 74(1):146–155

Polack FP, Teng MN, Collins PL, Prince GA, Exner M, Regele H, Lirman DD, Rabold R, Hoffman SJ, Karp CL, Kleeberger SR, Wills-Karp M, Karron RA (2002) A role for immune complexes in enhanced respiratory syncytial virus disease. J Exp Med 196(6):859–865

Polack FP, Irusta PM, Hoffman SJ, Schiatti MP, Melendi GA, Delgado MF, Laham FR, Thumar B, Hendry RM, Melero JA, Karron RA, Collins PL, Kleeberger SR (2005) The cysteine-rich region of respiratory syncytial virus attachment protein inhibits innate immunity elicited by the virus and endotoxin. Proc Natl Acad Sci U S A 102(25):8996–9001

Rawling J, Cano O, Garcin D, Kolakofsky D, Melero JA (2011) Recombinant Sendai viruses expressing fusion proteins with two furin cleavage sites mimic the syncytial and receptor-independent infection properties of respiratory syncytial virus. J Virol 85(6):2771–2780. doi:10.1128/JVI.02065-10 JVI.02065-10 [pii]

Rezaee F, Gibson LF, Piktel D, Othumpangat S, Piedimonte G (2011) Respiratory syncytial virus infection in human bone marrow stromal cells. Am J Respir Cell Mol Biol 45(2):277–286. doi:10.1165/rcmb.2010-0121OC

Roberts SR, Compans RW, Wertz GW (1995) Respiratory syncytial virus matures at the apical surfaces of polarized epithelial cells. J Virol 69(4):2667–2673

Rohwedder A, Keminer O, Forster J, Schneider K, Schneider E, Werchau H (1998) Detection of respiratory syncytial virus RNA in blood of neonates by polymerase chain reaction. J Med Virol 54(4):320–327

Sastre P, Oomens AG, Wertz GW (2007) The stability of human respiratory syncytial virus is enhanced by incorporation of the baculovirus GP64 protein. Vaccine 25(27):5025–5033. doi:10.1016/j.vaccine.2007.04.066 S0264-410X(07)00506-3 [pii]

Spann KM, Collins PL, Teng MN (2003) Genetic recombination during coinfection of two mutants of human respiratory syncytial virus. J Virol 77(20):11201–11211

Spann KM, Tran KC, Collins PL (2005) Effects of nonstructural proteins NS1 and NS2 of human respiratory syncytial virus on interferon regulatory factor 3, NF-kappaB, and proinflammatory cytokines. J Virol 79(9):5353–5362

Srinivasakumar N, Ogra PL, Flanagan TD (1991) Characteristics of fusion of respiratory syncytial virus with HEp-2 cells as measured by R18 fluorescence dequenching assay. J Virol 65(8):4063–4069

Surman SR, Collins PL, Murphy BR, Skiadopoulos MH (2007) An improved method for the recovery of recombinant paramyxovirus vaccine candidates suitable for use in human clinical

trials. J Virol Methods 141(1):30–33. doi:10.1016/j.jviromet.2006.11.024 S0166-0934(06)00409-5 [pii]
Swedan S, Andrews J, Majumdar T, Musiyenko A, Barik S (2011) Multiple functional domains and complexes of the two nonstructural proteins of human respiratory syncytial virus contribute to interferon suppression and cellular location. J Virol 85(19):10090–10100. doi:10.1128/JVI.00413-11
Tawar RG, Duquerroy S, Vonrhein C, Varela PF, Damier-Piolle L, Castagne N, MacLellan K, Bedouelle H, Bricogne G, Bhella D, Eleouet JF, Rey FA (2009) Crystal structure of a nucleocapsid-like nucleoprotein-RNA complex of respiratory syncytial virus. Science 326(5957):1279–1283. doi:10.1126/science.1177634 326/5957/1279 [pii]
Tayyari F, Marchant D, Moraes TJ, Duan W, Mastrangelo P, Hegele RG (2011) Identification of nucleolin as a cellular receptor for human respiratory syncytial virus. Nat Med 17:1132–1135
Teng MN, Collins PL (1998) Identification of the respiratory syncytial virus proteins required for formation and passage of helper-dependent infectious particles. J Virol 72(7):5707–5716
Teng MN, Collins PL (2002) The central conserved cystine noose of the attachment G protein of human respiratory syncytial virus is not required for efficient viral infection in vitro or in vivo. J Virol 76(12):6164–6171
Teng MN, Whitehead SS, Bermingham A, Clair MS, Elkins WR, Murphy BR, Collins PL (2000) Recombinant respiratory syncytial virus that does not express the NS1 or M2-2 protein is highly attenuated and immunogenic in chimpanzees. J Virol 74:9317–9321
Tran TL, Castagne N, Dubosclard V, Noinville S, Koch E, Moudjou M, Henry C, Bernard J, Yeo RP, Eleouet JF (2009) The respiratory syncytial virus M2-1 protein forms tetramers and interacts with RNA and P in a competitive manner. J Virol 83(13):6363–6374. doi:10.1128/JVI.00335-09 JVI.00335-09 [pii]
Tripp RA, Jones LP, Haynes LM, Zheng H, Murphy PM, Anderson LJ (2001) CX3C chemokine mimicry by respiratory syncytial virus G glycoprotein. Nat Immunol 2(8):732–738
Utley TJ, Ducharme NA, Varthakavi V, Shepherd BE, Santangelo PJ, Lindquist ME, Goldenring JR, Crowe JE Jr (2008) Respiratory syncytial virus uses a Vps4-independent budding mechanism controlled by Rab11-FIP2. Proc Natl Acad Sci U S A 105(29):10209–10214. doi:10.1073/pnas.0712144105 0712144105 [pii]
Venter M, van Niekerk S, Rakgantso A, Bent N (2011) Identification of deletion mutant respiratory syncytial virus strains lacking most of the G protein in immunocompromised children with pneumonia in South Africa. J Virol 85(16):8453–8457. doi:10.1128/JVI.00674-11
Villenave R, Thavagnanam S, Sarlang S, Parker J, Douglas I, Skibinski G, Heaney LG, McKaigue JP, Coyle PV, Shields MD, Power UF (2012) In vitro modeling of respiratory syncytial virus infection of pediatric bronchial epithelium, the primary target of infection in vivo. Proc Natl Acad Sci U S A 109(13):5040–5045. doi:10.1073/pnas.1110203109
Wang JT, McElvain LE, Whelan SP (2007) Vesicular stomatitis virus mRNA capping machinery requires specific cis-acting signals in the RNA. J Virol 81(20):11499–11506. doi:10.1128/JVI.01057-07
Whitehead SS, Juhasz K, Firestone CY, Collins PL, Murphy BR (1998) Recombinant respiratory syncytial virus (RSV) bearing a set of mutations from cold-passaged RSV is attenuated in chimpanzees. J Virol 72(5):4467–4471
Whitehead SS, Bukreyev A, Teng MN, Firestone CY, St Claire M, Elkins WR, Collins PL, Murphy BR (1999) Recombinant respiratory syncytial virus bearing a deletion of either the NS2 or SH gene is attenuated in chimpanzees. J Virol 73(4):3438–3442
Wright PF, Ikizler MR, Gonzales RA, Carroll KN, Johnson JE, Werkhaven JA (2005) Growth of respiratory syncytial virus in primary epithelial cells from the human respiratory tract. J Virol 79(13):8651–8654. doi:10.1128/JVI.79.13.8651-8654.2005 79/13/8651 [pii]
Wright PF, Karron RA, Belshe RB, Shi JR, Randolph VB, Collins PL, O'Shea AF, Gruber WC, Murphy BR (2007) The absence of enhanced disease with wild type respiratory syncytial virus infection occurring after receipt of live, attenuated, respiratory syncytial virus vaccines. Vaccine 25(42):7372–7378. doi:10.1016/j.vaccine.2007.08.014 S0264-410X(07)00937-1 [pii]

Yeo DS, Chan R, Brown G, Ying L, Sutejo R, Aitken J, Tan BH, Wenk MR, Sugrue RJ (2009) Evidence that selective changes in the lipid composition of raft-membranes occur during respiratory syncytial virus infection. Virology 386(1):168–182. doi:10.1016/j.virol.2008.12.017 S0042-6822(08)00799-X [pii]

Zhang L, Peeples ME, Boucher RC, Collins PL, Pickles RJ (2002) Respiratory syncytial virus infection of human airway epithelial cells is polarized, specific to ciliated cells, and without obvious cytopathology. J Virol 76(11):5654–5666

Clinical and Epidemiologic Features of Respiratory Syncytial Virus

Caroline B. Hall, Eric A. F. Simőes and Larry J. Anderson

Abstract Since its discovery in 1955, respiratory syncytial virus (RSV) has consistently been noted to be the single most important cause of lower respiratory tract illness in infants <1 year of age. RSV also causes repeat infections and significant disease throughout life. In addition to the young child, persons with compromised immune, pulmonary or cardiac systems, and the elderly have significant risk from infection. Though RSV causes the full spectrum of acute respiratory illnesses, it is most notably associated with signs and symptoms of increased airway resistance manifested as wheezing and, in the young child, diagnosed as bronchiolitis. In temperate climates, RSV occurs as yearly outbreaks usually between late fall and early spring lasting 3–4 months in a community. The timing of outbreaks varies between years and in the same year between regions and even between nearby communities. RSV can be a serious nosocomial pathogen in high risk individuals but nosocomial transmission that can often be prevented with meticulous attention to good infection control practices. High risk groups include the premature infants and persons of any age with compromised cardiac, pulmonary, or immune systems. Risk factors for infection include increased number of children in the household and day care center attendance. There are reasonable

C. B. Hall
Departments of Pediatrics and Medicine, University of Rochester, School of Medicine and Dentistry, Rochester, NY, USA

E. A. F. Simőes
Colorado School of Public Health, University of Colorado School of Medicine, Center for Global Health, Aurora, CO, USA

E. A. F. Simőes
Division of Infectious Diseases, Children's Hospital Colorado, Aurora, CO, USA

L. J. Anderson (✉)
Department of Pediatrics and Children's Healthcare of Atlanta, Emory University, 2015 Uppergate Dr, Atlanta, GA 30322, USA
e-mail: larry.anderson@emory.edu

L. J. Anderson and B. S. Graham (eds.), *Challenges and Opportunities for Respiratory Syncytial Virus Vaccines*, Current Topics in Microbiology and Immunology 372, DOI: 10.1007/978-3-642-38919-1_2,

estimates of the sizable burden of RSV disease in infants and young children and the elderly but less data on disease in older children, the role of RSV in later reactive airway disease (see chapter by M.T. Lotz et al., this volume), and RSV-associated mortality in developing countries. The available data on burden of disease suggests there are at least four potential target populations for a vaccine, the young infant, young children >4–6 months of age, pregnant women, and the elderly. A link between infection in the young infant and later reactive airway disease and mortality in developing countries is needed. Each target population has different vaccine safety and efficacy concerns and may warrant a different type of vaccine.

Contents

1 Introduction

Over a half century ago, Morris et al. (1956) reported an outbreak of colds and coryza among a colony of chimpanzees and the recovery of a new agent, the chimpanzee coryza agent (CCA). Their description of its clinical import and transmission is prescient of the epidemiologic and clinical mein of CCA in humans, now recognized as respiratory syncytial virus (RSV). In chimpanzees, CCA could be transmitted by inoculating respiratory secretions into the nose, it had a high attack rate, and it spread efficiently. The illness, URI symptoms and cough, became evident ~3 days after inoculation and lasted up to 2 weeks. Subsequent studies over the following decades found similar features in human RSV infections globally and have demonstrated the substantial healthcare burden RSV imposes in developed and developing countries. RSV causes repeat infections throughout life but severe disease is most notable in the very young. In this chapter, we summarize our understanding of the clinical and epidemiologic features of RSV disease with a focus on the relevance of these features to developing vaccines.

2 Clinical Features

Primary RSV Infection often occurs during the first encounter with RSV, which is usually in infancy, and essentially all become infected by 2 years of age (Glezen et al. 1986). The singular clinical features of RSV primary infection are that most children are symptomatic, lower respiratory tract involvement is frequent, wheezing is prominent, and the very young, those in the first 3 months of life, are most severely affected (Hall 2012; Kim et al. 1973; Ogra 2004).

Typically, RSV infection starts with several days of mild upper respiratory tract signs, cough, and low grade fever. A worsening cough is usually heralded by lower respiratory tract involvement, and the infant becomes tachypneic and may have progressively more labored breathing, with dyspnea and retractions of the chest wall (Fig. 1). The most common auscultatory signs are crackles and wheezes, but they are often variable over minutes to hours. Radiologic findings most frequently show hyperinflation and peribronchial thickening (Wright and Piedimonte 2011). Scattered interstitial infiltrates may be present, but more characteristic are areas of atelectasis, particularly in the right middle and upper lobes. The physical examination and radiographic findings commonly do not reflect the degree of illness, e.g., severely ill children may have little or no fever and minimal auscultatory findings.

The acute illness usually lasts about 5–10 days, but the cough may be prolonged for several weeks. The duration of hospitalization for 918 infants and children under 5 years of age with laboratory-confirmed RSV infection was a median of 2 days and no children died (Hall et al. 2009). The most frequent discharge diagnosis among RSV positive hospitalized infants <1 year of age was

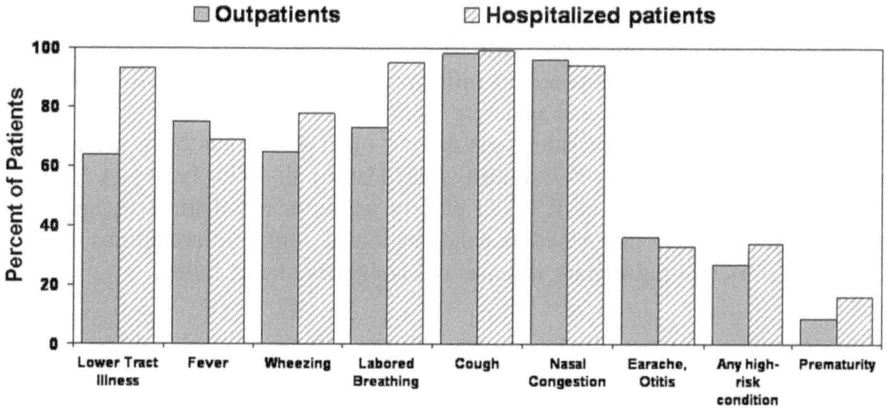

Fig. 1 The clinical characteristics of children less than 5 years of age with respiratory syncytial virus (RSV) infection who were outpatients compared to those who were hospitalized. RSV infection was laboratory confirmed during population-based surveillance of acute respiratory illnesses conducted in counties surrounding Nashville, TN, Rochester, NY, and Cincinnati, OH during 2000–2004. (Data form Hall et al. (2009))

bronchiolitis (85 %) but was asthma (60 %) and pneumonia (51 %) among older children 2–5 years of age. In this same study, RSV positive children seen in a clinician's office or the emergency department were most commonly diagnosed with upper respiratory tract infections (32 %), bronchiolitis (20 %), asthma (13 %), and pneumonia (8 %).

Apnea may develop in 1.2–23.8 % of infants and may be the initial manifestation before other respiratory signs are present (Ralston and Hill 2009). The apnea is generally self-limited, is most common in premature infants, and does not recur with subsequent respiratory infections.

In developed countries, simultaneous or secondary bacterial infection, other than otitis media, is uncommon with RSV infection (McIntosh 1991; Hall et al. 1988; Ralston et al. 2011). Urinary tract infections are the most frequent concurrent infections, identified in about 3 % of infants, and may be coincidental because the age of their initial occurrence is similar to that of RSV (Ralston et al. 2011). In developing countries, secondary bacterial infection or co-infection may be a more substantial contributor to RSV disease (Madhi and Klugman 2004).

Recurrent wheezing and long-term pulmonary sequelae have been reported in up to 30–50 % of infants hospitalized with RSV lower respiratory infection (Sigurs et al. 2010; Sly et al. 2010; Stein and Martinez 2010). However, it is unclear if RSV infection causes the predisposition to recurrent wheezing or a predisposition to recurrent wheezing increases the risk of RSV hospitalization (Stensballe et al. 2009).

RSV Infections among Older Children and Adults are frequent and occur multiple times throughout life independent of antigenic changes in the virus (Hall et al. 1991). Most recurrent infections are symptomatic upper respiratory tract illnesses that tend to be more severe and prolonged than the average cold (Falsey 2007; Hall et al. 1978, 2001). Low grade fever and upper respiratory tract complications, such as sinusitis and otitis media, are common. As many as one-fourth of RSV-infected, healthy adults will have lower respiratory tract signs such as wheezing and cough. Compared to influenza, RSV infection usually has a less acute onset and less fever and systemic symptoms.

The importance of RSV infections among older adults has been increasingly appreciated (Walsh et al. 2007; Walsh 2011; Hall et al. 1976; Falsey et al. 2005) The clinical presentation of RSV in elderly adults is not distinctive and more severe illness is often diagnosed as exacerbation of comorbid conditions, such as chronic obstructive pulmonary disease and congestive heart failure.

3 Transmission

Clinical observations of the spread of RSV within families, daycare, and other groups of children have noted that the transmission of RSV requires close contact with infected individuals or with their secretions (Hall 2007 and Lindsley et al. 2010). These observations and volunteer studies indicate that RSV is primarily spread by two mechanisms: (1) large particle droplet aerosols (10–100 μm) which are propelled short distances (≤0.9 m) by sneezing, coughing, and even quiet breathing; and (2) by infectious secretions contaminating environmental surfaces followed by self-inoculation (Lindsley et al. 2010; Hall and Douglas 1981; Hall et al. 1981). RSV in the nasal secretions of infants remains infectious on countertops for about 6 h and on cloth and paper tissue for about 30 min (Hall et al. 1980). Survival is also augmented when the humidity is low (≤30 %) as is usual during the winter respiratory season (Miller and Artenstein 1967 and Siegel et al. 2007). The likely portals of inoculation are the nose, eyes, and upper respiratory tract. Although detected in air samples in a fashion suggestive of having spread long distances by small particle aerosols (Lindsley et al. 2010), epidemiologic studies suggest this mode of transmission is uncommon.

Nosocomial spread of RSV is problematic in high risk populations such as pediatric wards and in patients with compromised immune, pulmonary, or cardiac systems (Bont and Nosocomial 2009; Englund et al. 1991; Hall 2000; El Saleeby et al. 2008). Risk of nosocomial RSV primary occurs during community outbreaks but periodic outbreaks can occur outside of the RSV season.

Recommendations for the control of RSV are primarily based on interrupting the assumed major modes of transmission by emphasizing the avoidance of close contact with infected individuals and preventing direct or indirect contact with infectious secretions. However, the ubiquity of infection during the RSV season among all ages and the fact that ill visitors and staff can contribute to nosocomial spread complicates control efforts (see chapter by H.Y. Chu and J.A. Englund, this volume). The U.S. Centers for Disease Control and Prevention (CDC) advises contact precautions in addition to standard precautions (Siegel et al. 2007). Integral to these recommendations is consistent and assiduous hand hygiene, which emphasizes the importance of fomite transmission and the pivotal role that personnel play in nosocomial spread of RSV.

4 Temporal and Geographic Patterns of Community Outbreaks

In temperate climates, RSV regularly causes community outbreaks in fall, winter, and spring months, i.e., November to April in many Northern hemisphere locations and March to October in many Southern hemisphere locations. Within this regularity, however, is substantial variability in the timing and duration of outbreaks. For example, in the USA, the timing of outbreaks for the same region or community between years and between different communities for the same year can vary substantially. Over a 10-year period, U.S. surveillance showed the onset of RSV season occurs between early November and late January in most communities. South census region tends to have earlier onset and the Midwest census region later onset of the RSV seasons (Mullins et al. 2003). Within regions and between communities there is additional variation. For example, RSV outbreaks in Florida occur as early as July and August in the Miami area and October or November in northern Florida (Light et al. 2008). Even communities within a few miles of each other can have substantially different timing of the onset of RSV outbreaks, e.g., in two communities 25 miles apart the onsets were >4 week different 20 % of the time (Mullins et al. 2003). The duration of RSV outbreaks is usually 12–20 weeks but can be substantially longer, especially in some southern regions. Surveillance studies from other countries have identified other patterns of RSV outbreaks such as alternating years with early and late onset of seasonal outbreaks. Seasonal patterns of community outbreaks in tropical regions are less consistent. In tropical regions, more distant from the equator, RSV outbreaks tend to occur in cool dry or cool wet seasons while in regions closer to the equator RSV tends to be detected throughout the year but with periods of increased activity (Stensballe et al. 2003). In other tropical regions with distinct yearly outbreaks, the timing of these outbreaks can vary more year to year than in temperate climates. Thus, local data is needed to most accurately predict timing of RSV outbreaks in a given community.

Understanding the temporal and geographic patterns of RSV outbreaks has been important for timing of RSV immune prophylaxis, implementing RSV infection control strategies, and estimating the burden of RSV disease. In temperate climates, the regularity of RSV seasonal outbreaks makes it possible to define the months during which RSV transmission is likely to occur and focus RSV immune prophylaxis (From the American Academy of Pediatrics 2009) (see chapter by H.Y. Chu and J.A. Englund, this volume). However, the variability of the time of RSV circulation that occurs between years and between communities makes such predictions imprecise even with local data (Panozzo et al. 2010).

5 Risk Factors for Infection and Disease

Age and sex are important risk factors for serious RSV disease with the very young and the elderly being the two highest risk groups for serious complications from infection (Sommer et al. 2011 and Langley and Anderson 2011). A review of recent studies of RSV hospitalization rates (Simoes 2003) revealed that approximately 10–28 % of infants hospitalized with RSV are aged below 6 weeks, 49–70 % below 6 months, and 66–100 % below 1 year. Consequently, young age, e.g., <6 months of age, at the onset of RSV season increases the risk of RSV hospitalization (Liese et al. 2003; Carbonell-Estrany et al. 2000; Figueras-Aloy et al. 2008; Law et al. 2004). Male sex has consistently been a risk factor for severe RSV LRTI and analysis of representative studies over the last 30 years found the risk ratio of boys to girls being 1.425:1 (Simoes 2003).

Much of the disease in elderly persons can be attributed to underlying conditions. In a comprehensive 4-year study of RSV in adults (Falsey et al. 2005), 608 healthy adults and 540 high risk adults were followed for one or more RSV seasons. In addition, 1388 adults 65 years or older or with physician diagnosed congestive heart failure or chronic lung disease hospitalized for an acute respiratory illness were studied. Among those followed prospectively, none of the 46 RSV-infected healthy elderly were hospitalized with their RSV infection while 9 (16 %) RSV-infected high risk adults were hospitalized. Among adults hospitalized with RSV, 80 % had an underlying cardiopulmonary condition. Of note, for many of the adults hospitalized with an RSV infection, the discharge diagnosis was exacerbation of underlying heart or lung disease. In this study, RSV was detected in 11 % of adults hospitalized with the diagnosis of pneumonia, 11 % with the diagnosis of chronic obstructive lung disease, 5 % with the diagnosis of congestive heart failure, and 7 % with the diagnosis of asthma. Over the four study years, 3–10 % of the prospectively followed adults became infected with RSV.

Underlying conditions such as prematurity in the infant and young child has been associated with a risk of hospitalization of 4–14 % depending on the study and gestational age at birth (Sommer et al. 2011 and Langley and Anderson 2011). Otherwise healthy infants <1 year of age have hospitalization rates of 1–3 % (Langley and Anderson 2011). Infants and children with chronic lung disease and prematurity are reported to have a 10–25 % risk of being hospitalized with RSV. Other chronic lung conditions such as cystic fibrosis have been associated with more severe illness with RSV infection (Garcia et al. 2007). Hospitalization rates among infants and children with Down syndrome have been reported as high as 7–11 % depending on presence of associated conditions such as congenital heart disease (Zachariah et al. 2012). The risk of hospitalization among children with congenital heart disease has been reported to be as high as 36 % but more recent estimates suggest lower rates of hospitalization. These lower recent estimates probably reflect better management of the patient's illness. At risk children also have more severe illness when hospitalized as indicated by longer hospital stays, more frequent admissions to the intensive care unit, and more frequent need for mechanical ventilation. Although several risk factors have been associated with a higher frequency of severe RSV-mediated disease, more than 50 % of hospitalizations caused by RSV infections are in infants and children with no known risk factors (Boyce et al. 2000).

RSV infection in immune compromised patients initially can be similar to infection in otherwise healthy persons but they can have a high rate of serious complications and death. Recipients of hematopoetic stem cell transplants (HSCT) have had especially high rates of severe disease and death from RSV infection with those receiving allogenic HSCT's having the greatest risk. The greatest risk is the post HSCT period before engraftment. Delay in engraftment and graft versus host disease also is associated with an increased risk of serious RSV disease. Up to 50 % of RSV infections in HSCT recipients progress to the lower respiratory tract and between 6 and 80 % of lower respiratory tract infections are reported to result in death (Ison 2009 and Shah and Chemaly 2011).

Recipients of other transplanted organs or immune suppressive therapy for cancer or other conditions with moderate levels of immune suppression may also have an increased risk of serious RSV disease, but this risk has not been well studied (Weigt et al. 2011). Severe immune suppression of any origin, however, likely is associated with a substantially increased risk of complications with RSV infection. Lung transplant recipients have been reported to have a high risk of fatal outcome with RSV infection, up to 10–15 %, as well as risk of new onset and progression of bronchiolitis obliterans syndrome [BOS] and the associated irreversible decline in lung function (Zamora et al. 2011 and Liu et al. 2010). HIV-infection can result in prolonged RSV shedding and may predispose to more severe RSV infection. A large prospective study in South Africa found nearly an eightfold increase in RSV hospitalization in HIV-infected children compared to those without HIV infection (Madhi et al. 2006).

Detection of RSV infection is one key to preventing and treating disease in high risk patients. In general, antigen detection assays and viral isolation are much less sensitive than molecular techniques but adequate for detecting infection in young children. In adults a sensitive assay, such as real time PCR, is necessary to reliably detect RSV and other viral respiratory infections (Falsey et al. 2002). Attention to good infection control practices inside and outside the health care setting can decrease the risk of infection (Danziger-Isakov et al. 2012) (see section on RSV Transmission).

Living at an altitude higher than 2500 m is associated with an increased risk of RSV hospitalization (relative risk [RR]: 1.30; $P < 0.018$ compared to living at moderate altitudes) and 1–4-year-old children exhibit an 80 % increase in their hospitalization rates (RR: 1.80; $P < 0.001$) (Choudhuri et al. 2006). Altitude could contribute to disease severity by lowering oxygen saturation, impairing respiratory airway ciliary activity, and causing hypoxia-related pulmonary vasoconstriction.

Malnutrition and small for gestational age have most often been studied as RSV risk factors in developing countries. A Kenyan cohort study found children with stunting (height for age z-scores <−2) had a higher rate of RSV ALRI (RR 1.73 [95 %CI 1.08–2.76]) (Okiro et al. 2008). A study from Philippines found that infants who were underweight (weight for age *z*-scores <−2) at 6 weeks of age had a significantly increased rate of subsequent RSV ALRI hospitalisation (RR 1.60 [95 %CI 1.07–2.41]) and were more likely to be hypoxemic than those who were not (30 % vs. 15 %, $p = 0.03$) (Paynter et al. 2013). Studies focusing on malnutrition in developing countries have concluded that malnutrition is less important to the severity of RSV than to bacterial infections and some studies suggested malnourished children may have less severe disease than well-nourished children (Simoes 2003). Intrauterine growth restriction was found to be an independent risk factor for RSV hospitalization in the Canadian PICNIC- Study (Law et al. 2004).

Day care attendance/older siblings in school or day-care and crowding are significant risk factors for LRTI. Liese et al. (2003) from the Munich RSV Study Group found that the presence of siblings in day-care attendance increased the risk factor for RSV rehospitalisation in preterm infants, while the Canadian PICNIC study (Law et al. 2004) showed that day-care-attendance of children was the single greatest risk factor for RSV hospitalization. Many studies demonstrated a significant effect of increased numbers of persons sharing a bedroom on RSV LRTI

(Simoes 2003). This effect was increased in families with low maternal education and even more in families with low maternal education who had not breast-fed their babies. School- and preschool-aged siblings impart an increased for the young infant to acquire RSV infection. In the Canadian PICNIC study (Law et al. 2004) the presence of preschool-aged siblings was significantly and independently associated with an increased risk for RSV related hospitalization, and a weaker association was found with the presence of school-aged siblings. Crowding, defined as five or more people living in one household, was also demonstrated to be a significant risk factor for RSV related hospitalization. The Spanish FLIP study (Carbonell-Estrany et al. 2000) revealed that only school-aged siblings and the presence of more than four additional residents and visitors at home were risk factors significantly associated with RSV related hospitalization. In the FLIP-2 study (Figueras-Aloy et al. 2008), the effect of school- aged siblings was confirmed but not crowding by use of the same definition. The Munich RSV Study Group (Liese et al. 2003) found that siblings at day care attendance significantly augmented the risk for RSV related hospitalization.

Multiple births carry an increased risk of RSV hospitalization. A study of twins and triplets in Colorado found a significantly higher risk of severe RSV LRTI and hospitalizations compared to matched singletons (Simoes et al. 1993). This finding was confirmed by a study on hospitalization rates in preterm infants aged 29–36 weeks by Resch et al. in (2005) that revealed multiple births being a risk factor for RSV related hospitalization (odds ratio 5.5, CI 95 % 1.439–21.028).

Other factors of uncertain association with disease have sometimes been linked to RSV disease. Lower socioeconomic status and parental education have been reported to be risk factors for RSV infection in some studies (Glezen et al. 1981 and Jansson et al. 2002) but not in other studies (Figueras-Aloy et al. 2008; Anderson et al. 1988; McConnochie and Roghmann 1960). In combination with other risk factors, maternal education was positively correlated with RSV related hospitalization in the Tucson study (Holberg et al. 1991). Lack of breast feeding was not found to be an independent risk factor for RSV disease or hospitalization, in epidemiologic studies of normal infants after accounting for other risk factors (Law et al. 2004 and Resch et al. 2005). In Denmark, the largest case control study of RSV hospitalization, that included all known risk factors in the model, did not include breastfeeding in any of the models of protection against RSV hospitalization (Stensballe et al. 2006). However, some studies, in more restricted populations (Carbonell-Estrany et al. 2000 and Bulkow et al. 2002) or older studies (Holberg et al. 1991) did find that the absence of breast-feeding in combination with other risk factors like crowding, passive smoke exposure or low socioeconomic status significantly increased the risk for development of RSV LRTI.

Maternal smoking and indoor smoke exposure contribute to lower respiratory tract infection (LRTI) in infants and young children (Simoes 2007), but it is not clear that it specifically contributes to risk of RSV LRTI (Carroll et al. 2007). An early case control study from the US showed an increased risk of bronchiolitis in families with $\geq$1 smoker (McConnochie and Roghmann 1960) but several other prospective case–control studies showed a significant effect in univariate analysis

($P = 0.018$ and 0.0004, respectively) but not in multivariate analysis (Bulkow et al. 2002; Sigurs et al. 1995; Juntti et al. 2003). A large cohort study in Arizona showed no significant effect of environmental tobacco smoke exposure in a multivariate analysis (Holberg et al. 1991) while a more recent large nested case–control study from the Danish birth cohort (2564 infants and children hospitalized with RSV and 12816 age-matched controls) (Stensballe et al. 2006), found an association between tobacco smoke exposure and an increased risk of hospitalization with RSV (odds ratio: 1.35; 95 % confidence interval: 1.20–1.52). A recent study implicated maternal smoking during pregnancy in ICU admission in infants and children with bronchiolitis (Mansbach et al. 2012). Among premature infants, several studies have found an association between maternal smoke exposure and RSV disease (Liese et al. 2003; Carbonell-Estrany et al. 2000; Figueras-Aloy et al. 2008; Law et al. 2004; Carbonell-Estrany and Quero 2001) but in only two was smoke exposure found to be an independent risk for RSV hospitalizations (Carbonell-Estrany et al. 2000 and Law et al. 2004).

A family history of atopy or asthma In the Canadian PICNIC—study (Law et al. 2004) a history of eczema in a first degree family member was found to be an independent protective factor for RSV hospitalization. However, no association was found for family history wheezing or any other allergic disorder. Data from the Spanish FLIP-2- study (Figueras-Aloy et al. 2008) confirmed these findings showing that a family history of wheezing did not reach statistical significance. In the previous FLIP—study, (Carbonell-Estrany et al. 2000) however, a history of wheezing in the family was found to be of statistical significance, whereas the interaction between a history of asthma or eczema in the family did not reach statistical significance in the multivariate logistic regression analysis. Finally the large Danish study established the role of an atopic disposition for hospitalization of infants with RSV bronchiolitis: the adjusted relative risk of RSV hospitalization in the offspring was 1.11 for maternal atopic dermatitis, 1.72 for maternal asthma, and 1.23 for paternal asthma (Stensballe et al. 2006).

6 Burden of RSV Disease: Industrialized Countries

Young Infants and Children are at higher risk for severe complications and hospitalization with RSV infection. RSV also accounts for significant outpatient disease and disease in older children and adults. Estimates from the U.S. National Hospital Discharge Survey (NHDS) from 1980 to 1996 indicated bronchiolitis was the leading cause of all hospitalizations for infants, and that RSV was the most frequent cause of bronchiolitis and all lower respiratory tract disease among young children (Shay et al. 1999). An estimated 74,000–1,26,000 hospitalizations for infants in the USA resulted from RSV each year between 1994 and 1996, and the number of RSV hospitalizations appeared to be increasing in both the USA and Canada, especially among those less than 6 months of age (Shay et al. 1999 and Langley et al. 2003). The annual RSV hospitalization rates estimated from national discharge and insurance databases generally have been 25–40 per 1,000 for infants and 6–10 times lower in the second year of life (Boyce et al. 2000; Zhou et al. 2012; Holman et al. 2004; Leader and Kohlhase 2003). More recent estimates for 1997–2006 showed the RSV-coded hospitalization rates were 26 per 1,000 infants and 1.8 per 1,000 children 1–5 years old and caused an estimated 24 % of all hospitalizations among children under 5 years of age (Stockman et al. 2012).

Prospective studies with population-based surveillance have provided more defined rates of RSV hospitalizations among young children (Hall et al. 2009 and Iwane et al. 2004). A 4-year study conducted by the CDC prospectively examined laboratory-confirmed RSV infections among children <5 years of age in counties in three states (Hall et al. 2009). Among over 5,000 enrolled children, 18 % had RSV infections. RSV was associated with 20 % of hospitalizations for acute respiratory illnesses during November–April, 18 % of emergency department visits, and 15 % of office practice visits. The annual rates of hospitalization were 3 per 1,000 for children under 5 years of age and 17 per 1,000 infants, among those under 6 months of age. In comparison to the hospitalization rates for influenza or parainfluenza viruses in this same population, rates for RSV were 3 times higher among children under 5 years of age and 6–8 times greater among infants. Similar rates of RSV-associated hospitalizations have been reported from other industrialized countries with rates reported of 9–28/1,000 children for the first year of life and 3–6/1,000 for children <5 years of age (Fjaerli et al. 2004; Forster et al. 2004; Nicholson et al. 2006; van Gageldonk-Lafeber et al. 2005; Eriksson et al. 2002).

Outpatient Disease with RSV infection confers appreciable clinical and economic burden, but it is less well estimated or appreciated. Information on the RSV burden from outpatients is limited and underappreciated. Few studies have defined and characterized the national healthcare impact from confirmed RSV illnesses among ambulatory patients, especially among those cared for in pediatric offices. In emergency departments in the USA, the average yearly rate of bronchiolitis has been estimated during 1992–2000 as 26 per 1,000 children under 2 years of age, and 64 % of ED visits for bronchiolitis were RSV positive (Mansbach et al. 2008) Among children under 8 years of age, the rate of emergency department visits for RSV during the winter has been estimated as 21.5 per 1,000 children (Bourgeois et al. 2009). The prospective, population-based CDC studies showed annual rates, 28 per 1,000 children under 5 years of age, and 55 per 1,000 under 6 months (Hall et al. 2009).

Visits for RSV illness among pediatric practices are notably greater. In the population-based study of German outpatients, the annual rate of RSV infection was 77 per 1,000 children under 3 years of age (Forster et al. 2004). In the USA, population-based surveillance showed rates of RSV visits to pediatric offices were 80 per 1,000 children less than 5 years old and 132 per 1,000 children under 6 months of age.(Hall et al. 2009) Extrapolating these results to the entire U.S. population suggests that among children <5 years of age, RSV results in 1 of 334 hospitalizations, 1 of 38 emergency department visits, and 1 of 13 pediatric office visits each year.

Older Children and Adults also suffer significant RSV associated disease but the impact on healthcare resources is least well recognized among older children and healthy adults. RSV infection is frequent in adults in families with young children, as high as 40 %, but usually undiagnosed (Hall et al. 1976). In a prospective virus surveillance study of healthy adults, the 211 RSV infected adults had substantial disease with infection. 40 % of infected adults missed work and the average duration of the RSV illness, 10 days, was twice that of acute respiratory infections from other viruses. Among military recruits, RSV is also a major cause of acute respiratory illness and causes ward confinement at rates similar to that of influenza (O'Shea et al. 2005). The morbidity and mortality associated with RSV in adults is greatest in those with comorbidities. In a 4-year study of 608 health community dwelling adults ≥65 years of age, 540 adults with cardiopulmonary disease, and 1,388 adults hospitalized with an acute respiratory illness, 3–13 % of the various cohorts were infected each year (Falsey et al. 2005). Among the prospective healthy cohort, 17 % saw their clinician during an RSV infection and none was hospitalized. In contrast, among the high-risk group, 29 % had office visits, 9 % visited the emergency department, and 16 % required hospitalization. In a recent study, it was estimated that yearly rates of RSV hospitalization among persons 50–64 years old was 0.82/1,000 and 2.5/1,000 persons ≥65 years of age or about 40,000 and 1,25,000 hospitalizations/year respectively in the USA (Widmer et al. 2012).

Mortality associated with RSV infection is relatively rare among young children in developed countries. Less than 500 fatal cases are estimated to occur in the USA per year (Leader and Kohlhase 2003; Shay et al. 2001; Thompson et al. 2003). Among older children and adults under 50 years of age mortality is generally the lowest. In the Netherlands, excess mortality during the winter related to RSV was not observed among those 1–18 years of age, was slight (0.3 per 1,00,000) among 18–49 year olds, and markedly increased among those 50–64 years (5.4) and those ≥65 years of age (98.7) (Jansen et al. 2007).

Economic burden of RSV disease in the USA and in other developed countries is appreciable. The greatest proportions of these costs are engendered by RSV illness among infants and the elderly. In the USA, the total annual RSV costs for infants have been estimated during 1997–2000 as $202 million for emergency department visits and $2.6 billion for hospitalizations (Leader and Kohlhase 2003). In comparison, estimates of the cost of RSV-associated disease in Australia were substantially lower, i.e., annual direct healthcare cost in 2005 of $24–$50 million for children <5 years of age and $20–$40 million for infants (Ranmuthugala et al. 2011). Among the elderly in the USA, the annual direct costs of RSV hospitalizations was estimated in 1995 as $150–$680 million, and more recently as exceeding $1 billion (Falsey et al. 2005 and Han et al. 1999).

7 Burden of RSV Disease: Developing Countries

Two recent estimates of the burden of RSV disease have been made, one estimating the global incidence of and mortality from episodes of acute lower respiratory infection due to RSV in children younger than 5 years in 2005 (Nair et al. 2010) and a more recent estimate of the Global Burden of Diseases (GBD) 2010, that included estimates of RSV specific mortality for all age groups (Lozano et al. 1990). The first estimate by Nair et al. utilized a systematic review of data published between January, 1995, and June, 2009, and ten unpublished population-based studies, utilizing the CHERG methodology, estimated that in 2005, 33·8 (95 % CI 19·3–46·2) million new episodes of RSV-associated ALRI occurred worldwide in children younger than 5 years. Of these about 10 % or 3·4 (2·8–4·3) million episodes required hospital admission. It was estimated that 66,000–1,99,000 children younger than 5 years died from RSV-associated ALRI in 2005, with 99 % of these deaths occurring in developing countries. The second estimate, by Lozano et al. (1990), is part of a GBD study of 291 causes of death. This cause of death analysis has been performed at the country level for 187 countries with models available from 1980 to 2010. The estimates for RSV used systematic reviews of published data and metanalysis using GBD Bayesian meta-regression methods. These generated region-age-sex specific estimates that were then applied to estimates of lower respiratory tract infections (the so called Aetiology Modeling). Using these methods, it has been estimated that RSV causes 234,000 deaths in children <5 years of age. Since the majority of acute respiratory deaths globally occur in the community without etiologic studies, it is difficult to validate these estimates. Better data on RSV mortality in the community in developing countries is needed.

8 Comment

The clinical and epidemiologic features of RSV disease foretell both difficulties and promise for developing an RSV vaccine. For example, repeat infections and serious disease throughout life suggest that inducing a protective immune response will be difficult. On the other hand, high titers of neutralizing antibodies correlate with protection and anti-F protein neutralizing antibody (palivizumab) prophylaxis protects the high risk infant and young child from serious disease. The substantial burden of disease throughout life suggests that there are at least four target populations that should be considered for vaccine development, the young infant to prevent the maximum amount of disease in children; the child over 4–6 months of age who is better able to respond to a vaccine but still suffers sufficient disease to warrant vaccination (especially in developing countries); pregnant women to offer protection to her infant during the period of highest risk—the first few months of life; and the elderly to protect them from serious complications of infection. Since each population has different safety and efficacy challenges, each may require a different approach to vaccination. Matching a candidate vaccine to the most appropriate target population is important to improve its chance of success. Understanding the epidemiology of RSV is also important to choosing a study population that is suited to efficient evaluation of candidate vaccines.

References

Anderson LJ, Parker RA, Strikas RA, Farrar JA, Gangarosa EJ, Keyserling HL, Sikes RK (1988) Day-care center attendance and hospitalization for lower respiratory tract illness. Pediatrics 82:300–308

Bont L (2009) Nosocomial RSV infection control and outbreak management. Paediatr Respir Rev 10(Suppl 1):16–17

Bourgeois FT, Valim C, McAdam AJ, Mandl KD (2009) Relative impact of influenza and respiratory syncytial virus in young children. Pediatrics 124:e1072–e1080

Boyce TG, Mellen BG, Mitchel EF Jr, Wright PF, Griffin MR (2000) Rates of hospitalization for respiratory syncytial virus infection among children in medicaid. J Pediatr 137:865–870

Bulkow LR, Singleton RJ, Karron RA, Harrison LH (2002) Risk factors for severe respiratory syncytial virus infection among Alaska native children. Pediatrics 109:210–216

Carbonell-Estrany X, Quero J (2001) Hospitalization rates for respiratory syncytial virus infection in premature infants born during two consecutive seasons. Pediatr Infect Dis J 20:874–879

Carbonell-Estrany X, Quero J, Bustos G, Cotero A, Domenech E, Figueras-Aloy J, Fraga JM, Garcia LG, Garcia-Alix A, Del Rio MG, Krauel X, Sastre JB, Narbona E, Roques V, Hernandez SS, Zapatero M (2000) Rehospitalization because of respiratory syncytial virus infection in premature infants younger than 33 weeks of gestation: a prospective study. IRIS study group. Pediatr Infect Dis J 19:592–597

Carroll KN, Gebretsadik T, Griffin MR, Dupont WD, Mitchel EF, Wu P, Enriquez R, Hartert TV (2007) Maternal asthma and maternal smoking are associated with increased risk of bronchiolitis during infancy. Pediatrics 119:1104–1112

Choudhuri JA, Ogden LG, Ruttenber AJ, Thomas DS, Todd JK, Simoes EA (2006) Effect of altitude on hospitalizations for respiratory syncytial virus infection. Pediatrics 117:349–356

Danziger-Isakov LA, Arslan D, Sweet S, Benden C, Goldfarb S, Wong J (2012) RSV prevention and treatment in pediatric lung transplant patients: a survey of current practices among the international pediatric lung transplant collaborative. Pediatr Transplant 16:638–644

El Saleeby CM, Somes GW, DeVincenzo JP, Gaur AH (2008) Risk factors for severe respiratory syncytial virus disease in children with cancer: the importance of lymphopenia and young age. Pediatrics 121:235–243

Englund JA, Anderson LJ, Rhame FS (1991) Nosocomial transmission of respiratory syncytial virus in immunocompromised adults. J Clin Microbiol 29:115–119

Eriksson M, Bennet R, Rotzen-Ostlund M, von Sydow M, Wirgart BZ (2002) Population-based rates of severe respiratory syncytial virus infection in children with and without risk factors, and outcome in a tertiary care setting. Acta Paediatr 91:593–598

Falsey AR (2007) Respiratory syncytial virus infection in adults. Seminars in respiratory and critical care medicine 28:171–181

Falsey AR, Formica MA, Walsh EE (2002) Diagnosis of respiratory syncytial virus infection: comparison of reverse transcription-PCR to viral culture and serology in adults with respiratory illness. J Clin Microbiol 40:817–820

Falsey AR, Hennessey PA, Formica MA, Cox C, Walsh EE (2005) Respiratory syncytial virus infection in elderly and high-risk adults. N Engl J Med 352:1749–1759

Figueras-Aloy J, Carbonell-Estrany X, Quero-Jimenez J, Fernandez-Colomer B, Guzman-Cabanas J, Echaniz-Urcelay I, Domenech-Martinez E (2008) FLIP-2 Study: risk factors linked to respiratory syncytial virus infection requiring hospitalization in premature infants born in Spain at a gestational age of 32 to 35 weeks. Pediatr Infect Dis J 27:788–793

Fjaerli HO, Farstad T, Bratlid D (2004) Hospitalisations for respiratory syncytial virus bronchiolitis in Akershus, Norway, 1993–2000: a population-based retrospective study. BMC Pediatr 4:25

Forster J, Ihorst G, Rieger CH, Stephan V, Frank HD, Gurth H, Berner R, Rohwedder A, Werchau H, Schumacher M, Tsai T, Petersen G (2004) Prospective population-based study of viral lower respiratory tract infections in children under 3 years of age (the PRI.DE study). Eur J Pediatr 163:709–716

From the American Academy of Pediatrics (2009) Policy statements–modified recommendations for use of palivizumab for prevention of respiratory syncytial virus infections. Pediatrics 124:1694–1701

Garcia DF, Hiatt PW, Jewell A, Schoonover SL, Cron SG, Riggs M, Grace S, Oermann CM, Piedra PA (2007) Human metapneumovirus and respiratory syncytial virus infections in older children with cystic fibrosis. Pediatr Pulmonol 42:66–74

Glezen WP, Paredes A, Allison JE, Taber LH, Frank AL (1981) Risk of respiratory syncytial virus infection for infants from low-income families in relationship to age, sex, ethnic group, and maternal antibody level. J Pediatr 98:708–715

Glezen WP, Taber LH, Frank AL, Kasel JA (1986) Risk of primary infection and reinfection with respiratory syncytial virus. Am J Dis Child 140:543–546

Hall C (2000) Nosocomial respiratory syncytial virus infections: the "cold war" has not ended. Clin Infect Dis 31:590–596

Hall CB (2007) The spread of influenza and other respiratory viruses: complexities and conjectures. Clin Infect Dis 45:353–359

Hall CB (2012) The burgeoning burden of respiratory syncytial virus among children. Infect Disord Drug Targets 12:92–97

Hall CB, Douglas RG Jr (1981) Modes of transmission of respiratory syncytial virus. J Pediatr 99:100–103

Hall CB, Douglas RG Jr, Geiman JM (1976) Respiratory syncytial virus infections in infants: quantitation and duration of shedding. J Pediatr 89:11–15

Hall WJ, Hall CB, Speers DM (1978) Respiratory syncytial virus infection in adults: clinical, virologic, and serial pulmonary function studies. Ann Intern Med 88:203–205

Hall CB, Douglas RG Jr, Geiman JM (1980) Possible transmission by fomites of respiratory syncytial virus. J Infect Dis 141:98–102

Hall CB, Douglas RG Jr, Schnabel KC, Geiman JM (1981) Infectivity of respiratory syncytial virus by various routes of inoculation. Infect Immun 33:779–783

Hall CB, Powell KR, Schnabel KC, Gala CL, Pincus PH (1988) Risk of secondary bacterial infection in infants hospitalized with respiratory syncytial viral infection. J Pediatr 113:266–271

Hall CB, Walsh EE, Long CE, Schnabel KC (1991) Immunity to and frequency of reinfection with respiratory syncytial virus. J Infect Dis 163:693–698

Hall CB, Long CE, Schnabel KC (2001) Respiratory syncytial virus infections in previously healthy working adults. Clin Infect Dis 33:792–796

Hall CB, Weinberg GA, Iwane MK, Blumkin AK, Edwards KM, Staat MA, Auinger P, Griffin MR, Poehling KA, Erdman D, Grijalva CG, Zhu Y, Szilagyi P (2009) The burden of respiratory syncytial virus infection in young children. N Engl J Med 360:588–598

Han LL, Alexander JP, Anderson LJ (1999) Respiratory syncytial virus pneumonia among the elderly: an assessment of disease burden. J Infect Dis 179:25–30

Holberg CJ, Wright AL, Martinez FD, Ray CG, Taussig LM, Lebowitz MD (1991) Risk factors for respiratory syncytial virus-associated lower respiratory illnesses in the first year of life. Am J Epidemiol 133:1135–1151

Holman RC, Curns AT, Cheek JE, Bresee JS, Singleton RJ, Carver K, Anderson LJ (2004) Respiratory syncytial virus hospitalizations among American Indian and Alaska Native infants and the general United States infant population. Pediatrics 114:e437–e444

Ison MG (2009) Respiratory syncytial virus and other respiratory viruses in the setting of bone marrow transplantation. Curr Opin Oncol 21:171–176

Iwane MK, Edwards KM, Szilagyi PG, Walker FJ, Griffin MR, Weinberg GA, Coulen C, Poehling KA, Shone LP, Balter S, Hall CB, Erdman DD, Wooten K, Schwartz B (2004) Population-based surveillance for hospitalizations associated with respiratory syncytial virus, influenza virus, and parainfluenza viruses among young children. Pediatrics 113:1758–1764

Jansen AG, Sanders EA, Hoes AW, van Loon AM, Hak E (2007) Influenza- and respiratory syncytial virus-associated mortality and hospitalisations. Eur Respir J 30:1158–1166
Jansson L, Nilsson P, Olsson M (2002) Socioeconomic environmental factors and hospitalization for acute bronchiolitis during infancy. Acta Paediatr 91:335–338
Juntti H, Kokkonen J, Dunder T, Renko M, Niinimaki A, Uhari M (2003) Association of an early respiratory syncytial virus infection and atopic allergy. Allergy 58:878–884
Kim HW, Arrobio JO, Brandt CD, Jeffries BC, Pyles G, Reid JL, Chanock RM, Parrott RH (1973) Epidemiology of respiratory syncytial virus infection in Washington, D.C. I. Importance of the virus in different respiratory tract disease syndromes and temporal distribution of infection. Am J Epidemiol 98:216–225
Langley GF, Anderson LJ (2011) Epidemiology and prevention of respiratory syncytial virus infections among infants and young children. Pediatr Infect Dis J 30:510–517
Langley JM, LeBlanc JC, Smith B, Wang EE (2003) Increasing incidence of hospitalization for bronchiolitis among Canadian children, 1980–2000. J Infect Dis 188:1764–1767
Law BJ, Langley JM, Allen U, Paes B, Lee DS, Mitchell I, Sampalis J, Walti H, Robinson J, O'Brien K, Majaesic C, Caouette G, Frenette L, Le Saux N, Simmons B, Moisiuk S, Sankaran K, Ojah C, Singh AJ, Lebel MH, Bacheyie GS, Onyett H, Michaliszyn A, Manzi P, Parison D (2004) The pediatric investigators collaborative network on infections in Canada study of predictors of hospitalization for respiratory syncytial virus infection for infants born at 33 through 35 completed weeks of gestation. Pediatr Infect Dis J 23:806–814
Leader S, Kohlhase K (2003) Recent trends in severe respiratory syncytial virus (RSV) among US infants, 1997 to 2000. J Pediatr 143:S127–S132
Liese JG, Grill E, Fischer B, Roeckl-Wiedmann I, Carr D, Belohradsky BH (2003) Incidence and risk factors of respiratory syncytial virus-related hospitalizations in premature infants in Germany. Eur J Pediatr 162:230–236
Light M, Bauman J, Mavunda K, Malinoski F, Eggleston M (2008) Correlation between respiratory syncytial virus (RSV) test data and hospitalization of children for RSV lower respiratory tract illness in Florida. Pediatr Infect Dis J 27:512–518
Lindsley WG, Blachere FM, Davis KA, Pearce TA, Fisher MA, Khakoo R, Davis SM, Rogers ME, Thewlis RE, Posada JA, Redrow JB, Celik IB, Chen BT, Beezhold DH (2010) Distribution of airborne influenza virus and respiratory syncytial virus in an urgent care medical clinic. Clin Infect Dis 50:693–698
Liu V, Dhillon GS, Weill D (2010) A multi-drug regimen for respiratory syncytial virus and parainfluenza virus infections in adult lung and heart-lung transplant recipients. Transpl Infect Dis: J Transplant Soc 12:38–44
Lozano R, Naghavi M, Foreman K, Lim S, Shibuya K, Aboyans V, Abraham J, Adair T, Aggarwal R, Ahn SY, Alvarado M, Anderson HR, Anderson LM, Andrews KG, Atkinson C, Baddour LM, Barker-Collo S, Bartels DH, Bell ML, Benjamin EJ, Bennett D, Bhalla K, Bikbov B, Bin Abdulhak A, Birbeck G, Blyth F, Bolliger I, Boufous S, Bucello C, Burch M, Burney P, Carapetis J, Chen H, Chou D, Chugh SS, Coffeng LE, Colan SD, Colquhoun S, Colson KE, Condon J, Connor MD, Cooper LT, Corriere M, Cortinovis M, de Vaccaro KC, Couser W, Cowie BC, Criqui MH, Cross M, Dabhadkar KC, Dahodwala N, De Leo D, Degenhardt L, Delossantos A, Denenberg J, Des Jarlais DC, Dharmaratne SD, Dorsey ER, Driscoll T, Duber H, Ebel B, Erwin PJ, Espindola P, Ezzati M, Feigin V, Flaxman AD, Forouzanfar MH, Fowkes FG, Franklin R, Fransen M, Freeman MK, Gabriel SE, Gakidou E, Gaspari F, Gillum RF, Gonzalez-Medina D, Halasa YA, Haring D, Harrison JE, Havmoeller R, Hay RJ, Hoen B, Hotez PJ, Hoy D, Jacobsen KH, James SL, Jasrasaria R, Jayaraman S, Johns N, Karthikeyan G, Kassebaum N, Keren A, Khoo JP, Knowlton LM, Kobusingye O, Koranteng A, Krishnamurthi R, Lipnick M, Lipshultz SE, Ohno SL, Mabweijano J, MacIntyre MF, Mallinger L, March L, Marks GB, Marks R, Matsumori A, Matzopoulos R, Mayosi BM, McAnulty JH, McDermott MM, McGrath J, Mensah GA, Merriman TR, Michaud C, Miller M, Miller TR, Mock C, Mocumbi AO, Mokdad AA, Moran A, Mulholland K, Nair MN, Naldi L, Narayan KM, Nasseri K, Norman P, O'Donnell M, Omer SB, Ortblad K, Osborne R, Ozgediz D, Pahari B, Pandian JD, Rivero AP, Padilla RP, Perez-Ruiz F, Perico N, Phillips D,

Pierce K, Pope CA, 3rd, Porrini E, Pourmalek F, Raju M, Ranganathan D, Rehm JT, Rein DB, Remuzzi G, Rivara FP, Roberts T, De Leon FR, Rosenfeld LC, Rushton L, Sacco RL, Salomon JA, Sampson U, Sanman E, Schwebel DC, Segui-Gomez M, Shepard DS, Singh D, Singleton J, Sliwa K, Smith E, Steer A, Taylor JA, Thomas B, Tleyjeh IM, Towbin JA, Truelsen T, Undurraga EA, Venketasubramanian N, Vijayakumar L, Vos T, Wagner GR, Wang M, Wang W, Watt K, Weinstock MA, Weintraub R, Wilkinson JD, Woolf AD, Wulf S, Yeh PH, Yip P, Zabetian A, Zheng ZJ, Lopez AD, Murray CJ (2013) Global and regional mortality from 235 causes of death for 20 age groups in 1990 and 2010: a systematic analysis for the global burden of disease study 2010. Lancet 380:2095–2128

Madhi SA, Klugman KP (2004) A role for Streptococcus pneumoniae in virus-associated pneumonia. Nat Med 10:811–813

Madhi SA, Kuwanda L, Cutland C, Klugman KP (2006) Five-year cohort study of hospitalization for respiratory syncytial virus associated lower respiratory tract infection in African children. J Clin Virol 36:215–221

Mansbach JM, McAdam AJ, Clark S, Hain PD, Flood RG, Acholonu U, Camargo CA Jr (2008) Prospective multicenter study of the viral etiology of bronchiolitis in the emergency department. Acad Emerg Med 15:111–118

Mansbach JM, Piedra PA, Stevenson MD, Sullivan AF, Forgey TF, Clark S, Espinola JA, Camargo CA Jr (2012) Prospective multicenter study of children with bronchiolitis requiring mechanical ventilation. Pediatrics 130:e492–e500

McConnochie KM, Roghmann KJ (1986) Parental smoking, presence of older siblings, and family history of asthma increase risk of bronchiolitis. Am J Dis Child (1960) 140:806–812

McIntosh K (1991) Pathogenesis of severe acute respiratory infections in the developing world: respiratory syncytial virus and parainfluenza viruses. Rev Infect Dis 13(Suppl 6):S492–S500

Miller WS, Artenstein MS (1967) Aerosol stability of three acute respiratory disease viruses. Proc Soc Exp Biol Med 125:222–227

Morris JA, Blount RE Jr, Savage RE (1956) Recovery of cytopathogenic agent from chimpanzees with coryza (22538). Proc Soc Exp Biol Med 92:544–549

Mullins JA, LaMonte AC, Bresee JS, Anderson LJ (2003) Substantial variability in community RSV season timing: An analysis using the National Respiratory and Enteric Viruses Sruveillance System (NREVSS). Pediatr Infect Dis J 22:857–862

Nair H, Nokes DJ, Gessner BD, Dherani M, Madhi SA, Singleton RJ, O'Brien KL, Roca A, Wright PF, Bruce N, Chandran A, Theodoratou E, Sutanto A, Sedyaningsih ER, Ngama M, Munywoki PK, Kartasasmita C, Simoes EA, Rudan I, Weber MW, Campbell H (2010) Global burden of acute lower respiratory infections due to respiratory syncytial virus in young children: a systematic review and meta-analysis. Lancet 375:1545–1555

Nicholson KG, McNally T, Silverman M, Simons P, Stockton JD, Zambon MC (2006) Rates of hospitalisation for influenza, respiratory syncytial virus and human metapneumovirus among infants and young children. Vaccine 24:102–108

O'Shea MK, Ryan MA, Hawksworth AW, Alsip BJ, Gray GC (2005) Symptomatic respiratory syncytial virus infection in previously healthy young adults living in a crowded military environment. Clin Infect Dis 41:311–317

Ogra PL (2004) Respiratory syncytial virus: the virus, the disease and the immune response. Paediatr Respir Rev 5(Suppl A):S119–S126

Okiro EA, Ngama M, Bett A, Cane PA, Medley GF, Nokes DJ (2008) Factors associated with increased risk of progression to respiratory syncytial virus-associated pneumonia in young Kenyan children. Trop Med Int Health 13:914–926

Panozzo CA, Stockman LJ, Curns AT, Anderson LJ (2010) Use of respiratory syncytial virus surveillance data to optimize the timing of immunoprophylaxis. Pediatrics 126:e116–e123

Paynter S, Ware RS, Lucero MG, Tallo V, Nohynek H, Weinstein P, Williams G, Sly PD, Simoes EAF (2013) Malnutrition is a risk factor for severe respiratory syncytial virus infection and hospitalization (Unpublished data)

Ralston S, Hill V (2009) Incidence of apnea in infants hospitalized with respiratory syncytial virus bronchiolitis: a systematic review. J Pediatr 155:728–733

Ralston S, Hill V, Waters A (2011) Occult serious bacterial infection in infants younger than 60 to 90 days with bronchiolitis: a systematic review. Arch Pediatr Adolesc Med 165:951–956

Ranmuthugala G, Brown L, Lidbury BA (2011) Respiratory syncytial virus–the unrecognised cause of health and economic burden among young children in Australia. Commun Dis Intell 35:177–184

Resch B, Pasnocht A, Gusenleitner W, Muller W (2005) Rehospitalisations for respiratory disease and respiratory syncytial virus infection in preterm infants of 29–36 weeks gestational age. J Infect 50:397–403

Shah JN, Chemaly RF (2011) Management of RSV infections in adult recipients of hematopoietic stem cell transplantation. Blood 117:2755–2763

Shay DK, Holman RC, Newman RD, Liu LL, Stout JW, Anderson LJ (1999) Bronchiolitis-associated hospitalizations among U.S. children, 1980–1995. JAMA 282:1440–1447

Shay D, Holman R, Roosevelt G, Clarke M, Anderson L (2001) Bronchiolitis-associated mortality and estimates of respiratory syncytial virus-associated deaths among U.S. children, 1979–1997. J Infect Dis 183:16–22

Siegel JD, Rhinehart E, Jackson M, Chiarello L (2007) 2007 Guideline for isolation precautions: preventing transmission of infectious agents in health care settings. Am J Infect Control 35:S65–S164

Sigurs N, Bjarnason R, Sigurbergsson F, Kjellman B, Bjorksten B (1995) Asthma and immunoglobulin E antibodies after respiratory syncytial virus bronchiolitis: a prospective cohort study with matched controls. Pediatrics 95:500–505

Sigurs N, Aljassim F, Kjellman B, Robinson PD, Sigurbergsson F, Bjarnason R, Gustafsson PM (2010) Asthma and allergy patterns over 18 years after severe RSV bronchiolitis in the first year of life. Thorax 65:1045–1052

Simoes EA (2003) Environmental and demographic risk factors for respiratory syncytial virus lower respiratory tract disease. J Pediatr 143:S118–S126

Simoes EA (2007) Maternal smoking, asthma, and bronchiolitis: clear-cut association or equivocal evidence? Pediatrics 119:1210–1212

Simoes EA, King SJ, Lehr MV, Groothuis JR (1993) Preterm twins and triplets. A high-risk group for severe respiratory syncytial virus infection. Am J Dis Child 147:303–306

Sly PD, Kusel M, Holt PG (2010) Do early-life viral infections cause asthma? J Allergy Clin Immunol 125:1202–1205

Sommer C, Resch B, Simoes EA (2011) Risk factors for severe respiratory syncytial virus lower respiratory tract infection. Open Microbiol J 5:144–154

Stein RT, Martinez FD (2010) Respiratory syncytial virus and asthma: still no final answer. Thorax 65:1033–1034

Stensballe LG, Devasundaram JK, Simoes EA (2003) Respiratory syncytial virus epidemics: the ups and downs of a seasonal virus. Pediatr Infect Dis J 22:S21–S32

Stensballe LG, Kristensen K, Simoes EA, Jensen H, Nielsen J, Benn CS, Aaby P (2006) Atopic disposition, wheezing, and subsequent respiratory syncytial virus hospitalization in Danish children younger than 18 months: a nested case-control study. Pediatrics 118:e1360–e1368

Stensballe LG, Ravn H, Kristensen K, Agerskov K, Meakins T, Aaby P, Simoes EA (2009) Respiratory syncytial virus neutralizing antibodies in cord blood, respiratory syncytial virus hospitalization, and recurrent wheeze. J Allergy Clin Immunol 123:398–403

Stockman LJ, Curns AT, Anderson LJ, Fischer-Langley G (2012) Respiratory syncytial virus-associated hospitalizations among infants and young children in the United States, 1997–2006. Pediatr Infect Dis J 31:5–9

Thompson WW, Shay DK, Brammer LT, Weintraub E, Nancy C, Anderson LJ, Fukuda K (2003) Mortality attributable to influenza and respiratory syncytial virus in the United States, 1990–1998. JAMA 289:179–186

van Gageldonk-Lafeber AB, Heijnen ML, Bartelds AI, Peters MF, van der Plas SM, Wilbrink B (2005) A case-control study of acute respiratory tract infection in general practice patients in The Netherlands. Clin Infect Dis 41:490–497

Walsh EE (2011) Respiratory syncytial virus infection in adults. Seminars in respiratory and critical care medicine 32:423–432
Walsh EE, Peterson DR, Falsey AR (2007) Is clinical recognition of respiratory syncytial virus infection in hospitalized elderly and high-risk adults possible? J Infect Dis 195:1046–1051
Weigt SS, Gregson AL, Deng JC, Lynch JP 3rd, Belperio JA (2011) Respiratory viral infections in hematopoietic stem cell and solid organ transplant recipients. Semin Respir Crit Care Med 32:471–493
Widmer K, Zhu Y, Williams JV, Griffin MR, Edwards KM, Talbot HK (2012) Rates of hospitalizations for respiratory syncytial virus, human metapneumovirus, and influenza virus in older adults. J Infect Dis 206:56–62
Wright M, Piedimonte G (2011) Respiratory syncytial virus prevention and therapy: past, present, and future. Pediatr Pulmonol 46:324–347
Zachariah P, Ruttenber M, Simoes EA (2012) Down syndrome and hospitalizations due to respiratory syncytial virus: a population-based study. J Pediatr 160(827–31):e1
Zamora MR, Budev M, Rolfe M, Gottlieb J, Humar A, Devincenzo J, Vaishnaw A, Cehelsky J, Albert G, Nochur S, Gollob JA, Glanville AR (2011) RNA interference therapy in lung transplant patients infected with respiratory syncytial virus. Am J Respir Crit Care Med 183:531–538
Zhou H, Thompson WW, Viboud CG, Ringholz CM, Cheng PY, Steiner C, Abedi GR, Anderson LJ, Brammer L, Shay DK (2012) Hospitalizations associated with influenza and respiratory syncytial virus in the United States, 1993–2008. Clin Infect Dis 54:1427–1436

Influence of Respiratory Syncytial Virus Strain Differences on Pathogenesis and Immunity

José A. Melero and Martin L. Moore

Abstract Molecular epidemiology studies have provided convincing evidence of antigenic and sequence variability among respiratory syncytial virus (RSV) isolates. Circulating viruses have been classified into two antigenic groups (A and B) that correlate with well-delineated genetic groups. Most sequence and antigenic differences (both inter- and intra-groups) accumulate in two hypervariable segments of the G-protein gene. Sequences of the G gene have been used for phylogenetic analyses. These studies have shown a worldwide distribution of RSV strains with both local and global replacement of dominant viruses with time. Although data are still limited, there is evidence that strain variation may contribute to differences in pathogenicity. In addition, there is some but limited evidence that RSV variation may be, at least partially, immune (antibody) driven. However, there is the paradox in RSV that, in contrast to other viruses (e.g., influenza viruses) the epitopes recognized by the most effective RSV-neutralizing antibodies are highly conserved. In contrast, antibodies that recognize strain-specific epitopes are poorly neutralizing. It is likely that this apparent contradiction is due to the lack of a comprehensive knowledge of the duration and specificities of the human antibody response against RSV antigens. Since there are some data supporting a group- (or clade-) specific antibody response after a primary infection

J. A. Melero
Unidad de Biología Viral, Centro Nacional de Microbiología, Instituto de Salud Carlos III, Madrid, Spain

J. A. Melero (✉)
Centro de Investigación Biomédica en Red de Enfermedades Respiratorias (CIBERES), Instituto de Salud Carlos III, Madrid, Spain
e-mail: melero@isciii.es

M. L. Moore
Department of Pediatrics, Emory University, Atlanta, GA, USA

M. L. Moore
Children's Healthcare of Atlanta, Atlanta, GA, USA

L. J. Anderson and B. S. Graham (eds.), *Challenges and Opportunities for Respiratory Syncytial Virus Vaccines*, Current Topics in Microbiology and Immunology 372, DOI: 10.1007/978-3-642-38919-1_3, © Springer-Verlag Berlin Heidelberg 2013

in humans, it may be wise to consider the incorporation of strains representative of groups A and B (or their antigens) in future RSV vaccine development.

Contents

1 Introduction

An early study of the seroepidemiology of RSV in Sendai, Japan found that patient sera did not differ in neutralization of a small number of homologous and heterologous RSV strains, as measured by reduction in tissue culture infectious dose ($TCID_{50}$) in HEp-2 cells (Suto et al. 1965). Using a methylcellulose overlay plaque assay developed in 1966, sera from infected ferrets detected limited strain antigenic variability, reflected in slightly different plaque reduction neutralization (PRN) titers for homologous (Long) versus heterologous (CH18537) strains (Coates et al. 1966). However, it was also found in those early days that children could be naturally infected in consecutive years with RSV strains indistinguishable by cross-PRN, and adults were naturally reinfected despite preexisting neutralizing antibodies (Abs) (Beem 1967).

Despite the previous comments, antigenic groups of RSV strains were definitively identified by enzyme-linked immunosorbent assay (ELISA) using a panel of 10 monoclonal Abs (mAbs) obtained from mice immunized with different RSV strains, such as A2, Long, and CH18537 (Anderson et al. 1985). In a separate study from the same year, RSV isolates from West Virginia were probed with a panel of mAbs generated against RSV Long (Mufson et al. 1985). RSV proteins recognized by the mAbs were identified by radioimmunoprecipitation assay (RIPA) and SDS-PAGE of ^{35}S-labeled infected cell extracts. When these mAbs were tested against RSV field isolates by RIPA, it was revealed that RSV separated into two antigenic groups, A and B, based on eight epitope differences in the attachment glycoprotein (G), one epitope difference in the fusion glycoprotein (F), and one epitope difference in the nucleoprotein (N). The antigenic groups correlated with genetic differences identified by sequencing cDNA clones of the G genes of RSV A2 (A group), Long (A group), and CH18537 (B group) strains. Thus, while the deduced G-protein sequences of A2 and Long strains shared 94 % amino acid identity, those of CH18537 and A2 strains shared only 53 % amino acid identity, with the majority of the diversity residing in the predicted extracellular domain (Johnson et al. 1987b). The classification of RSV isolates into A and B antigenic groups is now more often done via sequencing of variable region(s) of the G extracellular domain, rather than by mAb reactivity. The RSV A and B group designation is also referred to as antigenic "subgroups" in the literature, group A being more prevalent than group B (Hall et al. 1990; Matheson et al. 2006).

Sequence-based molecular epidemiology of RSV led to the identification of genetically distinct, cocirculating genotypic lineages. Evidence of RSV lineages within group A was revealed in isolates from Birmingham, U.K. (1989) using partial sequences of the small hydrophobic (SH) gene and restriction patterns of RSV nucleoprotein (N) gene PCR amplicons (Cane and Pringle 1991). RSV G gene sequences from 27 group A isolates from Montevideo, Uruguay and Madrid, Spain (1987–1993) were aligned with those of A2, Long, and six isolates from Birmingham, UK to analyze the phylogenic relatedness of group A strains, and distinct lineages were evident (Garcia et al. 1994). Similarly, lineages were observed by analyzing sequences of the two variable domains of the G gene from 48 group A RSV isolates collected from 1956 to 1993 in the US, Australia, UK, Norway, Sweden, and Finland (Cane and Pringle 1995). Both studies also found local cocirculation of group A lineages and a high ratio of nonsynonymous to synonymous (dN/dS) mutations in the C-terminal variable region of G, suggesting positive selection. Furthermore, both studies probed isolates with panels of mAbs to the G protein and found that the strength of reactivity roughly paralleled the position on the phylogenetic dendrogram, consistent with contribution of immune selection to RSV G variability (Cane and Pringle 1995; Garcia et al. 1994).

A more detailed picture of RSV genetic lineages emerged with additional sequences. It was determined that the C-terminal 270 nt of the G gene can serve as a proxy for full length G gene variability (Peret et al. 1998). Phylogenetic analysis of G sequences from 204 RSV isolates collected in Rochester, New York from winter 1990/1991 to winter 1994/1995 revealed a number of genetically distinct clusters of genotypes (clades) within A and B groups of RSV (Peret et al. 1998). These clades were designated GA1 to GA5 for group A and GB1 to GB4 for group B. This work provided a clade nomenclature and framework that was consistent with earlier observations of distinct RSV lineages and aided ongoing investigations of RSV molecular epidemiology. Subsequent RSV studies with large sequence datasets of RSV isolates over time from around the globe confirmed these cocirculating clades and identified additional clades (Gaunt et al. 2011; Matheson et al. 2006; Reiche and Schweiger 2009; Shobugawa et al. 2009; Venter et al. 2001; Zlateva et al. 2007; Botosso et al. 2009). Genetic relatedness of RSV group A strains is depicted in Fig. 1 by a phylogenetic tree composited from multiple studies, showing clades and representative isolates. Since the late 1990s, the GA2 and GA5 clades have dominated among group A RSV clades, with

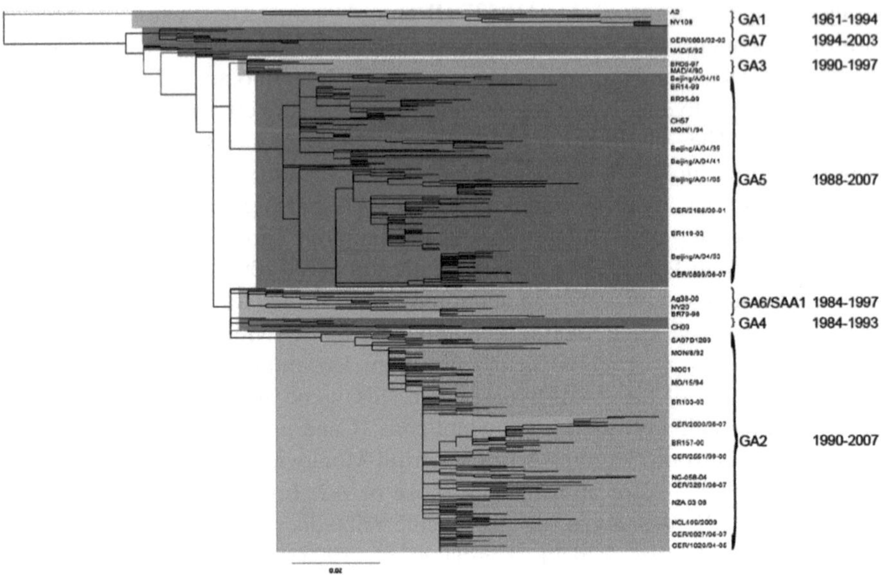

Fig. 1 Phylogenetic tree for RSV group A isolates. All group A G gene sequences from the following references were obtained from the NCBI database (Garcia et al. 1994; Peret et al. 1998, 2000; Venter et al. 2001, 2002; Botosso et al. 2009; Matheson et al. 2006; Zhang et al. 2007; Gaunt et al. 2011; Reiche and Schweiger 2009; Stokes et al. 2011; Yamaguchi et al. 2011). The isolates span years 1961–2007. Using Geneious Pro (Aukland, New Zealand) software, the sequences were trimmed to 270 nt corresponding to the second variable region of G, and identical, redundant sequences were removed. A total of 496 unique G sequences were aligned using ClustalW. Phylogenetic reconstruction was by the Maximum Likelihood (general time reversible model, 1000 bootstrap replicates) using MEGA5 software (Tamura et al. 2011). The tree was graphed and annotated using FigTree version 1.4. Representative isolates are shown to the right. Clades GA1 through GA7 are indicated, as well as the span of years of isolation for each clade. Branch lengths are measured in the number of substitutions per site (scale bar)

season-to-season fluctuation in relative rate of isolation. Group B RSV clades will be discussed in more detail below. In general, there is greater diversity between group A clades than there is between group B clades.

2 Temporal and Geographical Distribution of RSV Strains

Initial analysis of multisequence alignments of RSV G gene sequences showed clustering of strains temporally and not geographically. Phylogenetically related RSV strains could be found at similar times in different continents, whereas isolates collected in the same place from the same epidemic may show greater diversity (Garcia et al. 1994). Additional evidence for temporal and not geographic clustering came from a study comparing 106 group A and 38 group B RSV isolates from New Zealand (collected 1967–2004) with published isolates from around the world. The New Zealand isolates did not cluster with each other, and RSV isolates clustered by clade, not by country (Matheson et al. 2006). However, despite data on wide geographic dissemination of RSV strains, community-based transmission likely plays a role in RSV epidemiology. RSV isolates were collected from the 1994–1995 RSV season in the following cities, states/province: Birmingham, Alabama, Rochester, New York, Houston, Texas, St. Louis, Missouri, and Winnipeg, Mannitoba (Peret et al. 2000). The GA1 clade was most prevalent in Birmingham, Rochester, and St. Louis, whereas the GA5 clade was dominant in Houston, and the GA5 and GA7 clades codominated that season in Winnipeg. RSV isolates from Japan generally clustered with known clades, but some were genotypically unique, suggesting a role for geographic clustering and community-based spread (Kuroiwa et al. 2005).

One way to gauge community-specific RSV distribution is to compare RSV studies reporting RSV isolates from different locations spanning the same time frame. Table 1 compares the prevalence of RSV group A and group B as well as the prevalence of the dominant group A clades (GA2, GA5, and GA7) over time in different regions. Dominant group B clades were omitted in Table 1 because their classification is less consistent. Table 1 shows some patterns of widespread RSV distribution. For example, the relative A to B and relative GA2 to GA5 rates are roughly similar in Belgium, Germany, and Buenos Aries over three RSV seasons, 1998–1999, 1999–2000, and 2000–2001 (Table 1). Yet, site-specific group and clade restrictions are also evident. For example, Belgium and Germany differ in whether A or B is more prevalent in three of the eight overlapping seasons (Table 1). Also, GA5 was the dominant group A clade in Belgium, Germany, and Japan in the 2003–2004 season. However, GA2 was dominant the next season in Belgium and Japan, whereas GA5 remained dominant in Germany (Table 1). In tropical and subtropical regions, the relative prevalence of A and B differed between Buenos Aires and Kenya in two out of three overlapping seasons (Table 1). RSV seasonality in tropical regions is distinct, and epidemics occur during the rainy season from July to November, with a biennial pattern of low and high incidence seasons, as observed in Cambodia (Arnott et al. 2011). In Cambodia and China, the dominance of group A or group B in a particular season was strong (Table 1). In summary, factors determining RSV group and clade compositions of epidemics are complex. Region-specific factors play a role, and, as discussed in more detail below, specific genotypes can also spread globally.

Although RSV groups and clades cocirculate and can appear in successive years, year-to-year changes in the predominance of group A and group B and year-to-year changes in the predominance of clades were observed in a given location (Peret et al. 1998). Subsequent studies confirmed that clades co-circulate locally and alternate in predominance over time, potentially due to immune selection but without evidence of progressive evolution as defined by new strain emergence (Botosso et al. 2009; Gaunt et al. 2011; Matheson et al. 2006; Reiche and Schweiger 2009; Venter et al. 2001). Two key questions about the temporal distribution of RSV strains are: (i) what drives season-to-season changes in clade

Table 1 RSV group prevalence and group A clade prevalence by location, study, and epidemic season

	1997–1998	1998–1999	1999–2000	2000–2001	2001–2002	2002–2003	2003–2004	2004–2005	2005–2006	2006–2007	2007–2008
Temperate Region Winter Epidemic											
Leuven, Belgium[a]											
A/B[b]	64/34	25/75	75/24	82/18	15/85	55/45	61/37	22/76	84/15		
GA2/GA5[c]	70/30	42/58	26/74	41/59	83/17	20/80	0/100	75/25	22/78		
Germany[d]											
A/B		20/80	93/7	58/42	67/33	34/66	75/25	80/20	64/36	53/47	
GA2/GA5/GA7		50/50/–	37/58/5	38/62/–	50/50/–	4/48/48	28/72/–	13/87/–	26/74/–	57/43/–	
Niigata, Japan[e]											
A/B					94/6	22/78	89/11	80/20	70/30	70/30	
GA2[f]/GA5/GA7					–/97/3	–/69/31	–/100/–	81/19/–	91/1/–	100/–/–	
Lanzhou and Chongqing, China[g]											
A/B										100/–	97/3
GA2/GA5										100/–	100/–
Subtropical or Tropical Epidemic											
		1999	2000	2001	2002	2003	2004	2005	2006	2007	2008
Buenos Aires, Argentina[h]											
A/B		38/62	100/0	83/17	29/71	74/26	82/18				
GA2/GA5		64/36	30/70	43/57	–/100	78/22	50/50				
Kilifi District, Kenya[i]											
A/B				100/0	98/2	39/61					
GA2/GA5				5/95	27/73	–/100					
Cambodia[j]											

(continued)

Table 1 (continued)

	1997–1998	1998–1999	1999–2000	2000–2001	2001–2002	2002–2003	2003–2004	2004–2005	2005–2006	2006–2007	2007–2008
A/B								0/100	100/–	92/2	8/92
GA2[k]/GA5								–	100/–	98/2	100/–

[a] Data from Zlateva et al. (2007)
[b] % of isolates that were RSV group A/% of isolates that were RSV group B
[c] % of group A isolates that were clade GA2/% of group A isolates that were clade GA5
[d] Data from Reiche and Schweiger (2009)
[e] Data from Shobugawa et al. (2009)
[f] Niigata isolates assigned to NA1 and NA2 clades are represented as GA2 because these three clades are more closely related to each other than to GA5
[g] Data combined from Zhang et al. (2010a, c)
[h] Data from Viegas and Mistchenko (2005)
[i] Data from Scott et al. Scott et al. (2004)
[j] Data from Arnott et al. (2011)
[k] Cambodia isolates assigned to NA1 clade are represented as GA2

predominance? and (ii) do RSV clades impact natural infection and reinfection by, for example, providing some degree of immune evasion? A recurring hypothesis has been that alternating clade prevalence is a result of short-lived strain- or clade-specific herd immunity that favors circulation of a heterologous clade. Although it has yet to be shown definitively, there are published data consistent with the hypothesis of selection by herd immunity. Thus, several of the amino acid residues in the C-terminal hypervariable region of the G protein identified as having a high dN/dS ratio map within known Ab epitopes (Botosso et al. 2009; Garcia et al. 1994). Furthermore, a number of positively selected amino acid sites in the C-terminus of G show a reversion ("flip-flop") pattern of evolution, consistent with rising and waning strain-specific immunity (Botosso et al. 2009). The infant serum Ab response to primary RSV infection contains clade-specific Abs, as measured by ELISA using plates coated with polypeptides corresponding to the C-terminus of G (Scott et al. 2007). More recently, the RSV group and clade of infecting and reinfecting strains were identified in a birth cohort in Kenya. Excluding seven reinfections that occurred during the same epidemic, there were 46 reinfections documented (Agoti et al. 2012). Over 28 (61 %) of those reinfections were group heterologous (A then B, or B then A). Among the 46 reinfections, there were only six instances where group A infection was followed by group A in a subsequent season. Of those six, four were heterologous for GA2/GA5 clade. Although the numbers are small, the authors state that the majority of reinfections are heterologous at the group or clade level (Agoti et al. 2012). It would be more compelling if higher rates of heterologous clade reinfection could be documented when the reinfecting clade is subdominant.

A better understanding of the potential for global spread of RSV strains has been facilitated by the discovery, in 1999, of a novel group B RSV genotype. It was isolated for the first time in Buenos Aires and contained a 60 nt duplicated region in the C-terminal one-third of G, and the clade was named BA (Trento et al. 2003). These initial BA isolates had an exact 60 nt duplication, resulting in two tandem sequences of TERDTSTSQSTVLDTTTSKH from amino acids 240 to 280 in G (Trento et al. 2003). This insertion was then used as a "tag" to analyze the global distribution and evolution of a new RSV genotype since its initial emergence (Trento et al. 2006, 2010). The BA clade rapidly disseminated worldwide and became dominant, and these findings have been confirmed by a number of RSV molecular epidemiology studies with relatively large datasets (Baek et al. 2012; Gaunt et al. 2011; Zhang et al. 2007; Botosso et al. 2009). The BA clade continued to evolve, the two exact 60 nt repeats diverged slightly, and a new BA-IV lineage essentially replaced all other group B RSV G strains. The biological significance of the duplicated region in BA G is not known.

Besides evolutionary studies, analysis of sequence variation in the G gene may have other practical applications as, for instance, tracing the origin of infecting viruses during outbreaks in hospitalized patients (Mazzulli et al. 1999; Taylor et al. 2001).

3 Clinical Differences Between RSV Strains

Early work showed that group A RSV is associated with slightly greater clinical severity than group B RSV (Hall et al. 1990). RSV disease severity was correlated with RSV clades in small cohorts, but these studies are not definitive because clade GA2 was found to be less pathogenic in one study and more pathogenic in others (Gilca et al. 2006; Martinello et al. 2002). Thus, the role of RSV strain differences in disease remains to be elucidated.

4 Phenotypic Differences Between RSV Strains In Vitro and in Animal Models

Small changes in viral gene sequences may have a large impact on pathogenesis. For instance, the elevated virulence of 1918 and avian influenza strains hinges on few amino acids differences (Conenello et al. 2007; Tumpey et al. 2005). Passage of viruses in animal hosts can lead to adaptation. The RSV Long strain was adapted to mice (line 19) by serial intracranial inoculation, resulting in higher virus replication presumably due to mutations that have not been characterized (Cavallaro and Maassab 1966). Infection of BALB/c mice with the laboratory RSV strains A2 or Long results in a predominant T_H1-type antiviral response (Moore and Peebles 2006). In contrast, the line 19 strain of RSV induces a T_H2 type and IL-13-dependent airway hyperreactivity (AHR) and pulmonary mucus in BALB/c mice (Lukacs et al. 2006). The F protein of the line 19 RSV strain has a unique sequence and is a factor that can induce pulmonary IL-13 and mucin expression in RSV infection (Moore et al. 2009). In addition, six RSV isolates have been screened for lung IL-13 levels and airway mucin expression in BALB/c mice. Three of these isolates induced lung IL-13 and gob-5 (a marker of mucin) expression, and were found to be differentially mucogenic in BALB/c mice (Stokes et al. 2011). RSV clinical isolates also infect the airway epithelium of mice to a greater extent than the laboratory adapted A2 strain (Stokes et al. 2011). In differentiated primary pediatric airway epithelial cells cultured at air–liquid interface, a RSV clinical isolate exhibited enhanced infectivity (but not virus yield) and induced greater mucus production than the laboratory A2 strain (Villenave et al. 2012). The molecular bases for these strain-specific phenotypes are largely unknown.

5 Protective Immunity

Animal and human studies have provided a wealth of evidence indicating that protection against RSV infection is afforded mainly by neutralizing antibodies. In early experiments (Prince et al. 1983), it was found that cotton rats infected with

RSV developed complete resistance to pulmonary reinfection which lasted at least 18 months. Adaptive transfer studies with the convalescence blood showed that serum antibody, but not circulating lymphocytes, conferred resistance. Immune pregnant cotton rat females transmitted protective antibodies to their young mainly through colostrum and milk (Prince et al. 1983). Similar findings were observed in pregnant ferrets (Prince and Porter 1975) and guinea pigs (Buraphacheep and Sullender 1997). Human convalescent serum and human immunoglobulin (Ig) preparations were also found to confer pulmonary protection against RSV in the cotton rats. Serum neutralizating antibody titres of 1:380 or higher were required for complete protection in the lungs, whereas about 10-fold higher titres were required for protection in the nose (Prince et al. 1985). Therapeutic administration of neutralizing Abs reduced significantly the level of RSV replication in the lungs of cotton rats and owl monkeys but showed only a slight trend of beneficial effects in a limited number of RSV-infected children (Hemming and Prince 1990). In addition to polyclonal antibodies, administration of certain monoclonal antibodies (mAbs) directed against either G or F glycoproteins protected the lungs of mice (Taylor et al. 1984) and cotton rats (Walsh et al. 1984) against a RSV challenge. In one study, it was found that passive protection of mice afforded by a nonneutralizing mAb directed against the G glycoprotein was dependent on both the Fc fragment of the antibody and the host complement (Corbeil et al. 1996), explaining the lack of strict correlation between in vitro neutralization in the absence of complement and in vivo protection (Taylor et al. 1984; Walsh et al. 1984).

In humans, protection of adult volunteers to RSV challenge was correlated with high titres of preexisting serum neutralizing antibodies (Hall et al. 1991). Additionally, an inverse correlation was observed between high titres of RSV-neutralizing serum antibodies and risk of infection in children (Glezen et al. 1986). However, whereas in animal models the quantitative aspects of passive protection to RSV infection by neutralizing antibodies is well established, the situation in human is less certain. For instance, contradicting results were reported by the same group, about correlation (Falsey and Walsh 1998) or lack of correlation (Falsey and Walsh 1992) between neutralizing antibody titers and risk of RSV infection in the elderly. Nevertheless, the consensus inferred from the majority of data is that neutralizing antibodies protect against RSV infection and particularly against RSV-associated pathology. Perhaps the best evidence for this assertion was provided by the clinical studies carried out with a Ig preparation (RS-IVIG, Respigam^{TM}), selected for high titers of RSV-neutralizing antibodies which was administered prophylactically to high-risk infants (Groothuis et al. 1993). The beneficial effect of RS-IVIG was noticed in the reduction of hospitalizations (55 %) and days of intensive care (97 %) rather than frequency of RSV infections. These studies led to licensing of Respigam in 1996 for prophylaxis of RSV infections in high-risk infants. Respigam was replaced in 1998 by a humanized neutralizing mAb (MEDI-493, palivizumab) directed against the RSV F glycoprotein (Beeler and Van Wyke 1989) that showed similar efficacy but it was easier to administer than Respigam (see chapter by H.Y. Chu and J.A. Englund, this volume).

Despite the beneficial effects of antibodies in protection against RSV, some caution is needed because:

(1) Passive serum Abs have been shown to inhibit the antibody responses to F and G glycoproteins expressed by recombinant vaccinia viruses (Murphy et al. 1988) or administered as purified antigens adjuvanted with alum (Murphy et al. 1991); however, they did not suppress the T-cell response nor the priming for a strong secondary antibody response (Fisher et al. 1999; Crowe et al. 2001). Therefore, a well-balanced dose of prophylactic neutralizing antibodies may be required to avoid interference with the host immune response to either RSV infection or RSV vaccination.
(2) Weakly neutralizing Abs may have detrimental effects that contribute to the enhanced respiratory disease observed in seronegative children that were vaccinated in the 1960s with a formalin inactivated RSV preparation (Kim et al. 1969). It has been shown that antibodies may lead to formation of immune complexes that correlate with enhanced pathology in mice after an RSV challenge. Immune complex activation of complement was also observed in postmortem lung sections from children with enhanced RSV disease (Polack et al. 2002). Nonreplicating RSV vaccines that fail to promote antibody affinity maturation may prime for immune complex formation upon RSV infection (Delgado et al. 2009).

Although antibodies are important for resistance to infection, T cells are imperative for virus clearance. Thus, individuals with compromised T cell immunity can shed virus for months (Hall et al. 1986). Prolonged virus shedding is also observed in nude or irradiated mice (Cannon et al. 1987) and in mice depleted of both $CD4^+$ and $CD8^+$ lymphocytes (Graham et al. 1991). Furthermore, $CD8^+$ cytotoxic lymphocytes may provide some protection in mice against infection, but this effect is short-lived (Connors et al. 1991, 1992). In infants with severe RSV infection, the peak of activated $CD8^+$ T cell numbers in bronchoalveolar lavage (BAL) samples and in blood correlated with convalescence, consistent with a role for $CD8^+$ T cells in recovery (Heidema et al. 2007).

6 Protective Antigens

Identification of antigens able to induce a neutralizing and protective immune response was achieved initially by immunization of mice or cotton rats with recombinant vaccinia viruses encoding individual RSV gene products (Stott et al. 1986, 1987; Wertz et al. 1987; Olmsted et al. 1986). It was promptly found that only the external F and G glycoproteins were able to confer long-lasting protection against RSV infection and that this protection correlated with induction of neutralizing antibodies. The nucleoprotein (N) and the M2-1 (or 22k) proteins were

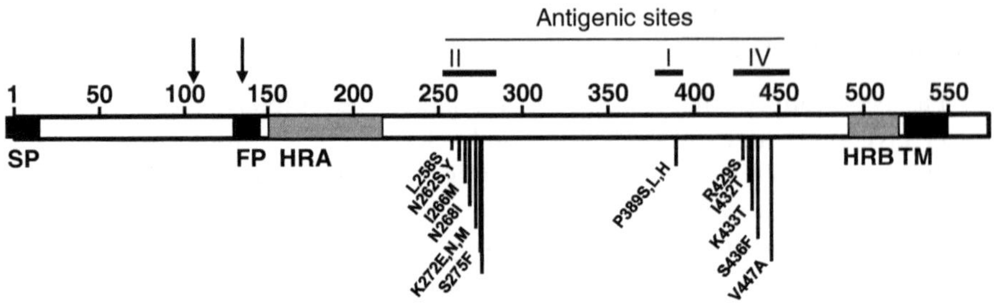

Fig. 2 Scheme of the F protein primary structure. The sequence length is indicated above the main *rectangle*. *Black boxes* denote the hydrophobic signal peptide (SP), fusion peptide (FP), and transmembrane region (TM). *Shaded boxes* symbolize heptad repeat sequences, HRA, and HRB and the two *arrows* indicate the location of the proteolytic cleavage sites. *Vertical lines* denote the location of the indicated amino acid changes selected in escape mutants described in the following articles: Arbiza et al. (1992), Lopez et al. (1998), Crowe et al. (1998), and Zhao et al. (2004a, b). These amino acid changes have been grouped in antigenic sites I, II, and IV as shown above the main *rectangle*

also able to induce partial and short-lived protection which, at least in the case of the M2-1 was mediated by cytotoxic T lymphocytes (Connors et al. 1992).

It was also noticed that: (i) recombinant vaccinia viruses expressing the F protein conferred higher level of protection against RSV than those expressing the G protein (Olmsted et al. 1986) and (ii) the neutralizing immune response against the F protein was cross-protective against viruses of a different antigenic group, whereas the neutralizing and protective response against the G protein was restricted to viruses of the same antigenic group (Johnson et al. 1987a; Stott et al. 1987). These differences reflect the dissimilar structural and antigenic characteristics of the F and G glycoproteins and their differences at the level of antigenic and genetic relatedness between RSV isolates.

In the case of the F glycoprotein (the viral glycoprotein that mediates fusion of the virus and cell membranes), there is 89 % amino acid sequence identity between the proteins of groups A and B of human RSV (Johnson and Collins 1988). This is reflected in the high level of antigenic relatedness observed with murine mAbs (Garcia-Barreno et al. 1989). Epitopes have been mapped in the F protein primary structure primarily by isolation and sequencing of mutants that grow in the presence of individual mAbs. These escape mutants normally contained single amino acid substitutions that obliterated the epitopes recognized by the antibodies used in their selection (see chapter by J.S. McLellan et al., this volume). Figure 2 summarizes the amino acid changes in escape mutants reported up to date by different groups (Arbiza et al. 1992; Lopez et al. 1998; Crowe et al. 1998; Zhao et al. 2004a, b). In most cases, selection was done with murine mAbs, except in the case of the mutant I266 M that was selected with a recombinant human Fab fragment (Crowe et al. 1998). It is worth stressing that frequently the same mutation could be repeatedly selected with different antibodies (e.g., K272N isolated with mAbs 151, 1200, 47F, and palivizumab). Occasionally, more than

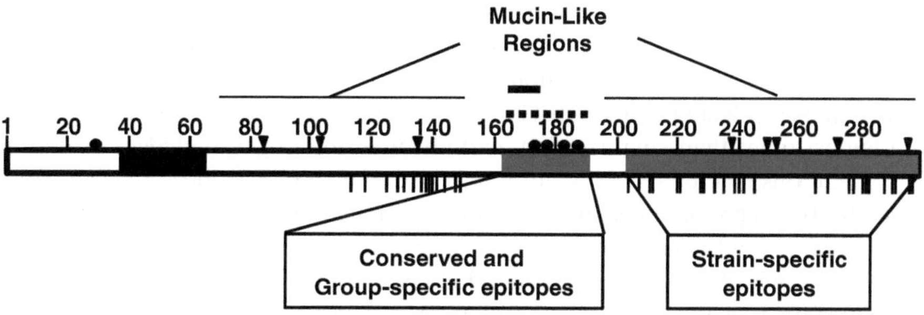

Fig. 3 Scheme of the G protein primary structure. The sequence length is indicated above the main *rectangle*. Two mucin-like variable regions of the G protein ectodomain are indicated. *Black dots* denote Cys residues, *black arrowheads* denote N-glycosylation sites, and short *vertical lines* (below the main *rectangle*) O-glycosylation sites. The *black box* delineates the transmembrane region. The *gray boxes* delineate the location of conserved and group-specific epitopes in the central segment and the strain-specific epitopes in the C-terminal mucin-like region, respectively (Garcia-Barreno et al. 1990; Rueda et al. 1991, 1994, 1995; Martinez et al. 1997; Walsh et al. 1998). The *continuous horizontal line* denotes the segment of identical sequence (amino acids 164–176) in all RSV isolates and the *broken horizontal line* the segment of identical sequence (amino acids 163–189) in all RSV group A isolates

one amino acid change was selected at the same position (e.g., N262S or N262Y) and one of the changes (N268I) was also selected after passing the virus in the presence of a polyclonal rabbit antiserum raised against purified F protein (Tome et al. 2012). These findings probably reflect the dominance of certain epitopes and the propensity of the F protein to incorporate certain mutations but not others. The amino acid changes depicted in Fig. 2 and therefore the corresponding antibody epitopes are clustered in three regions (antigenic sites) of the F protein primary structure, named sites II, I, and IV and corresponding to sites A, B, and C as designated by Beeler and Van Wyke (1989). The epitopes of antigenic sites II and IV (Anderson et al. 1985; Beeler and Van Wyke 1989; Garcia-Barreno et al. 1989) are very conserved among RSV strains, while those of antigenic site I are group-specific; i.e., they are conserved only in RSV strains of the same antigenic group (Garcia-Barreno et al. 1989).

In contrast to the antigenic and genetic conservation of the F protein, the G glycoproteins of RSV group A and B strains share less than 50 % amino acid sequence identity (Johnson et al. 1987b). This genetic variation is reflected in the high level of antigenic differences detected with anti-G mAbs (Garcia-Barreno et al. 1989). As with the F protein, the epitopes recognized by anti-G MAbs have been mapped in the protein primary structure mainly by selection and sequencing of escape mutants (Fig. 3). Three types of epitopes were identified: (i) epitopes ***conserved*** among all RSV strains that mapped in the central unglycosylated segment of the G protein ectodomain, (ii) ***group-specific*** epitopes that overlapped partially with the conserved epitopes and that were retained only in RSV strains of the same antigenic group, and (iii) a majority of ***strain-specific*** epitopes that were located in the C-terminal third of the G protein ectodomain and that were present only in some RSV strains of the same antigenic group (Martinez et al. 1997). Additionally, and in clear distinction with the escape mutants selected with anti-F MAbs, those selected with anti-G mAbs frequently contained drastic genetic alterations, other than single amino acid changes and which included: (i) frame-shift mutations that altered the C-terminal third of the G molecule (Garcia-Barreno et al. 1990), (ii) premature stop codons that shortened the G polypeptide between 1 and 42 amino acids (Rueda et al. 1991, 1995; Martinez et al. 1997), (iii) multiple A-G transitions (A-G hypermutations) that change several amino acids, including some cysteines in the G protein ectodomain (Rueda et al. 1994; Martinez et al. 1997; Walsh et al. 1998), and (iv) amino acid changes that prevent insertion of G in the viral membrane (Walsh et al. 1998). All these findings emphasize the extreme plasticity of the G molecule to adopt sequence changes, something that might be related to the fact that G is not required for RSV replication in Vero cells, although viruses lacking G are attenuated in HEp-2 cells and in vivo (Karron et al. 1997).

Another major difference between mAbs specific for the F and G glycoproteins is their potency and mechanism of neutralization. Whereas certain mAbs reacting with epitopes in sites II or IV of the F glycoprotein are potent neutralizers of RSV infectivity in vitro (Lopez et al. 1998) most mAbs specific of the G glycoprotein neutralize RSV very poorly, even if tested against the homologous strain used in their isolation. However, mixtures of anti-G mAbs show a synergistic effect in neutralization that is not observed with anti-F MAbs (Martinez and Melero 1998; Anderson et al. 1988). It has then been postulated that neutralizing antibodies directed against the F glycoprotein inhibit RSV infectivity by blocking the conformational changes that follow activation of RSV F to initiate the process of virus-cell membrane fusion (Magro et al. 2010). In contrast, results obtained with anti-G mAbs suggest that neutralization is afforded in this case by steric hindrance of G-protein interactions with cell surface components (likely proteoglycans). This steric inhibition requires simultaneous binding of several antibodies to the same G molecule, explaining the synergistic effect found with combinations of these antibodies (Martinez and Melero 1998). Despite these findings, it has been reported that a minority of human antibodies with high affinity for the central conserved region of RSV G are potent neutralizers of virus infectivity (Collarini et al. 2009) and may neutralize RSV infectivity by mechanisms others than steric hindrance.

In summary, most data indicate that F is more potent than G in inducing neutralizing and protective antibodies. The potent neutralizing antibody response afforded by anti-F antibodies is widely cross-reactive, while the poorer neutralizing anti-G response is rather strain or group-specific. These differences are probably deep-rooted in the level of genetic variability of F and G between RSV isolates and in the mechanism of virus neutralization afforded by different antibodies, particularly in vivo.

7 Human Antibody Response

As noted in previous sections, most studies carried out so far with neutralizing antibodies have been done with murine mAbs. Therefore, two relevant questions are: (i) what are the specificities of human neutralizing Abs? and (ii) are they the same as those of murine Abs?

Not much is known about the specificities of human anti-RSV neutralizing Abs. This gap between human and animal studies is due at least in part to the inherent difficulty of experimenting with humans. The use of technologies to clone and express human antibodies produced by individual B cells should facilitate in the future dissecting the repertoire of specificities represented in the human neutralizing antibody response against RSV.

Nevertheless, it has been reported already that human Ig preparations contain antibodies that compete with most murine anti-F antibodies (Sastre 2004); however, human neutralizing Abs have been identified that recognize F protein fragments laying outside of the antigenic sites demarcated by murine mAbs (Sastre et al. 2004). Additionally, neutralizing Abs specific for the untriggered form of RSV F, not represented in the collections of murine mAbs so far described were recently unveiled in human Ig preparations (Magro et al. 2012). Therefore, although the specificities of murine anti-F Abs are present in human sera, human Abs seem to embrace a broader range of specificities than murine Abs.

Similarly, certain specificities of anti-G mAbs also seem to be relevant in the human antibody response. Thus, human Igs were able to out compete binding of certain anti-G mAbs that recognized epitopes of the unglycosylated central segment of the G protein ectodomain (Sastre 2004). Furthermore, neutralizing antibodies could be purified from human Ig preparations by affinity chromatography with a fragment of the G glycoprotein corresponding to the conserved ectodomain segment (Sastre et al. 2004). In agreement with these results, group-specific serum antibodies directed against the central nonglycosylated G-protein segment have been found after primary infections of children (Murata et al. 2010a, b). These antibodies are likely responsible for the partially group-specific neutralizing antibody response reported in previous studies after primary RSV infections (Hendry et al. 1988; Muelenaer et al. 1991). However, it remains uncertain whether or not human sera contain Abs reacting with strain-specific epitopes of the C-terminal third of the RSV G glycoprotein. Detection of this type of antibodies in convalescent sera is likely dependent on the use of an appropriate G protein (or fragment thereof) that matches the antigenic properties of the infecting virus. Since detailed information about the infecting virus is not available for most sera, detection of strain-specific antibodies has been reported only in a few cases (Palomo et al. 2000; Cane et al. 1996; Jones et al. 2002; Scott et al. 2007)

In summary, it is apparent that the antibody specificities of murine mAbs are also represented in human convalescent sera, but human antibodies seem to recognize a broader range of F and G-protein epitopes than murine Abs. It will be important in the future to elucidate the actual relevance of the different antibody specificities for protection against infection and pathology. This might be particularly relevant in the case of anti-G Abs; for instance, it has been reported that certain murine Abs that block the interaction of the fractalkine-like motif of RSV_G with the CX3CR1 receptor reduced pulmonary inflammation and virus replication in mice (Zhang et al. 2010b). Likewise, other Abs may block additional, but still unrecognized activities of the F or G glycoproteins involved in virus replication and/or pathology.

8 Antigenic Variation, Immune Selection, and RSV Evolution

It has been observed that generally viral antigens accumulate amino acid changes at the sites that are recognized by neutralizing Abs as exemplified by the influenza virus haemaglutinin (HA) (Knossow and Skehel 2006). In this case, most of the sequence changes that are retained in viruses of the same subtype over the years involve residues on the surface of the HA head that are part of the epitopes recognized by neutralizing Abs. In other words, to escape neutralization by pre-existing antibodies new influenza strains are positively selected that can infect again the same human population.

The situation described for influenza virus is in apparent contradiction with that found in RSV and outlined in previous sections. Thus, in the case of human RSV, the most effective neutralizing antibodies are those directed against antigenic sites II and IV of the F glycoprotein (Fig. 2), which are highly conserved among RSV isolates. In contrast, the antibodies that recognize epitopes in the C-terminal third of the G glycoprotein are only weakly neutralizing (Fig. 3), although this segment of the G molecule represents an extreme example of sequence variation, compared with antigens of related viruses (e.g., influenza virus). In other words, the paradoxical situation in RSV is that the epitopes recognized by the most potent neutralizing antibodies are highly conserved, whereas those recognized by poorly neutralizing antibodies are extremely variable.

As mentioned before, two types of evidence support the notion that the sequence variation found in the G protein is the result of positive selection: (i) whereas synonymous nucleotide changes have a uniform distribution along the G-protein gene, nonsynonymous changes accumulate in the two variable mucin-like regions of G (Melero et al. 1997), in analogy with the accumulation of nonsynonymous changes at the antigenic sites of influenza HA due to positive Darwinian evolution (Fitch et al. 1991) and (ii) phylogenetic methods have identified several positively selected sites in group A and B of RSV isolates (Woelk and Holmes 2001; Botosso et al. 2009). The fact that some of the positively selected sites coincide with epitopes recognized by anti-G mAbs and that the antigenic changes detected with a panel of anti-G mAbs correlated with the position of RSV isolates in a phylogenetic tree (Garcia et al. 1994) support the notion that positive selection of changes in the G glycoprotein might be immune driven.

Data from human studies offer conflicting interpretations about the relevance of antigenic variation and immune selection in protection against infection and in RSV evolution, respectively. On the one hand, as noted before the epitopes recognized by the most efficient neutralizing anti-F mAbs are highly conserved among RSV isolates and antibodies with the same specificities are present in human Ig preparations (Sastre et al. 2004). These results would argue against immune selection playing any role in the generation of RSV variability and support the idea that RSV vaccines may require only one virus strain or its antigens.

On the other hand, studies conducted with convalescent sera have provided evidence that upon primary RSV infections there was some dominance of group-specific neutralizing antibodies (Muelenaer et al. 1991; Cane et al. 1996; Scott et al. 2007). This group-specific response was also observed by ELISA with sera from convalescent primary infections using segments of the central core of the G glycoprotein as antigens (Murata et al. 2010a; Scott et al. 2007; Cane et al. 1996). However, the group-specificity of the neutralizing and antigen-binding antibodies was blurred upon secondary infections (Murata et al. 2010a). Consequently, there is a critical need for detailed assessment of the antibody specificities (including strain-specific) induced upon RSV infection and reinfection and their actual contribution to the human neutralizing immune response.

It is possible that the immune selective pressure on RSV is not as strong as that operating in influenza virus; for instance, if the antibody response is short-lived in RSV in comparison with influenza virus, replacement of preexisting RSV strains by new variants may be a slower process than in influenza. Of note, the new variants of RSV group B with a 60-nucleotide duplication in the G-protein gene replaced the preexisting strains of the same antigenic group worldwide but after a 6 to 7-year period (Trento et al. 2010), which is clearly longer than the observed replacement of influenza A viruses by new variants (usually 2–3 years). A combination of immune selection and high plasticity of the G protein may be at the basis of the extreme sequence variation observed for this gene among RSV isolates. In consequence, until new information concerning the specificities of human Ab responses is available, it would be wise to include viruses (or antigens) representing the two RSV antigenic groups in vaccine development.

Acknowledgments Work in the Madrid laboratory is currently funded by grants GR09/0039 from Instituto de Salud Carlos III and SAF2009-11632 and SAF2012-31217 from Plan Nacional de I+D+i. Work in the Atlanta laboratory is supported by the following grants: NIH 1R01AI087798 and NIH 1U19AI095227.

References

Agoti CN, Mwihuri AG, Sande CJ, Onyango CO, Medley GF, Cane PA, Nokes DJ (2012) Genetic relatedness of infecting and reinfecting respiratory syncytial virus strains identified in a birth cohort from rural Kenya. J Infect Dis 206:1532–1541

Anderson LJ, Hierholzer JC, Tsou C, Hendry RM, Fernie BF, Stone Y, McIntosh K (1985) Antigenic characterization of respiratory syncytial virus strains with monoclonal antibodies. J Infect Dis 151:623–633

Anderson LJ, Bingham P, Hierholzer JC (1988) Neutralization of respiratory syncytial virus by individual and mixtures of F and G protein monoclonal antibodies. J Virol 62:4232–4238

Arbiza J, Taylor G, Lopez JA, Furze J, Wyld S, Whyte P, Stott EJ, Wertz G, Sullender W, Trudel M, Melero JA (1992) Characterization of two antigenic sites recognized by neutralizing monoclonal antibodies directed against the fusion glycoprotein of human respiratory syncytial virus. J Gen Virol 73(Pt 9):2225–2234

Arnott A, Vong S, Mardy S, Chu S, Naughtin M, Sovann L, Buecher C, Beaute J, Rith S, Borand L, Asgari N, Frutos R, Guillard B, Touch S, Deubel V, Buchy P (2011) A study of the genetic variability of human respiratory syncytial virus (HRSV) in Cambodia reveals the existence of a new HRSV group B genotype. J Clin Microbiol 49:3504–3513

Baek YH, Choi EH, Song MS, Pascua PN, Kwon HI, Park SJ, Lee JH, Woo SI, Ahn BH, Han HS, Hahn YS, Shin KS, Jang HL, Kim SY, Choi YK (2012) Prevalence and genetic characterization of respiratory syncytial virus (RSV) in hospitalized children in Korea. Arch Virol 157:1039–1050

Beeler JA, Van Wyke CK (1989) Neutralization epitopes of the F glycoprotein of respiratory syncytial virus: effect of mutation upon fusion function. J Virol 63:2941–2950

Beem M (1967) Repeated infections with respiratory syncytial virus. J Immunol 98:1115–1122

Botosso VF, Zanotto PM, Ueda M, Arruda E, Gilio AE, Vieira SE, Stewien KE, Peret TC, Jamal LF, Pardini MI, Pinho JR, Massad E, Sant'anna OA, Holmes EC, Durigon EL (2009) Positive selection results in frequent reversible amino acid replacements in the G protein gene of human respiratory syncytial virus. PLoS Pathog 5:e1000254

Buraphacheep W, Sullender WM (1997) The guinea pig as a model for the study of maternal immunization against respiratory syncytial virus infections in infancy. J Infect Dis 175:935–938

Cane PA, Pringle CR (1991) Respiratory syncytial virus heterogeneity during an epidemic: analysis by limited nucleotide sequencing (SH gene) and restriction mapping (N gene). J Gen Virol 72(Pt 2):349–357

Cane PA, Pringle CR (1995) Evolution of subgroup A respiratory syncytial virus: evidence for progressive accumulation of amino acid changes in the attachment protein. J Virol 69:2918–2925

Cane PA, Thomas HM, Simpson AF, Evans JE, Hart CA, Pringle CR (1996) Analysis of the human serological immune response to a variable region of the attachment (G) protein of respiratory syncytial virus during primary infection. J Med Virol 48:253–261

Cannon MJ, Stott EJ, Taylor G, Askonas BA (1987) Clearance of persistent respiratory syncytial virus infections in immunodeficient mice following transfer of primed T cells. Immunology 62:133–138

Cavallaro JJ, Maassab HF (1966) Adaptation of respiratory syncytial (RS) virus to brain of suckling mice. Proc Soc Exp Biol Med 121:37–41

Coates HV, Alling DW, Chanock RM (1966) An antigenic analysis of respiratory syncytial virus isolates by a plaque reduction neutralization test. Am J Epidemiol 83:299–313

Collarini EJ, Lee FE, Foord O, Park M, Sperinde G, Wu H, Harriman WD, Carroll SF, Ellsworth SL, Anderson LJ, Tripp RA, Walsh EE, Keyt BA, Kauvar LM (2009) Potent high-affinity antibodies for treatment and prophylaxis of respiratory syncytial virus derived from B cells of infected patients. J Immunol 183:6338–6345

Conenello GM, Zamarin D, Perrone LA, Tumpey T, Palese P (2007) A single mutation in the PB1-F2 of H5N1 (HK/97) and 1918 influenza A viruses contributes to increased virulence. PLoS Pathog 3:1414–1421
Connors M, Collins PL, Firestone CY, Murphy BR (1991) Respiratory syncytial virus (RSV) F, G, M2 (22 K), and N proteins each induce resistance to RSV challenge, but resistance induced by M2 and N proteins is relatively short-lived. J Virol 65:1634–1637
Connors M, Kulkarni AB, Collins PL, Firestone CY, Holmes KL, Morse HC III, Murphy BR (1992) Resistance to respiratory syncytial virus (RSV) challenge induced by infection with a vaccinia virus recombinant expressing the RSV M2 protein (Vac-M2) is mediated by CD8+ T cells, while that induced by Vac-F or Vac-G recombinants is mediated by antibodies. J Virol 66:1277–1281
Corbeil S, Seguin C, Trudel M (1996) Involvement of the complement system in the protection of mice from challenge with respiratory syncytial virus Long strain following passive immunization with monoclonal antibody 18A2B2. Vaccine 14:521–525
Crowe JE, Firestone CY, Crim R, Beeler JA, Coelingh KL, Barbas CF, Burton DR, Chanock RM, Murphy BR (1998) Monoclonal antibody-resistant mutants selected with a respiratory syncytial virus-neutralizing human antibody fab fragment (Fab 19) define a unique epitope on the fusion (F) glycoprotein. Virology 252:373–375
Crowe JE Jr, Firestone CY, Murphy BR (2001) Passively acquired antibodies suppress humoral but not cell-mediated immunity in mice immunized with live attenuated respiratory syncytial virus vaccines. J Immunol 167:3910–3918
Delgado MF, Coviello S, Monsalvo AC, Melendi GA, Hernandez JZ, Batalle JP, Diaz L, Trento A, Chang HY, Mitzner W, Ravetch J, Melero JA, Irusta PM, Polack FP (2009) Lack of antibody affinity maturation due to poor Toll-like receptor stimulation leads to enhanced respiratory syncytial virus disease. Nat Med 15:34–41
Falsey AR, Walsh EE (1992) Humoral immunity to respiratory syncytial virus infection in the elderly. J Med Virol 36:39–43
Falsey AR, Walsh EE (1998) Relationship of serum antibody to risk of respiratory syncytial virus infection in elderly adults. J Infect Dis 177:463–466
Fisher RG, Johnson JE, Dillon SB, Parker RA, Graham BS (1999) Prophylaxis with respiratory syncytial virus F-specific humanized monoclonal antibody delays and moderately suppresses the native antibody response but does not impair immunity to late rechallenge. J Infect Dis 180:708–713
Fitch WM, Leiter JM, Li XQ, Palese P (1991) Positive Darwinian evolution in human influenza A viruses. Proc Natl Acad Sci U S A 88:4270–4274
Garcia O, Martin M, Dopazo J, Arbiza J, Frabasile S, Russi J, Hortal M, Perez-Brena P, Martinez I, Garcia-Barreno B (1994) Evolutionary pattern of human respiratory syncytial virus (subgroup A): cocirculating lineages and correlation of genetic and antigenic changes in the G glycoprotein. J Virol 68:5448–5459
Garcia-Barreno B, Palomo C, Penas C, Delgado T, Perez-Brena P, Melero JA (1989) Marked differences in the antigenic structure of human respiratory syncytial virus F and G glycoproteins. J Virol 63:925–932
Garcia-Barreno B, Portela A, Delgado T, Lopez JA, Melero JA (1990) Frame shift mutations as a novel mechanism for the generation of neutralization resistant mutants of human respiratory syncytial virus. EMBO J 9:4181–4187
Gaunt ER, Jansen RR, Poovorawan Y, Templeton KE, Toms GL, Simmonds P (2011) Molecular epidemiology and evolution of human respiratory syncytial virus and human metapneumovirus. PLoS ONE 6:e17427
Gilca R, De Serres G, Tremblay M, Vachon ML, Leblanc E, Bergeron MG, Dery P, Boivin G (2006) Distribution and clinical impact of human respiratory syncytial virus genotypes in hospitalized children over 2 winter seasons. J Infect Dis 193:54–58
Glezen WP, Taber LH, Frank AL, Kasel JA (1986) Risk of primary infection and reinfection with respiratory syncytial virus. Am J Dis Child 140:543–546

Graham BS, Bunton LA, Wright PF, Karzon DT (1991) Role of T lymphocyte subsets in the pathogenesis of primary infection and rechallenge with respiratory syncytial virus in mice. J Clin Invest 88:1026–1033

Groothuis JR, Simoes EA, Levin MJ, Hall CB, Long CE, Rodriguez WJ, Arrobio J, Meissner HC, Fulton DR, Welliver RC (1993) Prophylactic administration of respiratory syncytial virus immune globulin to high-risk infants and young children. The Respiratory Syncytial Virus Immune Globulin Study Group. N Engl J Med 329:1524–1530

Hall CB, Powell KR, Macdonald NE, Gala CL, Menegus ME, Suffin SC, Cohen HJ (1986) Respiratory syncytial viral infection in children with compromised immune function. N Engl J Med 315:77–81

Hall CB, Walsh EE, Schnabel KC, Long CE, McConnochie KM, Hildreth SW, Anderson LJ (1990) Occurrence of groups A and B of respiratory syncytial virus over 15 years: associated epidemiologic and clinical characteristics in hospitalized and ambulatory children. J Infect Dis 162:1283–1290

Hall CB, Walsh EE, Long CE, Schnabel KC (1991) Immunity to and frequency of reinfection with respiratory syncytial virus. J Infect Dis 163:693–698

Heidema J, Lukens MV, van Maren WW, van Dijk ME, Otten HG, van Vught AJ, van der Werff DB, van Gestel SJ, Semple MG, Smyth RL, Kimpen JL, van Bleek GM (2007) CD8 + T cell responses in bronchoalveolar lavage fluid and peripheral blood mononuclear cells of infants with severe primary respiratory syncytial virus infections. J Immunol 179:8410–8417

Hemming VG, Prince GA (1990) Immunoprophylaxis of infections with respiratory syncytial virus: observations and hypothesis. Rev Infect Dis 12:S470–S475

Hendry RM, Burns JC, Walsh EE, Graham BS, Wright PF, Hemming VG, Rodriguez WJ, Kim HW, Prince GA, McIntosh K (1988) Strain-specific serum antibody responses in infants undergoing primary infection with respiratory syncytial virus. J Infect Dis 157:640–647

Johnson PR, Collins PL (1988) The fusion glycoproteins of human respiratory syncytial virus of subgroups A and B: sequence conservation provides a structural basis for antigenic relatedness. J Gen Virol 69(Pt 10):2623–2628

Johnson PR, Olmsted RA, Prince GA, Murphy BR, Alling DW, Walsh EE, Collins PL (1987a) Antigenic relatedness between glycoproteins of human respiratory syncytial virus subgroups A and B: evaluation of the contributions of F and G glycoproteins to immunity. J Virol 61:3163–3166

Johnson PR, Spriggs MK, Olmsted RA, Collins PL (1987b) The G glycoprotein of human respiratory syncytial viruses of subgroups A and B: extensive sequence divergence between antigenically related proteins. Proc Natl Acad Sci U S A 84:5625–5629

Jones LP, Zheng HQ, Karron RA, Peret TC, Tsou C, Anderson LJ (2002) Multiplex assay for detection of strain-specific antibodies against the two variable regions of the G protein of respiratory syncytial virus. Clin Diagn Lab Immunol 9:633–638

Karron RA, Buonagurio DA, Georgiu AF, Whitehead SS, Adamus JE, Clements-Mann ML, Harris DO, Randolph VB, Udem SA, Murphy BR, Sidhu MS (1997) Respiratory syncytial virus (RSV) SH and G proteins are not essential for viral replication in vitro: clinical evaluation and molecular characterization of a cold-passaged, attenuated RSV subgroup B mutant. Proc Natl Acad Sci U S A 94:13961–13966

Kim HW, Canchola JG, Brandt CD, Pyles G, Chanock RM, Jensen K, Parrott RH (1969) Respiratory syncytial virus disease in infants despite prior administration of antigenic inactivated vaccine. Am J Epidemiol 89:422–434

Knossow M, Skehel JJ (2006) Variation and infectivity neutralization in influenza. Immunology 119:1–7

Kuroiwa Y, Nagai K, Okita L, Yui I, Kase T, Nakayama T, Tsutsumi H (2005) A phylogenetic study of human respiratory syncytial viruses group A and B strains isolated in two cities in Japan from 1980–2002. J Med Virol 76:241–247

Lopez JA, Bustos R, Orvell C, Berois M, Arbiza J, Garcia-Barreno B, Melero JA (1998) Antigenic structure of human respiratory syncytial virus fusion glycoprotein. J Virol 72:6922–6928

Lukacs NW, Moore ML, Rudd BD, Berlin AA, Collins RD, Olson SJ, Ho SB, Peebles RS Jr (2006) Differential immune responses and pulmonary pathophysiology are induced by two different strains of respiratory syncytial virus. Am J Pathol 169:977–986

Magro M, Andreu D, Gomez-Puertas P, Melero JA, Palomo C (2010) Neutralization of human respiratory syncytial virus infectivity by antibodies and low-molecular-weight compounds targeted against the fusion glycoprotein. J Virol 84:7970–7982

Magro M, Mas V, Chappell K, Vazquez M, Cano O, Luque D, Terron MC, Melero JA, Palomo C (2012) Neutralizing antibodies against the preactive form of respiratory syncytial virus fusion protein offer unique possibilities for clinical intervention. Proc Natl Acad Sci U S A 109:3089–3094

Martinello RA, Chen MD, Weibel C, Kahn JS (2002) Correlation between respiratory syncytial virus genotype and severity of illness. J Infect Dis 186:839–842

Martinez I, Melero JA (1998) Enhanced neutralization of human respiratory syncytial virus by mixtures of monoclonal antibodies to the attachment (G) glycoprotein. J Gen Virol 79(Pt 9):2215–2220

Martinez I, Dopazo J, Melero JA (1997) Antigenic structure of the human respiratory syncytial virus G glycoprotein and relevance of hypermutation events for the generation of antigenic variants. J Gen Virol 78(Pt 10):2419–2429

Matheson JW, Rich FJ, Cohet C, Grimwood K, Huang QS, Penny D, Hendy MD, Kirman JR (2006) Distinct patterns of evolution between respiratory syncytial virus subgroups A and B from New Zealand isolates collected over thirty-seven years. J Med Virol 78:1354–1364

Mazzulli T, Peret TC, McGeer A, Cann D, MacDonald KS, Chua R, Erdman DD, Anderson LJ (1999) Molecular characterization of a nosocomial outbreak of human respiratory syncytial virus on an adult leukemia/lymphoma ward. J Infect Dis 180:1686–1689

Melero JA, Garcia-Barreno B, Martinez I, Pringle CR, Cane PA (1997) Antigenic structure, evolution and immunobiology of human respiratory syncytial virus attachment (G) protein. J Gen Virol 78(Pt 10):2411–2418

Moore ML, Peebles RS Jr (2006) Respiratory syncytial virus disease mechanisms implicated by human, animal model, and in vitro data facilitate vaccine strategies and new therapeutics. Pharmacol Ther 112:405–424

Moore ML, Chi MH, Luongo C, Lukacs NW, Polosukhin VV, Huckabee MM, Newcomb DC, Buchholz UJ, Crowe JE Jr, Goleniewska K, Williams JV, Collins PL, Peebles RS Jr (2009) A chimeric A2 strain of respiratory syncytial virus (RSV) with the fusion protein of RSV strain line 19 exhibits enhanced viral load, mucus, and airway dysfunction. J Virol 83:4185–4194

Muelenaer PM, Henderson FW, Hemming VG, Walsh EE, Anderson LJ, Prince GA, Murphy BR (1991) Group-specific serum antibody responses in children with primary and recurrent respiratory syncytial virus infections. J Infect Dis 164:15–21

Mufson MA, Orvell C, Rafnar B, Norrby E (1985) Two distinct subtypes of human respiratory syncytial virus. J Gen Virol 66(Pt 10):2111–2124

Murata Y, Lightfoote PM, Biear JN, Falsey AR, Walsh EE (2010a) Humoral response to the central unglycosylated region of the respiratory syncytial virus attachment protein. Vaccine 28:6242–6246

Murata Y, Lightfoote PM, Falsey AR, Walsh EE (2010b) Identification of and human serum reactogenicity to neutralizing epitopes within the central unglycosylated region of the respiratory syncytial virus attachment protein. Clin Vaccine Immunol 17:695–697

Murphy BR, Olmsted RA, Collins PL, Chanock RM, Prince GA (1988) Passive transfer of respiratory syncytial virus (RSV) antiserum suppresses the immune response to the RSV fusion (F) and large (G) glycoproteins expressed by recombinant vaccinia viruses. J Virol 62:3907–3910

Murphy BR, Prince GA, Collins PL, Hildreth SW, Paradiso PR (1991) Effect of passive antibody on the immune response of cotton rats to purified F and G glycoproteins of respiratory syncytial virus (RSV). Vaccine 9:185–189

Olmsted RA, Elango N, Prince GA, Murphy BR, Johnson PR, Moss B, Chanock RM, Collins PL (1986) Expression of the F glycoprotein of respiratory syncytial virus by a recombinant

vaccinia virus: comparison of the individual contributions of the F and G glycoproteins to host immunity. Proc Natl Acad Sci U S A 83:7462–7466
Palomo C, Cane PA, Melero JA (2000) Evaluation of the antibody specificities of human convalescent-phase sera against the attachment (G) protein of human respiratory syncytial virus: influence of strain variation and carbohydrate side chains. J Med Virol 60:468–474
Peret TC, Hall CB, Schnabel KC, Golub JA, Anderson LJ (1998) Circulation patterns of genetically distinct group A and B strains of human respiratory syncytial virus in a community. J Gen Virol 79(Pt 9):2221–2229
Peret TC, Hall CB, Hammond GW, Piedra PA, Storch GA, Sullender WM, Tsou C, Anderson LJ (2000) Circulation patterns of group A and B human respiratory syncytial virus genotypes in 5 communities in North America. J Infect Dis 181:1891–1896
Polack FP, Teng MN, Collins PL, Prince GA, Exner M, Regele H, Lirman DD, Rabold R, Hoffman SJ, Karp CL, Kleeberger SR, Wills-Karp M, Karron RA (2002) A role for immune complexes in enhanced respiratory syncytial virus disease. J Exp Med 196:859–865
Prince GA, Porter DD (1975) The pathogenesis of respiratory syncytial virus infection in infant ferrets. Am J Pathol 82:339–352
Prince GA, Horswood RL, Camargo E, Koenig D, Chanock RM (1983) Mechanisms of immunity to respiratory syncytial virus in cotton rats. Infect Immun 42:81–87
Prince GA, Horswood RL, Chanock RM (1985) Quantitative aspects of passive immunity to respiratory syncytial virus infection in infant cotton rats. J Virol 55:517–520
Reiche J, Schweiger B (2009) Genetic variability of group A human respiratory syncytial virus strains circulating in Germany from 1998 to 2007. J Clin Microbiol 47:1800–1810
Rueda P, Delgado T, Portela A, Melero JA, Garcia-Barreno B (1991) Premature stop codons in the G glycoprotein of human respiratory syncytial viruses resistant to neutralization by monoclonal antibodies. J Virol 65:3374–3378
Rueda P, Garcia-Barreno B, Melero JA (1994) Loss of conserved cysteine residues in the attachment (G) glycoprotein of two human respiratory syncytial virus escape mutants that contain multiple A-G substitutions (hypermutations). Virology 198:653–662
Rueda P, Palomo C, Garcia-Barreno B, Melero JA (1995) The three C-terminal residues of human respiratory syncytial virus G glycoprotein (Long strain) are essential for integrity of multiple epitopes distinguishable by antiidiotypic antibodies. Viral Immunol 8:37–46
Sastre P (2004) Caracterizaci6n de anticuerpos dirigidos frente al virus respiratorio sincitial humano presentes en preparaciones de inmunoglobulinas humanas. PhD Thesis
Sastre P, Melero JA, Garcia-Barreno B, Palomo C (2004) Comparison of antibodies directed against human respiratory syncytial virus antigens present in two commercial preparations of human immunoglobulins with different neutralizing activities. Vaccine 23:435–443
Scott PD, Ochola R, Ngama M, Okiro EA, Nokes DJ, Medley GF, Cane PA (2004) Molecular epidemiology of respiratory syncytial virus in Kilifi district, Kenya. J Med Virol 74:344–354
Scott PD, Ochola R, Sande C, Ngama M, Okiro EA, Medley GF, Nokes DJ, Cane PA (2007) Comparison of strain-specific antibody responses during primary and secondary infections with respiratory syncytial virus. J Med Virol 79:1943–1950
Shobugawa Y, Saito R, Sano Y, Zaraket H, Suzuki Y, Kumaki A, Dapat I, Oguma T, Yamaguchi M, Suzuki H (2009) Emerging genotypes of human respiratory syncytial virus subgroup A among patients in Japan. J Clin Microbiol 47:2475–2482
Stokes KL, Chi MH, Sakamoto K, Newcomb DC, Currier MG, Huckabee MM, Lee S, Goleniewska K, Pretto C, Williams JV, Hotard A, Sherrill TP, Peebles RS Jr, Moore ML (2011) Differential pathogenesis of respiratory syncytial virus clinical isolates in BALB/c mice. J Virol 85:5782–5793
Stott EJ, Ball LA, Young KK, Furze J, Wertz GW (1986) Human respiratory syncytial virus glycoprotein G expressed from a recombinant vaccinia virus vector protects mice against live-virus challenge. J Virol 60:607–613
Stott EJ, Taylor G, Ball LA, Anderson K, Young KK, King AM, Wertz GW (1987) Immune and histopathological responses in animals vaccinated with recombinant vaccinia viruses that express individual genes of human respiratory syncytial virus. J Virol 61:3855–3861

Suto T, Yano N, Ikeda M, Miyamoto M, Takai S, Shigeta S, Hinuma Y, Ishida N (1965) Respiratory syncytial virus infection and its serologic epidemiology. Am J Epidemiol 82:211–224
Tamura K, Peterson D, Peterson N, Stecher G, Nei M, Kumar S (2011) MEGA5: molecular evolutionary genetics analysis using maximum likelihood, evolutionary distance, and maximum parsimony methods. Mol Biol Evol 28:2731–2739
Taylor G, Stott EJ, Bew M, Fernie BF, Cote PJ, Collins AP, Hughes M, Jebbett J (1984) Monoclonal antibodies protect against respiratory syncytial virus infection in mice. Immunology 52:137–142
Taylor GS, Vipond IB, Caul EO (2001) Molecular epidemiology of outbreak of respiratory syncytial virus within bone marrow transplantation unit. J Clin Microbiol 39:801–803
Tome L, Frabasile S, Candia C, Pittini A, Farina N, Melero JA, Arbiza J (2012) Selection and characterization of human respiratory syncytial virus escape mutants resistant to a polyclonal antiserum raised against the F protein. Arch Virol 157(6):1071–1080
Trento A, Galiano M, Videla C, Carballal G, Garcia-Barreno B, Melero JA, Palomo C (2003) Major changes in the G protein of human respiratory syncytial virus isolates introduced by a duplication of 60 nucleotides. J Gen Virol 84:3115–3120
Trento A, Viegas M, Galiano M, Videla C, Carballal G, Mistchenko AS, Melero JA (2006) Natural history of human respiratory syncytial virus inferred from phylogenetic analysis of the attachment (G) glycoprotein with a 60-nucleotide duplication. J Virol 80:975–984
Trento A, Casas I, Calderon A, Garcia-Garcia ML, Calvo C, Perez-Brena P, Melero JA (2010) Ten years of global evolution of the human respiratory syncytial virus BA genotype with a 60-nucleotide duplication in the G protein gene. J Virol 84:7500–7512
Tumpey TM, Basler CF, Aguilar PV, Zeng H, Solorzano A, Swayne DE, Cox NJ, Katz JM, Taubenberger JK, Palese P, Garcia-Sastre A (2005) Characterization of the reconstructed 1918 Spanish influenza pandemic virus. Science 310:77–80
Venter M, Madhi SA, Tiemessen CT, Schoub BD (2001) Genetic diversity and molecular epidemiology of respiratory syncytial virus over four consecutive seasons in South Africa: identification of new subgroup A and B genotypes. J Gen Virol 82:2117–2124
Venter M, Collinson M, Schoub BD (2002) Molecular epidemiological analysis of community circulating respiratory syncytial virus in rural South Africa: comparison of viruses and genotypes responsible for different disease manifestations. J Med Virol 68:452–461
Viegas M, Mistchenko AS (2005) Molecular epidemiology of human respiratory syncytial virus subgroup A over a six-year period (1999–2004) in Argentina. J Med Virol 77:302–310
Villenave R, Thavagnanam S, Sarlang S, Parker J, Douglas I, Skibinski G, Heaney LG, McKaigue JP, Coyle PV, Shields MD, Power UF (2012) In vitro modeling of respiratory syncytial virus infection of pediatric bronchial epithelium, the primary target of infection in vivo. Proc Natl Acad Sci U S A 109:5040–5045
Walsh EE, Schlesinger JJ, Brandriss MW (1984) Protection from respiratory syncytial virus infection in cotton rats by passive transfer of monoclonal antibodies. Infect Immun 43:756–758
Walsh EE, Falsey AR, Sullender WM (1998) Monoclonal antibody neutralization escape mutants of respiratory syncytial virus with unique alterations in the attachment (G) protein. J Gen Virol 79(Pt 3):479–487
Wertz GW, Stott EJ, Young KK, Anderson K, Ball LA (1987) Expression of the fusion protein of human respiratory syncytial virus from recombinant vaccinia virus vectors and protection of vaccinated mice. J Virol 61:293–301
Woelk CH, Holmes EC (2001) Variable immune-driven natural selection in the attachment (G) glycoprotein of respiratory syncytial virus (RSV). J Mol Evol 52:182–192
Yamaguchi M, Sano Y, Dapat IC, Saito R, Suzuki Y, Kumaki A, Shobugawa Y, Dapat C, Uchiyama M, Suzuki H (2011) High frequency of repeated infections due to emerging genotypes of human respiratory syncytial viruses among children during eight successive epidemic seasons in Japan. J Clin Microbiol 49:1034–1040

Zhang Y, Xu W, Shen K, Xie Z, Sun L, Lu Q, Liu C, Liang G, Beeler JA, Anderson LJ (2007) Genetic variability of group A and B human respiratory syncytial viruses isolated from 3 provinces in China. Arch Virol 152:1425–1434

Zhang RF, Jin Y, Xie ZP, Liu N, Yan KL, Gao HC, Song JR, Yuan XH, Xiao NG, Guo MW, Zhou QH, Hou YD, Duan Z (2010a) Human respiratory syncytial virus in children with acute respiratory tract infections in China. J Clin Microbiol 48:4193–4199

Zhang W, Choi Y, Haynes LM, Harcourt JL, Anderson LJ, Jones LP, Tripp RA (2010b) Vaccination to induce antibodies blocking the CX3C-CX3CR1 interaction of respiratory syncytial virus G protein reduces pulmonary inflammation and virus replication in mice. J Virol 84:1148–1157

Zhang ZY, Du LN, Chen X, Zhao Y, Liu EM, Yang XQ, Zhao XD (2010c) Genetic variability of respiratory syncytial viruses (RSV) prevalent in Southwestern China from 2006 to 2009: emergence of subgroup B and A RSV as dominant strains. J Clin Microbiol 48:1201–1207

Zhao X, Chen FP, Megaw AG, Sullender WM (2004a) Variable resistance to palivizumab in cotton rats by respiratory syncytial virus mutants. J Infect Dis 190:1941–1946

Zhao X, Chen FP, Sullender WM (2004b) Respiratory syncytial virus escape mutant derived in vitro resists palivizumab prophylaxis in cotton rats. Virology 318:608–612

Zlateva KT, Vijgen L, Dekeersmaeker N, Naranjo C, Van Ranst M (2007) Subgroup prevalence and genotype circulation patterns of human respiratory syncytial virus in Belgium during ten successive epidemic seasons. J Clin Microbiol 45:3022–3030

Structure and Function of Respiratory Syncytial Virus Surface Glycoproteins

Jason S. McLellan, William C. Ray and Mark E. Peeples

Abstract The two major glycoproteins on the surface of the respiratory syncytial virus (RSV) virion, the attachment glycoprotein (G) and the fusion glycoprotein (F), control the initial phases of infection. G targets the ciliated cells of the airways, and F causes the virion membrane to fuse with the target cell membrane. The F protein is the major target for antiviral drug development, and both G and F glycoproteins are the antigens targeted by neutralizing antibodies induced by infection. In this chapter, we review the structure and function of the RSV surface glycoproteins, including recent X-ray crystallographic data of the F glycoprotein in its pre- and postfusion conformations, and discuss how this information informs antigen selection and vaccine development.

Contents

J. S. McLellan
Vaccine Research Center, NIAID, National Institutes of Health, Bethesda, MD, USA

W. C. Ray
Battelle Center for Mathematical Medicine, The Research Institute at Nationwide Children's Hospital, Columbus, OH, USA

M. E. Peeples (✉)
Center for Vaccines & Immunity, The Research Institute at Nationwide Children's Hospital, Columbus, OH, USA
e-mail: Mark.Peeples@nationwidechildrens.org

L. J. Anderson and B. S. Graham (eds.), *Challenges and Opportunities for Respiratory Syncytial Virus Vaccines*, Current Topics in Microbiology and Immunology 372,
DOI: 10.1007/978-3-642-38919-1_4,

1 F Glycoprotein

The F gene encodes a type I integral membrane protein that is synthesized as a 574 amino acid inactive precursor, F_0, decorated with 5–6 *N*-linked glycans, depending on the strain (Collins et al. 1984). It is also palmitoylated at a cysteine in its cytoplasmic domain (Arumugham et al. 1989). Three F_0 monomers assemble into a trimer and, as the trimer passes through the Golgi, the monomers are activated by a furin-like host protease (Bolt et al. 2000; Collins and Mottet 1991). The protease cleaves twice, after amino acids 109 and 136 (González-Reyes et al. 2001; Zimmer et al. 2001a), generating three polypeptides (Fig. 1). The N-terminal and C-terminal cleavage products are the F_2 and F_1 subunits (named in order of size),

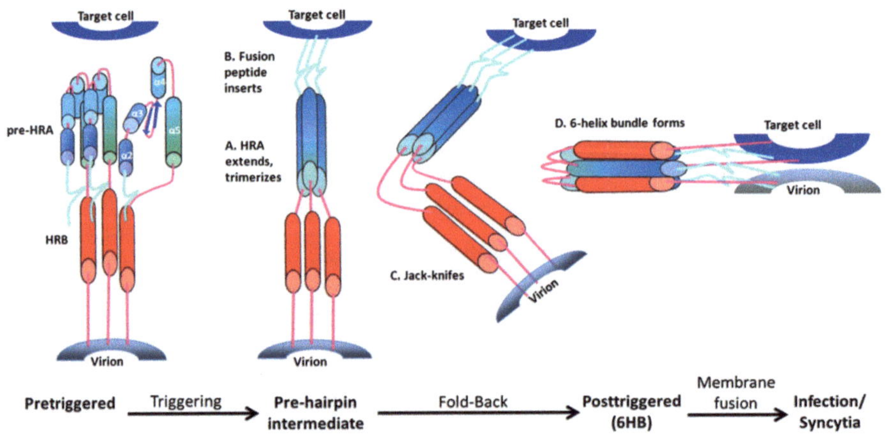

Fig. 1 Refolding the F protein to initiate fusion. In the prefusion form of the F_1 protein the fusion peptide (FP) at the N terminus of F_1 (turquoise) is followed by four short α-helices (*blue*) connected by three nonhelical peptides. Triggering causes these nonhelical connecting peptides to refold into α-helices, completing a single long HRA α-helix that thrusts the FP into the target cell membrane. The long HRA α-helices trimerize, the molecule folds in half, and the HRB α-helices (*red*) insert into the grooves between the HRA units forming a stable 6-helix bundle (6HB). As a result, the virion and cell membranes are brought together and initiate membrane fusion. The central region of the F protein does not rearrange during triggering and refolding and, therefore, is not represented here. It would be positioned at the bend in the molecule

respectively, and are covalently linked to each other by two disulfide bonds (Gruber and Levine 1983; Day et al. 2006). The intervening 27 amino acid peptide, pep27, contains two or three *N*-linked glycans, but dissociates after cleavage (Ruiz-Arguello et al. 2002). The F_2 subunit contains two *N*-linked glycans, whereas the larger F_1 subunit contains a single *N*-linked site. Unlike the others, this F_1 glycan is essential for the protein to cause membrane fusion (Li et al. 2007; Zimmer et al. 2001b).

During respiratory syncytial virus (RSV) replication, the F mRNA is produced in the cytoplasm, and is not exposed to the polyadenylation and splicing machinery of the nucleus. The F mRNA contains cryptic polyadenylation sites and splice sites (Ternette et al. 2007) that must be removed for transient expression of the RSV F protein from a plasmid in cultured cells.

The functional F protein trimer in the virion membrane is in a metastable, prefusion form. It is not yet clear what causes the F protein to trigger, but the result is a major refolding into its postfusion form (Fig. 1). At the N-terminus of each F_1 subunit is the fusion peptide (FP), a stretch of hydrophobic residues that insert into the target membrane (Collins et al. 1984). The FP is mirrored by the transmembrane (TM) domain near the C-terminus of F_1, and each is connected to a heptad repeat (HR) in this order: FP-HRA-HRB-TM. Upon triggering, the pre-HRA refolds into the long HRA helix and trimerizes. The F protein folds in the center as the target and viral membranes approach each other, enabling HRB to bind to the grooves in the HRA trimer, forming a hairpin 6-helix bundle (6HB) (Zhao et al. 2000).

The F glycoprotein is highly conserved among RSV isolates from both A and B subgroups, with amino acid sequence identities of 90 % or higher. Much of the variability in F ($\sim$ 25 %) is found within an antigenic site at the apex of the prefusion trimer (antigenic site Ø) composed of an α-helix from F_1 (aa 196–210) and a strand from F_2 (aa 62–69) and may be a site that determines subtype-specific immunity (McLellan et al. 2013). This high sequence conservation combined with its surface location on the virion, its obligatory role in viral entry, and antigenic sites associated with potent neutralization, make F an ideal target for neutralizing antibodies (Anderson et al. 1988; Walsh and Hruska 1983). Thus, F protein is being examined as a vaccine antigen (Costello et al. 2012), and is the target of antibodies used in, and being developed for, passive prophylaxis (The IMpact-RSV Study Group 1998; Wu et al. 2007b). In addition to these factors, the dramatic conformational changes that the F protein undergoes make it a major target for small molecule antiviral drug development (Costello et al. 2012).

1.1 Postfusion F Protein

Jose Melero's group was the first to produce and isolate a soluble form of the RSV F (sF) protein that lacked its transmembrane and cytoplasmic domains. They expressed this sF protein from a vaccinia virus vector in HEp-2 cells (Calder et al. 2000). Many of the sF protein molecules were cleaved at both furin sites and formed organized aggregates or 'rosettes' detected both by EM and velocity sucrose gradient analysis. The rosettes were a result of aggregation of the exposed, highly hydrophobic FPs, and since the FP is only exposed upon triggering, these molecules were in the postfusion form (Gonzalez-Reyes et al. 2001). Some of the sF proteins that were inefficiently cleaved remained as separate trimers, as did mutants whose furin cleavage sites were mutated (Ruiz-Arguello et al. 2002).

Melero's group also found that deletion of the first 10 amino acids of the RSV FP prevented rosette formation without inhibiting cleavage, confirming that the FP is responsible for rosette formation (Ruiz-Arguello et al. 2004). Two independent groups recently determined the crystal structure of a similar mutant (McLellan et al. 2011; Swanson et al. 2011). The structures revealed a cone-shaped molecule, with a globular head and an extended stalk (Fig. 2). The three F_2/F_1 subunits that make up the trimeric molecule are tightly intertwined, with threefold symmetry that runs the length of the molecule. The globular head contains both the F_2 and F_1 subunits, as well as the cysteine-rich region, and has both α-helices and β-sheets. The stalk region is almost entirely helical, composed of the 6HB that is characteristic of the postfusion state of many type I viral fusion proteins. Since the 6HB is composed of three HRA coiled-coils in the center, with three anti-parallel HRB

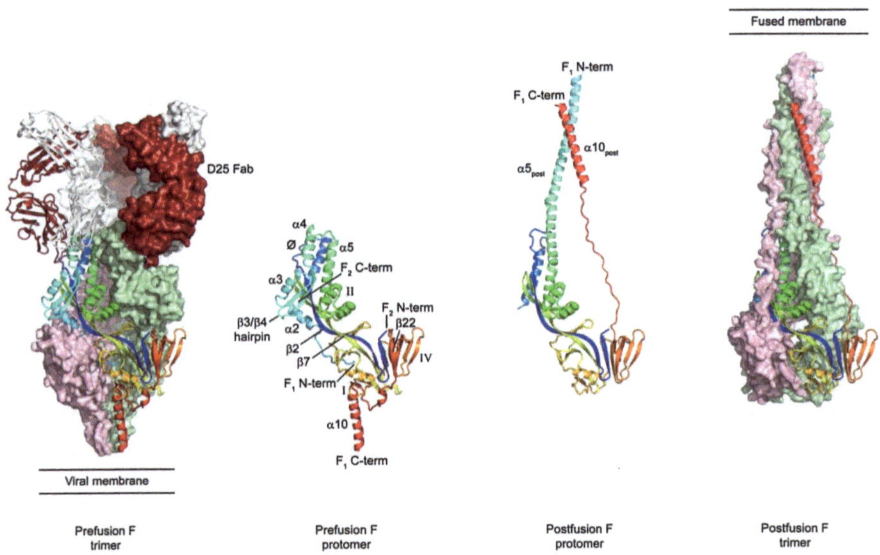

Fig. 2 Crystal structures of RSV F in pre- and postfusion conformations. Binding of antibody D25 locks the F glycoprotein in the prefusion conformation. Two of the prefusion F protomers are shown in surface representation and colored *pink* and *green*, while the third protomer is shown as ribbons colored *blue* to *red*, from the N-terminus of F_2 to the C-terminus of F_1, respectively. Three D25 Fabs are shown, with the heavy chain colored *dark red* and the light chain colored *white*. The Fab shown as ribbons is bound primarily to the F protomer shown as ribbons, while the other two Fabs are shown in surface representation. The two middle images show a pre- and postfusion protomer in ribbons, with labeled secondary structure elements. Antigenic sites Ø, I, II and IV are labeled

helices on the outside, the N- and C-termini of F_1 reside near each other in the postfusion state. It is the formation of this extremely stable 6HB that brings the viral and cellular membranes together to initiate fusion and ensures that the process is nonreversible.

Prior to the determination of the postfusion F structures, crystal structures of two RSV-neutralizing antibodies that target distinct antigenic sites on the F protein were determined in complex with their peptide epitopes (McLellan et al. 2010a, b). The first antibody structure was that of motavizumab (Wu et al. 2007a), a more potent derivative of palivizumab (Johnson et al. 1997), in complex with a peptide corresponding to F residues 254–277, known as antigenic site II. This structure revealed that antigenic site II exists as a helix-loop-helix, with motavizumab binding to one face of the epitope (McLellan et al. 2010b). Modeling of motavizumab bound to a prefusion F conformation based on the parainfluenza virus 5 (PIV5) crystal structure suggested that the epitope may be buried in the prefusion state, and that motavizumab would recognize an intermediate conformation. The second antibody structure was that of 101F in complex with a peptide corresponding to F_1 residues 422–438, known as antigenic site IV (McLellan et al. 2010a). The peptide epitope existed as an unstructured coil, lying in a groove that ran the length of the heavy and light chain interface. Using structures of related paramyxovirus F proteins in the pre- and postfusion states, it was suggested that 101F could bind both conformations.

When the postfusion F structure was determined, one of the unexpected findings was that both antigenic sites II and IV were in conformations that resembled the antibody-bound peptide structures, suggesting that antibodies targeting these two epitopes could bind, and be elicited by, the postfusion F conformation. Binding studies confirmed that palivizumab, motavizumab, 101F, and a site I-directed antibody bound with nanomolar affinity to the postfusion ectodomain F protein (McLellan et al. 2011). This result was surprising given that neutralizing antibodies are generally thought to bind the prefusion conformation and prevent transition to the postfusion state. Since 101F and palivizumab are known to block fusion (McLellan et al. 2010a; Magro et al. 2010; Huang et al. 2010), and because binding to the postfusion state should not prevent fusion, these antibodies very likely also bind to the prefusion form and intermediate states, and perhaps block fusion of the membranes due to steric effects. Additional structural and biochemical studies are needed to define the precise mechanism by which antibodies targeting antigenic sites II and IV disrupt the fusion process.

1.2 Prefusion F Protein

The Lamb/Jardetzky groups first stabilized a paramyxovirus F protein in the prefusion conformation by fusing a known self-trimerizing GCNt domain to the C-terminus of the PIV5 sF protein (Yin et al. 2006). They also mutated the cleavage site on the PIV5 sF protein to prevent cleavage by furin during its passage through the Golgi. This prefusion sF protein had a 'lollipop' shape by EM (Connolly et al. 2006), similar to the shape of the GCNt-stabilized sF protein determined from the crystal structure (Yin et al. 2006).

Prefusion RSV sF protein can be expressed in mammalian cells without the GCNt addition and without mutation of the furin cleavage sites. It does not form rosettes as determined by velocity sedimentation or EM (Chaiwatpongsakorn et al. 2011). However, this sF protein can be triggered by reducing the buffer molarity, resulting in the classical postfusion rosettes. Dialysis in the presence of liposomes resulted in sF associating with liposomes, confirming that triggering by reduced molarity results in exposure of the highly hydrophobic FP. Mutation of the FP-proximal furin cleavage site prevented liposome association under low molarity conditions, confirming the requirement for N-terminal FP for liposome insertion (Chaiwatpongsakorn et al. 2011). At physiological molarity this sF protein remains in its prefusion form (S. Chaiwatpongskorn and M. Peeples unpublished data), suggesting that a reduction in molarity could be involved in the physiological triggering of the F protein.

In addition to high osmolality, recently discovered neutralizing antibodies can stabilize the RSV F protein in the prefusion state. McLellan et al. (2013) isolated a mouse antibody, 5C4, that potently neutralized RSV but showed no binding to F proteins in the postfusion state. It was determined that 5C4 shared these properties with two human antibodies, D25 and AM22, which were shown to neutralize RSV with 100-fold greater potency than the prophylactic antibody palivizumab (Synagis®) (Kwakkenbos et al. 2010). Co-expression of these antibodies with an RSV sF construct containing residues 1-513 with a C-terminal fibritin T4 trimerization motif (Frank et al. 2001) allowed for purification of antibody-F complexes. The complex of D25 Fab bound to RSV F was crystallized, and its structure was solved to 3.6 Å resolution (McLellan et al. 2013).

The D25-bound RSV F structure (Fig. 2) resembled the lollipop shapes seen previously by EM, and although its structural elements were similar to the PIV5 sF structure, its shape is more oval and the spatial relationship of individual residues is quite different than could be predicted by modeling. Each F monomer is divided into two lobes separated by a seven-strand anti-parallel barrel. Two of those strands hydrogen bond for over 70 Å and make up portions of the two lobes and central barrel. Each lobe contains the F_2 and F_1 subunits, which are tightly intertwined, as observed in the postfusion structure. F_2 starts in the membrane-proximal lobe and then extends through the central barrel and into the membrane-distal lobe. The N-terminal portion of F_2 is mostly β-strand and the C-terminal portion is α-helical. A short unstructured region connects these two secondary structure elements, and this is the only portion of F_2 that moves more than 5 Å when compared to the postfusion conformation. The N-terminus of F_1, which contains the FP, is buried in the central trimer cavity, and is connected to two perpendicular helices ($\alpha2$ and $\alpha3$) followed by a β-hairpin ($\beta3$ and $\beta4$) and another helix ($\alpha4$). The FP and these five secondary structure elements undergo a dramatic conformational change upon triggering and refold into a single α-helix in the postfusion state. The next 240 amino acids of F_1, which includes antigenic sites I,

II and IV, show little conformational change between the pre- and postfusion states. The remaining F_1 residues, like the N-terminal F_1 residues, also undergo a dramatic conformational change that swings HRB (α10) around the molecule, bringing it near HRA to complete the 6HB observed in the postfusion state.

The epitope for D25, which resides at the apex of the prefusion F trimer, consists of the unstructured region in F_2 (residues 62–69) and helix α4 in F_1 (residues 196–210). Both of these regions move more than 5 Å between the pre- and postfusion state, with the helix changing orientations by ~180°. Minor contacts are also made between D25 and a neighboring protomer. Thus, the specificity of D25 for the prefusion state is because its epitope does not exist in the postfusion conformation. The D25 epitope, which is also targeted by AM22 and 5C4, is referred to as antigenic site Ø. Perhaps due to the potency of antibodies against this site, or its prominent location at the apex of the trimer, this region is the most variable portion of the prefusion sF protein, suggesting that it may be under immune pressure. Indeed, some antibodies against this epitope are subtype specific, whereas others can broadly neutralize RSV strains from both subtypes A and B (J. McLellan unpublished observations). Understanding the structural basis for this specificity will be important for designing vaccines.

1.3 F Protein Intermediates

It is not clear what triggering events initiate restructuring of the metastable prefusion F protein into the pre-hairpin intermediate. In this fully extended transient state, the viral heptad repeats are 180° apart, the viral and cellular membranes are parallel, and the F protein is inserted into both membranes. The pre-hairpin intermediate then 'jackknifes', folding at its center and bringing the HRBs closer to the HRA coiled-coil trimer, as the viral and cellular membranes come closer together (Fig. 1). As the three HRB helices lock into the grooves on the surface of the trimeric HRA coiled-coil to form the 6HB, their attached hydrophobic domains are pulled together thereby merging the membranes in which they are embedded to initiate membrane fusion.

It was discovered in the 1990s that peptides that target fusion intermediates and prevent formation of the 6HB are capable of inhibiting viral fusion. Wild et al. first demonstrated that synthetic peptides derived from the HRA or HRB regions of the HIV-1 gp41 fusion protein potently inhibited HIV-1 infection (Wild et al. 1992, 1993). The HRB peptides were found to be the most effective leading to the development of a peptide drug (T-20, Fuzeon® or enfuvirtide), the first licensed antiretroviral fusion inhibitor. This synthetic HRB peptide competes with the viral HRBs for binding to the HRA coiled-coil trimer thereby preventing 6HB formation. A similar approach was developed to inhibit fusion caused by paramyxovirus F proteins (Lambert et al. 1996). A synthetic HRB peptide from RSV F (residues 488–522) inhibited RSV infection with an EC_{50} of 50 nM. Although the HRB-derived peptides are potent and specific, their high cost and requirement for frequent subcutaneous injections are prohibitive.

Little structural information is available on the intermediates of any type I viral fusion glycoprotein, due in part to their instability and the spectrum of conformations that exist during the transition. Structures of a pre-hairpin intermediate state, stabilized by a peptide, antibody, drug or cross-link could greatly enhance our knowledge of the fusion process, and possibly identify new drug or vaccine targets. Neutralizing antibodies that target the intermediates are rare, and none have been found for the RSV F protein. But the peptide and drug inhibitors of the RSV F protein may provide an advantage.

1.4 F Protein Receptors, Triggering and Antiviral Drugs

Virions that contain the RSV F protein as their only glycoprotein are infectious, indicating that the RSV F protein can trigger without help from the viral attachment glycoprotein, unlike most other paramyxoviruses. It is not known what causes the RSV F protein to trigger. Bovine and human RSV infect cells of their respective host species preferentially and this species specificity has been traced to the F protein, particularly the F_2/pep27 region (Schlender et al. 2003). This result suggests that the F protein interacts with a receptor and that this receptor is species-specific. Several cell surface proteins that interact with the F protein and might function as receptors have been identified: ICAM-1 (Behera et al. 2001), TLR4 (Haynes et al. 2001), and nucleolin (Tayyari et al. 2011). One or more of these host proteins may be involved in attaching virions to target cells, triggering the F protein, or both. However, these molecules have been studied in immortalized cells or by ectopic expression in cells that are not susceptible to RSV infection. If they are available and functional on the airway epithelium in vivo is not known. Other factors such as exposure to low molarity may contribute to triggering (Chaiwatpongsakorn et al. 2011). An attachment function for the F protein would be especially important for virions lacking the G protein, but in complete virions, the G protein is required for efficient infection of primary well differentiated human airway epithelial (HAE) cultures (Kwilas et al. 2009). The key cellular receptors involved in attachment and required for F triggering have not been identified and cellular receptors that explain tropism have still not been determined.

Most small molecules that inhibit RSV infection in cell culture target the F protein, probably due to its metastable nature and the major rearrangements that it must make to initiate membrane fusion. These small molecules could cause premature F protein triggering, before the virion is close enough to a target cell to allow membrane fusion, or they could prevent triggering once the F protein is in contact with a target cell. The prefusion F protein, therefore, would seem to be the most likely target for antiviral drugs against the F protein. But another possibility is that an antiviral compound prevents one of the motions required during the refolding process. The antiviral peptides that represent a portion of the HRB sequence and compete for the F protein's own HRB binding to its HRA trimer during the 6HB formation would prevent this final, essential refolding step thereby preventing membrane fusion.

We have recently reviewed the small molecule drugs developed against RSV and we would refer readers to this review for a list and a more thorough discussion (Costello et al. 2012). The largest and best studied group of small molecule antiviral compounds against the RSV F protein bind to Y198 in the HRA domain (Cianci et al. 2004b; Douglas et al. 2003; Roymans et al. 2010). They share drug-resistant mutants, but none of these compounds select mutations in Y198 suggesting that Y198 plays an essential role in F protein function. BMS-433771 inhibited both RSV subgroups A and B with an average EC_{50} of 20 nM (Cianci et al. 2004b). Modeling based on the crystal structure of the RSV F 6HB (Zhao et al. 2000) suggested that BMS-433771 bound in a hydrophobic pocket in the HRA coiled-coil and prevented HRB from binding properly in that region (Cianci et al. 2004a). Crystal structure analysis revealed that TMC353121, a benzimidazole-based compound with an EC_{50} of 0.1 nM (Bonfanti et al. 2008), bound similarly (Roymans et al. 2010). This structure suggested that rather than completely preventing 6HB formation, these small molecule fusion inhibitors distort the membrane-distal structure of the postfusion 6HB.

2 G Glycoprotein

The RSV G protein was first described by Seymour Levine as a heavily glycosylated 80 kDa protein in purified virions produced in HeLa cells (Levine 1977). He later showed that rabbit antibodies to G protein, but not to F protein, prevented virions from binding to HeLa cells, indicating that the G protein is the major virus attachment protein (Levine et al. 1987). The G protein backbone contains 289–299 amino acids (32–33 kDa), depending on the strain, and is palmitoylated (Collins and Mottet 1992). It has no sequence homology with other paramyxovirus attachment proteins, and no hemagglutinating or neuraminidase functions. With 30–40 *O*-linked glycans and 4–5 *N*-linked glycans, the G protein is similar to mucins produced in the airways although much smaller in molecular mass (Satake et al. 1985; Wertz et al. 1985). Approximately 60 % of the G protein molecular mass is carbohydrate.

The size of the G protein varies depending on the cell type in which it is produced: 80–100 kDa in immortalized cell lines (Garcia-Beato et al. 1996) but 180 kDa in primary HAE cultures (Kwilas et al. 2009). This larger form is not a disulfide-linked dimer because it does not dissociate in reducing conditions. It is likely either a dimer held together by a different bond or a more heavily glycosylated monomer.

2.1 G Protein Domains

The central region of the G protein contains a 13-amino acid highly conserved domain (Fig. 3a), partially overlapping the cysteine noose domain with four cysteines linked 1–4 and 2–3 (Gorman et al. 1997), followed by a highly basic heparin-binding domain (HBD). The HBD is the likely attachment site for heparan sulfate (HS) found on the surface of most cells. A peptide from the G protein HBD (amino acids 184–198) binds efficiently to HEp-2 cells and inhibits RSV infection (Feldman et al. 1999).

Two large mucin-like domains flank the central region (Fig. 3b) and are highly variable in sequence, making the G protein the most variable RSV protein, a useful characteristic for RSV evolution studies. The overall Ser and Thr content of these two regions is relatively stable, suggesting that they may provide substrates for *O*-linked glycan decoration rather than any particular sequence or specific function. The appearance of a 20 amino acid repeat in the second mucin-like domain of the G protein of a B strain virus (Trento et al. 2003) and a 24 amino acid insertion in the same region of an A strain virus (Eshaghi et al. 2012) underscore the flexibility of this region. This B strain spread throughout the world in the decade since it appeared, suggesting that the repeat provides some advantage to the virus.

2.2 G Protein Receptor Candidates

The receptor for the RSV G protein on immortalized cells appears to be HS (Feldman et al. 1999, 2000; Hallak et al. 2000a, b; Krusat and Streckert 1997; Escribano-Romero et al. 2004), similar to a number of other viruses (Hallak et al. 2007). But HS is not detectable on the apical surface of HAE cultures (Zhang et al. 2005) suggesting that RSV most likely uses a different receptor to enter these cells. Furthermore, RSV infects nearly exclusively the ciliated cells in these HAE cultures (Villenave et al. 2012; Zhang et al. 2002), suggesting a more specific receptor.

Tripp et al. noticed that the third and fourth cysteines in the cysteine noose of the G protein are separated by three amino acids, similar to the CX3C motif in the chemokine CX3CL1, also called fractalkine (Tripp et al. 2001). G competes with CX3CL1 for binding to its receptor, CX3CR1, and like CX3CL1, G attracts neutrophils in modified Boyden chamber experiments (Tripp et al. 2001).

MAb 131-2G against the G protein prevents it from binding to CX3CR1 (Tripp et al. 2001). When this mAb is mixed with RSV, it does not neutralize infection of immortalized cells (Anderson et al. 1988). However, since the G protein likely uses a different receptor in vivo this mAb might block attachment to that receptor. In fact, we recently found that mAb 131-2G does neutralize RSV, reducing infectivity nearly 100-fold in HAE cultures (Johnson, S.M. and Peeples, M.E., manuscript in preparation). Since this mAb has been shown to prevent the G

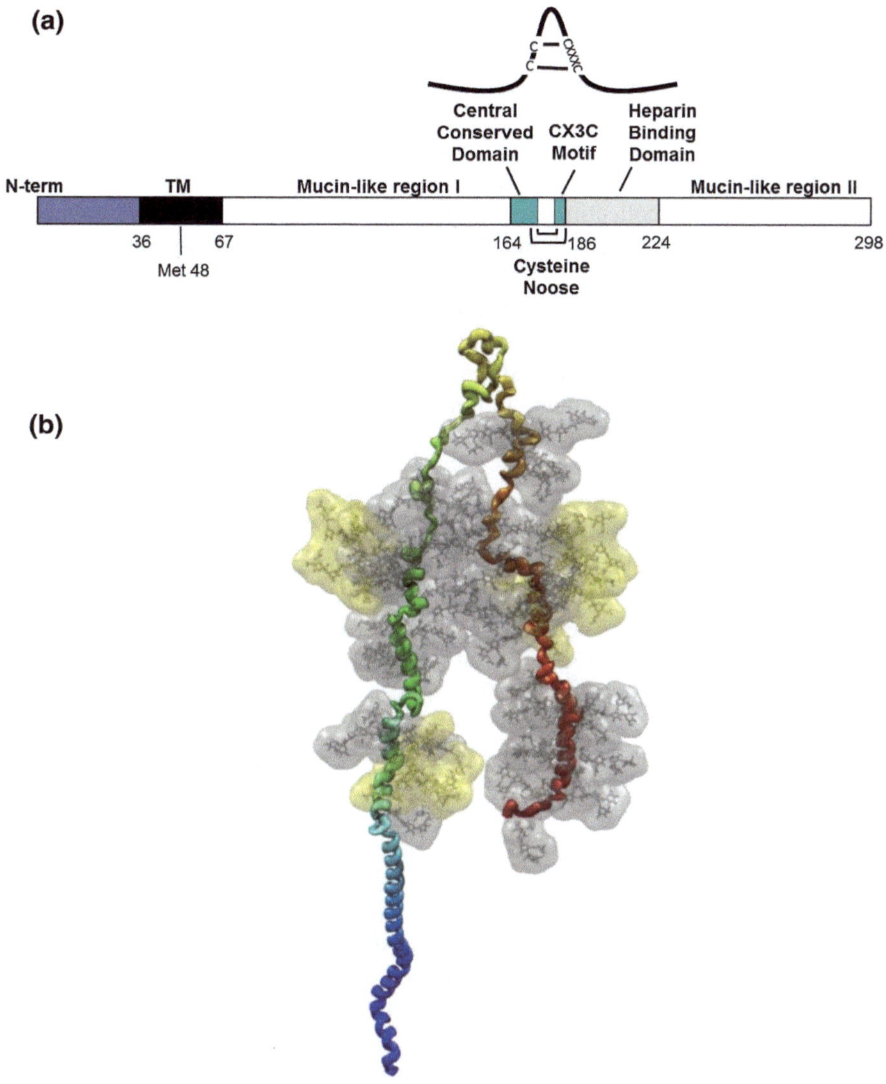

Fig. 3 Schematic of the RSV G protein. The RSV A2 strain G protein is 298 amino acids long and consists of two heavily glycosylated mucin-like regions, separated by a central conserved, unglycosylated cysteine noose (*yellow-green* loop at the *top*) that is stabilized by a pair of disulfide bonds. The unglycosylated N terminus (*blue*) includes the cytoplasmic and transmembrane domains. To approximate the structure, a linear α-helical prototype of the G protein was subjected to steered molecular dynamics (NAMD), pulling the central noose perpendicular to the backbone until mucin-like region 1 (*green*) and 2 (*orange*) arrived in a near-parallel arrangement. This simulation resulted in the loss of α-helical structure over much of each mucin domain, primarily due to the abundant prolines, without affecting the helical structure in the TM and N-terminal domains. Complex glycans were positioned at each of the four *N*-linked sites (*yellow* side chains), and simple glycans (*gray*) were positioned at each O-linked site predicted by NetOGlyc3.1. The glycans in this representation have slightly higher-than-biological mass to reflect the probable space they would occupy

protein from binding to CX3CR1 (Tripp et al. 2001), CX3CR1 might be a receptor for RSV on HAE cells. MAb 131-2G also reduces RSV production in mice (Haynes et al. 2009; Radu et al. 2010), suggesting that it does neutralize the virus in vivo.

Two other molecules have been identified as potential RSV G protein receptors: Surfactant Protein A (SP-A) and Annexin II. SP-A is a secreted, innate immune pattern recognizing collectin that has been shown to bind to RSV, enhancing infection of HEp-2 cells. SP-A binds to G in a Ca^{2+} and carbohydrate dependent manner (Barr et al. 2000; Hickling et al. 2000). SP-A has a receptor (p63) that enables it to bind to cells, although p63 has only been reported on type II pneumocytes (Gupta et al. 2006). An antibody that binds Annexin II inhibited HEp-2 cell infection, and Annexin II binds to G in a Ca^{2+} dependent manner (Malhotra et al. 2003). The roles of these G-binding proteins, as those of the F-binding proteins described above, need to be examined in HAE cultures or in vivo infection.

RSV G also interacts with the lectins DC-SIGN and L-SIGN on dendritic cells, but neither functions as a receptor for virus infection (Johnson et al. 2012). Instead, this interaction stimulates ERK1/2 phosphorylation which inhibits dendritic cell activation and could represent a partial explanation for the limited immunity against RSV reinfection. The addition of RSV virions to A549 cells also induces ERK1/2 phosphorylation (Kong et al. 2004), but these cells are nonetheless infected by RSV (Kwilas et al. 2009).

2.3 Soluble G Protein

A soluble form of the G protein (sG) is released from infected HEp-2 cells (Hendricks et al. 1987), and detected in the medium prior to the release of virions (Hendricks et al. 1988). The sG protein is 65–74 amino acids shorter at its N-terminus than the full-length G protein. Its translation begins with the second AUG (codon 48) in the G mRNA (Roberts et al. 1994) which deletes the cytoplasmic domain and a portion of the transmembrane domain. This remaining hydrophobic portion of the G protein is essential for translocating it into the lumen of the ER during translation and would likely maintain its membrane association until a proteolytic event releases sG into the medium. It is not clear why such a proteolytic event occurs only on the shortened G protein since the full-length G protein contains the same sequence. The sG protein produced in HEp-2 cells is a monomer, whereas the anchored G protein is an oligomer, perhaps a tetramer (Escribano-Romero et al. 2004).

To examine the role of the sG protein in vivo, the wild-type virus was compared to a recombinant virus lacking the second methionine in mice (Bukreyev et al. 2008). The sG protein occupied neutralizing antibody and, in addition, inhibited antibody-mediated antiviral effects of pulmonary macrophages and complement (Bukreyev et al. 2012).

3 SH Protein

The SH gene encodes a protein that is 64 (subgroup A) or 65 (subgroup B) amino acids in length. The protein contains a single transmembrane region, with an extracellular C-terminus and an intracellular N-terminus (Collins and Mottet 1993). The SH protein primarily localizes to the ER and Golgi complex in infected cells, though protein clusters are also observed on the plasma membrane. The SH protein is incorporated into mature RSV filaments at very low levels (Rixon et al. 2004). Several different species of the SH protein have been detected, including an N-terminally truncated form, and two *N*-linked glycan variants, one of which is further modified with polylactosamine (Olmsted and Collins 1989). The major species, however, is the full-length, nonglycosylated form.

The oligomeric state of the SH protein has been extensively investigated using a number of techniques. Early cross-linking experiments demonstrated oligomers as large as pentamers (Collins and Mottet 1993), and electron microscopy images of recombinant SH resuspended in liposomes revealed channel-like structures with five- or sixfold symmetry (Carter et al. 2010). Analytical ultracentrifugation data of SH protein solubilized in C14 betaine or dodecylphosphocholine (DPC) were best fit by a monomer-pentamer equilibrium (Gan et al. 2012).

NMR spectra of the SH protein solubilized in DPC was used to determine the structure of an SH monomer. The SH protein has an N-terminal α-helix co-planar with the membrane, connected by a linker to the transmembrane-spanning α-helix, which is connected by a linker to a C-terminal β-turn (Gan et al. 2012). The monomer structure was used to reconstruct a model of the pentameric SH protein using a number of additional experimental restraints. The convergent model revealed a funnel-like channel approximately 45 Å long, with a pore diameter of 3.5 Å at its narrowest. This structure suggests that the SH protein belongs to a class of channel-forming proteins called viroporins (Nieva et al. 2012). Indeed, the SH protein has been shown to induce membrane permeability in liposomes (Carter et al. 2010) and act as an acid-activated, nonselective cation channel in mammalian cells (Gan et al. 2012).

Unlike the F and G glycoproteins, the role of the SH protein in RSV replication and pathogenesis is not well understood. Serial cold-passaging of RSV in cell culture resulted in a virus, cp-52, lacking both the G and SH proteins that was infectious and replicated in vitro (Karron et al. 1997). In comparison to wild-type virus, recombinant RSV lacking the SH gene produced plaques that were 70 % larger in HEp-2 cells (Bukreyev et al. 1997). In some cell lines, the ΔSH virus replicated $>$ 12-fold better than wild-type virus. In mice, the ΔSH virus replicated in the lower respiratory tract as well as wild-type virus, but was tenfold lower in the upper respiratory tract (Bukreyev et al. 1997). In chimpanzees, however, the ΔSH virus replication was decreased 40-fold in the lower respiratory tract but was similar to wild-type in the upper respiratory tract (Whitehead et al. 1999). Collectively, these data demonstrate that the SH protein is not essential for RSV replication in cell culture, but is involved to some degree in RSV survival in vivo.

The RSV SH protein, like the SH proteins of PIV5 and mumps, inhibit TNF-α-induced apoptosis in the context of PIV5 missing its own SH protein (Fuentes et al. 2007). Inhibiting TNF-α production might enhance viral replication in vivo.

4 Vaccine Implications

As discussed above, the F and G glycoproteins are the target of neutralizing antibodies, and one or both glycoproteins are included in most vaccine modalities. In this section, we describe several vaccines that are being developed and summarize their attributes based on our knowledge of RSV glycoprotein structure and function (see chapters by R.A. Karron et al., and by T.G. Morrison and E.E. Walsh, this volume).

4.1 Attenuated Virus Vaccine

RSV lacking its G gene is viable, but replicates to lower titers than the complete virus in immortalized cells (Karron et al. 1997; Techaarpornkul et al. 2001; Teng et al. 2001), and is overly attenuated in human vaccination experiments (Karron et al. 1997). The attenuated RSV vaccines that are in development are produced in a World Health Organization approved Vero (African green monkey's kidney) cell line. Much of the G protein produced in Vero cells and inserted into virions is cleaved (Kwilas et al. 2009). As a result, these virions infect HAE cultures at least 10-fold less efficiently than the same virus grown in HEp-2 cells and similar to RSV lacking the G gene (Kwilas et al. 2009).

4.2 Experimental F Protein Vaccines

The F protein is highly conserved across the spectrum of RSV strains, making it likely that an F protein vaccine would protect against all strains of RSV. An experimental F protein vaccine produced by Wyeth-Lederle has been evaluated in adults (Munoz et al. 2003). This vaccine contained full-length F protein from disrupted RSV-infected cells that was purified by mAb-affinity chromatography. The vaccine was safe and induced antibodies to the F protein, but the antibodies were not very effective at neutralizing the virus. It is likely that the F protein in this vaccine was in the postfusion form.

Immunization of mice with the postfusion sF protein does induce antibodies to the F protein at a titer sufficient to neutralize RSV and protect cotton rats from RSV challenge (Swanson et al. 2011). Three neutralizing antigenic sites on F (I, II, and IV) are present in the postfusion F protein structures (McLellan et al. 2011;

Swanson et al. 2011), and mAbs to sites I, II, and IV do, in fact, bind the postfusion form (McLellan et al. 2011). However, postfusion F lacks antigenic site Ø that is uniquely found in the prefusion F protein, and thus would not elicit the remarkably potent antibodies that target this site (McLellan et al. 2013).

The location of antigenic site Ø at the apex of the prefusion conformation of F suggests that this epitope will be readily accessible to antibodies, and may be an immunodominant epitope. Indeed, Magro et al. determined that most neutralizing activity in *Respigam* (MedImmune), a high-titered antibody product from pooled human plasma, was specific for the prefusion conformation of RSV F (Magro et al. 2012). Furthermore, they found that neutralizing antibodies raised in rabbits against the complete F protein were not removed by exposure to the postfusion sF protein but did react with a disulfide-stabilized sF protein, indicating that these neutralizing antibodies targeted the prefusion F protein. Therefore, antibodies that uniquely recognize the prefusion sF protein are much more effective at neutralizing RSV than antibodies to the postfusion sF protein, suggesting that prefusion F would be the preferred vaccine antigen conformation.

4.3 Experimental G Protein Vaccines

The RSV G protein is the other major neutralizing antibody target on the surface of the RSV virion, and its expression from a vaccinia or Sendai virus vector induced a protective immune response in animals (Stott et al. 1986; Takimoto et al. 2004). Although G is highly variable and decorated with glycans that are in general poorly immunogenic, the central region is not glycosylated and is conserved in sequence, particularly a region on the upstream side of the cysteine noose. A large unglycosylated peptide that included the central region of the G protein (amino acids 130–230), linked to an albumin-binding domain of streptococcal protein G (BBG2Na) was able to induce a protective immune response in mice (Power et al. 1997). Immunization with this *Escherichia coli*-produced peptide did not cause enhanced disease in mice upon RSV challenge (Plotnicky-Gilquin et al. 1999). This vaccine progressed to phase III clinical trials but rare adverse events stopped the trials. These problems have been attributed to an Arthus reaction to the BB component (Libon et al. 2007). Without the BB component, the G2Na peptide induced protective immunity in cotton rats and, in previously immunized mice, it was recently shown to boost antibody titer to RSV (Nguyen et al. 2012), suggesting that this approach might work to boost immunity to RSV in older adults. Another group performed vaccination studies in naïve mice with a similar peptide (amino acids 131–230) and mucosal immunization, without the addition of an adjuvant. The immunized mice were nearly completely protected from an RSV challenge, without indications of enhanced disease (Kim et al. 2012).

The Tripp group confirmed the immunogenic value of the central region of the G protein by immunizing mice with a shorter peptide (amino acid 148–198). This synthetic peptide induced a greater neutralizing antibody response to RSV than did the peptides flanking it (Choi et al. 2012). These antibodies also inhibited G protein binding to CX3CR1 and reduced the effect of the G protein on lymphocyte migration. However, if CX3CR1 is the receptor for RSV on the ciliated cells of the airway epithelium as suggested above, these antibodies could also be neutralizing RSV in vivo.

4.4 Experimental SH Protein Vaccine

Antibodies to the SH protein are not neutralizing, but they can affect viral replication in vivo by ADCC (antibody dependent cellular cytotoxicity) (Schepens et al. 2012).

5 Conclusions

We are entering a new era in our understanding of the RSV glycoproteins, the major targets for vaccination strategies and for antiviral drug development. Solving the structures of the pre- and postfusion sF protein has been a major accomplishment that will allow us to evaluate and improve drugs that target the F protein and to design better vaccine antigens. The prefusion sF in a native, metastable form and in a stabilized form will also provide important reagents for understanding biochemically what triggers the F protein, identifying cellular receptors that determine tropism, and characterizing serological responses to natural infection and vaccines more precisely.

Another major advance has been the use of primary well differentiated HAE cultures for RSV entry studies. RSV enters these cells via a different receptor than it uses to enter immortalized cells. Identification of the cellular receptors on HAE cultures for the G and F proteins using new information from neutralizing mAbs against G and the structure of the prefusion F will provide additional targets for antiviral drug development and guide vaccine antigen design. The G protein that is produced in these HAE cells is dramatically different from the G protein produced in standard immortalized cells, perhaps providing another target for antiviral drug development and vaccine design efforts. All in all, this is an exciting time to be working with the RSV surface glycoproteins.

Acknowledgments The authors thank Supranee Chaiwatpongskorn and Jacqueline Corry for their help with the graphics. Authors are supported by NIH grants AI095684 (MEP and WCR) and AI093848 (MEP), Tibotec/Johnson & Johnson (MEP) and Cystic Fibrosis Foundation Therapeutics, Inc. (MEP).

References

Anderson LJ, Bingham P, Hierholzer JC (1988) Neutralization of respiratory syncytial virus by individual and mixtures of F and G protein monoclonal antibodies. J Virol 62(11):4232–4238

Arumugham RG, Seid RC, Doyle S, Hildreth SW, Paradiso PR (1989) Fatty acid acylation of the fusion glycoprotein of human respiratory syncytial virus. J Biol Chem 264(18):10339–10342

Barr FE, Pedigo H, Johnson TR, Shepherd VL (2000) Surfactant protein-A enhances uptake of respiratory syncytial virus by monocytes and U937 macrophages. Am J Respir Cell Mol Biol 23(5):586–592

Behera AK, Matsuse H, Kumar M, Kong X, Lockey RF, Mohapatra SS (2001) Blocking intercellular adhesion molecule-1 on human epithelial cells decreases respiratory syncytial virus infection. Biochem Biophys Res Commun 280(1):188–195

Bolt G, Pedersen LØ, Birkeslund HH (2000) Cleavage of the respiratory syncytial virus fusion protein is required for its surface expression: role of furin. Virus Res 68(1):25–33

Bonfanti J-F, Meyer C, Doublet F, Fortin J, Muller P, Queguiner L, Gevers T, Janssens P, Szel H, Willebrords R, Timmerman P, Wuyts K, van Remoortere P, Janssens F, Wigerinck P, Andries K (2008) Selection of a respiratory syncytial virus fusion inhibitor clinical candidate. 2. Discovery of a morpholinopropylaminobenzimidazole derivative (TMC353121). J Med Chem 51(4):875–896

Bukreyev A, Whitehead SS, Murphy BR, Collins PL (1997) Recombinant respiratory syncytial virus from which the entire SH gene has been deleted grows efficiently in cell culture and exhibits site-specific attenuation in the respiratory tract of the mouse. J Virol 71(12):8973–8982

Bukreyev A, Yang L, Collins PL (2012) The secreted G protein of human respiratory syncytial virus antagonizes antibody-mediated restriction of replication involving macrophages and complement. J Virol 86(19):10880–10884

Bukreyev A, Yang L, Fricke J, Cheng L, Ward JM, Murphy BR, Collins PL (2008) The secreted form of respiratory syncytial virus G glycoprotein helps the virus evade antibody-mediated restriction of replication by acting as an antigen decoy and through effects on Fc receptor-bearing leukocytes. J Virol 82(24):12191–12204

Calder LJ, Gonzalez-Reyes L, Garcia-Barreno B, Wharton SA, Skehel JJ, Wiley DC, Melero JA (2000) Electron microscopy of the human respiratory syncytial virus fusion protein and complexes that it forms with monoclonal antibodies. Virology 271(1):122–131

Carter SD, Dent KC, Atkins E, Foster TL, Verow M, Gorny P, Harris M, Hiscox JA, Ranson NA, Griffin S, Barr JN (2010) Direct visualization of the small hydrophobic protein of human respiratory syncytial virus reveals the structural basis for membrane permeability. FEBS Lett 584(13):2786–2790

Chaiwatpongsakorn S, Epand RF, Collins PL, Epand RM, Peeples ME (2011) Soluble respiratory syncytial virus fusion protein in the fully cleaved, pretriggered state is triggered by exposure to low-molarity buffer. J Virol 85(8):3968–3977

Choi Y, Mason CS, Jones LP, Crabtree J, Jorquera PA, Tripp RA (2012) Antibodies to the central conserved region of respiratory syncytial virus (RSV) G protein block RSV G protein CX3C-CX3CR1 binding and cross-neutralize RSV A and B strains. Viral Immunol 25(3):193–203

Cianci C, Langley DR, Dischino DD, Sun Y, Yu K-L, Stanley A, Roach J, Li Z, Dalterio R, Colonno R, Meanwell NA, Krystal M (2004a) Targeting a binding pocket within the trimer-of-hairpins: Small-molecule inhibition of viral fusion. Proc Natl Acad Sci USA 101(42):15046–15051

Cianci C, Yu KL, Combrink K, Sin N, Pearce B, Wang A, Civiello R, Voss S, Luo G, Kadow K, Genovesi EV, Venables B, Gulgeze H, Trehan A, James J, Lamb L, Medina I, Roach J, Yang Z, Zadjura L, Colonno R, Clark J, Meanwell N, Krystal M (2004b) Orally active fusion inhibitor of respiratory syncytial virus. Antimicrob Agents Chemother 48(2):413–422

Collins PL, Huang YT, Wertz GW (1984) Nucleotide sequence of the gene encoding the fusion (F) glycoprotein of human respiratory syncytial virus. Proc Natl Acad Sci USA 81(24):7683–7687

Collins PL, Mottet G (1991) Post-translational processing and oligomerization of the fusion glycoprotein of human respiratory syncytial virus. J Gen Virol 72(12):3095–3101

Collins PL, Mottet G (1992) Oligomerization and post-translational processing of glycoprotein G of human respiratory syncytial virus: altered O-glycosylation in the presence of brefeldin A. J Gen Virol 73(4):849–863

Collins PL, Mottet G (1993) Membrane orientation and oligomerization of the small hydrophobic protein of human respiratory syncytial virus. J Gen Virol 74(7):1445–1450

Connolly SA, Leser GP, Yin HS, Jardetzky TS, Lamb RA (2006) Refolding of a paramyxovirus F protein from prefusion to postfusion conformations observed by liposome binding and electron microscopy. Proc Natl Acad Sci USA 103(47):17903–17908

Costello HM, Ray WC, Chaiwatpongsakorn S, Peeples ME (2012) Targeting RSV with vaccines and small molecule drugs. Infect Disord Drug Targets 12(2):110–128

Day ND, Branigan PJ, Liu C, Gutshall LL, Luo J, Melero JA, Sarisky RT, Del Vecchio AM (2006) Contribution of cysteine residues in the extracellular domain of the F protein of human respiratory syncytial virus to its function. Virol J 3:34

Douglas JL, Panis ML, Ho E, Lin KY, Krawczyk SH, Grant DM, Cai R, Swaminathan S, Cihlar T (2003) Inhibition of respiratory syncytial virus fusion by the small molecule VP-14637 via specific interactions with F protein. J Virol 77(9):5054–5064

Escribano-Romero E, Rawling J, Garcia-Barreno B, Melero JA (2004) The soluble form of human respiratory syncytial virus attachment protein differs from the membrane-bound form in its oligomeric state but is still capable of binding to cell surface proteoglycans. J Virol 78(7):3524–3532

Eshaghi A, Duvvuri VR, Lai R, Nadarajah JT, Li A, Patel SN, Low DE, Gubbay JB (2012) Genetic variability of human respiratory syncytial virus A strains circulating in Ontario: a novel genotype with a 72 nucleotide G gene duplication. PLoS ONE 7(3):e32807

Feldman SA, Audet S, Beeler JA (2000) The fusion glycoprotein of human respiratory syncytial virus facilitates virus attachment and infectivity via an interaction with cellular heparan sulfate. J Virol 74(14):6442–6447

Feldman SA, Hendry RM, Beeler JA (1999) Identification of a linear heparin binding domain for human respiratory syncytial virus attachment glycoprotein G. J Virol 73(8):6610–6617

Frank S, Kammerer RA, Mechling D, Schulthess T, Landwehr R, Bann J, Guo Y, Lustig A, Bachinger HP, Engel J (2001) Stabilization of short collagen-like triple helices by protein engineering. J Mol Biol 308(5):1081–1089

Fuentes S, Tran KC, Luthra P, Teng MN, He B (2007) Function of the respiratory syncytial virus small hydrophobic protein. J Virol 81(15):8361–8366

Gan S-W, Tan E, Lin X, Yu D, Wang J, Tan GM-Y, Vararattanavech A, Yeo CY, Soon CH, Soong TW, Pervushin K, Torres J (2012) The small hydrophobic protein of the human respiratory syncytial virus forms pentameric ion channels. J Biol Chem 287(29):24671–24689

Garcia-Beato R, Martinez I, Franci C, Real FX, Garcia-Barreno B, Melero JA (1996) Host cell effect upon glycosylation and antigenicity of human respiratory syncytial virus G glycoprotein. Virology 221(2):301–309

Gonzalez-Reyes L, Ruiz-Arguello MB, Garcia-Barreno B, Calder L, Lopez JA, Albar JP, Skehel JJ, Wiley DC, Melero JA (2001) Cleavage of the human respiratory syncytial virus fusion protein at two distinct sites is required for activation of membrane fusion. Proc Natl Acad Sci USA 98(17):9859–9864

Gorman JJ, Ferguson BL, Speelman D, Mills J (1997) Determination of the disulfide bond arrangement of human respiratory syncytial virus attachment (G) protein by matrix-assisted laser desorption/ionization time-of-flight mass spectrometry. Protein Sci 6(6):1308–1315

Gruber C, Levine S (1983) Respiratory syncytial virus polypeptides. III. The envelope-associated proteins. J Gen Virol 64(4):825–832

Gupta N, Manevich Y, Kazi AS, Tao JQ, Fisher AB, Bates SR (2006) Identification and characterization of p63 (CKAP4/ERGIC-63/CLIMP-63), a surfactant protein A binding protein, on type II pneumocytes. Am J Physiol Lung Cell Mol Physiol 291(3):L436–L446

Hallak LK, Collins PL, Knudson W, Peeples ME (2000a) Iduronic acid-containing glycosaminoglycans on target cells are required for efficient respiratory syncytial virus infection. Virology 271(2):264–275

Hallak LK, Kwilas SA, Peeples ME (2007) Interaction between respiratory syncytial virus and glycosaminoglycans, including heparan sulfate. Methods Mol Biol 379:15–34

Hallak LK, Spillmann D, Collins PL, Peeples ME (2000b) Glycosaminoglycan sulfation requirements for respiratory syncytial virus infection. J Virol 74(22):10508–10513

Haynes LM, Caidi H, Radu GU, Miao C, Harcourt JL, Tripp RA, Anderson LJ (2009) Therapeutic monoclonal antibody treatment targeting respiratory syncytial virus (RSV) G protein mediates viral clearance and reduces the pathogenesis of RSV infection in BALB/c mice. J Infect Dis 200(3):439–447

Haynes LM, Moore DD, Kurt-Jones EA, Finberg RW, Anderson LJ, Tripp RA (2001) Involvement of toll-like receptor 4 in innate immunity to respiratory syncytial virus. J Virol 75(22):10730–10737

Hendricks DA, Baradaran K, McIntosh K, Patterson JL (1987) Appearance of a soluble form of the G protein of respiratory syncytial virus in fluids of infected cells. J Gen Virol 68(6):1705–1714

Hendricks DA, McIntosh K, Patterson JL (1988) Further characterization of the soluble form of the G glycoprotein of respiratory syncytial virus. J Virol 62(7):2228–2233

Hickling TP, Malhotra R, Bright H, McDowell W, Blair ED, Sim RB (2000) Lung surfactant protein A provides a route of entry for respiratory syncytial virus into host cells. Viral Immunol 13(1):125–135

Huang K, Incognito L, Cheng X, Ulbrandt ND, Wu H (2010) Respiratory syncytial virus-neutralizing monoclonal antibodies motavizumab and palivizumab inhibit fusion. J Virol 84(16):8132–8140

Johnson S, Oliver C, Prince GA, Hemming VG, Pfarr DS, Wang SC, Dormitzer M, O'Grady J, Koenig S, Tamura JK, Woods R, Bansal G, Couchenour D, Tsao E, Hall WC, Young JF (1997) Development of a humanized monoclonal antibody (MEDI-493) with potent in vitro and in vivo activity against respiratory syncytial virus. J Infect Dis 176(5):1215–1224

Johnson TR, McLellan JS, Graham BS (2012) Respiratory syncytial virus glycoprotein G interacts with DC-SIGN and L-SIGN to activate ERK1 and ERK2. J Virol 86(3):1339–1347

Karron RA, Buonagurio DA, Georgiu AF, Whitehead SS, Adamus JE, Clements-Mann ML, Harris DO, Randolph VB, Udem SA, Murphy BR, Sidhu MS (1997) Respiratory syncytial virus (RSV) SH and G proteins are not essential for viral replication in vitro: clinical evaluation and molecular characterization of a cold-passaged, attenuated RSV subgroup B mutant. Proc Natl Acad Sci USA 94(25):13961–13966

Kim S, Joo DH, Lee JB, Shim BS, Cheon IS, Jang JE, Song HH, Kim KH, Song MK, Chang J (2012) Dual role of respiratory syncytial virus glycoprotein fragment as a mucosal immunogen and chemotactic adjuvant. PLoS ONE 7(2):e32226

Kong X, San Juan H, Behera A, Peeples ME, Wu J, Lockey RF, Mohapatra SS (2004) ERK-1/2 activity is required for efficient RSV infection. FEBS Lett 559(1–3):33–38

Krusat T, Streckert HJ (1997) Heparin-dependent attachment of respiratory syncytial virus (RSV) to host cells. Arch Virol 142(6):1247–1254

Kwakkenbos MJ, Diehl SA, Yasuda E, Bakker AQ, van Geelen CMM, Lukens MV, van Bleek GM, Widjojoatmodjo MN, Bogers WMJM, Mei H, Radbruch A, Scheeren FA, Spits H, Beaumont T (2010) Generation of stable monoclonal antibody-producing B cell receptor-positive human memory B cells by genetic programming. Nat Med 16(1):123–128

Kwilas S, Liesman RM, Zhang L, Walsh E, Pickles RJ, Peeples ME (2009) Respiratory syncytial virus grown in Vero cells contains a truncated attachment protein that alters its infectivity and dependence on glycosaminoglycans. J Virol 83(20):10710–10718

Lambert DM, Barney S, Lambert AL, Guthrie K, Medinas R, Davis DE, Bucy T, Erickson J, Merutka G, Petteway SR (1996) Peptides from conserved regions of paramyxovirus fusion (F) proteins are potent inhibitors of viral fusion. Proc Natl Acad Sci USA 93(5):2186–2191

Levine S (1977) Polypeptides of respiratory syncytial virus. J Virol 21(1):427–431

Levine S, Klaiber-Franco R, Paradiso PR (1987) Demonstration that glycoprotein G is the attachment protein of respiratory syncytial virus. J Gen Virol 68(Pt 9):2521–2524

Li P, Mc LRHW, Brown G, Sugrue RJ (2007) Functional analysis of the *N*-linked glycans within the fusion protein of respiratory syncytial virus. Methods Mol Biol 379:69–83

Libon C, Desplanches F, Corbiere JC, Haeuw JF, Robert A, Nguyen T (2007) Identification of the component responsible for the type III hypersensitivity reaction (HS III) induced by BBG2Na, a subunit respiratory syncytial virus (RSV) vaccine. Paper presented at the 6th international respiratory syncytial virus symposium, Marco Island, Florida, 25–28 October 2007

Magro M, Andreu D, Gomez-Puertas P, Melero JA, Palomo C (2010) Neutralization of human respiratory syncytial virus infectivity by antibodies and low-molecular-weight compounds targeted against the fusion glycoprotein. J Virol 84(16):7970–7982

Magro M, Mas V, Chappell K, Vazquez M, Cano O, Luque D, Terron MC, Melero JA, Palomo C (2012) Neutralizing antibodies against the preactive form of respiratory syncytial virus fusion protein offer unique possibilities for clinical intervention. Proc Natl Acad Sci USA 109(8):3089–3094

Malhotra R, Ward M, Bright H, Priest R, Foster MR, Hurle M, Blair E, Bird M (2003) Isolation and characterisation of potential respiratory syncytial virus receptor(s) on epithelial cells. Microbes Infect 5(2):123–133

McLellan JS, Chen M, Chang J-S, Yang Y, Kim A, Graham BS, Kwong PD (2010a) Structure of a major antigenic site on the respiratory syncytial virus fusion glycoprotein in complex with neutralizing antibody 101F. J Virol 84(23):12236–12244

McLellan JS, Chen M, Kim A, Yang Y, Graham BS, Kwong PD (2010b) Structural basis of respiratory syncytial virus neutralization by motavizumab. Nat Struct Mol Biol 17(2):248–250

McLellan JS, Chen M, Leung S, Graepel KW, Du X, Yang Y, Zhou T, Baxa U, Yasuda E, Beaumont T, Kumar A, Modjarrad K, Zheng Z, Zhao M, Xia N, Kwong PD, Graham BS (2013) Structure of RSV fusion glycoprotein trimer bound to a prefusion-specific neutralizing antibody. Science 340(6136):1113–7

McLellan JS, Yang Y, Graham BS, Kwong PD (2011) Structure of respiratory syncytial virus fusion glycoprotein in the postfusion conformation reveals preservation of neutralizing epitopes. J Virol 85(15):7788–7796

Munoz FM, Piedra PA, Glezen WP (2003) Safety and immunogenicity of respiratory syncytial virus purified fusion protein-2 vaccine in pregnant women. Vaccine 21(24):3465–3467

Nguyen TN, Power UF, Robert A, Haeuw JF, Helffer K, Perez A, Asin MA, Corvaia N, Libon C (2012) The respiratory syncytial virus G protein conserved domain induces a persistent and protective antibody response in rodents. PLoS ONE 7(3):e34331

Nieva JL, Madan V, Carrasco L (2012) Viroporins: structure and biological functions. Nat Rev Micro 10(8):563–574

Olmsted RA, Collins PL (1989) The 1A protein of respiratory syncytial virus is an integral membrane protein present as multiple, structurally distinct species. J Virol 63(5):2019–2029

Plotnicky-Gilquin H, Huss T, Aubry JP, Haeuw JF, Beck A, Bonnefoy JY, Nguyen TN, Power UF (1999) Absence of lung immunopathology following respiratory syncytial virus (RSV) challenge in mice immunized with a recombinant RSV G protein fragment. Virology 258(1):128–140

Power UF, Plotnicky-Gilquin H, Huss T, Robert A, Trudel M, Stahl S, Uhlen M, Nguyen TN, Binz H (1997) Induction of protective immunity in rodents by vaccination with a prokaryotically expressed recombinant fusion protein containing a respiratory syncytial virus G protein fragment. Virology 230(2):155–166

Radu GU, Caidi H, Miao C, Tripp RA, Anderson LJ, Haynes LM (2010) Prophylactic treatment with a G glycoprotein monoclonal antibody reduces pulmonary inflammation in respiratory

syncytial virus (RSV)-challenged naive and formalin-inactivated RSV-immunized BALB/c mice. J Virol 84(18):9632–9636
Rixon HWML, Brown G, Aitken J, McDonald T, Graham S, Sugrue RJ (2004) The small hydrophobic (SH) protein accumulates within lipid-raft structures of the Golgi complex during respiratory syncytial virus infection. J Gen Virol 85(5):1153–1165
Roberts SR, Lichtenstein D, Ball LA, Wertz GW (1994) The membrane-associated and secreted forms of the respiratory syncytial virus attachment glycoprotein G are synthesized from alternative initiation codons. J Virol 68(7):4538–4546
Roymans D, De Bondt HL, Arnoult E, Geluykens P, Gevers T, Van Ginderen M, Verheyen N, Kim H, Willebrords R, Bonfanti JF, Bruinzeel W, Cummings MD, van Vlijmen H, Andries K (2010) Binding of a potent small-molecule inhibitor of six-helix bundle formation requires interactions with both heptad-repeats of the RSV fusion protein. Proc Natl Acad Sci USA 107(1):308–313
Ruiz-Arguello MB, Gonzalez-Reyes L, Calder LJ, Palomo C, Martin D, Saiz MJ, Garcia-Barreno B, Skehel JJ, Melero JA (2002) Effect of proteolytic processing at two distinct sites on shape and aggregation of an anchorless fusion protein of human respiratory syncytial virus and fate of the intervening segment. Virology 298(2):317–326
Ruiz-Arguello MB, Martin D, Wharton SA, Calder LJ, Martin SR, Cano O, Calero M, Garcia-Barreno B, Skehel JJ, Melero JA (2004) Thermostability of the human respiratory syncytial virus fusion protein before and after activation: implications for the membrane-fusion mechanism. J Gen Virol 85(12):3677–3687
Satake M, Coligan JE, Elango N, Norrby E, Venkatesan S (1985) Respiratory syncytial virus envelope glycoprotein (G) has a novel structure. Nucleic Acids Res 13(21):7795–7812
Schepens B, De Baets S, Sedyen K, Bogaert P, Gilbert B, Piedra PA, Fiers W, Saelens X (2012) SHe's a novel target for RSV vaccination. Paper presented at the 8th respiratory syncytial virus symposium, Santa Fe, USA, 28 September 2012
Schlender J, Zimmer G, Herrler G, Conzelmann KK (2003) Respiratory syncytial virus (RSV) fusion protein subunit F2, not attachment protein G, determines the specificity of RSV infection. J Virol 77(8):4609–4616
Stott EJ, Ball LA, Young KK, Furze J, Wertz GW (1986) Human respiratory syncytial virus glycoprotein G expressed from a recombinant vaccinia virus vector protects mice against live-virus challenge. J Virol 60(2):607–613
Swanson KA, Settembre EC, Shaw CA, Dey AK, Rappuoli R, Mandl CW, Dormitzer PR, Carfi A (2011) Structural basis for immunization with postfusion respiratory syncytial virus fusion F glycoprotein (RSV F) to elicit high neutralizing antibody titers. Proc Natl Acad Sci USA 108(23):9619–9624
Takimoto T, Hurwitz JL, Coleclough C, Prouser C, Krishnamurthy S, Zhan X, Boyd K, Scroggs RA, Brown B, Nagai Y, Portner A, Slobod KS (2004) Recombinant Sendai virus expressing the G glycoprotein of respiratory syncytial virus (RSV) elicits immune protection against RSV. J Virol 78(11):6043–6047
Tayyari F, Marchant D, Moraes TJ, Duan W, Mastrangelo P, Hegele RG (2011) Identification of nucleolin as a cellular receptor for human respiratory syncytial virus. Nat Med 17(9):1132–1135
Techaarpornkul S, Barretto N, Peeples ME (2001) Functional analysis of recombinant respiratory syncytial virus deletion mutants lacking the small hydrophobic and/or attachment glycoprotein gene. J Virol 75(15):6825–6834
Teng MN, Whitehead SS, Collins PL (2001) Contribution of the respiratory syncytial virus G glycoprotein and its secreted and membrane-bound forms to virus replication in vitro and in vivo. Virology 289(2):283–296
Ternette N, Stefanou D, Kuate S, Uberla K, Grunwald T (2007) Expression of RNA virus proteins by RNA polymerase II dependent expression plasmids is hindered at multiple steps. Virol J 4:51

The IMpact-RSV Study Group (1998) Palivizumab, a humanized respiratory syncytial virus monoclonal antibody, reduces hospitalization from respiratory syncytial virus infection in high-risk infants. Pediatrics 102(3):531–537

Trento A, Galiano M, Videla C, Carballal G, Garcia-Barreno B, Melero JA, Palomo C (2003) Major changes in the G protein of human respiratory syncytial virus isolates introduced by a duplication of 60 nucleotides. J Gen Virol 84(11):3115–3120

Tripp RA, Jones LP, Haynes LM, Zheng H, Murphy PM, Anderson LJ (2001) CX3C chemokine mimicry by respiratory syncytial virus G glycoprotein. Nat Immunol 2(8):732–738

Villenave R, Thavagnanam S, Sarlang S, Parker J, Douglas I, Skibinski G, Heaney LG, McKaigue JP, Coyle PV, Shields MD, Power UF (2012) In vitro modeling of respiratory syncytial virus infection of pediatric bronchial epithelium, the primary target of infection in vivo. Proc Natl Acad Sci USA 109(13):5040–5045

Walsh EE, Hruska J (1983) Monoclonal antibodies to respiratory syncytial virus proteins: identification of the fusion protein. J Virol 47(1):171–177

Wertz GW, Collins PL, Huang Y, Gruber C, Levine S, Ball LA (1985) Nucleotide sequence of the G protein gene of human respiratory syncytial virus reveals an unusual type of viral membrane protein. Proc Natl Acad Sci USA 82(12):4075–4079

Whitehead SS, Bukreyev A, Teng MN, Firestone C-Y, St. Claire M, Elkins WR, Collins PL, Murphy BR (1999) Recombinant respiratory syncytial virus bearing a deletion of either the NS2 or SH gene is attenuated in chimpanzees. J Virol 73(4):3438–3442

Wild C, Greenwell T, Matthews T (1993) A synthetic peptide from HIV-1 gp41 is a potent inhibitor of virus-mediated cell–cell fusion. AIDS Res Hum Retroviruses 9(11):1051–1053

Wild C, Oas T, McDanal C, Bolognesi D, Matthews T (1992) A synthetic peptide inhibitor of human immunodeficiency virus replication: correlation between solution structure and viral inhibition. Proc Natl Acad Sci USA 89(21):10537–10541

Wu H, Pfarr DS, Johnson S, Brewah YA, Woods RM, Patel NK, White WI, Young JF, Kiener PA (2007a) Development of motavizumab, an ultra-potent antibody for the prevention of respiratory syncytial virus infection in the upper and lower respiratory tract. J Mol Biol 368(3):652–665

Wu S-J, Schmidt A, Beil EJ, Day ND, Branigan PJ, Liu C, Gutshall LL, Palomo C, Furze J, Taylor G, Melero JA, Tsui P, Del Vecchio AM, Kruszynski M (2007b) Characterization of the epitope for anti-human respiratory syncytial virus F protein monoclonal antibody 101F using synthetic peptides and genetic approaches. J Gen Virol 88(10):2719–2723

Yin HS, Wen X, Paterson RG, Lamb RA, Jardetzky TS (2006) Structure of the parainfluenza virus 5 F protein in its metastable, prefusion conformation. Nature 439(7072):38–44

Zhang L, Bukreyev A, Thompson CI, Watson B, Peeples ME, Collins PL, Pickles RJ (2005) Infection of ciliated cells by human parainfluenza virus type 3 in an in vitro model of human airway epithelium. J Virol 79(2):1113–1124

Zhang L, Peeples ME, Boucher RC, Collins PL, Pickles RJ (2002) Respiratory syncytial virus infection of human airway epithelial cells is polarized, specific to ciliated cells, and without obvious cytopathology. J Virol 76(11):5654–5666

Zhao X, Singh M, Malashkevich VN, Kim PS (2000) Structural characterization of the human respiratory syncytial virus fusion protein core. Proc Natl Acad Sci USA 97(26):14172–14177

Zimmer G, Budz L, Herrler G (2001a) Proteolytic activation of respiratory syncytial virus fusion protein. Cleavage at two furin consensus sequences. J Biol Chem 276(34):31642–31650

Zimmer G, Trotz I, Herrler G (2001b) N-glycans of F protein differentially affect fusion activity of human respiratory syncytial virus. J Virol 75(10):4744–4751

Respiratory Syncytial Virus and Reactive Airway Disease

Matthew T. Lotz, Martin L. Moore and R. Stokes Peebles Jr.

Abstract Reactive airway disease (RAD) is a general term for respiratory illnesses manifested by wheezing. Respiratory syncytial virus (RSV) results in wheezing, either by causing bronchiolitis or by inducing acute exacerbations of asthma. There has been a long-standing interest in whether severe RSV bronchiolitis in infancy is a risk factor for the development of asthma later in childhood. While epidemiologic studies have suggested that such a link exists, a very recent study suggests that infants with greater airways responsiveness to methacholine instead have an increased prevalence of severe RSV bronchiolitis. Increased airways responsiveness to methacholine has been implicated as a key factor for loss of lung function in asthmatic subjects, suggesting that instead of being causal, severe RSV infection may instead be a marker of a predisposing factor for asthma. In this chapter, we will explore the evidence that RSV infection leads to RAD in infants and adults, and how these different forms of RAD may be linked.

Funding: National Institutes of Health R01 HL 090664, R01 AI 070672, R01 AI 059108, GM 015431, R21 HL106446, U19AI095227, and Veteran Affairs (1I01BX000624).

M. T. Lotz
Division of Allergy, Pulmonary and Critical Care Medicine, Department of Medicine, Vanderbilt University School of Medicine, Nashville, TN, USA

M. L. Moore
Division of Infectious Diseases, Department of Pediatrics, Emory University School of Medicine, Atlanta, GA, USA

R. S. Peebles Jr. (✉)
T-1218 Medical Center North, Vanderbilt University Medical Center, 1161 21st Avenue South, Nashville, TN 37232-2650, USA
e-mail: stokes.peebles@vanderbilt.edu

L. J. Anderson and B. S. Graham (eds.), *Challenges and Opportunities for Respiratory Syncytial Virus Vaccines*, Current Topics in Microbiology and Immunology 372,
DOI: 10.1007/978-3-642-38919-1_5,

Contents

1 Introduction

The relationship between respiratory syncytial virus (RSV) infection and reactive airway disease (RAD) has been of great interest for decades, mainly because of the question as to whether early life RSV infection predisposes to the development of asthma later in childhood. RAD is a general term for respiratory illnesses manifested by wheezing. Wheezing is produced by turbulent airflow as a result of airway narrowing, either by constriction or obstruction. The wheezing that occurs in asthma is characterized by reversible bronchial reactivity that is predominantly a result of smooth muscle constriction, airway mucus production, and inflammatory cell migration into the airway wall (NHLBI/WHO workshop report 1995). It is important to recognize that RAD constitutes more than just asthma, and therefore RAD is a broader concept to consider when exploring the relationship of RSV to lung disease. In contrast to asthma, the wheezing that occurs in infants with RSV infection is a manifestation of bronchiolitis (American Academy of Pediatrics Subcommittee on Diagnosis and Management of Bronchiolitis 2006). Bronchiolitis in its simplest definition is inflammation of the bronchioles, which are the smallest air passages in the lung. In this chapter, we will discuss the immunologic and physiologic aspects that lead to both bronchiolitis and asthma, and the supportive, but not definitive, evidence that there is a link between severe bronchiolitis in early life and the later development of asthma.

2 RSV Bronchiolitis

2.1 Epidemiology of Human RSV Bronchiolitis

Almost all children are infected with RSV by age 3 (Glezen and Denny 1973). Most RSV infections in these early years of life are mild and do not require medical attention. However, some infections are more severe and result in outpatient visits to emergency departments, or hospitalization. The percentage of children with mild symptoms versus those with more severe bronchiolitis was reported in infants enrolled in the Tennessee Medicaid program born between 1995 and 2000 (Carroll et al. 2009b). In that study, 82 % of infants did not require a health care visit, 9 % sought care in an outpatient clinic, 4 % required an emergency department visit, and 5 % were admitted to the hospital for bronchiolitis. A slightly decreased hospitalization rate of 26 per 1,000 (2.6 %) was reported in a retrospective analysis of hospital discharges in the USA from 1997 to 2006 (Stockman et al. 2012). There are several consistent risk factors that have been identified for severe RSV bronchiolitis. In an Italian study performed over four consecutive RSV seasons from 2000 to 2004, the seven predictors for hospitalization for RSV infection included the number of children in the family, the chronological age at the onset of the RSV season, birth weight and gestational age, being at least the second baby in the family, daycare attendance, and previous RSV infections (Rossi et al. 2007). This study suggested that there were specific host/environmental factors that identified children at greatest risk for hospitalization for RSV infection. This was supported by a different study using the Tennessee Medicaid data base previously mentioned in which infants who were 122 days of age (95 % CI, 118–126) at the winter virus peak had the highest risk of developing clinically significant bronchiolitis after adjusting for gender, race, number of living siblings, birth weight, gestational age, maternal smoking, marital status, maternal education, region of residence, and season (Wu et al. 2008). A more recent study in Canadian infants identified age less than 6 months at the time of infection, underlying heart disease, and household crowding as risk factors for RSV hospitalization (Papenburg et al. 2012). Of those hospitalized, age less than 6 months and prematurity were associated with more severe RSV disease. Interestingly, breast-feeding and viral co-infection were protective against hospitalization (Papenburg et al. 2012). In a study examining infants admitted to the hospital in St. Louis, exposure to post-natal cigarette smoke from the mother, younger age, and Caucasian race in contrast to black race were all predictors of more severe RSV bronchiolitis (Bradley et al. 2005).

2.2 Immunologic Determinants of Severe RSV Bronchiolitis in Human Infants

RAD and wheezing with RSV-induced bronchiolitis are most likely consequences of physical changes in the airway that are associated with the immune response to infection and the body's mechanism of clearing the virus. Airway obstruction is a key underlying factor in the severity of RSV infection. The extent of obstruction is determined by the size of the airway, the extent of the inflammatory infiltrate and mucus production, and the dynamic properties of the airway. The pathologic manifestations of bronchiolitis in infants include influx of monocytes,

CD3 + double-negative T-cells, CD8 + T-cells, and neutrophils into the airway smooth muscle, epithelium, and lumen (Johnson et al. 2007). The airway obstruction in RSV-induced bronchiolitis in infants is caused by luminal epithelial and inflammatory cell debris mixed with fibrin, mucus, and edema, in addition to extrinsic airway compression by hyperplastic lymphoid follicles (Johnson et al. 2007).

Several investigative teams have attempted to determine the immune response profile associated with severe RSV bronchiolitis in order to understand the pathogenic mechanisms leading to morbidity associated with airway obstruction during infection. Bronchoalveolar lavage (BAL) is a technique to sample the alveolar lining fluid and cells in the airway, while nasal lavage can be performed to measure the mediators and cells in the nose as a representation of the upper airway. BAL can be obtained by two methods. First, a bronchoscope is placed into the airway and directed into specific subsegments of the lung, where it is then wedged into a small airway, after which lavage can be performed to obtain samples distal to the location where the bronchoscope is wedged. The second technique is to perform BAL in intubated patients where a suction catheter is advanced until it meets resistance and then lavage fluid is introduced through the catheter and then suctioned. In this situation, the location of the catheter relative to the airways is not known and it is not clear that the catheter is necessarily wedged. Therefore, there may be differences in the results obtained with these techniques.

There are only a few studies in which bronchoscopy and BAL have been performed in infants with bronchiolitis because of the risk associated with hypoxemia during the procedure. One such study compared the BAL findings in six healthy children compared to 20 subjects with recurrent wheezing or prolonged wheezing of greater than 2 months over a 6-month period (Krawiec et al. 2001). In this report, the average age of the healthy children was 23 months compared to 15 months for the wheezing subjects. The wheezing subjects had a significantly greater number of total cells, lymphocytes, epithelial cells, neutrophils, and eosinophils compared to controls. Wheezing subjects also had increased BAL levels of leukotriene (LT)B_4 and LTE_4 than control subjects (Krawiec et al. 2001). LTB_4 is a neutrophil chemotactic factor (Wenzel 1997). LTE_4 is the final metabolite in the cysteinyl leukotriene pathway and has been shown to be an important mucus secretagogue, induce smooth muscle constriction, and be an eosinophil chemotactic factor (Wenzel 1997). While mucus is important in clearing foreign particles from the airway, excessive mucus secretion can be an important contributor to airway obstruction. Eosinophils are thought to have an important role in RAD because they produce proteins and enzymes (major basic protein, eosinophil cationic protein, and eosinophil peroxidase) that cause desquamation of airway epithelial cells, further contributing to airway obstruction. In addition, eosinophil major basic protein blocks the inhibitory muscarinic M2 receptor on the postganglionic parasympathetic nerve that normally inhibits acetylcholine release (Jacoby et al. 1993). This inhibition of M2 receptor function leads to increased acetylcholine release and airway smooth muscle contraction (Jacoby et al. 2001). There was no difference between the wheezing and healthy children in the BAL concentrations of either PGD_2 or β-tryptase, mediators released predominantly by mast cells, cells which are important in IgE-mediated hypersensitivity reactions that occur in allergic diseases such as asthma (Krawiec et al. 2001). Bronchoscopy has also been performed to determine if there were differences in mediators and cells present in the lower airway of children with either RSV-induced bronchiolitis (average age 9 months), acute asthma with no identifiable viral infection (average age 5 years), or controls who did not have asthma, but who did have indication for bronchoscopy (average age 6.4 years) (Kim et al. 2003). The subjects with bronchiolitis had significantly greater total cells that the other two groups. Asthmatic subjects had a greater median percentage of BAL eosinophils (3 %) than either the bronchiolitis group or control group (both 0 %). There was a small subgroup of bronchiolitis subjects with BAL eosinophils (6 subjects with eosinophils and 16 without). Asthmatic subjects and subjects with bronchiolitis and BAL eosinophils had a greater concentration of IL-5 in BAL fluid compared to the subjects with bronchiolitis and no BAL eosinophils or controls. IL-5 is the most important eosinophil growth, differentiation, and survival factor. Interestingly, in this study, there was no difference in interferon (IFN)-γ levels in the bronchiolitis, asthma, and control groups. Children with RSV bronchiolitis also had an increased median percentage of BAL neutrophils (37.5 %) compared to the asthma group (3.3 %) or the controls (2.3 %). The increase in total cells and BAL neutrophils in subjects with RSV-induced bronchiolitis was also confirmed by other investigators (McNamara et al. 2003).

Nasal lavage was used as a means to compare the immune profile of 63 children with RSV bronchiolitis and 22 controls in a study in Texas of children less than 2 years of age (Bennett et al. 2007). The RSV bronchiolitis group had a significantly greater number of inflammatory cells, as well as median nasal lavage concentrations of IL-8 (269.4 vs. 13.8) compared to controls. IL-8 is an important neutrophil survival and chemotactic factor and the increase demonstrated in this study suggests a mechanism for the increased neutrophils seen in the previously mentioned bronchoscopy studies. There were also significant increases in nasal lavage IL-6, GM-CSF, IFN-γ, TNF-α, IL-1β, G-CSF, and MIP-1β concentrations in the RSV group compared to controls. The subjects were stratified based on age at time of infection into four groups, $0 \leq 3$ months, $3 \leq 6$ months, $6 \leq 12$ months, $12 \leq 24$ months, and there was no difference between the groups and cytokine levels. Interestingly, there was an inverse correlation between nasal lavage levels of IL-6, IL-8, IL-10, and IFN-γ and the duration of supplemental oxygen necessary to prevent hypoxia in the RSV group (Bennett et al. 2007). Higher nasal lavage levels of IFN-γ were associated with protection against RSV illness in a British study of 197 infants admitted to the hospital with RSV bronchiolitis (Semple et al. 2007). Therefore, a deficient IFN-γ response may be a risk factor for more severe RSV disease. IFN-γ is the signature cytokine produced by CD4 + T helper (Th1) cells, and is also produced in abundance by CD8 + cytotoxic T lymphocytes, and Natural Killer (NK) cells which have potent anti-viral activity.

2.3 Primary Infection in Mice

Examining host immune response to naturally occurring RSV infection in humans is very complex. First, almost every host has a unique genetic composition. Second, there is little similarity in the environments in which people live or work. Third, there are genetic differences between RSV strains that may greatly affect the host immune response. These complexities are somewhat simplified by the use of the mouse model of RSV where at least a relatively constant environment can be maintained and utilizing genetically identical mice minimizes differences in host immune responses. However, mice are not perfect models of human disease. Beside the fact that some immune molecules present in humans have no murine counterpart, the airway epithelial desquamation seen in humans is virtually absent in mice. Some of the benefits of using the mouse model are the opportunity to test the inflammatory response of different RSV strains in hosts of the same strain of mouse, or to use mouse strains that vary in only one gene to determine the impact of that gene on the host immune response (see chapter by P.J. Openshaw, this volume).

The impact on viral genetics on airways responsiveness to methacholine, a surrogate of RAD has been extensively studied between RSV strains A2 and line 19 (Lukacs et al. 2006; Moore et al. 2009). Both A2 and line 19 are RSV group A viruses. There are two major antigenic groups of RSV, A and B, which are classified based on monoclonal antibodies to the major structural glycoproteins G and F (Anderson et al. 1985). A2 was isolated in Australia in the late 1950s and line 19 is a clinical isolate from the University of Michigan cultured in the 1960s. In BALB/cJ mice, line 19 infection induced lung IL-13 protein expression which was not detected in A2 infected mice (Lukacs et al. 2006). IL-13 is a central mediator of airway responsiveness, in part as a result of its direct effects on smooth muscle contraction and epithelial cell mucus metaplasia as well as mucus secretion (Wills-Karp et al. 1998). Indeed, line 19-infected mice had goblet cell metaplasia and hyperplasia, whereas neither occurred following A2 infection (Lukacs et al. 2006). Line 19 infection induced significant increases in Muc5ac, a mucus gene expressed in the airways, and gob-5, a member of the $Ca2^{+}$-activated chloride channel family which has selective expression in airway goblet cells. Mice infected with line 19 had significantly increased airways responsiveness to methacholine compared to mice similarly infected with A2. In this technique, mice are anesthetized, intubated, mechanically ventilated, and challenged with either aerosolized or intravenous methacholine. Methacholine is a synthetic derivative of acetylcholine, which causes airway smooth muscle constriction (Cockcroft 2010). The methacholine challenge test is performed in people to help make the diagnosis of asthma when other causes of dyspnea are being considered. People with asthma have increased airways obstruction with concomitant decrease in pulmonary function when they are challenged with aerosolized methacholine in comparison to persons without asthma. The meaning of methacholine reactivity in mice is not as clear as in humans, but the airways responsiveness induced by methacholine in mice is likely a result of a combination of heightened smooth muscle construction, the presence of epithelial mucus metaplasia narrowing the airway lumen, mucus present in the airway, and infiltration of inflammatory cells in the airway wall. In these experiments, the mucus airways responsiveness and mucus seen with line 19 infection was dependent upon the ability of this strain to induce lung IL-13 protein expression, as neither characteristic was present in IL-13 knockout mice. The differences seen between A2 and Line 19 were independent of viral load (Lukacs et al. 2006).

Further studies have revealed the importance of differences in the F protein between line 19 and A2 in modulating lung IL-13 protein expression, airway epithelial cell mucus metaplasia, and airways responsiveness. Generation of chimeric RSV strains using a reverse genetics approach where all genes are from the A2 strain with the exception of the line 19 F protein (rA2-line 19F) or an A2 F (rA2-A2F) revealed that rA2-line 19F infection recapitulated the lung IL-13 protein expression, epithelial cell mucus metaplasia, and airways responsiveness seen with line 19 infection, while these were absent from rA2-A2F. Studies are ongoing to determine which amino acids in the F protein regulate these immunologic and physiologic parameters (Moore et al. 2009).

Mice infected with different low passage clinical isolates also may have pathologic sequelae similar to those seen with line 19 and A2 infection (Stokes et al. 2011). Nasal secretions obtained from children presenting to the Vanderbilt Vaccine Clinic with RSV group A infection were used to infect HEp-2 cells. The isolates were passaged by limiting dilution and nine strains were then propagated. BALB/cJ mice infected with one of these strains, RSV A2001/2-20, had lung IL-13 protein expression, epithelial cell mucus metaplasia and increased airway mucus and airway responsiveness. In contrast, mice infected with another strain, RSV A2001/3-12, did not have lung IL-13 protein expression, increase airway responsiveness or mucus production. Interestingly, the child infected with RSV A2001/2-20 had a more severe lower respiratory tract illness than the child infected with RSV A2001/3-12. Studies are ongoing to define the role of RSV strain differences in RSV-induced disease severity and asthma development (Stokes et al. 2011). (see chapter by J.A. Melero and M.L. Moore, this volume).

3 RSV-Induced Bronchiolitis and Asthma Later in Childhood

In the 1980s, several retrospective studies suggested that severe RSV infection in infancy predisposed to the development of asthma later in childhood. However, other retrospective studies refuted these findings, leaving the role of RSV-induced bronchiolitis in asthma development in doubt. Two large prospective longitudinal studies published in the 1990s supported the concept that severe RSV lower respiratory tract infection in early life was an important risk factor for asthma or wheezing-related illness up to adolescence. The first report was a prospective cohort study of 47 Swedish children enrolled before 1 year of age with severe RSV bronchiolitis that was compared to age and gender matched controls in the same location as the cases (Sigurs et al. 1995). At 3 years of age, the children admitted to the hospital with RSV bronchiolitis had a significant increase in asthma compared to the control group (23 % vs. 1 %; $p < 0.001$). In addition, allergic sensitization as defined by a positive skin test to an aeroallergen was also significantly increased in the RSV bronchiolitis group compared to controls (32 % vs. 9 %; $p = 0.002$) In this study, RSV bronchiolitis was the most important risk factor for asthma and allergic sensitization, with a family history for asthma or allergic disease further increasing this risk. This cohort has now been followed to age 18 and the RSV bronchiolitis group continues to have an increased prevalence of asthma (39 % vs. 9 %) and allergic sensitization (41 % vs. 14 %) compared to the control group (Sigurs et al. 2010). In addition, persistent or relapsing wheezing at age 18 was significantly increased in the group that had RSV bronchiolitis in infancy compared to the controls (30 % vs. 1 %) and the RSV group had decreased lung function compared to controls, independent of a diagnosis of asthma. The second report examined 1,246 newborns were enrolled into the Tucson Children's Respiratory Study between May 1980 and January 1984 (Stein et al. 1999). Lower respiratory tract illnesses (LRTI) suffered by these children in the first 3 years of life were assessed for etiologic agents by means of culture and serology and the children were followed to age 13 for determination of wheezing. In this study, RSV LRTI was defined as deep or wet chest cough, wheezing, hoarseness, stridor, shortness of breath. RSV LTRI was associated with an increased risk of infrequent wheezing and frequent wheezing at age 6 and 11, but not at age 13. These results suggested that children might "outgrow" the effect of early life RSV LRTI on wheezing at the start of the teenage years. In this study, RSV LTRI before age 3 was not a risk factor for aeroallergen sensitization as defined by skin testing. A more recent population-based birth cohort study of 90,341 children further suggests a link between the severity of early life RSV infection and the later development of asthma (Carroll et al. 2009a). The Tennessee Asthma Bronchiolitis Study (TABS) examined the children enrolled in the state Medicaid program from 1995 to 2000 and found that infants requiring a health care visit for respiratory symptoms during the winter months when RSV is the dominant cause of bronchiolitis were significantly more likely to be diagnosed with asthma between the 4 and 5 ½ years of age (relative risk 1.98). Further investigation of the TABS database revealed that there was a severity-dependent relationship between infant bronchiolitis severity and childhood asthma. Of the children assessed in this study, 11.6 % were hospitalized for bronchiolitis, 8.2 % had an emergency department visit for bronchiolitis, 13.7 % had a clinic visit for bronchiolitis, and 66.5 % had no health care visit for bronchiolitis. Infants that had a bronchiolitis hospitalization during infancy had an adjusted odds ratio of 1.51 (1.26–1.80) of an asthma diagnosis at age 4–5 ½ years. One limitation of this study was a lack of virologic confirmation of the diagnosis of RSV being the causative agent of bronchiolitis. This is an important concern given the recent recognition that rhinovirus (RV) has also been determined to be a cause of infant bronchiolitis. In one study, wheezing with RV from birth to age 3 was associated with a greater risk of asthma at age 6 (odds ratio 9.8) compared to RSV (odds ratio 2.6), although wheezing from either virus conferred a greater risk of asthma compared to children who did not wheeze (Jackson et al. 2008).

Other prospective studies support the concept that children who suffer from severe RSV infection in infancy have a predisposition to the later development of asthma and eczema, both diseases associated with allergic diatheses. A total of 206 infants in St. Louis were enrolled in the RSV Bronchiolitis in Early Life (RBEL) cohort and followed prospectively through 6 years of life (Castro et al. 2008). Of these children with severe RSV bronchiolitis during infancy, 48 % developed asthma and 48 % were diagnosed with eczema at age 6. Thirty-two percent had allergic sensitization, as defined by positive skin tests to allergens prevalent in the St. Louis region (Castro et al. 2008). Independent determinants that were significantly linked to physician-diagnosed asthma at the seventh birthday included the mother having asthma, exposure to high levels of dog allergen, positive aeroallergen skin tests at age 3, recurrent wheezingduring the first 3 years of life, and CCL5 (RANTES) expression in nasal epithelia during the acute RSV infection during infancy. Of this cohort, white children and children who attended day care were significantly less likely to have physician-diagnosed asthma at the seventh birthday (Bacharier et al. 2012). Interestingly, when peripheral blood was drawn from the children diagnosed with asthma at age 6, the isolated T cells had decreased IL-13 expression following phorbol 12-myristate 13-acetate (PMA) and ionomycin stimulation compared to those without asthma (Castro et al. 2008).

While severe RSV bronchiolitis has been linked to the development of asthma later in life, it is possible that the severe RSV infection is not causative of asthma, but instead a marker of the true risk factor for asthma. This possibility was investigated in the Copenhagen Prospective Studies on Asthma in Childhood ($COPSAC_{2000}$) cohort, a prospective clinical study of a birth cohort of 411 neonates born to mothers with a history of asthma (Chawes et al. 2012). Methacholine challenge and pulmonary function measurements were performed on 1-month-old infants using the raised volume rapid thoracoabdominal compression technique before these children had experienced a lower respiratory tract infection. The infants were then prospectively followed and closely monitored for bronchiolitis signs and symptoms. A total of 34 subjects (8.5 %) were diagnosed with acute severe bronchiolitis before 2 years of age. A total of 21 of these children (62 %) required hospitalization and 23 of those diagnosed with bronchiolitis had detectable RSV infection. Those children who subsequently experienced severe bronchiolitis had a 2.5-fold increased responsiveness to methacholine at 1 month of age compared to control subjects. There was no difference in pulmonary function at 1 month between those who developed bronchiolitis and those who did not (Chawes et al. 2012). This study strongly suggests that baseline bronchial responsiveness to methacholine is an important risk factor for severe bronchiolitis. This study has two significant implications. First, children who are at greatest risk for bronchiolitis can be identified for prophylactic treatment with RSV immune globulin. Second, it is highly likely that these children with increased bronchial responsiveness to methacholine are those that are more likely to develop asthma later in childhood because bronchial responsiveness has long been recognized as the prime factor that determines longitudinal loss of lung function in asthmatic adults (Peat et al. 1987; Rijcken et al. 1995; Van Schayck et al. 1991).

4 RSV as a Cause for Asthma Exacerbations

4.1 Identification of RSV During Asthma Exacerbations in People

Asthma is a disease of intermittent, reversible airway obstruction (Lemanske and Busse 2010). During exacerbations of asthma, patients experience dyspnea, chest tightness, and wheezing, often associated with cough. Approximately 90 % of children with asthma have allergic sensitization defined as positive skin tests to common aeroallergens. Roughly 75 % of adults with allergies are allergically sensitized. While exposure to an allergen to which one is sensitive may result in an allergic reaction resulting in increased inflammation and subsequent airflow limitation, in actuality, viral infections are the predominant cause of exacerbations of symptoms and decreased pulmonary function. Approximately 80–85 % of children and about 50 % of adults with asthma exacerbations have detectable virus by polymerase chain reaction (PCR) in their respiratory tract secretions (Atmar et al. 1998; Johnston et al. 1995). The virus most commonly identified in asthma exacerbations in both children and adults is rhinovirus; however, RSV has also been attributed to be a cause of acute worsening asthma symptoms. In British children aged 9–11, RSV was detected in 5 % of those subjects who were identified as having a viral infection during an asthma exacerbation (Johnston et al. 1995). In a study of 49 Australian adults who were enrolled within 4 h of an emergency department visit for an asthma exacerbation, 19 (39 %) had RSV detected by either PCR or direct fluorescent antigen (Wark et al. 2002). To our knowledge, there are no other studies which report the RSV detection rate in adults with acute asthma exacerbations; however, a 39 % detection rate seems unusually high given that only 8 % had detection of rhinovirus, which in most studies is by far the most common virus detected in adult asthmatics. In the Australian study, the subjects were enrolled from February 2 to December 1, which should have included the peak times of both RSV and rhinovirus infection in the Southern hemisphere. In any case, RSV has been implicated as a cause of acute asthma exacerbation in both children and adults.

4.2 Mouse Model of RSV-Induced Asthma Exacerbation

A combined mouse model of allergic airways inflammation and RSV infection has to discern the immunologic determinants that result in virally-induced airways responsiveness. In this model, allergic airways inflammation was induced by sensitization and subsequent sequential challenges with ovalbumin followed by RSV infection during the allergen challenges (Peebles et al. 1999). RSV was chosen over rhinovirus as mice lack the receptor for the rhinovirus group, major receptor group, most often associated with asthma exacerbations (Bartlett et al. 2008). RSV challenge during ongoing allergic airway inflammation resulted in airway responsiveness and increased airway mucus production, but a decrease in lung levels of the Th2 a cytokines IL-5 and IL-13 (Peebles et al. 2001b). In contrast to the decrease in Th2 cytokines, RSV infection increased lung protein expression of IL-17A, a cytokine that has been associated with airway epithelial mucus metaplasia and inflammation that is resistant to corticosteroid treatment (Hashimoto et al. 2004). While mucus metaplasia and airway mucus obstruction accounts for some of the airways responsiveness witnessed in this model, there is a strong component of smooth muscle constriction as well as evidenced by the inhibition of airways responsiveness with a Rho kinase inhibitor (Hashimoto et al. 2002). The timing of RSV infection relative to allergic sensitization and challenge is critical to the inflammatory and physiologic phenotypes that result from this combined model (Hashimoto et al. 2002; Peebles et al. 2001a). RSV infection prior to allergen sensitization and challenge reduced airways responsiveness compared to sensitization and challenge alone, while RSV infection either after sensitization and challenge, or during sensitization and challenge increased airways responsiveness. We are currently in the process of determining the role of IL-17A in RSV-induced airways responsiveness and epithelial cell mucus metaplasia. This model should provide greater insight into other mechanisms by which RSV may exacerbate asthma in humans with underling allergic inflammation.

5 Conclusion

In this chapter, we have explored the evidence that RSV causes different forms of RAD, both bronchiolitis and acute asthma exacerbations. However, there are still unanswered questions. First, confirmatory studies need to be performed to determine if increased airways responsiveness to methacholine is truly a risk factor for severe RSV bronchiolitis, and whether this increased methacholine responsiveness is also a risk factor for the later development of asthma, independent of severe RSV bronchiolitis. Second, once a vaccine against RSV is developed, it will be important to determine if it reduces the prevalence of asthma or wheezing at age 6, since severe RSV bronchiolitis in infancy has been linked to asthma inception. This endpoint may be more difficult to assess in regard to the efficacy of a vaccine which may be targeted at children great than 6 months of age, since infant birth approximately 4 months before the winter virus peak carried the highest risk of developing childhood asthma. Third, more work needs to be performed in the area of RSV-induced acute asthma exacerbations in adults and the elderly and whether specific treatment with cytokine antagonists or prevention with an RSV vaccine can be safely and efficaciously utilized to prevent the sequelae of the symptoms and pulmonary function abnormalities resulting from these exacerbations.

References

American Academy of Pediatrics Subcommittee on Diagnosis and Management of Bronchiolitis (2006) Diagnosis and management of bronchiolitis. Pediatrics 118:1774–1793

Anderson LJ, Hierholzer JC, Tsou C, Hendry RM, Fernie BF, Stone Y, McIntosh K (1985) Antigenic characterization of respiratory syncytial virus strains with monoclonal antibodies. J Infect Dis 151:626–633

Atmar RL, Guy E, Guntupalli KK, Zimmerman JL, Bandi VD, Baxter BD, Greenberg SB (1998) Respiratory tract viral infections in inner-city asthmatic adults. Arch Intern Med 158:2453–2459

Bacharier LB, Cohen R, Schweiger T, Yin-Declue H, Christie C, Zheng J, Schechtman KB, Strunk RC, Castro M (2012) Determinants of asthma after severe respiratory syncytial virus bronchiolitis. J Allergy Clin Immunol 130:91–100

Bartlett NW, Walton RP, Edwards MR, Aniscenko J, Caramori G, Zhu J, Glanville N, Choy KJ, Jourdan P, Burnet J, Tuthill TJ, Pedrick MS, Hurle MJ, Plumpton C, Sharp NA, Bussell JN, Swallow DM, Schwarze J, Guy B, Almond JW, Jeffery PK, Lloyd CM, Papi A, Killington RA, Rowlands DJ, Blair ED, Clarke NJ, Johnston SL (2008) Mouse models of rhinovirus-induced disease and exacerbation of allergic airway inflammation. Nat Med 14:199–204

Bennett BL, Garofalo RP, Cron SG, Hosakote YM, Atmar RL, Macias CG, Piedra PA (2007) Immunopathogenesis of respiratory syncytial virus bronchiolitis. J Infect Dis 195:1532–1540

Bradley JP, Bacharier LB, Bonfiglio J, Schechtman KB, Strunk R, Storch G, Castro M (2005) Severity of respiratory syncytial virus bronchiolitis is affected by cigarette smoke exposure and atopy. Pediatrics 115:e7–e14

Carroll KN, Wu P, Gebretsadik T, Griffin MR, Dupont WD, Mitchel EF, Hartert TV (2009a) Season of infant bronchiolitis and estimates of subsequent risk and burden of early childhood asthma. J Allergy Clin Immunol 123:964–966

Carroll KN, Wu P, Gebretsadik T, Griffin MR, Dupont WD, Mitchel EF, Hartert TV (2009b) The severity-dependent relationship of infant bronchiolitis on the risk and morbidity of early childhood asthma. J Allergy Clin Immunol 123:1055–1061

Castro M, Schweiger T, Yin-Declue H, Ramkumar TP, Christie C, Zheng J, Cohen R, Schechtman KB, Strunk R, Bacharier LB (2008) Cytokine response after severe respiratory syncytial virus bronchiolitis in early life. J Allergy Clin Immunol 122:726–733

Chawes BL, Poorisrisak P, Johnston SL, Bisgaard H (2012) Neonatal bronchial hyperresponsiveness precedes acute severe viral bronchiolitis in infants. J Allergy Clin Immunol 130:354–361

Cockcroft DW (2010) Direct challenge tests: airway hyperresponsiveness in asthma: its measurement and clinical significance. Chest 138:18S–24S

Glezen P, Denny FW (1973) Epidemiology of acute lower respiratory disease in children. N Engl J Med 288:498–505

Hashimoto K, Graham BS, Ho SB, Adler KB, Collins RD, Olson SJ, Zhou W, Suzutani T, Jones PW, Goleniewska K, O'Neal JF, Peebles RS Jr (2004) Respiratory syncytial virus in allergic lung inflammation increases Muc5ac and gob-5. Am J Respir Crit Care Med 170:306–312

Hashimoto K, Peebles RS Jr, Sheller JR, Jarzecka K, Furlong J, Mitchell DB, Hartert TV, Graham BS (2002) Suppression of airway hyperresponsiveness induced by ovalbumin sensitisation and RSV infection with Y-27632, a Rho kinase inhibitor. Thorax 57:524–527

Jackson DJ, Gangnon RE, Evans MD, Roberg KA, Anderson EL, Pappas TE, Printz MC, Lee WM, Shult PA, Reisdorf E, Carlson-Dakes KT, Salazar LP, DaSilva DF, Tisler CJ, Gern JE, Lemanske RF Jr (2008) Wheezing rhinovirus illnesses in early life predict asthma development in high-risk children. Am J Respir Crit Care Med 178:667–672

Jacoby DB, Costello RM, Fryer AD (2001) Eosinophil recruitment to the airway nerves. J Allergy Clin Immunol 107:211–218

Jacoby DB, Gleich GJ, Fryer AD (1993) Human eosinophil major basic protein is an endogenous allosteric antagonist at the inhibitory muscarinic M2 receptor. J Clin Invest 91:1314–1318

Johnson JE, Gonzales RA, Olson SJ, Wright PF, Graham BS (2007) The histopathology of fatal untreated human respiratory syncytial virus infection. Mod Pathol 20:108–119

Johnston SL, Pattemore PK, Sanderson G, Smith S, Lampe F, Josephs L, Symington P, O'Toole S, Myint SH, Tyrrell DA, Holgate ST (1995) Community study of role of viral infections in exacerbations of asthma in 9–11 year old children [see comments]. BMJ 310:1225–1229

Kim CK, Kim SW, Park CS, Kim BI, Kang H, Koh YY (2003) Bronchoalveolar lavage cytokine profiles in acute asthma and acute bronchiolitis. J Allergy Clin Immunol 112:64–71

Krawiec ME, Westcott JY, Chu HW, Balzar S, Trudeau JB, Schwartz LB, Wenzel SE (2001) Persistent wheezing in very young children is associated with lower respiratory inflammation. Am J Respir Crit Care Med 163:1338–1343

Lemanske RF Jr, Busse WW (2010) Asthma: clinical expression and molecular mechanisms. J Allergy Clin Immunol 125:S95–S102

Lukacs NW, Moore ML, Rudd BD, Berlin AA, Collins RD, Olson SJ, Ho SB, Peebles RS Jr (2006) Differential immune responses and pulmonary pathophysiology are induced by two different strains of respiratory syncytial virus. Am J Pathol 169:977–986

McNamara PS, Ritson P, Selby A, Hart CA, Smyth RL (2003) Bronchoalveolar lavage cellularity in infants with severe respiratory syncytial virus bronchiolitis. Arch Dis Child 88:922–926

Moore ML, Chi MH, Luongo C, Lukacs NW, Polosukhin VV, Huckabee MM, Newcomb DC, Buchholz UJ, Crowe JE Jr, Goleniewska K, Williams JV, Collins PL, Peebles RS Jr (2009) A chimeric A2 strain of respiratory syncytial virus (RSV) with the fusion protein of RSV strain line 19 exhibits enhanced viral load, mucus, and airway dysfunction. J Virol 83:4185–4194

NHLBI/WHO workshop report (1995) Global strategy for asthma management and prevention. National Institutes of Health

Papenburg J, Hamelin ME, Ouhoummane N, Carbonneau J, Ouakki M, Raymond F, Robitaille L, Corbeil J, Caouette G, Frenette L, De SG, Boivin G (2012) Comparison of risk factors for human metapneumovirus and respiratory syncytial virus disease severity in young children. J Infect Dis 206:178–189

Peat JK, Woolcock AJ, Cullen K (1987) Rate of decline of lung function in subjects with asthma. Eur J Respir Dis 70:171–179

Peebles RS Jr, Sheller JR, Johnson JE, Mitchell DB, Graham BS (1999) Respiratory syncytial virus infection prolongs methacholine-induced airway hyperresponsiveness in ovalbumin sensitized mice. J Med Virol 57:186–192

Peebles RS Jr, Hashimoto K, Collins RD, Jarzecka K, Furlong J, Mitchell DB, Sheller JR, Graham BS (2001a) Immune interaction between respiratory syncytial virus infection and allergen sensitization critically depends on timing of challenges. J Infect Dis 184:1374–1379

Peebles RS Jr, Sheller JR, Collins RD, Jarzecka AK, Mitchell DB, Parker RA, Graham BS (2001b) Respiratory syncytial virus infection does not increase allergen- induced type 2 cytokine production, yet increases airway hyperresponsiveness in mice. J Med Virol 63:178–188

Rijcken B, Schouten JP, Xu X, Rosner B, Weiss ST (1995) Airway hyperresponsiveness to histamine associated with accelerated decline in FEV1. Am J Respir Crit Care Med 151:1377–1382

Rossi GA, Medici MC, Arcangeletti MC, Lanari M, Merolla R, Paparatti UD, Silvestri M, Pistorio A, Chezzi C (2007) Risk factors for severe RSV-induced lower respiratory tract infection over four consecutive epidemics. Eur J Pediatr 166:1267–1272

Semple MG, Dankert HM, Ebrahimi B, Correia JB, Booth JA, Stewart JP, Smyth RL, Hart CA (2007) Severe respiratory syncytial virus bronchiolitis in infants is associated with reduced airway interferon gamma and substance P. PLoS ONE 2:e1038

Sigurs N, Aljassim F, Kjellman B, Robinson PD, Sigurbergsson F, Bjarnason R, Gustafsson PM (2010) Asthma and allergy patterns over 18 years after severe RSV bronchiolitis in the first year of life. Thorax

Sigurs N, Bjarnason R, Sigurbergsson F, Kjellman B, Bjorksten B (1995) Asthma and immunoglobulin E antibodies after respiratory syncytial virus bronchiolitis: a prospective cohort study with matched controls. Pediatrics 95:500–505

Stein RT, Sherrill D, Morgan WJ, Holberg CJ, Halonen M, Taussig LM, Wright AL, Martinez FD (1999) Respiratory syncytial virus in early life and risk of wheeze and allergy by age 13 years. Lancet 354:541–545

Stockman LJ, Curns AT, Anderson LJ, Fischer-Langley G (2012) Respiratory syncytial virus-associated hospitalizations among infants and young children in the United States, 1997–2006. Pediatr Infect Dis J 31:5–9

Stokes KL, Chi MH, Sakamoto K, Newcomb DC, Currier MG, Huckabee MM, Lee S, Goleniewska K, Pretto C, Williams JV, Hotard A, Sherrill TP, Peebles RS Jr, Moore ML (2011) Differential pathogenesis of respiratory syncytial virus clinical isolates in BALB/c mice. J Virol 85:5782–5793

Van Schayck CP, Dompeling E, Van Herwaarden CL, Wever AM, Van WC (1991) Interacting effects of atopy and bronchial hyperresponsiveness on the annual decline in lung function and the exacerbation rate in asthma. Am Rev Respir Dis 144:1297–1301

Wark PA, Johnston SL, Moric I, Simpson JL, Hensley MJ, Gibson PG (2002) Neutrophil degranulation and cell lysis is associated with clinical severity in virus-induced asthma. Eur Respir J 19:68–75

Wenzel SE (1997) Arachidonic acid metabolites: mediators of inflammation in asthma. Pharmacotherapy 17:3S–12S

Wills-Karp M, Luyimbazi J, Xu X, Schofield B, Neben TY, Karp CL, Donaldson DD (1998) Interleukin-13: central mediator of allergic asthma. Science 282:2258–2261

Wu P, Dupont WD, Griffin MR, Carroll KN, Mitchel EF, Gebretsadik T, Hartert TV (2008) Evidence of a causal role of winter virus infection during infancy in early childhood asthma. Am J Respir Crit Care Med 178:1123–1129

Part II
Host Response to Infection and Genetics of Disease

Human Genetics and Respiratory Syncytial Virus Disease: Current Findings and Future Approaches

Eun Hwa Choi, Hoan Jong Lee and Stephen J. Chanock

Abstract Infection with respiratory syncytial virus (RSV) can result in a wide spectrum of pulmonary manifestations, from mild upper respiratory symptoms to severe bronchiolitis and pneumonia. Although there are several known risk factors for severe RSV disease, namely, premature birth, chronic lung disease, congenital heart disease, and T cell immunodeficiency, the majority of young children who develop severe RSV disease are otherwise healthy children. Genetic susceptibility to RSV infection is emerging as a complex trait, in which many different host genetic variants contribute to risk for distinct disease manifestations. Initially, host genetic studies focused on severe RSV disease using the candidate gene approach to interrogate common single nucleotide polymorphisms (SNPs). Many studies have reported genetic associations between severe RSV bronchiolitis and SNPs in genes within plausible biological pathways, such as in innate host defense genes (*SPA*, *SPD*, *TLR4*, and *VDR*), cytokine or chemokine response genes (*CCR5*, *IFN*, *IL6*, *IL10*, *TGFB1*), and altered Th1/Th2 immune responses (*IL4*, *IL13*). Due to the complexity of RSV susceptibility, genome studies done on a larger scale, such as genome-wide association studies have certainly identified more of the host factors that contribute to the development of severe RSV bronchiolitis or excessive pathology. Furthermore, whole-genome approaches can reveal robust associations between genetic markers and RSV disease susceptibility. Recent introduction of 'exome' genotyping or sequencing, which specifically analyzes the majority of coding variants, should be fruitful in sufficiently large, well-powered studies. The advent of new genomic technologies together with improved computational tools offer the promise of interrogating the host genome in search of genetic factors, rare, uncommon, or common that should give new insights into the underlying

E. H. Choi · H. J. Lee
Department of Pediatrics, Seoul National University College of Medicine, 101 Daehak-ro, Jongno-gu, Seoul 110-769, Korea

S. J. Chanock (✉)
Division of Cancer Epidemiology and Genetics, National Cancer Institute, National Institutes of Health, Bethesda, MD 20892, USA
e-mail: chanocks@mail.nih.gov

L. J. Anderson and B. S. Graham (eds.), *Challenges and Opportunities for Respiratory Syncytial Virus Vaccines*, Current Topics in Microbiology and Immunology 372,
DOI: 10.1007/978-3-642-38919-1_6,

biology of susceptibility to or protection from severe RSV infection. Careful assessment of novel pathways and further identification of specific genes could identify new approaches for vaccine development and perhaps lead to effective risk modeling.

Contents

1 Introduction

Respiratory syncytial virus (RSV) is a major respiratory virus in infants and young children worldwide. The spectrum of RSV infection ranges from mild upper respiratory symptoms to severe bronchiolitis and pneumonia. Risk factors that predispose infants and young children to severe RSV disease include premature birth, chronic lung disease, congenital heart disease, and T cell immunodeficiency. However, the majority of young children who develop severe RSV disease are healthy and have no identifiable risk factors (Hall et al. 2009). Since disease manifestations of RSV infection in infants vary substantially, it is likely that both host and viral factors contribute to RSV disease severity. The association of RSV disease susceptibility with family history of asthma and ethnicity, race, and gender support a genetic predisposition to severe RSV disease (Hull 2007) and host genetic variation playing a substantial role in severity of RSV disease.

Over the last century, investigators implicated human genetic variation as a major determinant of susceptibility to many infectious pathogens, beginning with twin and HLA studies (Hill 2012). Although the variability in disease severity of RSV remains inadequately explained, evidence exists that variations within genes involved in immune responses have been associated with the outcome of RSV infection, presumably due to alterations in the quality and strength of the immune response. The recent genomic revolution now provides the opportunity to identify genetic contributions to a complex disease, such as RSV infection.

In this article, we review the candidate gene studies focused on variability associated with RSV disease severity, specifically looking at pathogenesis, pilot studies examining common genetic polymorphisms, genetic contribution to critical pathogenesis or novel pathways, susceptibility, and outcome of the disease. We also summarize emerging issues in genome-wide approaches and highlight future implications of human genomics for the development of an effective and safe RSV vaccine.

2 Epidemiological Evidence for a Genetic Component of RSV Susceptibility

By the age of 2 years, nearly all children are infected with RSV at least once. Approximately 20 % of children infected with RSV develop lower respiratory tract infections (LRTIs) during their initial infection, and severe disease that requires hospitalization may develop in ~1 % of all infected infants under one year of age (Collins et al. 2001). In the decade, 1997–2006 in the United States, RSV-associated hospitalization was responsible for 24 % of an estimated 5.5 million LRTI hospitalizations among children <5 years of age. An estimated 132,000–172,000 RSV-associated hospitalizations occurred annually in children <5 years of age. The hospitalization rate was highest among infants <3 months old (48.9 per 1,000), followed by infants 3–5 months old (28.4 per 1,000), and lower among those >1 year old (1.8 per 1,000) (Stockman et al. 2012).

In a prospective, population-level hospital-based surveillance study, which was implemented during three RSV seasons (1997–2000) in the US, the overall seasonal RSV hospitalization rate among children younger than 2 years was 63.6 per 1,000 and 91.3 per 1,000 among children younger than 1 year. The Navajo and White Mountain Apache children were at a high risk for severe RSV disease that required hospitalization; the rates of RSV hospitalization were three times that of the general US population (Bockova et al. 2002). Similarly, higher hospitalization rates associated with bronchiolitis have been reported in American Indian and Alaska Native children (Lowther et al. 2000). Although these populations share several similar socio-environmental factors that may predispose them to severe RSV bronchiolitis, genetic factors, some general across populations, and some specific to populations with distinct histories likely contribute to the differences in the incidence of high risk RSV infection.

Studies of susceptibility to infections in identical and nonidentical twins have been instrumental in untangling the relative contributions of genetics and environmental factors. To determine differences in the severity of RSV infection contributed by genetic and environmental factors, a nationwide study analyzed linkage data on RSV hospitalizations for all twin pairs born in Denmark during a

10-year period (Thomsen et al. 2008). In this study, there were significantly higher concordance rates in the susceptibility to severe RSV infection in identical twin pairs, when compared with fraternal twin pairs (concordance rate 0.66 vs. 0.53, $P = 0.02$), which supports a role for genetic factors in risk for severe RSV disease. Approximately 20 % of the propensity to develop severe RSV infection was attributable to genetic differences in the twin study, whereas household, or shared environmental factors contributes nearly three-quarters of the risk for severe RSV infection.

3 Genetic Susceptibility in Animal Study

Animal models of human diseases have been used to unravel the underlying genetic architecture of infectious diseases. However, these approaches have important limitations, particularly for the identification of a novel class of host genes that confer resistance to RSV in humans. For instance, murine models of RSV infection are challenging to interpret because RSV is not normally pathogenic to mice. Despite the limitation, it has been shown that the susceptibility to RSV infection in mice can be genetically determined and more importantly, specific insights have been observed in mouse models (Isaacs et al. 1989; Stark et al. 2002; Jessen et al. 2011).

First, inbred mouse strains have been used to identify genetic factors that influence susceptibility to RSV infection (Stark et al. 2002; Jessen et al. 2011). In one study, disease susceptibility has been analyzed by comparing different virus titers and RSV-induced weight loss after RSV infection in 8 strains of mice. The 'resistant' C57BL/6 mice showed low viral titers and no disease, whereas the 'susceptible' AKR/J, 129P3, or BALB/c mice showed high virus replication and more prominent lung pathology, which implies that susceptibility to RSV infection is influenced by background genetic factors (Stark et al. 2002). In addition, this study of inbred mice suggested that susceptibility to RSV is likely to be a multigenic trait because the phenotypic distribution of the backcross progeny did not follow the predicted distribution.

Cytotoxic T lymphocyte (CTL) response has been identified as an important mechanism in virus control and disease severity. Both CD4 and CD8 T cells can eliminate virus and cause immunopathology independently. A recent study revealed that the Major Histocompatibility Complex (*MHC*) haplotype is an important factor of RSV susceptibility using *MHC*-congenic mice (Jessen et al. 2011). This study observed weight loss and pulmonary inflammatory response were greater in BALB/c (H-2^d) mice than in C57BL/6 (H-2^b) mice following RSV infection. The H-2^d allele can partially transfer disease susceptibility to C57BL/6 mice. The *MHC*–determined composition and efficacy of CD4 and CD8 T cell responses contributed to differences in disease susceptibility.

The contribution of the *MHC* locus has been difficult to demonstrate in RSV infections in humans, partly because expression of *MHC* class I alleles in human is extremely complex, whereas single homozygous *MHC* alleles are present on a constant non-*MHC* background in mice. CTL recognize virus-infected cells that are associated with the Human Leukocyte Antigen (*HLA*) class I antigens in humans. Although there is strong evidence for the importance of CTL in the occurrence of severe RSV infections, association between particular *HLA* class I antigens and severe RSV bronchiolitis has not yet been demonstrated in humans (Isaacs et al. 1989). Large population-based studies are needed to determine the role of *HLA* alleles for severity of RSV in children.

Important observations regarding host genetic factors on the development of RSV vaccine include sensitization with the attachment protein (G) of RSV, which leads to CD4 T cell-mediated lung eosinophilia after challenge with RSV in mice (Hussell et al. 1998). Nonpathogenic vaccines for RSV disease that are also protective against RSV infection should induce a strong CD8 T cell response. Although there may be a possible influence of non-*MHC* genes on the immune response, a mouse model demonstrated that *MHC* is the primary determinant of robust T cell responses and eosinophilia during subsequent RSV challenge.

4 Study Design to Detect Genetic Susceptibility to Severe RSV Infection

Studies of the host genetic determinants of severe RSV infection may lead to an understanding of the interaction of RSV and the individual host as well as identification of novel disease pathways to help development of specific therapeutic interventions and prevention measures. In order to identify the genetic contribution to the disease phenotype, there have been two main approaches to determine genetic associations in complex disease: association and linkage studies (Fig. 1).

Association studies initially searched for a relationship between one or more known polymorphisms within a "candidate gene" and the disease phenotype. The overall success of the candidate gene approach has been modest, partly due to inadequate sample sizes tested to find effects with small estimated sizes (odds ratio <1.5) and partly due to our limited understanding of the biology of complex diseases, such as severe RSV infection. However, in HIV infection, there have been a number of important candidate gene findings, mainly related to cytokine pathways (O'Brien and Nelson 2004). The approach has been to look at the most frequent variant, the SNP but also common insertion/deletions, such as *CCR5*. Association studies of SNPs or haplotypes are frequently used to investigate infectious diseases between cases and controls, or within families. On the other hand, genome-wide linkage studies, which identify nonrandom segregation of microsatellite markers from parent to child, can detect only large genetic effects, usually in family studies.

As a natural extension of the Human Genome Project (Venter et al. 2001; Lander et al. 2001), there has been an international effort to map common variants

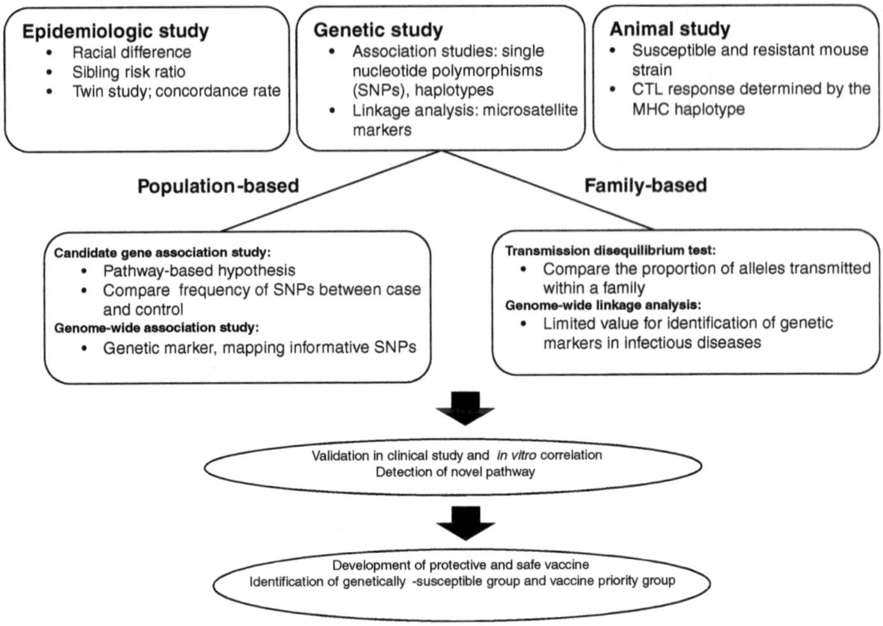

Fig. 1 Detection strategy for searching of genetic component for severe RSV disease

in the International HapMap Project (International HapMap Consortium et al. 2007) and less common variants in the 1,000 Genome Project (1,000 Genomes Project Consortium et al. 2010); the former has cataloged variants with minor allele frequencies (MAF) >5 % and the latter has annotated the genetic space between 0.5 and 5 % MAFs.

An extension of the annotation of variation across the genome has been the emergence of the genome-wide association studies (GWAS) as a powerful tool to scan more than 500,000 SNP genetic markers in thousands of individuals. It is predicated on the principle that it is possible to find highly correlated surrogates for the functional variants, due to linkage disequilibrium . This approach has been successful in the discovery of new regions associated with hundreds of diseases and traits (Hindorff et al. 2009). The GWAS approach is agnostic, namely it is statistically driven and can identify regions of interest in an unbiased manner. GWAS have effectively discovered variants that have a minor allele frequency of 5 % or greater in the test population, thus beginning to unravel the genetic architecture of complex diseases and traits (Park et al. 2010, 2011). Previous studies have demonstrated that GWAS can provide insights into crucial host–pathogen interaction in tuberculosis, malaria, and meningococcal disease (Khor and Hibberd 2012). Most notably, GWAS have identified a region on chromosome 19 that is important for both infections with hepatitis C and treatment response to interferon-alpha (Thomas et al. 2009a; Ge et al. 2009; Suppiah et al. 2009). It is notable that GWAS identify regions that require extensive follow-up work, beginning with fine mapping of the genetic region. This is required to nominate the optimal variants for laboratory studies, designed to illuminate the underlying functional basis of the direct association with susceptibility to a disease, such as severe RSV infection.

5 Preliminary Studies of Host Genetic Polymorphism Associated with Severe RSV Disease

The current understanding of genetic polymorphisms associated with increased (or decreased) risk of severe RSV disease illustrates that numerous host factors contribute to the variability in host susceptibility and pathogenesis (Table 1). There are two competing hypotheses on the pathogenesis of the variability of RSV disease, one of which is based on genes responsible for the dynamic interaction between the host and pathogen, whereas the other hypothesis focuses more on host immune response to the pathogen.

5.1 Host Genetics on Innate Defense

One of the most striking features of RSV disease is that primary RSV infection occurs before the development of specific acquired immunity and in the presence of maternal antibody. The early innate host response to primary infection is important. Thus, the first hypothesis emphasizes the primary host–pathogen interaction, which focuses on the innate defense system of the lung. This may play a particularly important role during infection in young infants before they acquire adequate immunity (Miyairi and DeVincenzo 2008).

Among the surfactant proteins, hydrophilic surfactant proteins A1, A2, and D (*SFPA1*, *SFPA2*, and *SFPD*) are considered to play a major role in the pulmonary defense mechanism, such as in the clearance of RSV infiltration from host pulmonary cells (Ghildyal et al. 1999; Hall et al. 2009; LeVine and Whitsett 2001). In a case–control study, genotype frequency of the surfactant protein A polymorphism was significantly different between severe RSV disease and control, which indicates that there is an association between severe RSV disease and surfactant protein A gene locus (Lofgren et al. 2002; Thomas et al. 2009b; El Saleeby et al. 2009). In comparison, the mannose-binding lectin (*MBL2*), which is a plasma protein that binds to pathogens in order to facilitate the immune destruction pathway, is postulated to limit the disease severity of RSV infection (Ribeiro et al. 2008; Nielsen et al. 2003). Yet, in contrast to other viral or bacterial pathogens, association of the *MBL2* polymorphism and RSV appears to be less conclusive, and further studies are needed (Kristensen et al. 2004).

Table 1 Association studies of host genetic polymorphism and candidate genes with severe respiratory syncytial virus disease

Study design	Pathway	Gene	Variants	Reference, country
Candidate gene association study	Innate immunity pulmonary defense	*SPA*	*SPA1*; Val19Ala, *SPA2*; Val91Pro, Gln223Lys	Lofgren et al. (2002), Finland
		SPD	Met11Thr	Lahti et al. (2002), Finland
	Innate immunity pattern recognition	*TLR4*	Asp299Gly, Thr399Ile	Tal et al. (2004), Israel
		CD14	C-550T	Inoue et al. (2007), Japan
	Innate immunity signal production	*VDR*	Thr1Met	Kresfelder et al. (2011), South Africa
	Cytokine Th2 pathway	*IL4*	T-589C, T-33C G8375A, A8412C	Choi et al. (2002), Korea
		IL4R	Q551R	Hoebee et al. (2003), The Netherlands
		IL13	C-1112T	Puthothu et al. (2006), Germany
		IL1RL1	Intron rs1921622	Faber et al. (2012), The Netherlands
	Cytokine proinflammatory	*IL18*	G-133C	Puthothu et al. (2007) Germany
		IL6	G-174C	Gentile et al. (2003), US
		IFN	T874A	
	Cytokine antiinflammatory	*IL10 family*	*IL10* Promoter variant or haplotype	Gentile et al. (2003), US Hoebee et al. (2004), The Netherlands
			IL19/IL20 haplotype	Ermers et al. (2011), The Netherlands
		TGFB1	Haplotype, codons 10, and 25	Gentile et al., (2003), US
Genome-wide association study	Innate immunity signal production	*VDR*	Thr1Met	Janssen et al. (2007), The Netherlands
		JUN	G750A	
	Innate immunity proinflammatory	*IFNA5*	C453T	
		NOS2A	G2757A	
	Allergic asthma	*FCER1A*	T-66C	
Transmission disequilibrium test	Chemotaxis Chemokine	*IL8*	T-251A	Hull et al. (2000), UK
Transmission disequilibrium test and candidate gene association study	Chemotaxis Chemokine receptor	*CCR5*	-2459G, -2554T	Hull et al. (2003), UK

Toll-like receptor 4 (*TLR4*) plays a crucial role in regulating innate and adaptive immune responses by recognizing pathogen-associated molecular patterns. Interaction between *TLR4* and the RSV F protein is shown to mediate cytokine production during innate immune response. *TLR4* mutations, Asp299Gly and Thr399Ile, either alone or in combination, are overrepresented in hospitalized infants for RSV bronchiolitis compared to infants who have mild RSV bronchiolitis. Functional significance of these *TLR4* SNPs has shown that human bronchial epithelial cells transfected with 299Gly and 399Ile alleles fail to efficiently translocate TLR4 to the cell surface (Tulic et al. 2007). Peripheral blood mononuclear cells from children expressing exonic *TLR4* variants demonstrated blunted responses to RSV, thus emphasizing the likelihood that this polymorphism is relevant to RSV pathogenesis. One of the promoter polymorphisms of *CD14*, another pattern recognition receptor, has also been found to be associated with the risk for hospitalized RSV bronchiolitis (Inoue et al. 2007).

In a recent study of prospective birth cohort from Netherlands, low levels of cord blood 25-hydroxy vitamin D were associated with increased risk of RSV LRTI in the first year of life (Belderbos et al. 2011). Vitamin D has important immunomodulatory effects; therefore varying levels of vitamin D receptor *(VDR*) may influence the mechanism of VDR in immune-mediated diseases such as RSV bronchiolitis. Allele frequency of the SNP in the first codon of the *VDR* gene, which may have resulted in changed transcriptional activity of the *VDR* gene, was associated with susceptibility to RSV infection (Kresfelder et al. 2011).

5.2 Th1/Th2 Cytokine Pathways

A tipping of the balance between Th1 and Th2 response has been postulated to cause disease variability among individuals, such as mild upper respiratory disease or severe RSV bronchiolitis. Measurement of the IL-4/IFN-γ and IL-10/IL-12 ratios revealed a higher IL-10/IL-12 ratio in those with acute bronchiolitis (Legg et al. 2003). In addition, reduced IL-18 mRNA levels, which elevated the IL-4/IFN-γ ratio, were demonstrated in stimulated peripheral blood mononuclear cells from infants with acute bronchiolitis in this study. This suggests the importance of excess type 2 and/or deficient type 1 immune responses in the pathogenesis of RSV bronchiolitis, which may be due to inter-individual genetic differences that are associated with variation in cytokine response.

Many genetic association studies have sought associations between SNPs in Th1 or Th2 cytokine genes and RSV disease severity. Common candidate variants in three Th2 cytokine genes, *IL4*, *IL13*, and *IL5,* which are clustered on chromosome 5q31.1 were investigated in a Korean cohort of hospitalized RSV bronchiolitis (Choi et al. 2002). In that study, a common *IL4* haplotype, which includes the -589T promoter variant, previously shown to be associated with increased IL-4 transcriptional activity and predisposition to asthma, was overrepresented in patients with severe RSV disease. A subsequent study has found an association of

IL4R alpha chain polymorphism (Q551R) with severe RSV disease in infants (Hoebee et al. 2003). The R551 allele, which was overrepresented in infants with RSV bronchiolitis requiring hospitalization, was associated with functionally enhanced signaling of IL-4 receptor. Studies illustrating genetic associations of *IL4* and *IL4R* polymorphisms with RSV disease severity, highly suggest a potential contribution of Th2 cytokine genes to RSV disease severity. Thus, in conjunction with the previous argument of asthma being associated with an increased level of IgE, supports the hypothesis that the similar genetic susceptibility factors for hyper-reactive airway disease may also play a key role in initiating severe RSV disease (Humbert et al. 1997; Webb et al. 2000).

IL-10 is a central anti-inflammatory cytokine that is known to promote Th2 and suppress Th1 immune responses. Contribution of promoter variants in the *IL10* gene to higher risk of hospitalization for RSV bronchiolitis has been shown by both a transmission/disequilibrium test (Hoebee et al. 2004) and two case–control approaches (Gentile et al. 2003; Wilson et al. 2005). Moreover, *IL10* family member genes, *IL19* and *IL20*, are clustered together with *IL10* on chromosome 1q31-32 and have similar protein structures. Genetic variations in the *IL19* and *IL20* genes have been evaluated in a large cohort of hospitalized infants with RSV LRTIs that was prospectively followed to evaluate the occurrence of recurrent wheeze (Ermers et al. 2011). A specific haplotype of the combined *IL19/IL20* genotypes demonstrated an association with the development of recurrent wheeze after RSV LRTIs. In another context related to gender difference in disease susceptibility, a polymorphism in *IL9* was found to have an opposite effect in males and females, an intriguing observation that needs to be confirmed (Schuurhof et al. 2010).

5.3 Other Cytokine- and Chemokine-Related Genes

RSV-infected airway epithelial cells secrete high levels of neutrophil-recruiting chemoattractant, IL-8, and other proinflammatory cytokines. The association of *IL8* promoter allele with RSV disease was analyzed by the transmission disequilibrium test in UK families (Hull et al. 2000). This promoter allele has been shown to be associated with increased levels of IL-8 production by LPS-stimulated whole blood cells and a significant increase in IL-8 transcript levels in human respiratory epithelial A549 cells.

It is known that the binding of the RSV G protein to leukocytes involves the host CX3C receptor (*CX3CR*), a leukocyte receptor for the chemokine fractalkine. The RSV G glycoprotein may compete with fractalkine for binding to CX3CR. Polymorphisms in the *CX3CR* gene, Val249Ile, and Thr280Met that disrupt fractalkine's affinity to its CX3CR have been identified. Of these, the 280Met-containing genotypes were overrepresented in children hospitalized with RSV LRTIs compared to sex-matched healthy adults without a history of LRTIs (Amanatidou et al. 2006). This finding demonstrates an association between the common variation (Thr280Met) of the *CX3CR* gene and an increased risk of severe RSV bronchiolitis.

6 Human Genetic Susceptibility and Vaccine Research

6.1 Association of HLA Allele with Vaccine Response

Genetic variation can have an important influence on disease susceptibility, progression, and resolution, as well as on vaccine-induced responses. However, the mechanism involved in vaccine responses will vary by pathogen and host immune response pathways. A number of studies have focused on the contribution of polymorphic alleles in the *HLA* system that could significantly influence immune responses to viral vaccines. There is an association of the *HLA*-DRB1*13 allele with the measles vaccine response (Hayney et al. 1996) and the *HLA*-DRB1*0701 allele with a poor neutralizing antibody response to the influenza vaccine (Lambkin et al. 2004). A study of influenza vaccine has found associations between H1- and H3-specific hemagglutination inhibition antibody titers and *HLA* gene polymorphisms or polymorphisms of cytokine and cytokine receptor genes (Poland et al. 2008). This suggests that the SNPs present in *HLA*, cytokine, and cytokine receptor genes may influence humoral responses to seasonal influenza vaccination.

It has been demonstrated that the level of anti-HBs antibody varies widely in different inbred strains of mice (Milich 1991). The response was determined to be a dominant trait related to genes within the class II region of the *MHC*. In familial studies in humans, the HBs-specific antibody response was dominantly inherited and closely linked to the *MHC* (Salazar et al. 1995). As in the mice, particular haplotypes (*HLA*-B8, SC01, DR3 and *HLA*-B44, FC31, DR7) were overrepresented among those who did not respond to the hepatitis B vaccine. When homozygotes and heterozygotes for the *HLA*-B8, SC01, and DR3 haplotypes were prospectively immunized with HBsAg, the antibody response was significantly lower in the homozygotes than in the heterozygotes (Alper et al. 1989). However, homozygotes do respond occasionally, which suggests that non-*MHC* genes could also be involved.

The GWAS approach has been employed to look for markers of genes which could map to known as well as novel immune responses to a single particular vaccine. A pilot GWAS has been recently reported to look at the cytokine response, characteristic of a major toxicity of the smallpox vaccine (Kennedy et al. 2012). In this study of approximately 1,000 well-characterized, healthy adult recipients of the smallpox vaccine, multiple SNPs were associated with variations in cytokine response. These SNPs were found in genes with known immune function, as well as in genes encoding for proteins involved in signal transduction, cytoskeleton, membrane channels and ion transport, and in others with no previously identified role in immune responses. The observation of a large number of significant SNP associations implies that cytokine secretion in response to vaccinia virus is a complex process that is controlled by multiple genes in multiple gene families.

6.2 Genetic Variability in Immune Response to RSV Vaccine

Development of an effective and safe vaccine to prevent RSV infection has been hampered by the fear of causing vaccine-enhanced disease. The formalin-inactivated RSV vaccine (FI-RSV), which was developed in the 1960s, had led to an increased risk of severe forms of RSV disease among the vaccine recipients (Kapikian et al. 1969). In the immunized mouse model, prominent Th2 response has been documented to enhance RSV pulmonary disease after the virus challenge (Connors et al. 1994). Moreover, the mice that were previously immunized with the G glycoprotein of RSV were observed to develop markedly enhanced RSV disease, which is characterized by pulmonary eosinophilia, when challenged with live RSV, probably mediated by the CD4 T cells that were G protein-specific and had characteristics of Th2 cells. Following immunization with the F protein from the formalin-inactivated RSV, a similar enhancement in disease severity with pathognomonic changes and bias towards a Th2 response were also observed (Graham et al. 1999).

It has been suggested that genetic influence on Th2 response for the vaccine-enhanced severe RSV disease has important implications for designing RSV vaccines. Hussell et al. demonstrated that different effect of G protein vaccination on pulmonary eosinophilia in mice with different *MHC* haplotype (Hussell et al. 1998). This study showed that *MHC* is the primary determinant of extensive T cell responses and eosinophilia during RSV challenge in G protein-primed mice. Despite the contribution of non-*MHC* genes in different mouse strains, this result may explain the variability in response to vaccination and RSV infection in humans. Thus, better understanding of genetic variability in immune responses to RSV vaccines and continuing analysis of human genetic susceptibility of the RSV disease severity will provide further instructive knowledge on the development of safe and effective RSV vaccines.

7 Future Aspects

The pursuit of dissecting the genetic basis of susceptibility and resistance to RSV infection, especially in the young or immunocompromised host will move forward when the new genomic technologies can be applied to large case–control and population-based studies, including maternal-infant cohorts. The ability to apply GWAS towards RSV-related questions will require large scale collaborations across centers and studies to accumulate the sample sizes needed to conclusively discover regions in the genome- worthy of further pursuit (Chanock et al. 2007).

It is also important to continue to conduct studies to validate the candidate gene approach results. With new genes being associated with RSV disease, there is a need to understand mechanisms and interactions related to this association. Thus, future studies should also focus on investigating physiological mechanism to explain functional correlation in animal models, in vitro laboratory studies, or clinical studies. These studies need to account for human gene–gene interactions and gene-environment interactions.

It is also possible that genetic variants associated with disease are only markers for a disease-modifying locus. It is likely that GWAS will uncover other cryptic genetic factors associated with RSV disease and 'exome-wide' genotyping identify additional genetic factors. The majority of nonsynonymous coding genetic variants detected by exome-wide re-sequencing add another level of detail which might be missed by standard GWAS analysis misses. GWAS only analyzes common and noncoding genetic variants. Lastly, future studies on the mechanisms by which genetic polymorphisms and/or epigenetic modifications regulate gene expression critical to RSV disease pathogenesis that influence immune responses to vaccine antigens will be important.

Recently, a relatively new approach has emerged, which is known as 'personalized vaccinology' and is related to precision medicine, namely therapy guided by genomic characterization of either a disease or cancer (Lahti et al. 2002). To date, the major efforts in vaccinology have focused at the population level, but with personal genomes not far in the future, it is plausible to consider how and in what way individual variability in immune response could influence decisions related to a range of vaccinations (Poland et al. 2008). However, as previously discussed, many studies have identified associations between immune response gene polymorphisms and variations in immune response to several vaccines in genetically heterogeneous populations. Recently, a new paradigm of 'individual-level' vaccinology, which is also called 'vaccinomics' , emphasizes the personalized approach. Though the potential benefits from vaccinomics, e.g., personalized vaccines, novel vaccines, and novel vaccine adjuvants, much research is needed before it can be used to guide use of vaccines in humans.

8 Conclusion

General conclusions that can be drawn from observations on host genetic determinants of severe RSV diseases are important confirmations of the link between inter-individual differences and population-based assessments of major viral pathogens, such as RSV. The challenge that we are now confronted with is to determine the role of genetic variations in the severity of RSV disease in a more systematic way, by employing more advanced genetic concepts. Thus, the progress in preventing severe RSV infection should be focused on vaccine development, while the new genomic approaches may uncover novel biological insights that can guide novel vaccine strategies. Future genetic studies on RSV disease should aim at identifying the genetic contributions to severity of RSV infection (including common and rare variants) and the follow-up studies to understand the mechanisms or underlying functional features of these associations.

References

1000 Genomes Project Consortium (2010) A map of human genome variation from population-scale sequencing. Nature 467:1061–1073

Alper CA, Kruskall MS, Marcus-Bagley D, Craven DE, Katz AJ, Brink SJ, Dienstag JL, Awdeh Z, Yunis EJ (1989) Genetic prediction of nonresponse to hepatitis B vaccine. N Engl J Med 321:708–712

Amanatidou V, Sourvinos G, Apostolakis S, Tsilimigaki A, Spandidos DA (2006) T280 M variation of the CX3C receptor gene is associated with increased risk for severe respiratory syncytial virus bronchiolitis. Pediatr Infect Dis J 25:410–414

Belderbos ME, Houben ML, Wilbrink B, Lentjes E, Bloemen EM, Kimpen JL, Rovers M, Bont L (2011) Cord blood vitamin D deficiency is associated with respiratory syncytial virus bronchiolitis. Pediatrics 127:e1513–e1520

Bockova J, O'Brien KL, Jane Oski J, Croll J, Reid R, Weatherholtz RC, Santosham M, Karron RA (2002) Respiratory syncytial virus infection in Navajo and White Mountain Apache children. Pediatrics 110:e20–e22

Choi EH, Lee HJ, Yoo T, Chanock SJ (2002) A common haplotype of interleukin-4 gene *IL4* is associated with severe respiratory syncytial virus disease in Korean children. J Infect Dis 186:1207–1211

Collins PL, Chanock RM, Murphy BR (2001) Respiratory syncytial virus. In: Knipe DM, Howley PM (eds) Fields virology, 4th edn. Lippincott Williams & Wilkins, Philadelphia, pp 1443–1486

Connors M, Giese NA, Kulkarni AB, Firestone CY, Morse HC 3rd, Murphy BR (1994) Enhanced pulmonary histopathology induced by respiratory syncytial virus (RSV) challenge of formalin-inactivated RSV-immunized BALB/c mice is abrogated by depletion of interleukin-4 (IL-4) and IL-10. J Virol 68:5321–5325

El Saleeby CM, Li R, Somes GW, Dahmer MK, Quasney MW, DeVincenzo JP (2009) Surfactant protein A2 polymorphisms and disease severity in a respiratory syncytial virus-infected population. J Pediatr 156:409–414

Ermers MJ, Janssen R, Onland-Moret NC, Hodemaekers HM, Rovers MM, Houben ML, Kimpen JL, Bont LJ (2011) IL10 family member genes IL19 and IL20 are associated with recurrent wheeze after respiratory syncytial virus bronchiolitis. Pediatr Res 70:518–523

Faber TE, Schuurhof A, Vonk A, Koppelman GH, Hennus MP, Kimpen JL, Janssen R, Bont LJ (2012) IL1RL1 gene variants and nasopharyngeal IL1RL-a levels are associated with severe RSV bronchiolitis: a multicenter cohort study. PLoS ONE 7:e34364

Ge D, Fellay J, Thompson AJ, Simon JS, Shianna KV, Urban TJ, Heinzen EL, Qiu P, Bertelsen AH, Muir AJ, Sulkowski M, McHutchison JG, Goldstein DB (2009) Genetic variation in IL28B predicts hepatitis C treatment-induced viral clearance. Nature 461:399–401

Gentile DA, Doyle WJ, Zeevi A, Howe-Adams J, Kapadia S, Trecki J, Skoner DP (2003) Cytokine gene polymorphisms moderate illness severity in infants with respiratory syncytial virus infection. Hum Immunol 64:338–344

Ghildyal R, Hartley C, Varrasso A, Meanger J, Voelker DR, Anders EM, Mills J (1999) Surfactant protein A binds to the fusion glycoprotein of respiratory syncytial virus and neutralizes virion infectivity. J Infect Dis 180:2009–2013

Graham BS, Henderson GS, Tang YW, Lu X, Neuzil KM, Colley DG (1999) Priming immunization determines T helper cytokine mRNA expression patterns in lungs of mice challenged with respiratory syncytial virus. J Immunol 151:2032–2040

Hall CB, Weinberg GA, Iwane MK, Blumkin AK, Edwards KM, Staat MA, Auinger P, Griffin MR, Poehling KA, Erdman D, Grijalva CG, Zhu Y, Szilagyi P (2009) The burden of respiratory syncytial virus infection in young children. N Engl J Med 360:588–598

Hayney MS, Poland GA, Jacobson RM, Schaid DJ, Lipsky JJ (1996) The influence of the *HLA*-DRB1*13 allele on measles vaccine response. J Investig Med 44:261–263

Hill AV (2012) Evolution, revolution and heresy in the genetics of infectious disease susceptibility. Phil Trans R Soc B 367:840–849

Hindorff LA, Sethupathy P, Junkins HA, Ramos EM, Mehta JP, Collins FS, Manolio TA (2009) Potential etiologic and functional implications of genome-wide association loci for human diseases and traits. Proc Natl Acad Sci U S A 106:9362–9367

Hoebee B, Rietveld E, Bont L, Oosten M, Hodemaekers HM, Nagelkerke NJ, Neijens HJ, Kimpen JL, Kimman TG (2003) Association of severe respiratory syncytial virus bronchiolitis with interleukin-4 and interleukin-4 receptor alpha polymorphisms. J Infect Dis 187:2–11

Hoebee B, Bont L, Rietveld E, van Oosten M, Hodemaekers HM, Nagelkerke NJ, Neijens HJ, Kimpen JL, Kimman TG (2004) Influence of promoter variants of interleukin-10, interleukin-9, and tumor necrosis factor-alpha genes on respiratory syncytial virus bronchiolitis. J Infect Dis 189:239–247

Hull J (2007) Genetic susceptibility to RSV disease. In: Cane PA (ed) Respiratory syncytial virus, vol 14. Elsevier, Amsterdam, pp 115–140

Hull J, Thomson A, Kwiatkowski D (2000) Association of respiratory syncytial virus bronchiolitis with the interleukin 8 gene region in UK families. Thorax 55:1023–1027

Hull J, Rowlands K, Lockhart E, Moore C, Sharland M, Kwiatkowski D (2003) Variants of the chemokine receptor CCR5 are associated with severe bronchiolitis caused by respiratory syncytial virus. J Infect Dis 188:904–907

Humbert M, Corrigan CJ, Kimmitt P, Till SJ, Kay AB, Durham SR (1997) Relationship between IL-4 and IL-5 mRNA expression and disease severity in atopic asthma. Am J Respir Crit Care Med 156:704–708

Hussell T, Georgiou A, Sparer TE, Matthews S, Pala P, Openshaw PJ (1998) Host genetic determinants of vaccine-induced eosinophilia during respiratory syncytial virus infection. J Immunol 161:6215–6222

Inoue Y, Shimojo N, Suzuki Y, Campos Alberto EJ, Yamaide A, Suzuki S, Arima T, Matsuura T, Tomiita M, Aoyagi M, Hoshioka A, Honda A, Hata A, Kohno Y (2007) CD14-550 C/T, which is related to the serum level of soluble CD14, is associated with the development of respiratory syncytial virus bronchiolitis in the Japanese population. J Infect Dis 195:1618–1624

International HapMap Consortium, Frazer KA, Ballinger DG, Cox DR, Hinds DA, Stuve LL et al (2007) A second generation human haplotype map of over 3.1 million SNPs. Nature 449:851–861

Isaacs D, Taylor CJ, Ting A, McMichael AJ (1989) HLA class I antigens in severe RSV bronchiolitis. Tissue Antigens 34:210–212

Janssen R, Bont L, Siezen CL, Hodemaekers HM, Ermers MJ, Doornbos G, van't Slot R, Wijmenga C, Goeman JJ, Kimpen JL, van Houwelingen HC, Kimman TG, Hoebee B (2007) Genetic susceptibility to respiratory syncytial virus bronchiolitis is predominantly associated with innate immune genes. J Infect Dis 196:826–834

Jessen B, Faller S, Krempl CD, Ehl S (2011) Major histocompatibility complex-dependent cytotoxic T lymphocyte repertoire and functional avidity contribute to strain-specific disease susceptibility after murine respiratory syncytial virus infection. J Virol 85:10135–10143

Kapikian AZ, Mitchell RH, Chanock RM, Shvedoff RA, Stewart CE (1969) An epidemiologic study of altered clinical reactivity to respiratory syncytial (RS) virus infection in children previously vaccinated with an inactivated RS virus vaccine. Am J Epidemiol 89:405–421

Kennedy RB, Ovsyannikova IG, Pankratz VS, Haralambieva IH, Vierkant RA, Jacobson RM, Poland GA (2012) Genome-wide genetic associations with IFNgamma response to smallpox vaccine. Hum Genet 131:1433–1451

Khor CC, Hibberd ML (2012) Host-pathogen interactions revealed by human genome-wide surveys. Trends Genet 28:233–243

Kresfelder TL, Janssen R, Bont L, Pretorius M, Venter M (2011) Confirmation of an association between single nucleotide polymorphisms in the VDR gene with respiratory syncytial virus related disease in South African children. J Med Virol 83:1834–1840

Kristensen IA, Thiel S, Steffensen R, Madhi S, Sorour G, Olsen J (2004) Mannan-binding lectin and RSV lower respiratory tract infection leading to hospitalization in children: a case-control study from Soweto, South Africa. Scand J Immunol 60:184–188

Lahti M, Lofgren J, Marttila R, Renko M, Klaavuniemi T, Haataja R, Ramet M, Hallman M (2002) Surfactant protein D gene polymorphism associated with severe respiratory syncytial virus infection. Pediatr Res 51:696–699

Lambkin R, Novelli P, Oxford J, Gelder C (2004) Human genetics and responses to influenza vaccination: clinical implications. Am J Pharmacogenomics 4:293–298

Lander ES, Linton LM, Birren B, Nusbaum C, Zody MC, Baldwin J et al (2001) Initial sequencing and analysis of the human genome. Nature 409:860–921

Legg JP, Hussain IR, Warner JA, Johnston SL, Warner JO (2003) Type 1 and type 2 cytokine imbalance in acute respiratory syncytial virus bronchiolitis. Am J Respir Crit Care Med 168:633–639

LeVine AM, Whitsett JA (2001) Pulmonary collectins and innate host defense of the lung. Microbes Infect 3:161–166

Lofgren J, Ramet M, Renko M, Marttila R, Hallman M (2002) Association between surfactant protein A gene locus and severe respiratory syncytial virus infection in infants. J Infect Dis 185:283–289

Lowther SA, Shay DK, Holman RC, Clarke MJ, Kaufman SF, Anderson LJ (2000) Bronchiolitis-associated hospitalizations among American Indian and Alaska Native children. Pediatr Infect Dis J 19:11–17

Milich DR (1991) Immune response to hepatitis B virus proteins: relevance of the murine model. Semin Liver Dis 11:93–112

Miyairi I, DeVincenzo JP (2008) Human genetic factors and respiratory syncytial virus disease severity. Clin Microbiol Rev 21:686–703

NCI-NHGRI Working Group on Replication in Association Studies, Chanock SJ, Manolio T, Boehnke M, Boerwinkle E, Hunter DJ et al (2007) Replicating genotype-phenotype associations. Nature 447:655–660

Nielsen HE, Siersma V, Andersen S, Gahrn-Hansen B, Mordhorst CH, Nørgaard-Pedersen B, Røder B, Sørensen TL, Temme R, Vestergaard BF (2003) Respiratory syncytial virus infection–risk factors for hospital admission: a case-control study. Acta Paediatr 92:1314–1321

O'Brien SJ, Nelson GW (2004) Human genes that limit AIDS. Nat Genet 36:565–574

Park JH, Wacholder S, Gail MH, Peters U, Jacobs KB, Chanock SJ, Chatterjee N (2010) Estimation of effect size distribution from genome-wide association studies and implications for future discoveries. Nat Genet 42:570–575

Park JH, Gail MH, Weinberg CR, Carroll RJ, Chung CC, Wang Z, Chanock SJ, Fraumeni JF Jr, Chatterjee N (2011) Distribution of allele frequencies and effect sizes and their interrelationships for common genetic susceptibility variants. Proc Natl Acad Sci U S A 108:18026–18031

Poland GA, Ovsyannikova IG, Jacobson RM (2008) Personalized vaccines: the emerging field of vaccinomics. Expert Opin Biol Ther 8:1659–1667

Puthothu B, Krueger M, Forster J, Heinzmann A (2006) Association between severe respiratory syncytial virus infection and IL13/IL4 haplotypes. J Infect Dis 193:438–441

Puthothu B, Krueger M, Forster J, Heinze J, Weckmann M, Heinzmann A (2007) Interleukin (IL)-18 polymorphism 133C/G is associated with severe respiratory syncytial virus infection. Pediatr Infect Dis J 26:1094–1098

Ribeiro LZ, Tripp RA, Rossi LM, Palma PV, Yokosawa J, Mantese OC, Oliveira TF, Nepomuceno LL, Queiróz DA (2008) Serum mannose-binding lectin levels are linked with respiratory syncytial virus (RSV) disease. J Clin Immunol 28:166–173

Salazar M, Deulofeut H, Granja C, Deulofeut R, Yunis DE, Marcus-Bagley D, Awdeh Z, Alper CA, Yunis EJ (1995) Normal HBsAg presentation and T-cell defect in the immune response of nonresponders. Immunogenetics 41:366–374

Schuurhof A, Bont L, Siezen CL, Hodemaekers H, van Houwelingen HC, Kimman TG, Hoebee B, Kimpen JL, Janssen R (2010) Interleukin-9 polymorphism in infants with respiratory syncytial virus infection: an opposite effect in boys and girls. Pediatr Pulmonol 45:608–613

Stark JM, McDowell SA, Koenigsknecht V, Prows DR, Leikauf JE, Le Vine AM, Leikauf GD (2002) Genetic susceptibility to respiratory syncytial virus infection in inbred mice. J Med Virol 67:92–100

Stockman LJ, Curns AT, Anderson LJ, Fischer-Langley G (2012) Respiratory syncytial virus-associated hospitalizations among infants and young children in the United States, 1997–2006. Pediatr Infect Dis J 31:5–9

Suppiah V, Moldovan M, Ahlenstiel G, Berg T, Weltman M, Abate ML, Bassendine M, Spengler U, Dore GJ, Powell E, Riordan S, Sheridan D, Smedile A, Fragomeli V, Müller T, Bahlo M, Stewart GJ, Booth DR, George J (2009) IL28B is associated with response to chronic hepatitis C interferon-alpha and ribavirin therapy. Nat Genet 41:1100–1104

Tal G, Mandelberg A, Dalal I, Cesar K, Somekh E, Tal A, Oron A, Itskovich S, Ballin A, Houri S, Beigelman A, Lider O, Rechavi G, Amariglio N (2004) Association between common Toll-like receptor 4 mutations and severe respiratory syncytial virus disease. J Infect Dis 189:2057–2063

Thomas DL, Thio CL, Martin MP, Qi Y, Ge D, O'Huigin C, Kidd J, Kidd K, Khakoo SI, Alexander G, Goedert JJ, Kirk GD, Donfield SM, Rosen HR, Tobler LH, Busch MP, McHutchison JG, Goldstein DB, Carrington M (2009a) Genetic variation in IL28B and spontaneous clearance of hepatitis C virus. Nature 461:798–801

Thomas NJ, DiAngelo S, Hess JC, Fan R, Ball MW, Geskey JM, Willson DF, Floros J (2009b) Transmission of surfactant protein variants and haplotypes in children hospitalized with respiratory syncytial virus. Pediatr Res 66:70–73

Thomsen SF, Stensballe LG, Skytthe A, Kyvik KO, Backer V, Bisgaard H (2008) Increased concordance of severe respiratory syncytial virus infection in identical twins. Pediatrics 121:493–496

Tulic MK, Hurrelbrink RJ, Prele CM, Laing IA, Upham JW, Le Souef P, Sly PD, Holt PG (2007) TLR4 polymorphisms mediate impaired responses to respiratory syncytial virus and lipopolysaccharide. J Immunol 179:132–140

Venter JC, Adams MD, Myers EW, Li PW, Mural RJ, Sutton GG et al (2001) The sequence of the human genome. Science 291:1304–1351

Webb DC, McKenzie AN, Koskinen AM, Yang M, Mattes J, Foster PS (2000) Integrated signals between IL-13, IL-4, and IL-5 regulate airways hyperreactivity. J Immunol 165:108–113

Wilson J, Rowlands K, Rockett K, Moore C, Lockhart E, Sharland M, Kwiatkowski D, Hull J (2005) Genetic variation at the IL10 gene locus is associated with severity of respiratory syncytial virus bronchiolitis. J Infect Dis 191:1705–1709

Innate Immune Responses to Respiratory Syncytial Virus Infection

Sumanta Mukherjee and Nicholas W. Lukacs

Abstract The innate immune response has a critical role in the initial stages of respiratory syncytial virus (RSV) infection and provides important instructional control that determines the direction of the acquired immune response and the severity of subsequent disease. Contributions to innate immunity include responses initiated in epithelial cells, dendritic cells, and macrophages. The initiation and the intensity of the response depends upon the recognition of pathogen-associated molecular patterns (PAMPs) that activate various pattern recognition receptors (PRRs) such as toll-like receptors (TLR), RIG-I–like receptors (RLR), and NOD-like receptors (NLR), that induce innate cytokines and chemokines that promote inflammation and direct the recruitment of immune cells as well as promote anti-viral responses. In this review, we summarize the results of numerous studies that have characterized the innate immune responses that contribute to the RSV-induced responses and may be important considerations for the development of efficacious vaccine strategies.

Contents

S. Mukherjee · N. W. Lukacs (✉)
Department of Pathology, University of Michigan Medical School, Ann Arbor, MI, USA
e-mail: nlukacs@med.umich.edu

L. J. Anderson and B. S. Graham (eds.), *Challenges and Opportunities for Respiratory Syncytial Virus Vaccines*, Current Topics in Microbiology and Immunology 372,
DOI: 10.1007/978-3-642-38919-1_7,

1 Introduction

RSV infection causing acute bronchiolitis is most commonly associated with lower respiratory tract infections in young children and infants. Although newborn infants are partially protected from the disease due to presence of maternal RSV specific neutralizing antibodies, the protection is short lived. Infants between 2 and 4 months are at very high risk due the diminishing levels of maternal antibodies of RSV specific antibodies (Hurwitz 2011). In the USA, it is estimated that nearly every child before they are 3 years old will get infected at least once with RSV. It is estimated that over 34 million cases of RSV lower respiratory tract infection of young children occur annually and about 10 % of these require hospitalization. Further, hospitalization of reinfected infants is common and exponentially increases the health care costs that are over $400 million in the USA alone and it is estimated that over 175,000 deaths of young children occur worldwide due to RSV infection (Paramore et al. 2010). RSV also infects adults with the elderly and immunocompromised patients most susceptible to this infection. In addition, several epidemiological studies have shown that RSV infection is an independent risk factor for the development of severe chronic asthma exacerbations. Thus, there continues to be a need for both therapeutic intervention, but more importantly, a vaccine that can address the overall RSV burden in the community to reduce the risk of infant and elderly disease acquisition. Thus, in this chapter, we will focus on the current concepts of innate immune recognition of RSV and consider whether they can be exploited specifically for vaccination strategies that will lead to more efficacious and safer vaccines against RSV.

2 RSV-Associated Immunopathology

Severe RSV infections are often associated with mucus production and accumulation of several leukocyte populations in the airways (Johnson et al. 2007). Mucus production can provide a beneficial innate immune response by preventing reinfection and forming a protective environment for the epithelial cell layer. However, uncontrolled increases in mucus production can inhibit gas exchange mechanisms in the lung, cause airways obstruction, and result in difficulty in breathing both mechanically and biochemically. The over production of mucus is particularly detrimental in infants whose airways are small and can be easily obstructed. While the activation of a differential cell response can be beneficial for mounting a robust innate antiviral response, an inappropriately activated adaptive immune cell response can lead to disease. This is observed in severely RSV-infected children. The array of leukocyte populations includes neutrophils, lymphocytes, macrophages, and eosinophils (Graham et al. 2000). Animal modeling of the RSV disease has shown that substantial leukocyte infiltration appears as early as 4-day post infection and consists of mainly macrophages, neutrophils and NK cells (Hussell and Openshaw 2000). Depending on the type of inflammatory cytokine production by the innate cells, the adaptive immune cells are activated. At this stage of the infection, it appears that the most appropriate immune response is the activation and migration of cytotoxic CD8+T cells to clear the virus-infected cells. However, delayed establishment of an effective antiviral environment by the innate immune response to activate adaptive CD8+T cells may lead to the initiation of other mechanisms to inhibit the virus, such as overproduction of mucus. Thus, severe RSV disease sequelae may be associated with an inappropriate immune environment. Although the host appears to clear RSV, it leaves behind a dysregulated cytokine environment with aberrant Th cell response further potentiating the disease. The specific impact of RSV on Type I IFN will be discussed in the chapter by S. Barik (this volume) and the adaptive immune response to RSV will be covered in the chapter S.M. Varga and T.J. Braciale (this volume). Thus, the most detrimental aspects of RSV-induced disease is related to immunopathology that may lead to long-term pulmonary dysfunction due to a reinforced pre-existing or newly established mucosal immune environment.

3 Epithelial Cell Responses During RSV Infection

The airway epithelium forms a mechanical barrier between the environment and the host. It also functions as a sentinel system to detect pathogens entering via the airways and initiates the initial host immune response. In the upper airway, epithelial cells consist of ciliated cells, secretory goblet cells, and cells with microvilli that provide mechanisms for mucocilliary clearance. In the bronchioles the secretory clara cells line the airways, while in the alveolar space the alveolar type I cells and type II cells constitute the alveolar epithelial boundaries. The alveolar type I cells make up 90–95 % of the internal lung surface and are fused with the endothelial cells. Together with the endothelial cells, they form the gas exchange barrier. The other prominent cell in the alveolar space is the alveolar type II cells that are involved in regulating the lung surfactant system, alveolar fluid content, and are important for the replacement of injured type I cells. Clinical observations suggest that infection and inflammation in the lower airways are associated with the most severe disease phenotypes. Infection and activation of the epithelial cells in both the upper and lower airways are involved in the immune response, as they are able to secrete various chemokines and cytokines that direct the immune response by selectively recruiting specific leukocyte populations.

The epithelial cells are one of the primary targets of RSV infection. Thus, viral attachment and fusion may be two critical steps in the initial RSV recognition response. RSV can initiate attachment via either the G or F (fusion) glycoproteins with subsequent entry mediated by F (see chapter by J.S. McLellan et al., this volume). While the cell surface targets for these two proteins have remained elusive for many years, recent data has shown that the G and F proteins can interact with immune-associated receptors, CX3CR1 and TLR4, respectively, as well as glycosaminoglycan (GAGs) and C-type lectins on the cellular surface. Indeed, studies have recognized that in patients with genetic alterations in TLR4 there is an association with more severe disease (Rallabhandi et al. 2006). While the implications for interacting with these innate immune receptors have been widely speculative and associative, they may directly alter the responsiveness of epithelial cells and the ongoing inflammatory response. Recently, cell surface nucleolin was identified as a functional RSV entry receptor using a virus overlay protein-binding assay (Tayyari et al. 2011). Nucleolin was shown to co-precipitate with RSV-F protein. Studies using neutralization antibody for nucleolin or RNA-mediated knockdown of lung nucleolin in mice demonstrated a significant reduction in RSV infection. Since nucleolin is present on most cells, cellular tropism may be specified by some other yet unknown receptor, tissue-specific isoforms on nucleolin, or intracellular cofactors. Another mechanism by which the lung epithelium is exploited by RSV for infection is by decreasing Na+ channel dependent increase in UTP production. This leads to reduced alveolar fluid clearance and oxygen saturation and causes increased lung water volume and hypoxemia.

RSV entry into host epithelial cells is also increased by surfactant protein A (SP-A), another important component of the pulmonary innate immune system, that is reported to bind to RSV G-protein and thereby enhances attachment of RSV and subsequent entry (Hickling et al. 2000). Surfactant proteins (SP-A, B, C, D) are mainly involved in maintenance of pulmonary homeostasis and lung mechanics. Changes in their expression can contribute to the development of lower airway obstruction and enhancement of respiratory workload and efficiencies of surfactant proteins can lead to the development of severe RSV bronchiolitis, and may be related to impairment in viral clearance and exacerbated inflammatory response. Genetic polymorphisms of SP-A, SP-B, and SP-D have been correlated to the development of severe forms of RSV bronchiolitis (El Saleeby et al. 2010). Further, the surfactant proteins are involved in the modulation of the pulmonary inflammatory process, regulating the release of oxygen radicals, as well as pro and anti-inflammatory cytokines (van de Wetering et al. 2004). Pulmonary surfactant replacement may be a promising treatment alternative in patients on mechanical ventilation to maintain airway patency.

4 RSV Disease Regulation through Innate Immune Surveillance

Innate immune cells are responsible for recognizing pathogen-associated molecular patterns (PAMPs) unique to various classes of pathogens, with a limited set of fixed germline-encoded receptors. The existence of invariant structures like lipopolysaccharide and zymosan in bacteria and fungi respectively, make them an easier target for recognition than viruses; which have the heterogeneity of viral glycoproteins, along with their ability to drift genetically from season to season. However, the innate immune system has evolved mechanisms to detect characteristics of viral nucleic acids including single-stranded RNA (ssRNA) or double-stranded RNA (dsRNA). Since, RSV is a negative-sense virus, both ssRNA and dsRNA species are formed during viral replication and available for detection by pathogen recognition receptors (PRRs).

Host immune responses have been implicated in both the protection and immunopathological mechanisms. For example, PRRs expressed on innate immune cells recognize the RSV-associated molecular patterns and activate a proinflammatory innate immune cell response, which is protective. However, the same receptors also mediate airway inflammation and pulmonary immunopathology. This characteristic of dual PRR driven function may be utilized judiciously for vaccine development.

There are several types of innate mononuclear cells that appear to play important roles in the development of the RSV-induced responses in the lung and can serve as antigen presenting cells (APC) that influence the adaptive responses. The myeloid (CD11b/CD11c) subset of dendritic cells (DC) appear to promote a more pathogenic inflammatory response, especially in the absence of the plasmacytoid DC (CD11c/B220) population that is a primary source of IFN-α. Studies have begun to dissect what functions each of these subsets perform during RSV infection. Other subsets of DC have also been described in the lung, including CD103^{+} DC, that are only beginning to be explored. Each of these subsets express different cytokine profiles that can activate immune responses associated with either viral clearance or immunopathology. A complete understanding of these complex regulatory activation pathways and the different subsets that are involved may be very helpful for targeted vaccine development. Here we will highlight the recent advances using experimental models of RSV disease in the innate immune recognition of and responses to RSV through PRRs, including Toll-like receptors (TLRs), RIG-I like receptors (RLRs) and NOD like receptors (NLRs).

4.1 TLRs

A number of TLRs have been linked to RSV infection, including TLR2, TLR3, TLR4, and TLR7. Murawski et al. suggested that TLR2 is a functional receptor for RSV and demonstrated that TLR2 and TLR6 signaling in leukocytes could activate innate immune response to RSV by promoting proinflammatory cytokines and chemokine production, neutrophil migration, and DCs activation as well as control viral replication in the lung (Murawski et al. 2009).

TLR3 is an intracellular receptor that recognizes dsRNA. TLR3-mediated response is channeled through TRIF (Toll-interleukin (IL)-1 receptor domain-containing adaptor inducing interferon-β) which activates both NFκB and IRF-3, and subsequently drives production of IFN-α, IP10/CXCL10, CCL12, and CCL5 (RANTES) (Bermejo-Martin et al. 2007). TLR3 detects dsRNA generated during the RSV replication cycle (Aeffner et al. 2011). TLR3 activation might be necessary to regulate the immune environment that contributes to RSV-associated pulmonary immunopathology, although it has little or no effect on RSV clearance (Rudd et al. 2005). Rudd et al. showed that RSV infection triggers the activation of the TLR3 signaling pathways that regulate the expression of MyD88-independent chemokines, such as IP-10/CXCL10 and CCL5, and further upregulate TLR3 expression in RSV-infected cells (Rudd et al. 2005). The activation of TLR3 during RSV infection promotes a predominant Th1-type response, whereas the deletion of TLR3 leads to increased pathogenic Th2-biased responses, resulting in IL-13 and IL-5 production, mucus overproduction, and an accumulation of eosinophils in the airways (Rudd et al. 2006). RSV infection increases TLR3 expression in the respiratory epithelial cells, which sensitizes the epithelial cells to subsequent extracellular dsRNA exposure through activation of the inflammation-related transcription factor NF-κB and production of the chemokine CXCL8 (IL-8) (Groskreutz et al. 2006). These results are consistent with the notion that the persistence of RSV-inducing inflammatory responses may provide a substratum for later challenges with other pathogens. The TLR3 signaling pathway may be an attractive therapeutic target for preventing RSV-induced lung inflammation and subsequent allergic asthma.

TLR4 was first shown to have an effect on RSV infection through its interaction with RSV F protein using CD14 as a co-receptor (Kurt-Jones et al. 2000). Further, TLR4 null mice challenged with RSV showed reduced pulmonary NK and CD14+ cell trafficking, deficient NK cell function, impaired IL-12 expression and delayed clearance of RSV compared with TLR4-positive mice (Kurt-Jones et al. 2000). Later it was shown that activation of TLR4 and CD14 by RSV F protein leads to NFκB-mediated innate immune responses and inflammation (Haeberle et al. 2002). Another study in 2007 showed that in infants TLR4 response is central to the development of an efficacious innate immune response to natural RSV infection (Awomoyi et al. 2007). Later it was shown that RSV-elicited production of proinflammatory cytokines IL-6 and IL-8 by epithelial cells is suppressed through inhibition of the RSV-TLR4/CD14 interaction (Numata et al. 2010). Two single nucleotide polymorphisms (SNPs) encoding Asp299Gly and Thr399Ile substitutions in the TLR4 ectodomain have been linked epidemiologically with an increased risk of severe RSV bronchiolitis and increased risk for hospitalization in previously healthy infants (Mandelberg et al. 2006; Tal et al. 2004). It was shown that TLR4 polymorphisms contribute to enhanced susceptibility to RSV infection in these individuals. In contrast to these reports, a clinical study has shown that upregulated TLR4 expression on blood monocytes in infants is linked closely to disease severity (Gagro et al. 2004). However, recent findings with TLR4 SNPs in humans and TLR4$^{-/-}$ mice showed little evidence of altered sensitivity to RSV infection has resulted in some controversy about the role of TLR4 in RSV recognition (Ehl et al. 2004; Douville et al. 2010).

Similar to other ssRNA viruses RSV can also be recognized via TLR7. Lindell et al., revealed that TLR7 mediates responses to RSV that regulate DC activation resulting in reduced IL-12 while promoting IL-23, two important T cell differentiation factors that determine Th1 versus Th17 development, respectively (Lindell et al. 2011). The study showed that TLR7 deficiency is associated with an alteration in T-cell responses with increases in mucogenic cytokines IL-4, IL-13, and IL-17 that enhanced immunopathogenesis and mucus production in RSV-infected TLR7−/− mice. Thus, the TLR7-mediated response is important for regulating the host defense against RSV and preventing RSV-associated pulmonary immunopathology.

4.2 RLRs

RIG-I and MDA5 are the two RLRs that are critical sensors of viral infection in most cell types including epithelial cells, macrophages, and conventional DCs. These are cytoplasmic receptors for dsRNA. MDA5 and RIG-I are upregulated during RSV infections and a significant positive correlation exists between RSV viral load and RIG-I mRNA levels (Scagnolari et al. 2009). Recent in vitro investigations have shown that RSV infection detected by RIG-I activates the downstream NF-κB and IRF3 pathways by complexing with the MAVS adaptor that localizes to the mitochondrial membrane. This is important for generating IFNβ, IP-10, and CCL5 expression in airway epithelial cells. Further activation of NFκB, IRF3, and cytokine expression is significantly inhibited in RIG-I-silenced cells during the early phase of infection (Liu et al. 2007). Subsequently, studies demonstrated that RSV is a poor inducer of type I IFNs and it was mediated by the RSV proteins- NS1 and NS2 that directly decreased RIG-I interaction with MAVS (Ling et al. 2009). These data suggest that the RIG-I-MAVS pathway plays an important role in RSV-induced innate immune responses and inflammatory pathology. The regulation of the RLR system and type I IFN by RSV NS1 and NS2 may be the most critical aspect of immune deviation during RSV-induced immunopathogenesis (see chapter by S. Barik, this volume, for more details).

4.3 NLRs

Recent evidence has indicated that NOD2 recognizes ssRNA virus in addition to detection of bacterial products, and is involved in innate antiviral responses by activation of IRF3 and IFN production. Sabbah et al. found that RSV infection results in increased NOD2 expression as well as activated IRF3 and IFN production within 2 h post-infection. NOD2, activated by RSV ssRNA, translocated to the mitochondria where it could interact with MAVS to induce activation of both IRF3 and NFκB (Sabbah et al. 2009). RSV infection of NOD2-KO mice results in markedly higher body weight loss and severe lung pathology with higher concentrations of pro-inflammatory cytokines, as well as chemokines, than wild-type counterparts. These results demonstrate that, like the RLR receptors, NOD2 can function as a cytoplasmic PRR for RSV and is important for host defense against RSV infection.

4.4 RSV G and F Proteins

The innate immune response is also regulated by some of the RSV proteins such as G and F, which are major integral proteins on the viral membrane and serve as excellent targets for neutralizing antibodies for treatment. However, since the G protein also exists as a secreted protein it could act as a decoy to block the host G-specific neutralizing antibody response. On the other hand, the F protein apart from facilitating virus fusion, is also capable of signaling through TLR4 and CD14. Whereas F-TLR4 signaling induces proinflammatory cytokines including interferon with antiviral activities the G, NS-1 and NS-2 proteins down regulate inflammatory events. In addition, the G-proteins also bind to C-type lectins such as surfactant protein and dendritic cell-specific intercellular adhesion molecule-3-grabbing Non Integrin (DC-SIGN) and DC-SIGN/R (L-SIGN) on DCs. The signaling events mediated by RSV G interactions with DC/L-SIGN are immunomodulatory and diminish DC activation, which may limit induction of RSV-specific immunity (Johnson et al. 2012).

5 RSV and Chemokine Production

Clinical studies have suggested that the severity of RSV-induced disease correlates with the influx of leukocytes that lead to the damage in the airways (Welliver 2003). Chemokines have been shown to correlate directly to the intensity of the inflammatory response and can be induced by viral infection in multiple pulmonary cell populations. In particular, a number of studies demonstrated that (El-Sahly et al. 2000), CCL3 (MIP-1α), and CCL5 (RANTES) released during the RSV infection correlates to the most severe cases of RSV infection in infants (Bonville et al. 1999). RSV can directly induce cytokines and chemokines from both airway epithelial cells and resident macrophage populations, such as TNF, IL-1, IL-6, and IL-8 (Becker et al. 1991; Mellow et al. 2004). RSV-infected epithelial cells appear to be a rich source of a number of different chemokines, including CCL2, CCL3, CCL5, and CxCL8, in vitro analyses and are induced via activation of NFkB (Mastronarde et al. 1996). Thus, macrophages and epithelial cells directly induce an amplified system of activating chemokines. It is not known, however, whether individuals who are more susceptible to severe RSV infections versus those that are less susceptible have different profiles of cytokines and chemokines. However, it is clear that samples from infants with severe RSV infections have a high level of chemokines that relate to the severity of the disease (Garofalo et al. 2001). Several of these chemokines, including CCL3 and CCL5, have also been shown to be associated with inflammatory responses in animal models of RSV infection. CCL3 has been linked to the severity of primary RSV-induced inflammation as well as with multiple infections with RSV. Studies examining CCL5 indicate a significant impact on the pathophysiologic responses in primary RSV infection as well as the pattern of leukocyte recruitment and leukotriene release in the lungs during RSV-induced allergen exacerbated disease (John et al. 2003). Overproduction of mucus and the development of airway hyperreactivity is directly related to the expression and activation of CXCR2 (an IL-8 receptor homolog) in mice that is a common receptor system associated with neutrophilia (Miller et al. 2003). In addition, CXCL10, the interferon-inducible CXC chemokine, is induced during RSV infection and has been identified to cause increased allergic pulmonary disease (Medoff et al. 2002). Later, CXCL10 (IP-10) was associated with RSV clearance (Lindell et al. 2008). Overall, the profile of chemokine production may dictate whether immune cell recruitment leads to severe disease, efficient clearance of the virus, and/or subsequent development of chronic pulmonary disease. Thus, the induction of the proper chemokine milieu and the magnitude of chemokine production may be critical for developing the most efficacious vaccine.

6 RSV and Autophagy

The autophagy system is a critical cellular response that facilitates self-digestion of misfolded or unused protein and cellular debris that impacts development, aging, and normal function of cellular processes. Induction of autophagy can be induced by amino acid starvation that entails the sequential activation of Atg proteins that direct assembly of autophagosomes and capture cytoplasmic components (Swanson 2006). The autophagy-associated proteins interact with important signaling pathways, including Beclin 1 (atg6) and BCL-2 family members, and these are often targeted for destruction (Xie and Klionsky 2007). Another autophagy related target is the LC3 protein (an atg8 protein) that is modified to LC3b and binds to the autophagosome membrane during its formation. LC3b is involved with late autophagosome formation, lysosome interaction and fusion that lead to degradation of the autophagosome contents (Munafo and Colombo 2001). Defective autophagy responses have been identified in numerous diseases including Alzheimer's and Parkinson's. More recent identification of the role of autophagy in Crohn's disease as well as in bacterial clearance has highlighted an important activation pathway in innate immune responses (Deretic 2009; Cario 2008). Thus, this conserved and relatively well-defined process has a role for anti-pathogen responses in all cells, including a role for alerting the immune system during pathogen invasion. RSV infects cells via a mechanism of cell membrane fusion and cytoplasmic entry that does not use an endosomal pathway and therefore must rely on other process, such as autophagy, to shuttle pathogenic signals to the lysosome compartment for TLR activation. This delay in TLR-mediated signaling along with the ability of RSV to alter cytoplasmic PAMP pathway activation via NS1 and NS2 may allow RSV the opportunity to delay and/or avoid appropriate innate immune responses. Recent evidence has demonstrated that RSV induces autophagy in DC and enhances appropriate anti-viral innate cytokines, especially type I IFN. Previous data have demonstrated that RSV-induced cytokine production from DCs is partially or completely dependent on activation of TLRs found predominately in the endosome compartment. Moreover, induction of autophagy can directly impact the immune system by optimizing the activation of APC and other innate immune cells for immediate cytokine responses as well as for activation of T cells through increased antigen presentation (Li et al. 2009; Zhou et al. 2005). In fact, autophagy has been identified not only in the delivery of antigens for MHC class II presentation to CD4 T cells, but also for cross presentation of antigens for CD8 T cell activation. Interestingly, immune responses differentially regulate autophagy with Th1 cytokines promoting and Th2 cytokines inhibiting the process for activation in innate immune cells (Harris et al. 2009). Recent studies have demonstrated that induction of autophagy in the lung can promote enhanced vaccination against M. tuberculosis in animal models. Understanding how autophagy is induced, how it is regulated during RSV infection, and its role in mediating the immune responses may provide insight into how the process can be used for optimizing vaccines.

7 RSV and Notch

Pathogen-induced Th2 responses tend to arise in the absence of TLR signaling leading some to suggest that Th2 differentiation is a default pathway that is triggered in the absence of IL-12. However, not all stimuli default to Th2 responses in the absence of IL-12, indicating that other signals may exist on antigen-presenting cells to elicit a Th1 response. The strongest evidence for this possibility comes from studies demonstrating that the expression of Notch ligands delta-like and jagged can provide instructional signals for the development of Th1 and Th2 cells, respectively (Dallman et al. 2003). Notch signaling regulates a wide range of cellular developmental and activation events in multi-celled organisms. In mammals there are five notch ligands (Jagged 1 and 2, Delta like 1, 3, and 4) and four Notch receptors (Notch 1–4) with a common pathway. In the mature immune system the Notch pathway has been described as a signaling mechanism involved in regulating cell lineage choices for T cells. A number of studies have demonstrated that Notch activation is responsible for the generation of Th1, Th2, or Treg cell generation, suggesting that the role of Notch may be a contextual mechanism and depend upon the immune environment that is associated with specific disease conditions (Tu et al. 2005). RSV infection of innate immune DC upregulate expression of notch ligand Dll4 and it was shown to modulate the Th2 response without affecting the Th1 responses (Schaller et al. 2007). The absence of Dll4 was associated with polarized Th2 differentiation during RSV infection along with induction of mucus over-production. Interestingly, Dll4 and not other notch ligands have been demonstrated to augment Th17 cell responses suggesting that it is responsible for skewing the immune response in a particular direction (Mukherjee et al. 2009). Thus, while the relationship of RSV and Notch activation are presently still being explored, it appears from preliminary reports that specific Notch activation pathways may provide an interesting avenue to explore in the pathogenesis of the disease process and for developing immune-modulatory strategies during disease and for directing vaccination efficiency.

8 Novel Strategies of Vaccination Targeting Innate Immune Responses

A number of traditional and innovative strategies have been employed for developing vaccines against RSV, including vector delivery systems, attenuated virus, subunit vaccines, and novel adjuvant approaches for optimizing immunization. While inactivated RSV vaccines have been avoided due to the original formalin-inactivated vaccine immunopotentiation problems, live-attenuated vaccines suffer from fear of reversion or insufficient immunogenicity. Although interest continues to grow in the field where limited alternatives exist for virus protection, the complications of RSV infection intricacies and immunopathology responses continue to hamper the development. Recent attempts using a RSV F protein containing Sendai virus, a virus that readily infects humans but rarely causes disease, appears to provide very promising pre-clinical results (Jones et al. 2012). This latter approach offers protection against both PIV1 and RSV. Several live-attenuated vaccines are also being developed that either utilize defective viruses or temperature-sensitive mutants that only grow at lower temperatures allowing replication of the virus in only the upper respiratory track [recently reviewed (Collins and Melero 2011)]. These latter viruses may offer superior protection to inactivated virus vaccines due to their ability to infect and drive the proper anti-viral responses. However, from an immunologic standpoint, if an infant is prone to a detrimental anti-viral response the initial viral response may only further sensitize the individual to more severe pathogenesis upon subsequent pulmonary disease, no matter if the initial RSV infection is attenuated or not. Thus, a goal in vaccine strategy may need to shift from trying to induce an enhanced innate immune response for modulation of the adaptive response to a more desirable adaptive anti-viral response with reduced pathogenic consequences.

One of the complications of RSV is that it encodes additional proteins that have inhibitory functions within the innate immune system. Two of these–NS1 and NS2, inhibit the host type I and type III interferon (IFN) responses and appear to promote some of the most complicated aspects of RSV-induced immunopathogenesis. Thus, one strategy has been to boost anti-viral responses by modifying the immune responses through the activation of type I IFN. Prior attempts to enhance vaccination by activating TLR pathways have been addressed by adding PAMPs such as CpG, PolyI:C or ssRNA into inactivated vaccines. Studies in animal models have been mixed, as not all PAMPs promote protection. For example, it appears that TLR9 (CpG), but not TLR7/8 (ssRNA), activation could alleviate certain aspects of FI-RSV vaccine-induced disease in mice (Johnson et al. 2009). Another study utilizing mucosal immunization with inactivated RSV supplemented with innate receptor ligands, CpG-ODN and L18-MDP demonstrated induction of neutralizing RSV-specific IgG antibodies and IFN-γ-producing T cell responses, indicating an anti-viral Th1-skewed response. Other novel adjuvant strategies are being attempted to specifically boost the innate immune response and promote immune system maturation without causing pathogenic immunopotentiation (recently reviewed) (Collins and Melero 2011). These strategies include MPL that is an immunomodulatory TLR4 agonist that has shown to inhibit the development of Th2 cytokine responses with inactivated RSV vaccination. Likewise, Protollin comprised of proteosomes (hydrophobic outer membrane proteins or OMPs derived from *Neisseria meningitidis*) non-covalently complexed with LPS isolated from *Shigella flexneri 2a* or *Pleisiomonas shigelloides*, provides a similar adjuvant activity as MPL by activating TLR4. More recently, a novel "nanoemulsion" that inactivates live virus in a lipid bilayer and provides immunization by intranasal administration has been used to provide a strong anti-viral response characterized by low Th2 responses and enhanced anti-viral clearance (Lindell et al. 2011). Yet another strategy that holds promise for activation of immunity through innate activation is using virus-like particles (VLP) coupled with RSV surface proteins as a vaccine strategy for long-lived anti-viral vaccine development (McGinnes et al. 2011). Thus, there are now a number of new adjuvant strategies being employed to activate the innate immune system and enhance the appropriate vaccine immune response to mitigate immunopotentiation.

While the focus of present vaccine strategies has been primarily on the RSV F protein due to the success of passively administered palivizumab, a significant effort has been made to also examine the surface G glycoprotein. However, the attachment (G) glycoprotein of RSV exhibits unusual features such as high sequence variability, extensive glycosylation, cytokine mimicry, blockade of TLR activation pathways, and a shed form that helps the virus evade neutralizing antibodies and has immunomodulatory properties. Interestingly, RSV G protein mimics CX3C chemokine (a.k.a. fractalkine) and binds to the CX3C receptor (CX3CR1). Using G protein polypeptides or peptides that would essentially block G protein CX3C-CX3CR1 interaction, Zhang et al. have shown that the strategy

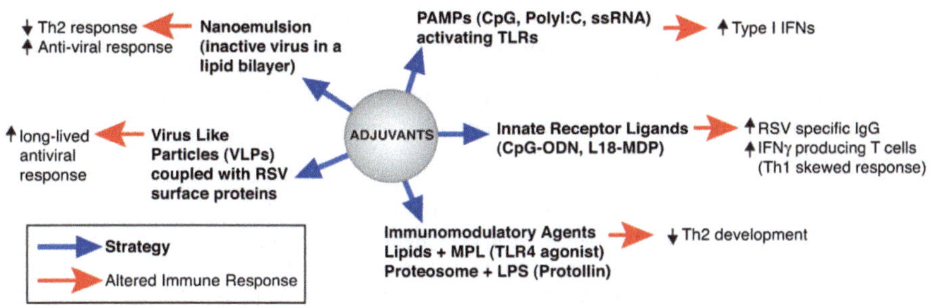

Fig. 1 Adjuvant Strategy for RSV Vaccine development

blocks RSV-mediated disease pathogenesis (Zhang et al. 2010). Thus, combination vaccines directed against multiple surface proteins may provide increased protection upon challenge and effectively alter disease pathogenesis.

9 Conclusions

The field of RSV research has never been more exciting with new strategies for therapy and vaccination being explored. It is becoming ever clearer that targeting both RSV infection and the detrimental immune responses to the virus that contribute to disease may be needed to achieve a safe and an effective vaccine. While strategies to combat viral replication will continue to be important for reducing viral burden, by themselves they may not be sufficient to reduce the disease severity in susceptible patient populations that have underlying disease and/or an altered immune environment that is prone to adverse responses. The most recent strategies, summarized in Fig. 1, highlight the belief that adjuvants may provide a promising avenue to promote an efficacious vaccine with minimal pathologic consequence. Thus, a better understanding of the innate immune response and how to best modulate the environment may be extremely important for advancing both therapeutic and vaccination strategies.

References

Aeffner F et al (2011) Double-stranded RNA induces similar pulmonary dysfunction to respiratory syncytial virus in BALB/c mice. Am J Physiol Lung Cell Mol Physiol 301(1):L99–L109

Awomoyi AA et al (2007) Association of TLR4 polymorphisms with symptomatic respiratory syncytial virus infection in high-risk infants and young children. J Immunol 179(5):3171–3177

Becker S, Quay J, Soukup J (1991) Cytokine (tumor necrosis factor, IL-6, and IL-8) production by respiratory syncytial virus-infected human alveolar macrophages. J Immunol 147(12):4307–4312

Bermejo-Martin JF et al (2007) Predominance of Th2 cytokines, CXC chemokines and innate immunity mediators at the mucosal level during severe respiratory syncytial virus infection in children. Eur Cytokine Netw 18(3):162–167

Bonville CA, Rosenberg HF, Domachowske JB (1999) Macrophage inflammatory protein-1alpha and RANTES are present in nasal secretions during ongoing upper respiratory tract infection. Pediatr Allergy Immunol 10(1):39–44

Cario E (2008) Innate immune signalling at intestinal mucosal surfaces: a fine line between host protection and destruction. Curr Opin Gastroenterol 24(6):725–732

Collins PL, Melero JA (2011) Progress in understanding and controlling respiratory syncytial virus: still crazy after all these years. Virus Res 162(1–2):80–99

Dallman MJ, Champion B, Lamb JR (2003) Notch signalling in the peripheral immune system. Novartis Found Symp 252:268–276; discussion 276-278

Deretic V (2009) Links between autophagy, innate immunity, inflammation and Crohn's disease. Dig Dis 27(3):246–251

Douville RN et al (2010) TLR4 Asp299Gly and Thr399Ile polymorphisms: no impact on human immune responsiveness to LPS or respiratory syncytial virus. PLoS ONE 5(8):e12087

Ehl S et al (2004) The role of toll-like receptor 4 versus interleukin-12 in immunity to respiratory syncytial virus. Eur J Immunol 34(4):1146–1153

El Saleeby CM et al (2010) Surfactant protein A2 polymorphisms and disease severity in a respiratory syncytial virus-infected population. J Pediatr 156(3):409–414

El-Sahly HM et al (2000) Spectrum of clinical illness in hospitalized patients with "common cold" virus infections. Clin Infect Dis 31(1):96–100

Gagro A et al (2004) Increased toll-like receptor 4 expression in infants with respiratory syncytial virus bronchiolitis. Clin Exp Immunol 135(2):267–272

Garofalo RP et al (2001) Macrophage inflammatory protein-1alpha (not T helper type 2 cytokines) is associated with severe forms of respiratory syncytial virus bronchiolitis. J Infect Dis 184(4):393–399

Graham BS, Johnson TR, Peebles RS (2000) Immune-mediated disease pathogenesis in respiratory syncytial virus infection. Immunopharmacology 48(3):237–247

Groskreutz DJ et al (2006) Respiratory syncytial virus induces TLR3 protein and protein kinase R, leading to increased double-stranded RNA responsiveness in airway epithelial cells. J Immunol 176(3):1733–1740

Haeberle HA et al (2002) Respiratory syncytial virus-induced activation of nuclear factor-kappaB in the lung involves alveolar macrophages and toll-like receptor 4-dependent pathways. J Infect Dis 186(9):1199–1206

Harris J et al (2009) Th1-Th2 polarisation and autophagy in the control of intracellular mycobacteria by macrophages. Vet Immunol Immunopathol 128(1–3):37–43

Hickling TP et al (2000) Lung surfactant protein A provides a route of entry for respiratory syncytial virus into host cells. Viral Immunol 13(1):125–135

Hurwitz JL (2011) Respiratory syncytial virus vaccine development. Expert Rev Vaccines 10(10):1415–1433

Hussell T, Openshaw PJ (2000) IL-12-activated NK cells reduce lung eosinophilia to the attachment protein of respiratory syncytial virus but do not enhance the severity of illness in CD8 T cell-immunodeficient conditions. J Immunol 165(12):7109–7115

John AE, Berlin AA, Lukacs NW (2003) Respiratory syncytial virus-induced CCL5/RANTES contributes to exacerbation of allergic airway inflammation. Eur J Immunol 33(6):1677–1685

Johnson JE et al (2007) The histopathology of fatal untreated human respiratory syncytial virus infection. Mod Pathol 20(1):108–119

Johnson TR et al (2009) TLR9 agonist, but not TLR7/8, functions as an adjuvant to diminish FI-RSV vaccine-enhanced disease, while either agonist used as therapy during primary RSV infection increases disease severity. Vaccine 27(23):3045–3052

Johnson TR, McLellan JS, Graham BS (2012) Respiratory syncytial virus glycoprotein G interacts with DC-SIGN and L-SIGN to activate ERK1 and ERK2. J Virol 86(3):1339–1347

Jones BG et al (2012) Sendai virus-based RSV vaccine protects African green monkeys from RSV infection. Vaccine 30(5):959–968
Kurt-Jones EA et al (2000) Pattern recognition receptors TLR4 and CD14 mediate response to respiratory syncytial virus. Nat Immunol 1(5):398–401
Li Y et al (2009) Cross-presentation of tumor associated antigens through tumor-derived autophagosomes. Autophagy 5(4):576–577
Lindell DM, Lane TE, Lukacs NW (2008) CXCL10/CXCR3-mediated responses promote immunity to respiratory syncytial virus infection by augmenting dendritic cell and CD8(+) T cell efficacy. Eur J Immunol 38(8):2168–2179
Lindell DM et al (2011) A novel inactivated intranasal respiratory syncytial virus vaccine promotes viral clearance without Th2 associated vaccine-enhanced disease. PLoS ONE 6(7):e21823
Ling Z, Tran KC, Teng MN (2009) Human respiratory syncytial virus nonstructural protein NS2 antagonizes the activation of beta interferon transcription by interacting with RIG-I. J Virol 83(8):3734–3742
Liu P et al (2007) Retinoic acid-inducible gene I mediates early antiviral response and Toll-like receptor 3 expression in respiratory syncytial virus-infected airway epithelial cells. J Virol 81(3):1401–1411
Mandelberg A et al (2006) Lipopolysaccharide hyporesponsiveness as a risk factor for intensive care unit hospitalization in infants with respiratory syncitial virus bronchiolitis. Clin Exp Immunol 144(1):48–52
Mastronarde JG et al (1996) Induction of interleukin (IL)-8 gene expression by respiratory syncytial virus involves activation of nuclear factor (NF)-kappa B and NF-IL-6. J Infect Dis 174(2):262–267
McGinnes LW et al (2011) Assembly and immunological properties of Newcastle disease virus-like particles containing the respiratory syncytial virus F and G proteins. J Virol 85(1):366–377
Medoff BD et al (2002) IFN-gamma-inducible protein 10 (CXCL10) contributes to airway hyperreactivity and airway inflammation in a mouse model of asthma. J Immunol 168(10):5278–5286
Mellow TE et al (2004) The effect of respiratory synctial virus on chemokine release by differentiated airway epithelium. Exp Lung Res 30(1):43–57
Miller AL et al (2003) CXCR2 regulates respiratory syncytial virus-induced airway hyperreactivity and mucus overproduction. J Immunol 170(6):3348–3356
Mukherjee S et al (2009) Regulation of T cell activation by Notch ligand, DLL4, promotes IL-17 production and Rorc activation. J Immunol 182(12):7381–7388
Munafo DB, Colombo MI (2001) A novel assay to study autophagy: regulation of autophagosome vacuole size by amino acid deprivation. J Cell Sci 114(Pt 20):3619–3629
Murawski MR et al (2009) Respiratory syncytial virus activates innate immunity through Toll-like receptor 2. J Virol 83(3):1492–1500
Numata M et al (2010) Pulmonary surfactant phosphatidylglycerol inhibits respiratory syncytial virus-induced inflammation and infection. Proc Natl Acad Sci U S A 107(1):320–325
Paramore LC, Mahadevia PJ, Piedra PA (2010) Outpatient RSV lower respiratory infections among high-risk infants and other pediatric populations. Pediatr Pulmonol 45(6):578–584
Rallabhandi P et al (2006) Analysis of TLR4 polymorphic variants: new insights into TLR4/MD-2/CD14 stoichiometry, structure, and signaling. J Immunol 177(1):322–332
Rudd BD et al (2005) Differential role for TLR3 in respiratory syncytial virus-induced chemokine expression. J Virol 79(6):3350–3357
Rudd BD et al (2006) Deletion of TLR3 alters the pulmonary immune environment and mucus production during respiratory syncytial virus infection. J Immunol 176(3):1937–1942
Sabbah A et al (2009) Activation of innate immune antiviral responses by Nod2. Nat Immunol 10(10):1073–1080

Scagnolari C et al (2009) Gene expression of nucleic acid-sensing pattern recognition receptors in children hospitalized for respiratory syncytial virus-associated acute bronchiolitis. Clin Vaccine Immunol 16(6):816–823

Schaller MA et al (2007) Notch ligand delta-like 4 regulates disease pathogenesis during respiratory viral infections by modulating Th2 cytokines. J Exp Med 204(12):2925–2934

Swanson MS (2006) Autophagy: eating for good health. J Immunol 177(8):4945–4951

Tal G et al (2004) Association between common toll-like receptor 4 mutations and severe respiratory syncytial virus disease. J Infect Dis 189(11):2057–2063

Tayyari F et al (2011) Identification of nucleolin as a cellular receptor for human respiratory syncytial virus. Nat Med 17(9):1132–1135

Tu L et al (2005) Notch signaling is an important regulator of type 2 immunity. J Exp Med 202(8):1037–1042

van de Wetering JK, van Golde LM, Batenburg JJ (2004) Collectins: players of the innate immune system. Eur J Biochem 271(7):1229–1249

Welliver RC (2003) Respiratory syncytial virus and other respiratory viruses. Pediatr Infect Dis J 22(2 Suppl):S6–S10; discussion S10-2

Xie Z, Klionsky DJ (2007) Autophagosome formation: core machinery and adaptations. Nat Cell Biol 9(10):1102–1109

Zhang W et al (2010) Vaccination to induce antibodies blocking the CX3C-CX3CR1 interaction of respiratory syncytial virus G protein reduces pulmonary inflammation and virus replication in mice. J Virol 84(2):1148–1157

Zhou D et al (2005) Lamp-2a facilitates MHC class II presentation of cytoplasmic antigens. Immunity 22(5):571–581

The Adaptive Immune Response to Respiratory Syncytial Virus

Steven M. Varga and Thomas J. Braciale

Abstract Respiratory syncytial virus (RSV) causes severe respiratory disease in children, the elderly and immunocompromised individuals. The combined actions of CD4 and CD8 T cells play a critical role in terminating an acute RSV infection whereas antibodies can provide protection from re-infection. Despite eliciting an immune response that mediates clearance of the virus, immunity to the virus appears to wane over time and individuals remain susceptible to reinfection with RSV throughout their lifetime. The ineffectiveness of the natural infection to induce long-term immunity has hampered vaccine efforts and there is currently no licensed RSV vaccine. In this review, we summarize our current understanding of the adaptive immune response to RSV and its contribution to disease.

S. M. Varga (✉)
Interdisciplinary Graduate Program in Immunology, University of Iowa, Iowa City, IA 52242, USA
e-mail: steven-varga@uiowa.edu

S. M. Varga
Department of Pathology, University of Iowa, Iowa City, IA 52242, USA

S. M. Varga
Department of Microbiology, University of Iowa, Iowa City, IA 52242, USA

T. J. Braciale
Beirne B. Carter Center for Immunology Research, University of Virginia Health Sciences Center, Charlottesville, VA 22908, USA

T. J. Braciale
Department of Pathology, University of Virginia Health Sciences Center, Charlottesville, VA 22908, USA

T. J. Braciale
Department of Microbiology, University of Virginia Health Sciences Center, Charlottesville, VA 22908, USA

L. J. Anderson and B. S. Graham (eds.), *Challenges and Opportunities for Respiratory Syncytial Virus Vaccines*, Current Topics in Microbiology and Immunology 372,
DOI: 10.1007/978-3-642-38919-1_8,

Contents

1 Respiratory Syncytial Virus-Induced Disease

Respiratory syncytial virus (RSV) is a leading cause of virus-induced respiratory tract infections in young children and the elderly. In addition, RSV causes respiratory disease in immune compromised individuals. Although the majority of individuals are infected with RSV at an early age, recurrent infections with RSV are common throughout life. Thus, a natural RSV infection fails to induce long-term sterilizing immunity. At present, we still lack a thorough understanding of the pulmonary pathophysiology of severe RSV infection in infants and children, in part due to the lack of post-mortem evaluation of children that have died following an acute RSV infection. Modern medical interventions have significantly reduced the number of RSV-induced deaths in industrialized nations. Such advances in medical care have made it difficult to obtain further information regarding the pathologic features of a natural RSV infection in humans because current interventions such as mechanical ventilation and oxygen therapy can induce pulmonary changes that may in some cases overlap with virus-induced alterations. In a recent report, a post-mortem histologic analysis was carried out on lung tissue from an untreated RSV-infected child who died in an automobile accident along with tissue from three archived cases of fatal RSV infection in children collected at Vanderbilt Medical Center between 1931 and 1949 where only supportive care was provided (i.e., no mechanical ventilation was employed) (Johnson et al. 2007). The histology revealed viral antigen present in bronchiolar and alveolar epithelial cells with extensive obstruction of small airways caused by inflammatory cells and cellular debris, edema and small airway compression caused by peri-bronchiolar lymphoid aggregates (Johnson et al. 2007). Inflammatory cells included macrophages, neutrophils as well as CD4 and CD8 T cells as identified using immunohistochemistry (Johnson et al. 2007). A separate study from Santiago, Chile examined lung histopathology in a group of infants with fatal RSV infection (Welliver et al. 2007, 2008). RSV infection caused substantial damage to the respiratory epithelium including cell sloughing and apoptotic epithelial cell death (Welliver et al. 2007). In contrast to the Vanderbilt study, histology performed on several children from Chile that succumbed to fatal RSV-induced respiratory disease revealed few CD4 and CD8 T cells in the lung calling into question a major role for the adaptive immune response in contributing to disease severity (Welliver et al. 2007). Thus, the relative contribution of tissue damage to the respiratory epithelium caused by infection of respiratory epithelial cells versus damage caused by virus-induced inflammation and the host immune response in contributing to RSV-induced disease severity remains unclear.

More recent studies have examined the correlation between increased viral load and relative disease severity. A study examining natural RSV infection in children <24 months of age found that higher viral loads in nasal washes correlated with increased severity of disease as measured by the length of hospital stay, the requirement for intensive care, or the development of respiratory failure (El Saleeby et al. 2011). A similar correlation between higher viral replication and increased disease symptoms was observed in adult volunteers experimentally infected with RSV (DeVincenzo et al. 2010). However, the same study also observed a strong correlation between disease severity and the levels of several pro-inflammatory cytokines and chemokines in the nasal washes of adult volunteers experimentally infected with RSV suggesting a correlation between the host immune response and disease severity (DeVincenzo et al. 2010). In addition, several studies have demonstrated a predominance of neutrophils in the airway secretions of human infants infected with RSV with increased neutrophil numbers correlating with disease severity (Everard et al. 1994; Lukens et al. 2010). Thus, multiple factors likely contribute to RSV-induced disease including a combination of direct virus-induced tissue damage as well as virus-induced pulmonary inflammation. Based on the above work and studies in animal models, RSV-induced disease is best characterized by small airway obstruction, airway hyper-reactivity, increased mucus production and pulmonary inflammation.

Given the prevalence and severity of RSV-induced disease in infants and the elderly, much effort has been placed into the development of an RSV vaccine. However, a number of obstacles must be overcome in order to develop a successful RSV vaccine including the ability to successfully vaccinate immunologically immature infants, avoiding interference by maternal antibodies and eliciting more effective and longer lasting immunity than that induced by natural RSV infection. A previous vaccine attempt using a formalin-inactivated RSV vaccine during the 1960s met with failure when the vaccinated children experienced enhanced morbidity and mortality following a natural RSV infection (Castilow and Varga 2008). The failure of the formalin-inactivated vaccine trial has hampered efforts to develop a vaccine, in part, because the underlying immunological mechanisms responsible were never clarified. In order to develop a safe and an effective RSV vaccine, we need to develop a better understanding of the immune response to the virus and how the virus evades various components of the innate and adaptive immune responses.

2 Initiation of the Adaptive Immune Response

Dendritic cells (DCs) are antigen-presenting cells that provide a critical link between the innate immune response and the induction of adaptive immunity. DCs are positioned at mucosal interfaces and other potential portals of entry for pathogens where they serve as initial sentinels to detect infection. Immature DCs internalize virus-derived antigens, become activated and further mature and migrate to the local draining lymph nodes where they present antigen to naïve T cells. The signals provided by DCs during the activation of naïve T cells direct the type of effector T cell response that subsequently develops.

Following acute RSV infection, DCs present in the respiratory tract can acquire virus-derived antigens either through direct infection or indirectly by phagocytizing virus particles or necrotic/apoptotic virus-infected epithelial cells. Several subsets of DCs have been identified in both humans and mice based on differences in cell surface phenotype and function. Prior to infection, the lung contains predominantly two major subsets of respiratory DCs, conventional $CD11b^+$ (cDCs; sometimes referred to as myeloid DCs) as well as $CD103^+$ DCs. The $CD103^+$ DCs are largely restricted to mucosal sites and known in the mouse to be linked developmentally and functionally to the $CD8\alpha^+$ DC found in secondary lymphoid organs, such as lymph nodes and spleen. The $CD103^+$ DCs are located in the basal lamina of small and large airways allowing them to extend their dendrites into the airway lumen and sample the antigen environment in the airways. Similar to $CD8\alpha^+$ DCs, $CD103^+$ DCs exhibit the capacity to cross-present antigens taken up from the extracellular space allowing them to activate CD8 T-cell responses from exogenously acquired antigens without direct infection. In contrast to the $CD103^+$ DCs, cDCs are located primarily within the lung parenchyma/interstitium where they aid in the recruitment of cells into the lung via the production of chemokines.

Acute RSV infection in the murine model results in an increase in the number of DCs in the lung for several weeks following infection due to an influx of DCs and DC precursors from the peripheral blood (Beyer et al. 2004; Smit et al. 2006; Wang et al. 2006a). The RSV-induced increase in pulmonary DC numbers is comprised of a variety of DC subsets including cDCs as well as $CD103^+$ DCs and plasmacytoid DCs (pDCs). Simultaneous to the increasing number of lung DCs, a substantial number of DCs present in the lung at the time of initial RSV infection migrate from the lung to the lung-draining lymph nodes via the lymphatics system (Lukens et al. 2009). Both cDCs and $CD103^+$ DCs rapidly migrate to the lung-draining lymph nodes and play a role in the activation of naïve CD8 T cells. $CD103^+$ DCs can either directly initiate the activation of naïve CD8 T cells expressing the appropriate T-cell receptors or transfer antigen to lymph node-resident $CD8\alpha^+$ DCs, which are also potent activators of CD8 T-cell responses. The relative role of $CD103^+$ DCs versus $CD8\alpha^+$ DCs in the induction of RSV-specific CD8 T-cell response remains to be elucidated, although evidence from other models of respiratory virus infection implicate $CD103^+$ DCs as the critical activator of naïve CD8 T cells. In addition, both CD103$^+$ DCs as well as CD8α^+ DCs and migrant cDCs are also capable of stimulating naïve CD4 T cells.

pDCs represent an important source of during many viral infections. Although infected at a much lower frequency than primarily myeloid DCs, RSV infection of pDCs in vitro results in type I IFN production (Johnson et al. 2011). In vivo depletion of pDCs abrogates increased type I IFN levels following RSV infection in mice. Furthermore, pDCs have been shown to be important in limiting pulmonary immunopathology during acute RSV infection in the mouse model as well as promoting viral clearance (Smit et al. 2008; Smit et al. 2006; Wang et al. 2006b). Depletion of pDC results in an increase in mucus production and increased Th2 cytokine production in the lung (Smit et al. 2006), consistent with a role for pDCs in limiting pulmonary Th2 responses. Recent work has indicated that the relative balance of pDCs versus cDCs may be critical in the outcome of acute RSV infection with increased pDC numbers accelerating viral clearance and limiting disease severity (Smit et al. 2008).

The impact of RSV infection on the activation and migration of DCs in humans is currently unclear. A number of studies examining in vitro infection of human monocyte-derived DCs suggest that RSV infection may modulate the function and T cell stimulatory capacity of DCs (Boogaard et al. 2007; de Graaff et al. 2005; Le Nouen et al. 2010; Le Nouen et al. 2011). Only a small fraction of monocyte-derived DCs exposed to RSV become productively infected in vitro, yet RSV exposure appears to affect DC function at the population level (de Graaff et al. 2005). RSV infection of human DCs in vitro results in maturation of the entire DC population, both infected and uninfected DCs in the culture (de Graaff et al. 2005). RSV infection of DCs induces an increase in cell surface expression of both MHC class I and class II as well as the costimulatory molecules CD80 and CD86 (de Graaff et al. 2005). However, RSV infection of DCs results in a decreased capacity to induce the activation and proliferation of autologous CD4 T cells in vitro (de Graaff et al. 2005; Le Nouen et al. 2010). This is thought to be due to either an unknown soluble factor released by the RSV-infected DC or through disruption of synapse formation between the DC and the naïve T cell. In addition, though RSV infection induces some maturation of the DCs, a recent report demonstrated that RSV infection fails to induce decreased expression of the chemokine receptors CCR1, CCR2 and CCR5 as well as an increase in the expression of CCR7 (Le Nouen et al. 2011). This failure to modulate chemokine receptor expression by RSV-infected DCs suggests that RSV may inhibit the trafficking of human DCs from the lung to the draining lymph nodes thereby preventing the induction of a robust T-cell response. More work needs to be done to understand the interaction of RSV with the distinct subsets of DCs present in the human respiratory tract and the impact of RSV infection on the DC response in humans in vivo (see chapters by P.L. Collins et al., S. Barik, S. Mukherjee and N.W. Lukacs, and by A.M.W. Malloy et al., this volume).

3 T-Cell Response

The cellular immune response plays a critical role in host defense against viral infections. In particular, induction of an appropriate T-cell response is required to clear a primary RSV infection. CD8 T cells control acute virus infections through the secretion of cytokines and the lysis of infected host cells. CD8 T cells possess multiple pathways that can induce apoptosis of virus-infected host cells including the lytic granule (i.e., perforin and granzymes) pathway as well as through the cell-surface up-regulation of the death-inducing molecule FasL (Braciale et al. 2012). Recent studies have also implicated a role for TRAIL in CD8 T cell-mediated killing of virus-infected target cells (Brincks et al. 2008). The role of adaptive cellular immunity in reducing RSV viral titers is especially evident in children with defective T-cell responses that demonstrate prolonged virus shedding and experience increased disease severity (Fishaut et al. 1980; Hall et al. 1986). However, despite their critical role in viral clearance, CD8 T cells may also contribute to disease. In vivo depletion of CD8 T cells prior to acute RSV infection results in a significant reduction in weight loss in mice (Graham et al. 1991c). In contrast to the histological analysis of fatal RSV cases that indicated few CD4 and CD8 T cells could be observed in the lung as described above, virus-specific CD8 T cells can be readily detected in the bronchial alveolar lavage and peripheral blood of RSV-infected infants (Heidema et al. 2007). However, no comprehensive human studies to date have evaluated the relationship between the magnitude of the virus-specific CD8 T-cell response during a primary RSV infection and disease severity. Nonetheless, as in mice, CD8 T cells can contribute to protective immunity as an increased frequency of CD8 T cells in young children correlates with a decreased risk of secondary infection with RSV (Mbawuike et al. 2001).

Work in animal models indicates that induction of a T-cell response is required to completely eliminate RSV. Depletion of CD8 T cells in mice delays viral clearance indicating that CD8 T cells are required to clear a primary RSV infection (Graham et al. 1991c). In BALB/c mice, the acute RSV-specific CD8 T-cell response is focused on the immunodominant $M2_{82-90}$ epitope, with 30–50 % of the CD8 T cells in the lungs at the peak of the acute infection recognizing this epitope (Chang and Braciale 2002). RSV $M2_{82-90}$-specific CD8 T cells have been demonstrated to mediate viral clearance. However, the effector mechanisms employed by RSV-specific CD8 T cells to terminate an acute RSV infection is currently unclear. Adoptive transfer of RSV-specific memory CD8 T cells deficient in IFN-γ failed to protect from a subsequent RSV challenge (Ostler et al. 2002). These data suggest that IFN-γ production by RSV-specific CD8 T cells is a critical determinant for their ability to eliminate virus. Data from other studies suggest that FasL may also play a role in viral clearance during acute RSV infection. FasL-deficient mice exhibit delayed clearance of RSV as compared to wild-type controls (Rutigliano and Graham 2004). Taken together, the above studies indicate that CD8 T cells contribute to the control and elimination of virus-infected cells during acute

RSV infection. However, more work needs to be done to determine what effector molecules are required under physiological conditions.

RSV-induced pulmonary pathology shares some characteristics with asthma such as the induction of airway hyperresponsiveness and mucus production. Asthma is characterized by the development of morphologic changes in the lung and cytokine signatures characteristic of Th2 responses, including the development of pulmonary eosinophilia and the production of cytokines such as IL-4, IL-5, and IL-13. Th2-derived cytokines such as IL-4, IL-5 and, most prominently, IL-13 has been shown to play important roles in mediating mucus production and airway hyperreactivity in various models of allergy and asthma. Since RSV-induced disease is associated with increased mucus production and airway hyperreactivity, Th2 responses have been implicated in contributing to the pathogenesis of disease associated with severe RSV infection. Because of the similarities between RSV-induced disease and asthma, the role of various CD4 T-cell subsets in contributing to RSV disease severity has received much attention. The risk of severe RSV-induced disease peaks between 2 and 6 months of age, a timeframe in which the infant immune system appears biased towards Th2 immune responses. The Th2 bias observed in children may be caused by functional differences in neonatal-derived DCs resulting in insufficient signals provided by DCs to newly activated naïve T cells. DCs isolated from human cord blood produce less IL-12, which is important for skewing CD4 T cells into IFN-γ-secreting Th1 cells (Hunt et al. 1994). There have been several reports demonstrating a correlation between the levels of Th2 cytokines and RSV-induced disease severity in young children (Legg et al. 2003; Roman et al. 1997). However, other studies failed to detect any correlation between the levels of Th2-derived cytokines and the severity of RSV-induced disease (Brandenburg et al. 2000; Garofalo et al. 2001). A number of factors may account for these discrepancies in the literature including differences in the age of the patient population, the source of the samples analyzed, and the time following initial infection and when the samples were obtained. It is possible that both Th1 and Th2 cells contribute to various manifestations of RSV-induced disease that may account for the disparate results in humans. In animals, there appears to be a strong correlation between the induction of a Th2 response and an increase in mucus production and increased airway hyperreactivity following acute RSV infection. For example, IL-13 production is associated with increased mucus production in RSV infected BALB/c mice (Tekkanat et al. 2001). Thus, it remains unclear what if any effect the relative balance between Th1 and Th2 cytokine signatures has on RSV-induced disease severity.

Recent data have indicated that Th17 cells may also impact RSV-induced disease severity (Kallal et al. 2010; Lukacs et al. 2010; Mukherjee et al. 2011). Interleukin-17 belongs to a family of cytokines that currently contains six known members IL-17A, IL-17B, IL-17C, IL-17D, IL-17E (i.e., IL-25), and IL-17F. IL-17A and IL-17F share the strongest homology among the family members and both are pro-inflammatory and have been shown to play a role in various autoimmune diseases such as rheumatoid arthritis, multiple sclerosis and inflammatory bowel disease. In addition, IL-17A plays a critical role in host defense against bacterial and fungal infections whereas IL-17F is involved in the development of asthma and airway inflammation. Elevated levels of IL-17A protein were observed in tracheal aspirate samples obtained from RSV-infected infants (Mukherjee et al. 2011). In mice, neutralization of IL-17A results in a significant decrease in mucus production following acute RSV infection (Mukherjee et al. 2011). Thus, IL-17 may play an important role in RSV-induced disease.

Although numerous studies have examined the levels of T cell-derived cytokines in RSV-infected children, there is currently little information available regarding the frequency and magnitude of RSV-specific CD4 T-cell responses. In contrast to influenza virus, there appears to be significantly fewer RSV-specific CD4 T cells in the peripheral blood of healthy adults (de Bree et al. 2007). The significance of these findings is presently unclear and will require more extensive analysis of T-cell frequencies in larger human cohorts to solidify these initial observations.

If the balance between Th1 and Th2 cytokine responses by CD4 T cells responding to RSV infection is a critical factor in the pathogenesis of RSV-induced disease, it would be equally critical to understand where the Th1 or Th2 lineage "commitment" of the CD4 T cells occurs, i.e., is it in the draining lymph nodes where naïve CD4 T cells activate in response to RSV infection or is it in the RSV infected respiratory tract? This is an important area for future investigations. Also it remains formally possible that CD8 T cells secreting IL-4 and/or IL-5, so called Tc2 cells, contribute to the Th2 cytokine signature of the cytokine profile detected in the respiratory tract during primary RSV infection. In this connection it is important to appreciate the potential contribution of the newly described "innate helper cells" (or nuocytes) as major producers of Th2 cytokines (Neill et al. 2010). These cells have been identified both in experimental models of respiratory virus infection as well as in the human respiratory tract (Monticelli et al. 2011). The contribution of this innate immune cell type to the severity of RSV infection and disease development awaits further analysis.

4 Regulation of the T-Cell Response

Tight control of the host immune response is critical to prevent bystander damage to host tissue during an infection. In particular, severe inflammation and tissue damage in the lungs can result in respiratory failure and death. Thus, the immune system employs a number of overlapping mechanisms to regulate the host immune response during respiratory infections. Because the lung is constantly exposed to foreign substances that are often inert (i.e., not infectious), an anti-inflammatory basal state (i.e., tolerogenic) is maintained by lung-resident immune cells such as alveolar macrophages and DCs to avoid induction of unnecessary or inappropriate immune responses. This anti-inflammatory state is orchestrated by environmental cues provided by respiratory epithelial cells. As a consequence, the activation threshold for initiating the innate and adaptive immune responses is increased requiring strong stimuli, such as pathogenic microorganisms to overcome this increased threshold. Once the activation threshold is exceeded and a T-cell response is induced, the immune system utilizes a combination of inhibitory cytokines, such as IL-10, IL-35 and TGF-β, as well as specialized regulatory T cells (Tregs) with inhibitory functions to limit inflammation and prevent damage to the lung tissue.

Tregs represent a specialized subset of CD4 T cells that are responsible for limiting tissue damage and inflammation during immune responses. Two major subsets of Tregs have been identified: naturally occurring Tregs (nTregs) that develop in the thymus and inhibit the activation of conventional T-cell responses to both self- and foreign antigens and induced (also referred to as adaptive) Tregs (iTregs) that develop in the periphery by stimulation of uncommitted naïve CD4 T cells with antigen in the presence of regulatory cytokines or other signals provided by antigen-presenting cells. Although originally characterized by their expression of CD25 (i.e., IL-2Rα chain), Tregs can be distinguished from other T-cell subsets based on the expression of the transcription factor forkhead box P3 (Foxp3) (Fontenot et al. 2003; Hori et al. 2003). Foxp3 expression is upregulated in a subset of thymocytes that display increased affinity for self peptide-MHC complexes and leave the thymus as nTregs. In contrast, iTregs arise from the induction of Foxp3 expression in Foxp3$^-$ CD4 T cells activated under certain conditions including low antigen concentrations, low-to-moderate levels of costimulation and TGF-β. The continued expression of Foxp3 is critical in maintaining the regulatory capacity of both nTregs and iTregs. In mice, expression of Foxp3 appears to be limited primarily to Tregs making it an excellent marker to identify these cells. However, Foxp3 can be transiently expressed by conventional human CD4 T cells following activation, preventing the use of Foxp3 expression alone as a marker for human Tregs. In addition, both nTregs and iTregs express Foxp3 making it difficult to distinguish these two subsets from one another. Both nTregs and iTregs have been reported to suppress the proliferation, cytokine production and cytotoxic activity of both CD4 and CD8 effector T cells, although the exact mechanism(s) of this suppression in vivo remain poorly understood.

Acute RSV infection results in an increase in the number of Tregs in the lung (Fulton et al. 2010; Ruckwardt et al. 2009). It is currently unclear if the majority of Tregs induced during acute RSV infection represent nTregs activated in response to the release of self-antigens by the death of virus-infected host epithelial cells or if they are iTregs responding to RSV-derived antigens. The kinetics of the Treg response in the lung appears to coincide with the kinetics of the RSV-specific CD4 and CD8 T-cell response (Fulton et al. 2010). In mice, a fraction of RSV-specific CD4 T cells identified through MHC class II tetramer staining at the peak of the CD4 T-cell response also stained Foxp3$^+$ indicating that at least some virus-specific iTregs are generated following infection (Liu et al. 2009). Consistent with the capacity of Tregs to modulate various aspects of adaptive immunity, elimination of Tregs following acute RSV infection impacts multiple aspects of the adaptive immune response. Treg depletion resulted in increased production of the pro-inflammatory cytokine IL-6 in the lung airways and resulted in increased cellular infiltration into the lungs including an increase in the numbers of macrophages, lymphocytes, and neutrophils (Fulton et al. 2010; Lee et al. 2010; Ruckwardt et al. 2009). The increased inflammation in Treg depleted mice resulted in increased disease severity following acute RSV infection (Fulton et al. 2010; Lee et al. 2010; Loebbermann et al. 2012b). Importantly, Treg depletion did not impact viral clearance suggesting that Tregs can serve to prevent immunopathology without compromising host defense (Lee et al. 2010; Loebbermann et al. 2012b; Ruckwardt et al. 2009). In addition, a recent study has implicated a critical role for granzyme B expressed by the Tregs in the modulation of the adaptive immune response following acute RSV infection (Loebbermann et al. 2012b). These studies all indicate that Tregs play a vital role in modulating disease severity following acute RSV infection.

Inhibitory cytokines also help regulate the adaptive immune response. Recent work has demonstrated that the inhibitory cytokine IL-10 is produced by multiple cell types during acute RSV infection including Tregs and RSV-specific CD4 and CD8 T cells (Loebbermann et al. 2012a; Sun et al. 2011a; Weiss et al. 2011). By both frequency and total number, CD4 T cells make up the majority of the IL-10-producing cells in the lung following acute RSV infection (Loebbermann et al. 2012a; Sun et al. 2011a; Weiss et al. 2011). The IL-10-producing CD4 T cells can be subdivided into three distinct populations, $Foxp3^{+}$ Tregs, Tr1 cells that make IL-10 but do not co-produce IFN-γ and Th1 cells that make IFN-γ and have gained the capacity to make IL-10 (Weiss et al. 2011). IL-10 production by CD4 T cells appears to peak at the same time as the peak of the RSV-induced T-cell response in the lung, suggesting that the production of IL-10 plays an important role in limiting the development of pulmonary immunopathology. IL-10-deficient mice or wild-type mice treated with blocking anti-IL-10R antibodies exhibit increased pro-inflammatory cytokine and chemokine production in the lung resulting in increased pulmonary inflammation (Loebbermann et al. 2012a; Sun et al. 2011a; Weiss et al. 2011). Disruption of IL-10 did not alter peak RSV titers in the lung or affect the rate of viral clearance (Loebbermann et al. 2012a; Sun et al. 2011a; Weiss et al. 2011). However, as a consequence of the increased inflammation, disruption of IL-10 resulted in significantly increased severity of the RSV-induced disease demonstrating the importance of IL-10 in limiting the induction of RSV-induced pulmonary injury caused by the host immune response (Loebbermann et al. 2012a; Sun et al. 2011a; Weiss et al. 2011). Therefore, in the experimental murine model of RSV infection, IL-10 produced in the infected respiratory tract during the acute phase of infection by several distinct cell types including CD8 and several distinct subpopulations of CD4 T cells may play a critical role into controlling excess inflammation during the host response to infection. It is currently unclear what signals induce the production of IL-10 by CD4 T cells during acute RSV infection. However, recent data examining the induction of IL-10 by CD8 T cells during acute influenza virus infection suggests that IL-2 made by CD4 T cells in conjunction with IL-27 produced by innate cells in the lung may induce potent IL-10 production by virus-specific CD4 and CD8 T cells (Sun et al. 2011b).

5 B Cell Response and Antibodies

Antibodies play an essential role in preventing re-infection with viruses by either directly neutralizing or aiding in the opsonization of extracellular virus particles. RSV infection elicits an antibody response following infection that rapidly decays over time. The RSV attachment (G) and fusion (F) proteins are the major targets of RSV-specific neutralizing antibodies. Both IgA and IgG play important roles in preventing re-infection with RSV with both IgA and IgG providing protection against upper respiratory tract infection and primarily IgG protecting against lower respiratory tract infection. Consistent with this notion, B cell depletion does not affect the clearance of a primary RSV infection but significantly impacts the rate of viral clearance after secondary RSV infection (Graham et al. 1991a). Following acute RSV infection, the frequency of antibody-secreting plasma cells rapidly declines in the upper respiratory tract (Singleton et al. 2003). RSV-immune mice with lower serum titers of RSV-specific IgG exhibit decreased protection from a secondary RSV challenge (Graham et al. 1991b).

In humans, both IgA and IgG titers rapidly wane following acute RSV infection, particularly in young children. In adults, lower nasal IgA titers were correlated with increased RSV infection rates (Walsh and Falsey 2004). In addition, within 1 year following a natural RSV infection, approximately 75 % of adults exhibit a >fourfold decrease in neutralizing anti-RSV IgG titers (Falsey et al. 2006). This decline likely contributes to re-infection with RSV as lower serum neutralizing antibody titers have been shown to correlate with an increased susceptibility to RSV infection in the elderly (Falsey and Walsh 1998). In children, the presence of maternal antibodies combined with the immaturity of the immune system likely contributes to diminished antibody responses induced by acute RSV infection (Kasel et al. 1987). Neutralizing RSV-specific antibody responses can only be detected in 50–75 % of children less than six months of age (Brandenburg et al. 1997; Murphy et al. 1986). Thus, RSV infection elicits an antibody response that fails to establish long-lasting immunity and prevent periodic reinfections throughout life.

The induction of primary antibody responses to RSV likely occurs in the lymph nodes draining the respiratory tract. In these draining lymph nodes, virus-specific extrafollicular and marginal zone B cells encounter viral constituents, such as RSV antigens, and initiate engagement of their surface immunoglobulin B cell receptor. At the same time, naïve CD4 T cells encounter respiratory DCs that have entered the lymph nodes and become activated and in response to engagement of critical costimulatory receptors (e.g., ICOS) and encounter critical cytokines (e.g., IL-6) to differentiate into RSV-specific T follicular helper (Tfh) cells. Tfh cells support the differentiation of extrafollicular B cells into antibody-secreting plasma cells by promoting germinal center formation and affinity maturation.

6 Modulation of the Adaptive Immune Response by RSV

Several RSV-encoded proteins have been shown to modulate aspects of the innate and adaptive immune responses. For example, the two nonstructural proteins NS1 and NS2 have been shown to act individually as well as in concert to inhibit type I IFN production as well as signaling (Ramaswamy et al. 2006; Spann et al. 2004). Inhibition of the type I IFN response by NS1 and NS2 also negatively impacts the magnitude of the CD8 T-cell response (Kotelkin et al. 2006; and chapter by J.S. McLellan et al., this volume). Recent work has indicated that naïve CD8 T cells require 3 signals in order to become activated, proliferate and differentiate into effector cytotoxic lymphocytes: (1) antigen; (2) costimulation; and (3) cytokines such as IL-12 and/or IFNs. Thus, the inhibition of type I IFN by NS1 and NS2 may deprive activated CD8 T cells of type I IFN during differentiation that serves as signal 3, providing critical cell-survival signals to the CD8 T cell.

The RSV G protein has been shown to contain a number of structural features that serve to modulate the adaptive immune response. The G protein features both a membrane anchored as well as a soluble secreted form. Because the G protein is a major target of neutralizing antibodies, one important role of the soluble form is to act as a decoy to bind neutralizing antibodies and prevent neutralization and/or opsonization of the viral particles. The RSV G protein also contains a highly conserved central region that bears high amino acid sequence homology with the CX3C chemokine fractalkine (Tripp et al. 2001). Purified G protein has been shown to bind cells expressing the fractalkine receptor and mediate their chemotaxis (Tripp et al. 2001). Thus, by secreting the G protein that can serve as a soluble chemokine receptor agonist, RSV can modulate the local inflammatory environment in the lung by altering the infiltrating inflammatory cells. Consistent with this, a report comparing infection of mice with a recombinant RSV lacking the G protein to wild-type virus indicated that the presence of the G protein inhibited the trafficking of $CD11b^{+}$ cells as well as NK cells into the lung (Harcourt et al. 2006). In addition, the expression of various chemokines was reduced in the lung in the presence of the G protein and reduced Th1 and increased Th2 responses were observed as well as a reduced CD8 T-cell response (see chapters by J.S. McLellan et al., S. Mukherjee and N.W. Lukacs, and by S. Barik, this volume).

Polymorphisms in the pattern recognition receptor toll-like receptor 4 (TLR4) are associated with increased susceptibility to RSV infection (Awomoyi et al. 2007; Mandelberg et al. 2006; Puthothu et al. 2006; Tal et al. 2004; Tulic et al. 2007). Consistent with a role for TLR4 in RSV infection, the RSV F protein has been shown to be a TLR4 agonist (Kurt-Jones et al. 2000). However, it is currently unclear how TLR4 activation by the RSV F protein is beneficial for the virus or how it may modulate the adaptive immune response.

7 Summary and Conclusions

The induction of long-lived T and B cell memory upon primary exposure to a foreign antigen serves as the primary basis for vaccination. However, RSV causes recurrent infections throughout life suggesting that the virus either evades or compromises the generation of long-lived memory. Studies of RSV-induced immune responses continue to uncover novel mechanisms by which RSV modulates the host immune response. More work is needed to gain a better understanding of how RSV prevents the development of long-lived T and B cell immunity in humans. Thus, it will be important to firmly establish the immunological correlates that provide protective immunity in order to design a vaccine that is capable of eliciting a protective immune response of greater magnitude and duration than that elicited by natural RSV infection. The relative role of the host immune response in contributing to RSV-induced disease in humans remains unclear. More work in humans is needed to gain a better understanding of RSV-induced host responses and their contribution to disease.

References

Awomoyi AA, Rallabhandi P, Pollin TI, Lorenz E, Sztein MB, Boukhvalova MS, Hemming VG, Blanco JC, Vogel SN (2007) Association of TLR4 polymorphisms with symptomatic respiratory syncytial virus infection in high-risk infants and young children. J Immunol 179(5):3171–3177 179/5/3171 [pii]

Beyer M, Bartz H, Horner K, Doths S, Koerner-Rettberg C, Schwarze J (2004) Sustained increases in numbers of pulmonary dendritic cells after respiratory syncytial virus infection. J Allergy Clin Immunol 113(1):127–133

Boogaard I, van Oosten M, van Rijt LS, Muskens F, Kimman TG, Lambrecht BN, Buisman AM (2007) Respiratory syncytial virus differentially activates murine myeloid and plasmacytoid dendritic cells. Immunology 122(1):65–72. doi:10.1111/j.1365-2567.2007.02613.x

Braciale TJ, Sun J, Kim TS (2012) Regulating the adaptive immune response to respiratory virus infection. Nat Rev Immunol 12(4):295–305. doi:10.1038/nri3166

Brandenburg AH, Groen J, van Steensel-Moll HA, Claas EC, Rothbarth PH, Neijens HJ, Osterhaus AD (1997) Respiratory syncytial virus specific serum antibodies in infants under six months of age: limited serological response upon infection. J Med Virol 52(1):97–104

Brandenburg AH, Kleinjan A, Land BV, Moll HA, Timmerman HH, de Swart RL, Neijens HJ, Fokkens W, Osterhaus ADME (2000) Type 1-like immune response is found in children with respiratory syncytial virus infection regardless of clinical severity. J Med Virol 62(2):267–277

Brincks EL, Katewa A, Kucaba TA, Griffith TS, Legge KL (2008) CD8 T cells utilize TRAIL to control influenza virus infection. J Immunol 181(7):4918–4925 181/7/4918 [pii]

Castilow EM, Varga SM (2008) Overcoming T cell-mediated immunopathology to achieve safe RSV vaccination. Future Virol 3(5):445–454. doi:10.2217/17460794.3.5.445

Chang J, Braciale TJ (2002) Respiratory syncytial virus infection suppresses lung $CD8^+$ T-cell effector activity and peripheral $CD8^+$ T-cell memory in the respiratory tract. Nat Med 8(1):54–60. doi:10.1038/nm0102-54

de Bree GJ, Daniels H, Schilfgaarde M, Jansen HM, Out TA, van Lier RA, Jonkers RE (2007) Characterization of $CD4^+$ memory T cell responses directed against common respiratory pathogens in peripheral blood and lung. J Infect Dis 195(11):1718–1725. doi:10.1086/517612 de Graaff PM, de Jong EC, van Capel TM, van Dijk ME, Roholl PJ, Boes J, Luytjes W, Kimpen JL, van Bleek GM (2005) Respiratory syncytial virus infection of monocyte-derived dendritic cells decreases their capacity to activate CD4 T cells. J Immunol 175(9):5904–5911

DeVincenzo JP, Wilkinson T, Vaishnaw A, Cehelsky J, Meyers R, Nochur S, Harrison L, Meeking P, Mann A, Moane E, Oxford J, Pareek R, Moore R, Walsh E, Studholme R, Dorsett P, Alvarez R, Lambkin-Williams R (2010) Viral load drives disease in humans experimentally infected with respiratory syncytial virus. Am J Respir Crit Care Med 182(10):1305–1314. doi:10.1164/rccm.201002-0221OC

El Saleeby CM, Bush AJ, Harrison LM, Aitken JA, Devincenzo JP (2011) Respiratory syncytial virus load, viral dynamics, and disease severity in previously healthy naturally infected children. J Infect Dis 204(7):996–1002. doi:10.1093/infdis/jir494

Everard ML, Swarbrick A, Wrightham M, McIntyre J, Dunkley C, James PD, Sewell HF, Milner AD (1994) Analysis of cells obtained by bronchial lavage of infants with respiratory syncytial virus infection. Arch Dis Child 71(5):428–432

Falsey AR, Walsh EE (1998) Relationship of serum antibody to risk of respiratory syncytial virus infection in elderly adults. J Infect Dis 177(2):463–466

Falsey AR, Singh HK, Walsh EE (2006) Serum antibody decay in adults following natural respiratory syncytial virus infection. J Med Virol 78(11):1493–1497. doi:10.1002/jmv.20724

Fishaut M, Tubergen D, McIntosh K (1980) Cellular response to respiratory viruses with particular reference to children with disorders of cell-mediated immunity. J Pediatr 96(2):179–186

Fontenot JD, Gavin MA, Rudensky AY (2003) Foxp3 programs the development and function of $CD4^{+}CD25^{+}$ regulatory T cells. Nat Immunol 4(4):330–336. doi:10.1038/ni904

Fulton RB, Meyerholz DK, Varga SM (2010) $Foxp3^{+}$ CD4 regulatory T cells limit pulmonary immunopathology by modulating the CD8 T cell response during respiratory syncytial virus infection. J Immunol 185(4):2382–2392. doi:10.4049/jimmunol.1000423

Garofalo RP, Patti J, Hintz KA, Hill V, Ogra PL, Welliver RC (2001) Macrophage inflammatory protein-1α (not T helper type 2 cytokines) is associated with severe forms of respiratory syncytial virus bronchiolitis. J Infect Dis 184(4):393–399. doi:10.1086/322788

Graham BS, Bunton LA, Rowland J, Wright PF, Karzon DT (1991a) Respiratory syncytial virus infection in anti-μ-treated mice. J Virol 65(9):4936–4942

Graham BS, Bunton LA, Wright PF, Karzon DT (1991b) Reinfection of mice with respiratory syncytial virus. J Med Virol 34(1):7–13

Graham BS, Bunton LA, Wright PF, Karzon DT (1991c) Role of T lymphocyte subsets in the pathogenesis of primary infection and rechallenge with respiratory syncytial virus in mice. J Clin Invest 88(3):1026–1033. doi:10.1172/JCI115362

Hall CB, Powell KR, MacDonald NE, Gala CL, Menegus ME, Suffin SC, Cohen HJ (1986) Respiratory syncytial viral infection in children with compromised immune function. N Engl J Med 315(2):77–81. doi:10.1056/NEJM198607103150201

Harcourt J, Alvarez R, Jones LP, Henderson C, Anderson LJ, Tripp RA (2006) Respiratory syncytial virus G protein and G protein CX3C motif adversely affect $CX3CR1^{+}$ T cell responses. J Immunol 176(3):1600–1608 176/3/1600 [pii]

Heidema J, Lukens MV, van Maren WW, van Dijk ME, Otten HG, van Vught AJ, van der Werff DB, van Gestel SJ, Semple MG, Smyth RL, Kimpen JL, van Bleek GM (2007) CD8+ T cell responses in bronchoalveolar lavage fluid and peripheral blood mononuclear cells of infants with severe primary respiratory syncytial virus infections. J Immunol 179(12):8410–8417

Hori S, Nomura T, Sakaguchi S (2003) Control of regulatory T cell development by the transcription factor Foxp3. Science 299(5609):1057–1061. doi:10.1126/science.1079490

Hunt DW, Huppertz HI, Jiang HJ, Petty RE (1994) Studies of human cord blood dendritic cells: evidence for functional immaturity. Blood 84(12):4333–4343

Johnson JE, Gonzales RA, Olson SJ, Wright PF, Graham BS (2007) The histopathology of fatal untreated human respiratory syncytial virus infection. Mod Pathol 20(1):108–119. doi:10.1038/modpathol.3800725

Johnson TR, Johnson CN, Corbett KS, Edwards GC, Graham BS (2011) Primary human mDC1, mDC2, and pDC dendritic cells are differentially infected and activated by respiratory syncytial virus. PLoS ONE 6(1):e16458. doi:10.1371/journal.pone.0016458

Kallal LE, Hartigan AJ, Hogaboam CM, Schaller MA, Lukacs NW (2010) Inefficient lymph node sensitization during respiratory viral infection promotes IL-17-mediated lung pathology. J Immunol 185(7):4137–4147. doi:10.4049/jimmunol.1000677

Kasel JA, Walsh EE, Frank AL, Baxter BD, Taber LH, Glezen WP (1987) Relation of serum antibody to glycoproteins of respiratory syncytial virus with immunity to infection in children. Viral Immunol 1(3):199–205

Kotelkin A, Belyakov IM, Yang L, Berzofsky JA, Collins PL, Bukreyev A (2006) The NS2 protein of human respiratory syncytial virus suppresses the cytotoxic T-cell response as a consequence of suppressing the type I interferon response. J Virol 80(12):5958–5967. doi:10.1128/JVI.00181-06

Kurt-Jones EA, Popova L, Kwinn L, Haynes LM, Jones LP, Tripp RA, Walsh EE, Freeman MW, Golenbock DT, Anderson LJ, Finberg RW (2000) Pattern recognition receptors TLR4 and CD14 mediate response to respiratory syncytial virus. Nat Immunol 1(5):398–401. doi:10.1038/80833

Le Nouen C, Hillyer P, Munir S, Winter CC, McCarty T, Bukreyev A, Collins PL, Rabin RL, Buchholz UJ (2010) Effects of human respiratory syncytial virus, metapneumovirus, parainfluenza virus 3 and influenza virus on CD4+ T cell activation by dendritic cells. PLoS ONE 5(11):e15017. doi:10.1371/journal.pone.0015017

Le Nouen C, Hillyer P, Winter CC, McCarty T, Rabin RL, Collins PL, Buchholz UJ (2011) Low CCR7-mediated migration of human monocyte derived dendritic cells in response to human respiratory syncytial virus and human metapneumovirus. PLoS Pathog 7(6):e1002105. doi:10.1371/journal.ppat.1002105

Lee DC, Harker JA, Tregoning JS, Atabani SF, Johansson C, Schwarze J, Openshaw PJ (2010) $CD25^+$ natural regulatory T cells are critical in limiting innate and adaptive immunity and resolving disease following respiratory syncytial virus infection. J Virol 84(17):8790–8798. doi:10.1128/JVI.00796-10

Legg JP, Hussain IR, Warner JA, Johnston SL, Warner JO (2003) Type 1 and type 2 cytokine imbalance in acute respiratory syncytial virus bronchiolitis. Am J Respir Crit Care Med 168(6):633–639. doi:10.1164/rccm.200210-1148OC

Liu J, Ruckwardt TJ, Chen M, Johnson TR, Graham BS (2009) Characterization of respiratory syncytial virus M- and M2-specific CD4 T cells in a murine model. J Virol 83(10):4934–4941. doi:10.1128/JVI.02140-08

Loebbermann J, Schnoeller C, Thornton H, Durant L, Sweeney NP, Schuijs M, O'Garra A, Johansson C, Openshaw PJ (2012a) IL-10 regulates viral lung immunopathology during acute respiratory syncytial virus infection in mice. PLoS ONE 7(2):e32371. doi:10.1371/journal.pone.0032371

Loebbermann J, Thornton H, Durant L, Sparwasser T, Webster KE, Sprent J, Culley FJ, Johansson C, Openshaw PJ (2012b) Regulatory T cells expressing granzyme B play a critical role in controlling lung inflammation during acute viral infection. Mucosal Immunol 5(2):161–172. doi:10.1038/mi.2011.62

Lukacs NW, Smit JJ, Mukherjee S, Morris SB, Nunez G, Lindell DM (2010) Respiratory virus-induced TLR7 activation controls IL-17-associated increased mucus via IL-23 regulation. J Immunol 185(4):2231–2239. doi:10.4049/jimmunol.1000733

Lukens MV, Kruijsen D, Coenjaerts FE, Kimpen JL, van Bleek GM (2009) Respiratory syncytial virus-induced activation and migration of respiratory dendritic cells and subsequent antigen presentation in the lung-draining lymph node. J Virol 83(14):7235–7243. doi:10.1128/JVI.00452-09 JVI.00452-09 [pii]

Lukens MV, van de Pol AC, Coenjaerts FE, Jansen NJ, Kamp VM, Kimpen JL, Rossen JW, Ulfman LH, Tacke CE, Viveen MC, Koenderman L, Wolfs TF, van Bleek GM (2010) A systemic neutrophil response precedes robust $CD8^+$ T-cell activation during natural respiratory syncytial virus infection in infants. J Virol 84(5):2374–2383. doi:10.1128/JVI.01807-09

Mandelberg A, Tal G, Naugolny L, Cesar K, Oron A, Houri S, Gilad E, Somekh E (2006) Lipopolysaccharide hyporesponsiveness as a risk factor for intensive care unit hospitalization in infants with respiratory syncitial virus bronchiolitis. Clin Exp Immunol 144(1):48–52. doi:10.1111/j.1365-2249.2006.03030.x CEI3030 [pii]

Mbawuike IN, Wells J, Byrd R, Cron SG, Glezen WP, Piedra PA (2001) HLA-restricted CD8$^+$ cytotoxic T lymphocyte, interferon-γ, and interleukin-4 responses to respiratory syncytial virus infection in infants and children. J Infect Dis 183(5):687–696. doi:10.1086/318815

Monticelli LA, Sonnenberg GF, Abt MC, Alenghat T, Ziegler CG, Doering TA, Angelosanto JM, Laidlaw BJ, Yang CY, Sathaliyawala T, Kubota M, Turner D, Diamond JM, Goldrath AW, Farber DL, Collman RG, Wherry EJ, Artis D (2011) Innate lymphoid cells promote lung-tissue homeostasis after infection with influenza virus. Nat Immunol 12(11):1045–1054. doi:10.1031/ni.2131

Mukherjee S, Lindell DM, Berlin AA, Morris SB, Shanley TP, Hershenson MB, Lukacs NW (2011) IL-17-induced pulmonary pathogenesis during respiratory viral infection and exacerbation of allergic disease. Am J Pathol 179(1):248–258. doi:10.1016/j.ajpath.2011.03.003

Murphy BR, Graham BS, Prince GA, Walsh EE, Chanock RM, Karzon DT, Wright PF (1986) Serum and nasal-wash immunoglobulin G and a antibody response of infants and children to respiratory syncytial virus F and G glycoproteins following primary infection. J Clin Microbiol 23(6):1009–1014

Neill DR, Wong SH, Bellosi A, Flynn RJ, Daly M, Langford TK, Bucks C, Kane CM, Fallon PG, Pannell R, Jolin HE, McKenzie AN (2010) Nuocytes represent a new innate effector leukocyte that mediates type-2 immunity. Nature 464(7293):1367–1370. doi:10.1038/nature08900

Ostler T, Davidson W, Ehl S (2002) Virus clearance and immunopathology by CD8$^+$ T cells during infection with respiratory syncytial virus are mediated by IFN-gamma. Eur J Immunol 32(8):2117–2123. doi:10.1002/1521-4141(200208)32:8<2117:AID-IMMU2117>3.0.CO;2-C

Puthothu B, Forster J, Heinzmann A, Krueger M (2006) TLR-4 and CD14 polymorphisms in respiratory syncytial virus associated disease. Dis Markers 22(5–6):303–308

Ramaswamy M, Shi L, Varga SM, Barik S, Behlke MA, Look DC (2006) Respiratory syncytial virus nonstructural protein 2 specifically inhibits type I interferon signal transduction. Virology 344(2):328–339. doi:10.1016/j.virol.2005.09.009

Roman M, Calhoun WJ, Hinton KL, Avendano LF, Simon V, Escobar AM, Gaggero A, Diaz PV (1997) Respiratory syncytial virus infection in infants is associated with predominant Th-2-like response. Am J Respir Crit Care Med 156(1):190–195

Ruckwardt TJ, Bonaparte KL, Nason MC, Graham BS (2009) Regulatory T cells promote early influx of CD8$^+$ T cells in the lungs of respiratory syncytial virus-infected mice and diminish immunodominance disparities. J Virol 83(7):3019–3028. doi:10.1128/JVI.00036-09

Rutigliano JA, Graham BS (2004) Prolonged production of TNF-α exacerbates illness during respiratory syncytial virus infection. J Immunol 173(5):3408–3417 173/5/3408 [pii]

Singleton R, Etchart N, Hou S, Hyland L (2003) Inability to evoke a long-lasting protective immune response to respiratory syncytial virus infection in mice correlates with ineffective nasal antibody responses. J Virol 77(21):11303–11311

Smit JJ, Rudd BD, Lukacs NW (2006) Plasmacytoid dendritic cells inhibit pulmonary immunopathology and promote clearance of respiratory syncytial virus. J Exp Med 203(5):1153–1159. doi:10.1084/jem.20052359

Smit JJ, Lindell DM, Boon L, Kool M, Lambrecht BN, Lukacs NW (2008) The balance between plasmacytoid DC versus conventional DC determines pulmonary immunity to virus infections. PLoS ONE 3(3):e1720. doi:10.1371/journal.pone.0001720

Spann KM, Tran KC, Chi B, Rabin RL, Collins PL (2004) Suppression of the induction of alpha, beta, and lambda interferons by the NS1 and NS2 proteins of human respiratory syncytial virus in human epithelial cells and macrophages. J Virol 78(8):4363–4369 (corrected)

Sun J, Cardani A, Sharma AK, Laubach VE, Jack RS, Muller W, Braciale TJ (2011a) Autocrine regulation of pulmonary inflammation by effector T-cell derived IL-10 during infection with respiratory syncytial virus. PLoS Pathog 7(8):e1002173. doi:10.1371/journal.ppat.1002173

Sun J, Dodd H, Moser EK, Sharma R, Braciale TJ (2011b) CD4$^+$ T cell help and innate-derived IL-27 induce Blimp-1-dependent IL-10 production by antiviral CTLs. Nat Immunol 12(4):327–334. doi:10.1038/ni.1996

Tal G, Mandelberg A, Dalal I, Cesar K, Somekh E, Tal A, Oron A, Itskovich S, Ballin A, Houri S, Beigelman A, Lider O, Rechavi G, Amariglio N (2004) Association between common Toll-like receptor 4 mutations and severe respiratory syncytial virus disease. J Infect Dis 189(11):2057–2063. doi:10.1086/420830 JID31682 [pii]

Tekkanat KK, Maassab HF, Cho DS, Lai JJ, John A, Berlin A, Kaplan MH, Lukacs NW (2001) IL-13-induced airway hyperreactivity during respiratory syncytial virus infection is STAT6 dependent. J Immunol 166(5):3542–3548

Tripp RA, Jones LP, Haynes LM, Zheng H, Murphy PM, Anderson LJ (2001) CX3C chemokine mimicry by respiratory syncytial virus G glycoprotein. Nat Immunol 2(8):732–738. doi:10.1038/90675

Tulic MK, Hurrelbrink RJ, Prele CM, Laing IA, Upham JW, Le Souef P, Sly PD, Holt PG (2007) TLR4 polymorphisms mediate impaired responses to respiratory syncytial virus and lipopolysaccharide. J Immunol 179(1):132–140 179/1/132 [pii]

Walsh EE, Falsey AR (2004) Humoral and mucosal immunity in protection from natural respiratory syncytial virus infection in adults. J Infect Dis 190(2):373–378. doi:10.1086/421524

Wang H, Peters N, Laza-Stanca V, Nawroly N, Johnston SL, Schwarze J (2006a) Local CD11c$^+$ MHC class II$^-$ precursors generate lung dendritic cells during respiratory viral infection, but are depleted in the process. J Immunol 177(4):2536–2542

Wang H, Peters N, Schwarze J (2006b) Plasmacytoid dendritic cells limit viral replication, pulmonary inflammation, and airway hyperresponsiveness in respiratory syncytial virus infection. J Immunol 177(9):6263–6270

Weiss KA, Christiaansen AF, Fulton RB, Meyerholz DK, Varga SM (2011) Multiple CD4$^+$ T cell subsets produce immunomodulatory IL-10 during respiratory syncytial virus infection. J Immunol 187(6):3145–3154. doi:10.4049/jimmunol.1100764

Welliver TP, Garofalo RP, Hosakote Y, Hintz KH, Avendano L, Sanchez K, Velozo L, Jafri H, Chavez-Bueno S, Ogra PL, McKinney L, Reed JL, Welliver RC Sr (2007) Severe human lower respiratory tract illness caused by respiratory syncytial virus and influenza virus is characterized by the absence of pulmonary cytotoxic lymphocyte responses. J Infect Dis 195(8):1126–1136. doi:10.1086/512615

Welliver TP, Reed JL, Welliver RC Sr (2008) Respiratory syncytial virus and influenza virus infections: observations from tissues of fatal infant cases. Pediatr Infect Dis J 27(10 Suppl):S92–S96. doi:10.1097/INF.0b013e318168b706

Respiratory Syncytial Virus Mechanisms to Interfere with Type 1 Interferons

Sailen Barik

Abstract Respiratory syncytial virus (RSV) is a member of the Paramyxoviridae family that consists of viruses with nonsegmented negative-strand RNA genome. Infection by these viruses triggers the innate antiviral response of the host, mainly type I interferon (IFN). Essentially all other viruses of this family produce IFN suppressor functions by co-transcriptional RNA editing. In contrast, RSV has evolved two unique nonstructural proteins, NS1 and NS2, to effectively serve this purpose. Together, NS1 and NS2 degrade or sequester multiple signaling proteins that affect both IFN induction and IFN effector functions. While the mechanism of action of NS1 and NS2 is a subject of active research, their effect on adaptive immunity is also being recognized. In this review, we discuss various aspects of NS1 and NS2 function with implications for vaccine design.

Contents

S. Barik (✉)
Center for Gene Regulation in Health and Disease, Department of Biological, Geological and Environmental Sciences, Cleveland State University, 2121 Euclid Avenue, Cleveland, OH 44115, USA
e-mail: s.barik@csuohio.edu

L. J. Anderson and B. S. Graham (eds.), *Challenges and Opportunities for Respiratory Syncytial Virus Vaccines*, Current Topics in Microbiology and Immunology 372,
DOI: 10.1007/978-3-642-38919-1_9,

1 Introduction

Respiratory syncytial virus (RSV) is a member of the *Paramyxoviridae* family that includes a number of clinically significant human and animal pathogens, such as measles, mumps, parainfluenza, Sendai, and metapneumovirus. RSV belongs to the *Pneumovirus* genus of this family, and is the most significant cause of severe lower respiratory tract infection in infants and an important cause of acute respiratory illness in the elderly (Collins and Melero 2011). Unfortunately, despite intense research, there is no clinically approved vaccine or reliable therapeutics against RSV. Palivizumab (Synagis®), a composite humanized monoclonal antibody (95 % human, 5 % murine) against RSV fusion protein (F), and ribavirin, a non-specific antiviral with significant toxicity, are currently used in severe cases only (see chapter by J.A. Melero and M.L. Moore, this volume). An intranasal RNA interference-based therapy (Bitko et al. 2005), currently targeting the viral N sequence (ALN-RSV01), has finished Phase IIb clinical trial, and appears promising (Zamora et al. 2011 and references therein). Ribavirin and RNAi inhibit viral gene expression and function as therapeutics, and palivizumab, an F-specific antibody is used prophylactically.

Like all members of this family, RSV contains a single-stranded, nonsegmented RNA genome of negative sense (anti-mRNA sense); the RSV genome itself contains 10 genes encoding 11 proteins (as the M2 gene contains two coding sequences), (see chapter by P.L. Collins et al., this volume). Two intriguing gene products of RSV are its nonstructural (NS) proteins that play a cardinal role in suppressing cellular innate immunity, mainly represented by type I interferons (IFN). The interaction between RSV and IFN involves two pathways: IFN *synthesis* and IFN *response*. Viral infection activates the synthesis of IFN, and the liberated IFN then acts upon the neighboring uninfected cells to activate the IFN response pathway, which makes the target cell virus-resistant. However, RNA viruses have evolved to encode specialized proteins, such as the NS proteins of RSV, which suppress these pathways to allow optimal virus growth. An understanding of this exciting new area of host-virus interaction will require an introduction to IFN and related pathways, which is provided below (see chapters by S. Mukherjee and N.W. Lukacs, and by S.M. Varga and T.J. Braciale, this volume).

2 Type I IFN-Based Innate Immunity and Its Suppression by RSV

Highly comprehensive reviews of type I IFN have appeared recently (Fensterl and Sen 2009). In brief, type I IFN genes are found in mammals, birds, fish, and perhaps in reptiles and amphibians (Fensterl and Sen 2009). They are specialized cytokines, 165–208 amino acids long, that are synthesized and released by host cells in response to various pathogens, such as viruses, bacteria, and parasites. IFN are divided into three types, based on amino acid sequence homology and the receptors they use. Type I IFN mainly refers to the many subtypes of IFN-α (e.g., 13 in human), and a single IFN-β. The remaining, lesser studied type I members are not ubiquitous, but expressed in specific cell-types and species, such as IFN-ζ, which is found only in mouse and shares 32 and 26 % amino acid sequence identity with mouse IFNα and mouse IFNβ, respectively. Type II IFN is represented by its only member, IFN-γ, which is unrelated to type I IFN, and is mainly produced by activated T cells and natural killer (NK) cells (Fensterl and Sen 2009). Type III IFNs are the most recently identified, and comprises of IFN-λ1, IFN-λ2, IFN-λ3, also known as interleukin-29 (IL-29), IL-28A, and IL-28B, respectively. IFNs of a given type are highly similar in sequence and also located in a single cluster on the same chromosome, perhaps indicating type-specific gene duplication. The possible functional redundancy of multiple type I and type III IFNs may have relevance in the immunopathology of the RSV disease, but this has remained under-explored.

2.1 IFN Synthesis Triggered by RSV Infection

The epithelial cells, lining the inner surface of the alveoli, serve as the initial primary target of lower respiratory tract infection in the RSV disease. Much of our knowledge of RSV in fact originated from the use of the alveolar carcinoma cell line, A549, which reproduces most of the properties of type II epithelial cells of the alveoli. These cells are major producers of type I IFN, and thus, studies of innate immunity towards RSV has focused primarily on IFN-α and IFN-β.

Recent studies have revealed that the A549 and nasal epithelial cells (as well as dendritic cells/macrophages) also produce type III IFN (IFN-λ) (Spann et al. 2004; Okabayashi et al. 2011). More recently, RSV infection of cells of the upper respiratory tract (URT), represented by nasal epithelial cells, was shown to trigger the induction of type III IFN (IFN-λ), but not type I (Okabayashi et al. 2011). Nonetheless, like type I IFN, type III IFN also rendered respiratory epithelial cells resistant against a broad spectrum of viruses including RSV (Mordstein et al. 2010). Lastly, T cell proliferation, suppressed by RSV infection, was only modestly restored by blocking either type I or type III IFN response, but strongly restored when both types are blocked. Together, these studies documented important roles of type III as well as type I IFN in RSV innate immunity. Nonetheless, as our knowledge of type III IFN is still relatively meager, we will concentrate on IFN-α and IFN-β here, referring to IFN-λ only when relevant information is available.

The innate immune pathway that leads to the induction of type I (and type III) IFN begins with the recognition of pathogen-associated molecular patterns (PAMP) by cognate receptors of the host cell, known as pattern-recognition receptors (PRR) (comprehensively reviewed by Kumar et al. 2011; Lee et al. 2012). The major PRRs, in order of relevance to innate immune detection of RSV, are: (i) the cytoplasmic DEAD/DEAH-box RNA helicase receptors or RIG-I-like receptors (RLR) (Liu et al. 2007; Sasai et al. 2006), (ii) Toll-like receptors (TLR), and (iii) the comparatively less important NOD-like receptors (NLR) (Sabbah et al. 2009; Vissers et al. 2012).

2.1.1 RLR-Mediated Innate Immune Response Against RSV

The two major members of the RLR family are RIG-I (retinoic acid-inducible gene I) and Melanoma Differentiation-Associated protein 5 (MDA5) (Yoneyama et al. 2005; Takeuchi and Akira 2009). Both are expressed basally at low levels in all cell types including epithelial cells, but are induced by type I IFN in a positive feedback loop. RIG-I, specifically, is rapidly induced following RSV infection (Bitko et al. 2008). Both RIG-I and MDA5 possess multiple caspase activation and recruitment domains (CARDs) (Yoneyama et al. 2005).

Each paramyxovirus shows various degrees of preference for using either RIG-I or MDA5 (Kato et al. 2006; Loo et al. 2008). While both are activated by dsRNA, there are other features of the dsRNA that determine specificity (Belgnaoui et al. 2011). RIG-I can be activated by short 5′-triphosphated (5′ppp) RNA that also contains double-stranded regions, whereas MDA5 generally prefers long dsRNA molecules that do not need to have 5′ppp (Kato et al. 2006, 2008; Schmidt et al. 2009). Regardless, RIG-I serves as the major RLR for IFN activation induction by RSV, promoting the induction of IRF3 and NF-κB and the resultant induction of type I and type III IFN (Liu et al. 2007; Bitko et al. 2008; Yoboua et al. 2010; Okabayashi et al. 2011). However, the physiological activator of RIG-I (or MDA5) is yet to be identified for most RNA viruses. In Sendai virus, the defective interfering (DI) virus genome has been shown to bind and activate RIG-I and activate IFN, likely because it contains both the features mentioned above (Strähle et al. 2007). In contrast, the DI particles of RSV or their possible roles in IFN induction have not been characterized. Naked, 47-nucleotide long RSV leader RNA is also capable of binding RIG-I and activating type I IFN (Bitko et al. 2008), perhaps due to its 5′-triphosphate and short, yet undefined, hairpin regions. However, in the RSV-infected cell, the leader RNA is hardly ever naked, but is shielded by the host La antigen; and in the genome replication mode, the nascent leader sequence is also covered by the viral N protein and the resultant RNP neither binds RIG-I nor activates IFN (Bitko et al. 2008). The full-length viral genome or antigenome in the RSV-infected cell also does not bind RIG-I, likely due to encapsidation with N protein (Bitko et al. 2008; Liu et al. 2007).

The first step in the RLR pathway is the binding of the viral ligand, likely a dsRNA of some form, to the RNA-binding domain of RIG-I or MDA5, which changes the protein conformation such that CARD dissociates from a negative regulatory domain and dimerizes via homotypic CARD–CARD interaction. The dimer binds to the strategically placed mitochondrial CARD-containing protein, MAVS (also known as IPS-1, VISA, Cardif), which in turn dimerizes to create a signaling platform (Belgnaoui et al. 2011). Docking of RIG-I/MDA5 with MAVS leads to the recruitment of members of the TRAF adapter family, resulting in the bifurcation of signaling into two branches. One branch utilizes TRAF3 and activates the NEMO-TANK-TBK1/IKKε kinase complex that phosphorylates IRF3 and IRF7. The other branch signals through a TRAF2/TRAF6 complex and activates three kinases, IKK, p38, and JNK, leading to the activation of three transcription factors, NF-κB, ATF-2 and c-Jun, respectively. These three, along with phosphorylated IRF3 and IRF7 in various combinations, bind to the IFN-stimulated response element (ISRE) sequence at the type I IFN promoters and promote the expression of IFN-α and IFN-β (Belgnaoui et al. 2011; Fensterl and Sen 2009). Available evidence indicates that essentially the same pathway activates type III IFN (Onoguchi et al. 2007). As we will see (Sect. 2.3.2), the RSV NS proteins inhibit the RIG-I signaling pathway.

2.1.2 TLR-Mediated Innate Immunity to RSV

While the RLRs are major players in epithelial innate immunity to RSV, the TLRs may be more important in the adaptive immune aspects of the RSV disease and in vaccine development (See later). Humans possess ten functional TLRs (TLR 1 through 10) that exhibit diverse locations and equally diverse preference for ligands (PAMPs) (Lee et al. 2012). Unlike RLR, the TLRs are located either on the cell membrane or endosomal membranes; also unlike RLR, they are found mainly in cells of immune origin, such as DC/macrophages and neutrophils, but relatively scarce in epithelial cells.

TLR3 is activated by double-stranded RNA (dsRNA); however, currently there is no published report of innate immunity activating dsRNA species generated by nonsegmented negative-strand RNA viruses (Weber et al. 2006). An interesting early study identified TLR4 as the PRR for purified RSV fusion protein (F), the significance of which remains unresolved (Kurt-Jones et al. 2000; Marr and

Turvey 2012). Regardless, activation of TLR-signaling leads to antiviral cytokines, TNF-α and IFNs, and multiple TLRs seem to activate this arm of innate immune response to RSV (Klein Klouwenberg et al. 2009). Moreover, our ability to use specific synthetic ligands of TLRs (such as polyI:C for TLR3) may allow manipulation of this arm to regulate adaptive immunity. Finally, RSV NS proteins may also block these pathways, thereby contributing to immune evasion and RSV survival (Sect. 2.3.2).

2.1.3 NLR-Mediated Innate Immunity to RSV

Nod2, a member of the NLR family is rapidly induced (within 2 h) after RSV infection and plays a role in the induction of IFN-β in 293 cells and PBMCs (Sabbah et al. 2009; Vissers et al. 2012). The induction requires RSV gene expression, since UV-inactivated virus fails to activate. The exact nature of the viral trigger, the relative importance of the pathway in comparison to the TLR and RLR pathways, and whether NS proteins can suppress it—all remain to be elucidated.

2.2 IFN Response, Promoting Antiviral Effect

Once IFNs are released from the initially infected cells, they induce an antiviral state in the neighboring uninfected cells (Aaronson and Horvath 2002; Sen and Sarkar 2007). To accomplish this, IFN-α and IFN-β bind to a shared receptor causing the activation of the Janus-activated kinase (JAK)/signal transducer and activator of transcription (STAT) pathway. This leads to the formation of the ISGF3 complex consisting of three proteins: IRF-9 (previously p48), STAT1, and STAT2. The complex translocates to the nucleus, binds to the ISRE and induces transcription of a plethora of interferon-stimulated genes (ISGs). However, the exact antiviral roles of the majority of the ISGs remain undefined (Fensterl and Sen 2011). Recent studies have shown that three ISGs of the IFIT (IFN-induced protein with tetratricopeptide repeat/TPR) family, namely IFIT1 (ISG56), IFIT2 (ISG54) and IFIT3 (ISG60), are robustly induced upon RSV infection of mice and/or A549 cells (Janssen et al. 2007; Hastie et al. 2012). Although their anti-RSV roles are still unknown, individual IFITs have been shown to inhibit closely related RNA viruses, such as Sendai. Interestingly, these IFIT genes can be directly induced by viral PAMPs independently of IFN action (Fensterl and Sen 2011), which is in part due to the IRF3/IRF7-binding sites near their promoters.

As detailed below, RSV NS proteins suppress IFN response as well as IFN synthesis, and may directly suppress some ISGs.

2.3 Suppression of Innate Immunity by RSV

2.3.1 The Unique Nonstructural Proteins of RSV

RSV, like all other *Paramyxoviridae* members studied to date, has evolved specialized arsenals to defeat the host cell's innate immune response, specifically the synthesis and action of type I IFNs. However, RSV appeared to be superbly efficient in this regard. The earliest studies observed that IFN-α or IFN-β concentrations of 10,000 U/ml reduced RSV growth in lung epithelial cell culture by only 10- to 20-fold, whereas the same concentrations of IFN reduced the growth of parainfluenza virus type 3 by 1000 to 10,000-fold (Atreya and Kulkarni 1999).

It is perhaps not surprising that RSV makes unique departures from the general theme used by all other Paramyxoviruses to subvert IFN, which makes its mechanism all the more intriguing and challenging. In all other viruses, the viral P gene gives rise to multiple mRNA sequences through "RNA editing", producing accessory proteins such as V, W and C, which inhibit various members of the type I IFN pathways. In contrast, RNA editing does not occur in RSV, and thus the P gene of RSV produces the P protein only. Instead, it was recognized early on that RSV has delegated the task of IFN suppression to its two NS proteins, NS1 and NS2. The NS proteins exhibit no discernible sequence similarity with any V or W protein or any other sequence in the biological world, except the NS1 and NS2 homologs of other RSV strains such as bovine RSV (Bossert et al. 2003; Schlender et al. 2000; Valarcher et al. 2003) and those of the closely related mouse virus, Pneumonia Virus of Mice (PVM) (Buchholz et al. 2009; Heinze et al. 2011), considered by many as the mouse RSV-equivalent (Brock et al. 2012).

The first indications of the anti-IFN role of RSV NS genes came from the discovery that recombinant human (as well as bovine) RSV with single or double deletions of NS genes (ΔNS1, ΔNS2, and ΔNS1/2) exhibited reduced replication in cultured cells that are competent to produce type I IFN as well as in animals; however, growth was significantly restored in IFN-negative Vero cells (Jin et al. 2000, 2003; Schlender et al. 2000; Spann et al. 2004, 2005; Teng and Collins 1999, Teng et al. 2000; Valarcher et al. 2003). Studies so far have shown that the NS proteins also suppress type III IFNs, such as IFN-λ (Spann et al. 2004), suggesting common signaling pathways and suppression mechanisms.

Why the pneumoviruses encode two unique NS proteins of apparently unprecedented sequence remains an evolutionary mystery; however, as we will see under Mechanism (Sect. 2.3.2), this may entail division of labor, in which NS1 and NS2 work separately and together, targeting multiple IFN pathways and cellular factors, and providing both versatility and synergistic efficiency. The high pathogenicity of RSV may in part derive from this extraordinary efficiency.

It should be mentioned that in a novel detour (Lifland et al. 2012), subcellular imaging of RSV-infected cells was conducted, which revealed that both RIG-I and MDA5 colocalize with viral genomic RNA and the nucleoprotein (N) within large viral inclusion bodies (IB). Moreover, expression of N alone substantially inhibited induction of IFN by Newcastle disease virus. Clearly, further studies are needed to evaluate how the contributions of N (or other structural proteins of RSV, if any) compare to those of NS proteins.

2.3.2 Mechanisms of IFN Antagonism by the RSV NS Proteins

The molecular mechanisms by which NS1 and NS2 suppress IFN are currently under intense investigation, but provocative insights have been obtained already. Early results showed that while wild-type RSV suppressed activation of IRF3 by preventing its nuclear translocation, both NS1- and NS2-deleted RSV failed to do so, which suggested that both the NS proteins may affect the IFN induction pathway by targeting IRF3 (Bossert et al. 2003; Spann et al. 2005). Accumulating evidence thus far has revealed that NS1 and NS2, working singularly and together, target various factors of the IFN synthesis as well as response pathway.

NS1 has recently been shown to bind and sequester IRF3 (Ren et al. 2011) and degrade TRAF3 (Swedan et al. 2009, 2011), the two key players of the IFN synthesis pathway (see Sect. 2.1.1). NS1, to a greater extent than NS2, can also regulate IRF7 (our unpublished result). Since these factors are common to both the RLR and the TLR pathways, their subversion ensures a stringent suppression of IFN synthesis. NS2 interacts with RIG-I, preventing IFN gene induction by this pathway (Ling et al. 2009). NS2 achieves this by binding to the N-terminal CARD of RIG-I and inhibiting its interaction with the downstream component MAVS. Since RIG-I and IRF3 are also important for induction of IFN-λ (Okabayashi et al. 2011), their subversion may also underlie the mechanism by which RSV suppresses type III IFN (Spann et al. 2004). The NS proteins, singly and together, also affect STAT2, likely through a combination of ubiquitin-promoted proteasomal mechanism and a general protease-mediated degradation (Lo et al. 2005; Ramaswamy et al. 2006; Elliott et al. 2007; Swedan et al. 2009, 2011), leading to abrogation of IFN response. Thus, the short list of the collective targets of RSV NS1 and NS2 currently include: RIG-I, TRAF3, IRF3, IRF7 and STAT2. Whereas the first four factors are involved in IFN synthesis, STAT2 is essential for IFN response, which explains the ability of RSV to suppress both pathways and the high IFN resistance of the virus.

The expanding set of NS substrates suggests that other cellular targets may still be waiting to be discovered, which would extend the contribution of NS proteins in IFN suppression or in other areas of host-RSV interaction. While the targeting of multiple members of IFN pathways may help explain the superior ability of RSV to resist IFN (Atreya and Kulkarni 1999), it also begs the mechanistic question of how NS1/NS2 regulates and/or degrades this apparently diverse substrate repertoire. It seems likely that NS1 and NS2 form a heterogenous multi-subunit degradasome that recruits proteolytic activities and diverse accessory subunits. Of note, select paramyxoviral V proteins target specific STAT proteins for proteasomal degradation that involves the assembly of a "V-dependent degradation complex" (VDC) into spherical macromolecular particles (Ulane et al. 2005). At least three cellular proteins have been identified in different VDCs: the DDB1, Cul4A and Roc1, all of which are components of cellular ubiquitin-proteasome system (Ulane et al. 2005; Andrejeva et al. 2002). In analogy, NS1 can degrade STAT2 in the presence of Elongin-Cullin E3 ligase (Elliott et al. 2007). It thus appears likely that the "NS1/2 degradasome" minimally contains the host proteasome machinery; however, since proteasome inhibitors incompletely inhibit the degradasome, full degradation may require subsequent action of proteases (Swedan et al. 2009; Ye and Maniatis 2011).

As indicated earlier, ISGs of the IFIT family may be inhibitory to RSV, and if so, one would expect that NS proteins may suppress their induction as well, which is in fact supported by recent results. Specifically, infection of A549 cells with ΔNS1-RSV generated a much greater induction of IFIT1 and IFIT3, in comparison to infection with wild type RSV (Hastie et al. 2012), although ΔNS2-RSV was not tested. Overall, the anti-RSV role of IFITs and whether or how NS proteins may degrade them deserve further investigation.

2.4 Other Functions of RSV NS Proteins

Although limited, available evidence points to roles of NS proteins in virus growth and cellular regulation that have little or no obvious relevance to their IFN-suppressive function. These areas are as follows.

2.4.1 Regulation of Viral Replication

One of the earliest studies (Jin et al. 2000) quantified the replication of recombinant, NS-deleted RSV in IFN-proficient HEp-2 cells versus IFN-negative Vero cells. RSVΔNS2 was found to grow poorly in Vero cells but growth was largely restored to wild-type levels in Vero. In contrast, RSVΔNS1 grew equally poorly in both cells, about 50-fold less efficiently than wild-type RSV. Replacement of RSV NS1 and NS2 by the V protein of PIV5 restored some of the IFN-inhibitory functions but did not fully restore viral replication (Tran et al. 2007), suggesting that NS1 has additional functions required for optimal RSV growth. In a minigenome-based reconstitution analysis, recombinantly expressed NS1, but not NS2, was found to be a potent inhibitor of viral transcription and replication (Atreya et al. 1998). Overall, NS1 appears to fine-tune viral RNA-dependent RNA polymerase through a mechanism yet unknown.

2.4.2 Regulation of Cellular Apoptosis

RSV-infection ultimately kills the infected cell, but NS1 and NS2 inhibit premature apoptosis, allowing the virus more time to replicate (Bitko et al. 2007).

This occurred in both A549 and Vero cells, suggesting an IFN-independent mechanism. NS1 and NS2 induced or activated a number of cellular anti-apoptotic factors at early stages of infection, including NF-kappa B (NF-κB) and phosphorylated forms of protein kinases AKT, PDK, and GSK (Bitko et al. 2007). Of note, the V protein of PIV5 also inhibits apoptosis (Sun et al. 2004), suggesting that this may be a common property of RNA viral IFN-antagonists.

2.4.3 Miscellaneous and Yet Undiscovered Functions of NS

Research to elucidate the composition of the NS degradasome has led to the finding that the host factor Microtubule-Associated Protein 1B (MAP1B) binds to both NS proteins, and this is important for functionality, at least as tested for the ability of NS2 to degrade STAT2. However, a more comprehensive proteomic analysis has revealed that more than 200 cellular proteins co-precipitate with NS1 alone, which include many nuclear proteins and complexes, such as cyclin C, mediator complex, RNA polymerase II, and ATR (Wu et al. 2012), involved in cell cycle regulation and DNA repair. Studies also revealed a role of NS1 in G1-phase arrest of the cell cycle (Wu et al. 2012), in agreement with previous studies reporting G0/G1-phase preference of RSV-infected A549 cells (Gibbs et al. 2009), suggesting that NS1 is a key player in the arrest. Clearly, it will be important to determine how the nuclear and cell cycle regulatory roles of NS1 dovetail with its IFN-suppressive and viral replicative function.

2.5 The Intriguing Similarities between NS1 and NS2

It is relevant to summarize the fascinating similarities between the two NS proteins of RSV, which may shed light on their commonality of action: (1) As described in this article in detail, they are both devoted to subverting the IFN response. (2) They exhibit some degree of shared substrate specificities; for example, both may degrade STAT2, TBK1, and IKK to various extents under appropriate experimental conditions (Elliott et al. 2007; Swedan et al. 2009, 2011). (3) They translocate to mitochondria, but are also found in the cytosol and nucleus (Swedan et al. 2011; Spann et al. 2005; Boyapalle et al. 2012), and it is possible that the relative distribution is dynamic and regulated. (4) They are both located at the 3′ end (promoter-proximal end) of the viral genome, next to each other. Thus, they are transcribed abundantly and in close succession (see chapter by P.L. Collins et al., this volume). (5) Although their short sequences have no discernible similarity with each other, the very C-terminal of both contains the intriguing pentapeptide sequence (F/Y)DLNP, which was found to be important for binding of MAP1B, as mentioned above. (6) Like the paramyxoviral V proteins, NS1 and NS2 form homo- as well as heteromers (Swedan et al. 2009, 2011), and a substantial fraction ($\sim$80 %) of NS1 and NS2 can be found in the mitochondria (Swedan et al. 2011; Boyapalle et al. 2012).

In summary, all evidence suggests that NS1 and NS2 work together as team players, sharing subunits, activities, and cellular locales, to serve the common goal of IFN antagonism. In addition, being separate proteins (as opposed to two domains of a single chimeric protein) allows them to perform specific individual roles, such as the replicative role of NS1, and inhibition of RIG-I by NS2. This one-two punch endows the NS proteins with efficiency, synergism, versatility and flexibility, unique to RSV.

3 Capitalizing on the Role of NS in RSV Vaccine Design or Treatment

The obstacles to RSV vaccine are covered in detail elsewhere in this publication. A major hurdle is the weak immunity of the target population groups, i.e., the infants and the elderly, residing at the two ends of the lifespan. While the infants have immature immunity, the elderly suffer from immunosenescence (see chapter by A.M.W. Malloy et al., this volume).

In brief, the human fetus begins to acquire maternal IgG through placental transfer at about 28 weeks of gestation; the process is so efficient that maternal antibody titer at birth can match or even exceed that in the adults, protecting the neonate against a variety of pathogens. In an ironic immunological trade-off, the maternally acquired passive antiviral immunity also neutralizes the vaccine virus and thus inhibits the induction of active immunity. The situation reverses after 3–4 months of age, at which time the maternal antibodies disappear; thus, slightly older infants with lower antibody levels and higher vulnerability to RSV should respond to an RSV vaccine. Immunosenescence of the elderly, on the other hand, entails a multifactorial, age-related loss of systemic immunity (Agarwal and Busse 2010). Although the humoral immunity is mostly preserved throughout life, old age reduces the ability of B cells to produce antibodies. At the same time, thymic atrophy occurs that reduces the generation of naïve T lymphocytes, anergic memory cells accumulate, and antigen presentation is also affected. Together, they dampen both innate and active immune responses. A possible concern in live ΔNS-RSV vaccine is whether it will trigger high IFN levels locally that may lead to pulmonary toxicity, especially in these vulnerable populations. Systemic IFN administration is generally well-tolerated, for example, in the treatment of hepatitis C (Iwasaki et al. 2006); however various adverse reactions have been reported.

Another RSV-vulnerable population is the recipients of solid organ transplants (such as kidney, lung, liver), as they undergo lifelong immunosuppressant therapy, often with a powerful three-drug regimen consisting of mycophenolate, sirolimus (rapamycin), and prednisolone, a cocktail that strongly inhibits proliferation of B cells and T cells by targeting multiple pathways. RSV infection is common and frequently lethal in this group. Although the siRNA-based therapy mentioned earlier (Bitko et al. 2005) has recently shown promising result in a small cohort of lung transplant recipients (Zamora et al. 2011), the difficulties of vaccination continues to apply to this group. One should also be cognizant of the fact that immunosuppressants affect multiple signaling pathways downstream of their targets, which may include IFN pathways as well. In a pioneering recent study (Kaur et al. 2012), rapamycin, a specific inhibitor of *m*ammalian or *m*echanistic *t*arget *o*f *r*apamycin (mTOR) also inhibited type I IFN responses because mTOR is essential for IFN-induced phosphorylation of AKT and ultimately required for ISG transcription. It remains to be seen how such suppression of IFN response impacts on the outcome of live RSV vaccine.

It is now also well known that denatured, whole-inactivated RSV fails as a vaccine in test children and in fact triggers a more severe disease in subsequent RSV challenge, variously called 'vaccine-enhanced disease', 'immunopotentiation', or 'enhanced respiratory disease' (ERD). This likely stemmed from a combination of factors (Graham 2011; also see chapter by P.L. Collins et al., this volume), including: possible disruption of critical antigenic epitopes by denaturation or premature triggering of F; failure of the nonreplicating virus to activate DC-mediated proliferation of virus-specific memory CD4+ T cells; lack of antibody affinity maturation due to poor TLR stimulation; and lack of CD8 T cell induction, all of which promoted a Th2-biased CD4 T cell response and allergic inflammation. In summary, it appears that all the RSV-vulnerable populations are also immunologically challenged and vulnerable or less responsive to traditional vaccines, and hence require novel approaches.

For reasons discussed above, the current emphasis is on live-attenuated RSV vaccines generated by reverse genetics, in which essential viral functions have been compromised (Graham 2011). The balancing act here is to reduce virus replication sufficiently to eliminate the associated pathology but at the same time retain antigenicity. Since NS proteins function effectively as viral virulence factors, a variation of this theme is to use RSV with mutated NS proteins, the various aspects of which are detailed in this section. Alternatively, the RSV vaccine antigens can be expressed in a gene-based vector to eliminate NS1 and NS2 from the initial immunizing event (see chapter by R.J. Loomis and P.R. Johnson, this volume).

3.1 Balancing the Immune-Antagonistic and Non-immune Roles of NS

A straightforward approach is the use of NS-deleted recombinant RSV as vaccine, since they have been shown to be highly attenuated in cell culture and in non-human primates, namely chimpanzees (Jin et al. 2000, 2003; Teng and Collins 1999; Teng et al. 2000; Whitehead et al. 1999). However, it should be borne in mind that NS proteins are also important for RSV replication and suppression of apoptosis, and that a productive infection of the vaccine virus is required for a robust antibody response via proper antigen presentation, activation of T cells, and finally, activation and differentiation of the antigen-producing B cells by the antiviral $CD4^+$ T cells (Braciale et al. 2012). Early on, recombinant ΔNS1 RSV virus was found to be more than 2,000-fold restricted in replication in the upper and lower respiratory tracts, and tenfold more restricted than the *cpts*248/404 virus, a cold-passaged, temperature-sensitive vaccine candidate (see chapter by R.A. Karron et al., this volume). In spite of the reduced growth, the RSV-neutralizing antibody induced in the serum by the ΔNS1 virus was close to the level induced by wild-type RSV, and the replication of challenge RSV in the respiratory tract of these immunized chimpanzees was reduced more than 10,000-fold (Teng et al. 2000; Whitehead et al. 1999). Apparently in this case, even if loss of NS1 contributed to the highly attenuated growth, robust antibody response occurred. It is also possible that NS2 had compensated for the loss of NS1 in this regard.

Following the ΔNS1 lead, recombinant ΔNS2 RSV, alone and in combination with various other cold-passaged mutations (*cpts*248/404 and *cpts*530/1009), was evaluated in adults as well as in RSV-seropositive and RSV-seronegative children (Wright et al. 2006). Viruses lacking the NS2 gene were indeed highly attenuated in adults, but surprisingly, were under-attenuated in children. Although ΔNS2 in combination with the *cpts* mutations were better attenuated in children, the ΔNS2 viruses were not pursued further. It would be interesting to see if ΔNS1/2 RSV would be more severely attenuated and be a viable vaccine. The one redeeming feature of all live-attenuated vaccine studies was that no immunopotentiation occurred.

Another approach would be to first dissect the anti-IFN and replication functions of NS1 by screening site-directed NS1 mutants by reverse genetics, using replicon assays in parallel with IFN suppression assays. The mapping must be conducted in an empirical manner, as the lack of homology of NS with other proteins makes prediction of any functional domain virtually impossible. If the sequences are indeed found to be different, however, recombinant RSV could then be generated incorporating the IFN-defective NS1 mutations, leaving the replication sequences intact. The level of attenuation and vaccine potential of this recombinant could then be tested. Lastly, NS1 and NS2 can interact with a multitude of cellular proteins (Sect. 2.4.3). If the functionality of each interaction is unraveled in the future, the cognate NS domain(s) could also be mapped and the knowledge utilized to further optimize the recombinant vaccine.

Genetic stability of the vaccine strain is important for not only efficacy but also for safety. To minimize reversion of point mutations of NS, synonymous mutant codons can be carefully selected that will contain two nucleotide changes instead of one, an approach recently used successfully for RSV L protein to stabilize the *ts*248 mutation at amino acid residue 831 (Luongo et al. 2009).

3.2 Roles of NS Proteins in Regulating Adaptive Immunity: Crosstalk between Pathways

Intriguing new findings implicate NS proteins in the regulation of aspects of the adaptive immunity that hold the key to a successful vaccine. RSV is also known to infect DC/macrophages, which then elaborates IFN (Kumagai et al. 2007). In a regulatory loop, IFN activates adaptive immunity by promoting DC maturation and T cell activation, and thus ΔNS RSV vaccine may be useful in this respect. However, in the RSV-infected DC, the NS1 protein tends to suppress proliferation and activation of two of the protective cell populations ($CD103^{+}CD8^{+}$ T cells and Th17 cells) and promotes proliferation and activation of Th2 cells that can enhance RSV disease (Munir et al. 2011). Besides, infection of DC with NS1-deleted RSV induced a strong induction of multiple genes, positively involved in DC maturation and T cell activation, much of which was suppressed by wild-type RSV (Munir et al. 2011 and references therein). Thus, a properly engineered NS1-mutant RSV devoid of Th2-promoting effect but retaining DC maturation and T cell activation properties would clearly be a safer vaccine.

Since some of the RSV target populations, e.g. the infant or the elderly, have skewed or insufficient immune response, pharmacological activation of the immune cells is worth testing. Specifically, PRR agonists can be used to enhance the activation, an approach that offers tools for novel adjuvant development. TLR agonists, such as CpG, LPS, poly(I:C), and the NLR activator, alum, could be pursued as adjuvants. These ligands should improve affinity maturation as well as neutralizing capacity of antibodies in immunized individuals. Care should be taken to ensure that these ligands preferably activate Th1 but not Th2 response. Recently, the adjuvant Protollin, derived from the *Neisseria meningitidis* outer membrane has been shown to improve RSV subunit vaccine candidates, by conferring a balanced Th1/Th2 immune response via TLR4 activation (Cyr et al. 2009). A candidate RSV vaccine was developed that contained "epitope-enhanced" recombinant RSV G protein, a fusion between G residues 128–229 with thioredoxin, with Protollin as adjuvant. It conferred strong protection against subsequent RSV challenge with markedly reduced eosinophilia and IL-13 (Huang et al. 2009). Similarly, CpG administration during immunization also induced a Th1 response and decreased pulmonary disease following RSV challenge (Johnson et al. 2009). Novel Th-response regulators are continually being discovered, some of which can be tested as well. For example, in PVM infection, IL-21 is important for producing pathology (Spolski et al. 2012), since pathology is reduced in IL21R KO mice and in those in which a soluble receptor-Fc fusion protein is administered. As PVM is generally regarded as the natural mouse equivalent of human RSV, it may be worth testing if a soluble IL-21 receptor can be a part of the human RSV therapeutic regimen, or used to reduce the potential for Th2-biased vaccine-induced responses, although this would have to be balanced against the potential effects this would have on Tfh cells and antibody production.

Lastly, enhanced apoptosis in ΔNS RSV infection may limit virus replication and spread (Domachowske et al. 2000; Bitko et al. 2007). However, DCs may internalize apoptotic cells and process them for presentation by MHC molecules. RSV infection has been shown to sensitize tracheal and bronchial epithelial cells to exogenously added TNF-related apoptosis-inducing ligand (TRAIL) (Kotelkin et al. 2003). Thus, it may be worth testing if the ΔNS2 RSV vaccine virus can be further attenuated in children through the use of TRAIL. However, care should be taken to ensure that TRAIL also does not sensitize RSV-infected DCs that must activate T cells for full antibody response.

4 Conclusion and Future Directions

The available evidence suggests that NS proteins subvert or degrade multiple members of the type I IFN induction and response pathways, likely by recruiting a degradasome, using mitochondria as scaffold. Our search for the molecular mechanisms of this suppression is still in its infancy but already promises to yield fascinating insights. In addition, however, NS proteins also regulate multiple steps of the adaptive immune response that are highly relevant to the design of a successful attenuated RSV vaccine. In creating such a vaccine, however, NS mutants must be selected to encode the desired balance of properties by reverse genetic analysis. Boosting of both B cell and T cell responses in the infants and the elderly with supplementary means such as Th1-promoting adjuvants or cytokine modulators may also help the creation of a robust yet safe RSV vaccine.

References

Aaronson DS, Horvath CM (2002) A road map for those who don't know JAK-STAT. Science 296:1653–1655

Agarwal S, Busse PJ (2010) Innate and adaptive immunosenescence. Ann Allergy Asthma Immunol 104:183–190

Andrejeva J, Poole E, Young DF, Goodbourn S, Randall RE (2002) The p127 subunit (DDB1) of the UV-DNA damage repair binding protein is essential for the targeted degradation of STAT1 by the V protein of the paramyxovirus simian virus 5. J Virol 76:11379–11386

Atreya PL, Kulkarni S (1999) Respiratory syncytial virus strain A2 is resistant to the antiviral effects of type I interferons and human MxA. Virology 261:227–241

Atreya PL, Peeples ME, Collins PL (1998) The NS1 protein of human respiratory syncytial virus is a potent inhibitor of minigenome transcription and RNA replication. J Virol 72:1452–1461

Belgnaoui SM, Paz S, Hiscott J (2011) Orchestrating the interferon antiviral response through the mitochondrial antiviral signaling (MAVS) adapter. Curr Opin Immunol 23:564–572

Bitko V, Musiyenko A, Bayfield MA, Maraia RJ, Barik S (2008) Cellular La protein shields nonsegmented negative-strand RNA viral leader RNA from RIG-I and enhances virus growth by diverse mechanisms. J Virol 82:7977–7987

Bitko V, Musiyenko A, Shulyayeva O, Barik S (2005) Inhibition of respiratory viruses by nasally administered siRNA. Nat Med 11:50–55

Bitko V, Shulyayeva O, Mazumder B, Musiyenko A, Ramaswamy M, Look DC, Barik S (2007) Nonstructural proteins of respiratory syncytial virus suppress premature apoptosis by an NF-kappaB-dependent, interferon-independent mechanism and facilitate virus growth. J Virol 81:1786–1795

Bossert B, Marozin S, Conzelmann KK (2003) Nonstructural proteins NS1 and NS2 of bovine respiratory syncytial virus block activation of interferon regulatory factor 3. J Virol 77:8661–8668

Braciale TJ, Sun J, Kim TS (2012) Regulating the adaptive immune response to respiratory virus infection. Nat Rev Immunol 12:295–305

Boyapalle S, Wong T, Garay J, Teng M, San Juan-Vergara H, Mohapatra S, Mohapatra S (2012) Respiratory syncytial virus NS1 protein colocalizes with mitochondrial antiviral signaling protein MAVS following infection. PLoS ONE 7:e29386

Brock LG, Karron RA, Krempl CD, Collins PL, Buchholz UJ (2012) Evaluation of pneumonia virus of mice as a possible human pathogen. J Virol 86:5829–5843

Buchholz UJ, Ward JM, Lamirande EW, Heinze B, Krempl CD, Collins PL (2009) Deletion of nonstructural proteins NS1 and NS2 from pneumonia virus of mice attenuates viral replication and reduces pulmonary cytokine expression and disease. J Virol 83:1969–1980

Collins PL, Melero JA (2011) Progress in understanding and controlling respiratory syncytial virus: still crazy after all these years. Virus Res 162:80–99

Cyr SL, Angers I, Guillot L, Stoica-Popescu I, Lussier M, Qureshi S, Burt DS, Ward BJ (2009) TLR4 and MyD88 control protection and pulmonary granulocytic recruitment in a murine intranasal RSV immunization and challenge model. Vaccine 27:421–430

Domachowske JB, Bonville CA, Mortelliti AJ, Colella CB, Kim U, Rosenberg HF (2000) Respiratory syncytial virus infection induces expression of the anti-apoptosis gene IEX-1L in human respiratory epithelial cells. J Infect Dis 181:824–830

Elliott J, Lynch OT, Suessmuth Y, Qian P, Boyd CR, Burrows JF, Buick R, Stevenson NJ, Touzelet O, Gadina M, Power UF, Johnston JA (2007) Respiratory syncytial virus NS1 protein degrades STAT2 by using the Elongin-Cullin E3 ligase. J Virol 81:3428–3436

Fensterl V, Sen GC (2009) Interferons and viral infections. BioFactors 35:14–20

Fensterl V, Sen GC (2011) The ISG56/IFIT1 gene family. J Interferon Cytokine Res 31:71–78

Gibbs JD, Ornoff DM, Igo HA, Zeng JY, Imani F (2009) Cell cycle arrest by transforming growth factor beta1 enhances replication of respiratory syncytial virus in lung epithelial cells. J Virol 83:12424–12431

Graham BS (2011) Biological challenges and technological opportunities for respiratory syncytial virus vaccine development. Immunol Rev 239:149–166

Hastie ML, Headlam MJ, Patel NB, Bukreyev AA, Buchholz UJ, Dave KA, Norris EL, Wright CL, Spann KM, Collins PL, Gorman JJ (2012) The human respiratory syncytial virus nonstructural protein 1 regulates type I and type II interferon pathways. Mol Cell Proteomics 11:108–127

Heinze B, Frey S, Mordstein M, Schmitt-Gräff A, Ehl S, Buchholz UJ, Collins PL, Staeheli P, Krempl CD (2011) Both nonstructural proteins NS1 and NS2 of pneumonia virus of mice are inhibitors of the interferon type I and type III responses in vivo. J Virol 85:4071–4084

Huang Y, Cyr SL, Burt DS, Anderson R (2009) Murine host responses to respiratory syncytial virus (RSV) following intranasal administration of a Protollin-adjuvanted, epitope-enhanced recombinant G protein vaccine. J Clin Virol 44:287–291

Iwasaki Y, Ikeda H, Araki Y, Osawa T, Kita K, Ando M, Shimoe T, Takaguchi K, Hashimoto N, Kobatake T, Tomita M, Kawaguchi M, Kobashi H, Sakaguchi K, Shiratori Y (2006) Limitation of combination therapy of interferon and ribavirin for older patients with chronic hepatitis C. Hepatology 43:54–63

Janssen R, Pennings J, Hodemaekers H, Buisman A, van Oosten M, de Rond L, Ozturk K, Dormans J, Kimman T, Hoebee B (2007) Host transcription profiles upon primary respiratory syncytial virus infection. J Virol 81:5958–5967

Jin H, Zhou H, Cheng X, Tang R, Munoz M, Nguyen N (2000) Recombinant respiratory syncytial viruses with deletions in the NS1, NS2, SH, and M2–2 genes are attenuated in vitro and in vivo. Virology 273:210–218

Jin H, Zhou H, Cheng X, Tang R, Munoz M, Nguyen N (2003) Evaluation of recombinant respiratory syncytial virus gene deletion mutants in African green monkeys for their potential as live attenuated vaccine candidates. Vaccine 21:3647–3652

Johnson TR, Rao S, Seder RA, Chen M, Graham BS (2009) TLR9 agonist, but not TLR7/8, functions as an adjuvant to diminish FI-RSV vaccine-enhanced disease, while either agonist used as therapy during primary RSV infection increases disease severity. Vaccine 27:3045–3052

Kato H, Takeuchi O, Sato S, Yoneyama M, Yamamoto M, Matsui K, Uematsu S, Jung A, Kawai T, Ishii KJ, Yamaguchi O, Otsu K, Tsujimura T, Koh CS, Reis e Sousa C, Matsuura Y, Fujita T, Akira S (2006) Differential roles of MDA5 and RIG-I helicases in the recognition of RNA viruses. Nature 441:101–105

Kato H, Takeuchi O, Mikamo-Satoh E, Hirai R, Kawai T, Matsushita K, Hiiragi A, Dermody TS, Fujita T, Akira S (2008) Length-dependent recognition of double-stranded ribonucleic acids by retinoic acid-inducible gene-I and melanoma differentiation-associated gene 5. J Exp Med 205:1601–1610

Kaur S, Sassano A, Majchrzak-Kita B, Baker DP, Su B, Fish EN, Platanias LC (2012) Regulatory effects of mTORC2 complexes in type I IFN signaling and in the generation of IFN responses. Proc Natl Acad Sci USA 109:7723–7728

Klein Klouwenberg P, Tan L, Werkman W, van Bleek GM, Coenjaerts F (2009) The role of Toll-like receptors in regulating the immune response against respiratory syncytial virus. Crit Rev Immunol 29:531–550

Kotelkin A, Prikhod'ko EA, Cohen JI, Collins PL, Bukreyev A (2003) Respiratory syncytial virus infection sensitizes cells to apoptosis mediated by tumor necrosis factor-related apoptosis-inducing ligand. J Virol 77:9156–9172

Kumagai Y, Takeuchi O, Kato H, Kumar H, Matsui K, Morii E, Aozasa K, Kawai T, Akira S (2007) Alveolar macrophages are the primary interferon-alpha producer in pulmonary infection with RNA viruses. Immunity 27:240–252

Kumar H, Kawai T, Akira S (2011) Pathogen recognition by the innate immune system. Int Rev Immunol 30:16–34

Kurt-Jones EA, Popova L, Kwinn L, Haynes LM, Jones LP, Tripp RA, Walsh EE, Freeman MW, Golenbock DT, Anderson LJ, Finberg RW (2000) Pattern recognition receptors TLR4 and CD14 mediate response to respiratory syncytial virus. Nat Immunol 1:398–401

Lee CC, Avalos AM, Ploegh HL (2012) Accessory molecules for toll-like receptors and their function. Nat Rev Immunol 12:168–179

Lifland AW, Jung J, Alonas E, Zurla C, Crowe JE Jr, Santangelo PJ (2012) Human respiratory syncytial virus nucleoprotein and inclusion bodies antagonize the innate immune response mediated by MDA5 and MAVS. J Virol 86:8245–8258

Ling Z, Tran KC, Teng MN (2009) Human respiratory syncytial virus nonstructural protein NS2 antagonizes the activation of beta interferon transcription by interacting with RIG-I. J Virol 83:3734–3742

Liu P, Jamaluddin M, Li K, Garofalo RP, Casola A, Brasier AR (2007) Retinoic acid-inducible gene I mediates early antiviral response and Toll-like receptor 3 expression in respiratory syncytial virus-infected airway epithelial cells. J Virol 81:1401–1411

Lo MS, Brazas RM, Holtzman MJ (2005) Respiratory syncytial virus nonstructural proteins NS1 and NS2 mediate inhibition of Stat2 expression and alpha/beta interferon responsiveness. J Virol 79:9315–9319

Loo YM, Fornek J, Crochet N, Bajwa G, Perwitasari O, Martinez-Sobrido L, Akira S, Gill MA, García-Sastre A, Katze MG, Gale M Jr (2008) Distinct RIG-I and MDA5 signaling by RNA viruses in innate immunity. J Virol 82:335–345

Luongo C, Yang L, Winter CC, Spann KM, Murphy BR, Collins PL, Buchholz UJ (2009) Codon stabilization analysis of the "248" temperature sensitive mutation for increased phenotypic stability of respiratory syncytial virus vaccine candidates. Vaccine 27:5667–5676

Marr N, Turvey SE (2012) Role of human TLR4 in respiratory syncytial virus-induced NF-κB activation, viral entry and replication. Innate Immun 18:856–865

Mordstein M, Neugebauer E, Ditt V, Jessen B, Rieger T, Falcone V, Sorgeloos F, Ehl S, Mayer D, Kochs G, Schwemmle M, Günther S, Drosten C, Michiels T, Staeheli P (2010) Lambda interferon renders epithelial cells of the respiratory and gastrointestinal tracts resistant to viral infections. J Virol 84:5670–5677

Munir S, Hillyer P, Le Nouën C, Buchholz UJ, Rabin RL, Collins PL, Bukreyev A (2011) Respiratory syncytial virus interferon antagonist NS1 protein suppresses and skews the human T lymphocyte response. PLoS Pathog 7:e1001336

Onoguchi K, Yoneyama M, Takemura A, Akira S, Taniguchi T, Namiki H, Fujita T (2007) Viral infections activate types I and III interferon genes through a common mechanism. J Biol Chem 282:7576–7581

Okabayashi T, Kojima T, Masaki T, Yokota S, Imaizumi T, Tsutsumi H, Himi T, Fujii N, Sawada N (2011) Type-III interferon, not type-I, is the predominant interferon induced by respiratory viruses in nasal epithelial cells. Virus Res 160:360–366

Ramaswamy M, Shi L, Varga SM, Barik S, Behlke MA, Look DC (2006) Respiratory syncytial virus nonstructural protein 2 specifically inhibits type I interferon signal transduction. Virology 344:328–339

Ren J, Liu T, Pang L, Li K, Garofalo RP, Casola A, Bao X (2011) A novel mechanism for the inhibition of interferon regulatory factor-3-dependent gene expression by human respiratory syncytial virus NS1 protein. J Gen Virol 92:2153–2159

Sabbah A, Chang TH, Harnack R, Frohlich V, Tominaga K, Dube PH, Xiang Y, Bose S (2009) Activation of innate immune antiviral responses by Nod2. Nat Immunol 10:1073–1080

Sasai M, Shingai M, Funami K, Yoneyama M, Fujita T, Matsumoto M, Seya T (2006) NAK-associated protein 1 participates in both the TLR3 and the cytoplasmic pathways in type I IFN induction. J Immunol 177:8676–8683

Schlender J, Bossert B, Buchholz U, Conzelmann KK (2000) Bovine respiratory syncytial virus nonstructural proteins NS1 and NS2 cooperatively antagonize alpha/beta interferon-induced antiviral response. J Virol 74:8234–8242

Schmidt A, Schwerd T, Hamm W, Hellmuth JC, Cui S, Wenzel M, Hoffmann FS, Michallet MC, Besch R, Hopfner KP, Endres S, Rothenfusser S (2009) 5′-triphosphate RNA requires base-paired structures to activate antiviral signaling via RIG-I. Proc Natl Acad Sci USA 106:12067–12072

Sen GC, Sarkar SN (2007) The interferon-stimulated genes: targets of direct signaling by interferons, double-stranded RNA, and viruses. Curr Top Microbiol Immunol 316:233–250

Spann KM, Tran KC, Chi B, Rabin RL, Collins PL (2004) Suppression of the induction of alpha, beta, and lambda interferons by the NS1 and NS2 proteins of human respiratory syncytial virus in human epithelial cells and macrophages. J Virol 78:4363–4369 (corrected)

Spann KM, Tran KC, Collins PL (2005) Effects of nonstructural proteins NS1 and NS2 of human respiratory syncytial virus on interferon regulatory factor 3, NF-kappaB, and proinflammatory cytokines. J Virol 79:5353–5362

Spolski R, Wang L, Wan CK, Bonville CA, Domachowske JB, Kim HP, Yu Z, Leonard WJ (2012) IL-21 promotes the pathologic immune response to pneumovirus infection. J Immunol 188:1924–1932

Strähle L, Marq JB, Brini A, Hausmann S, Kolakofsky D, Garcin D (2007) Activation of the beta interferon promoter by unnatural Sendai virus infection requires RIG-I and is inhibited by viral C proteins. J Virol 81:12227–12237

Sun M, Rothermel TA, Shuman L, Aligo JA, Xu S, Lin Y, Lamb RA, He B (2004) Conserved cysteine-rich domain of paramyxovirus simian virus 5 V protein plays an important role in blocking apoptosis. J Virol 78:5068–5078

Swedan S, Andrews J, Majumdar T, Musiyenko A, Barik S (2011) Multiple functional domains and complexes of the two nonstructural proteins of human respiratory syncytial virus contribute to interferon suppression and cellular location. J Virol 85:10090–10100

Swedan S, Musiyenko A, Barik S (2009) Respiratory syncytial virus nonstructural proteins decrease levels of multiple members of the cellular interferon pathways. J Virol 83:9682–9693

Takeuchi O, Akira S (2009) Innate immunity to virus infection. Immunol Rev 227:75–86

Teng MN, Collins PL (1999) Altered growth characteristics of recombinant respiratory syncytial viruses which do not produce NS2 protein. J Virol 73:466–473

Teng MN, Whitehead SS, Bermingham A, St. Claire M, Elkins WR, Murphy BR, Collins PL (2000) Recombinant respiratory syncytial virus that does not express the NS1 or M2–2 protein is highly attenuated and immunogenic in chimpanzees. J Virol 74:9317–9321

Tran KC, He B, Teng MN (2007) Replacement of the respiratory syncytial virus nonstructural proteins NS1 and NS2 by the V protein of parainfluenza virus 5. Virology 368:73–82

Ulane CM, Kentsis A, Cruz CD, Parisien JP, Schneider KL, Horvath CM (2005) Composition and assembly of STAT-targeting ubiquitin ligase complexes: paramyxovirus V protein carboxyl terminus is an oligomerization domain. J Virol 79:10180–10189

Valarcher JF, Furze J, Wyld S, Cook R, Conzelmann KK, Taylor G (2003) Role of alpha/beta interferons in the attenuation and immunogenicity of recombinant bovine respiratory syncytial viruses lacking NS proteins. J Virol 77:8426–8439

Vissers M, Remijn T, Oosting M, de Jong DJ, Diavatopoulos DA, Hermans PW, Ferwerda G (2012) Respiratory syncytial virus infection augments NOD2 signaling in an IFN-ß-dependent manner in human primary cells. Eur J Immunol 42:2727–2735

Weber F, Wagner V, Rasmussen SB, Hartmann R, Paludan SR (2006) Double-stranded RNA is produced by positive-strand RNA viruses and DNA viruses but not in detectable amounts by negative-strand RNA viruses. J Virol 80:5059–5064

Whitehead SS, Bukreyev A, Teng MN, Firestone CY, St Claire M, Elkins WR, Collins PL, Murphy BR (1999) Recombinant respiratory syncytial virus bearing a deletion of either the NS2 or SH gene is attenuated in chimpanzees. J Virol 73:3438–3442

Wright PF, Karron RA, Madhi SA, Treanor JJ, King JC, O'Shea A, Ikizler MR, Zhu Y, Collins PL, Cutland C, Randolph VB, Deatly AM, Hackell JG, Gruber WC, Murphy BR (2006) The interferon antagonist NS2 protein of respiratory syncytial virus is an important virulence determinant for humans. J Infect Dis 193:573–581

Wu W, Tran KC, Teng MN, Heesom KJ, Matthews DA, Barr JN, Hiscox JA (2012) The interactome of the human respiratory syncytial virus NS1 protein highlights multiple effects on host cell biology. J Virol 86:7777–7789

Ye J, Maniatis T (2011) Negative regulation of interferon-β gene expression during acute and persistent virus infections. PLoS ONE 6:e20681

Yoboua F, Martel A, Duval A, Mukawera E, Grandvaux N (2010) Respiratory syncytial virus-mediated NF-kappa B p65 phosphorylation at serine 536 is dependent on RIG-I, TRAF6, and IKK beta. J Virol 84:7267–7277

Yoneyama M, Kikuchi M, Matsumoto K, Imaizumi T, Miyagishi M, Taira K, Foy E, Loo YM, Gale M Jr, Akira S, Yonehara S, Kato A, Fujita T (2005) Shared and unique functions of the DExD/H-Box helicases RIG-I, MDA5, and LGP2 in antiviral innate immunity. J Immunol 175:2851–2858

Zamora MR, Budev M, Rolfe M, Gottlieb J, Humar A, Devincenzo J, Vaishnaw A, Cehelsky J, Albert G, Nochur S, Gollob JA, Glanville AR (2011) RNA interference therapy in lung transplant patients infected with respiratory syncytial virus. Am J Respir Crit Care Med 183:531–538

Host Gene Expression and Respiratory Syncytial Virus Infection

Ralph A. Tripp, Asuncion Mejias and Octavio Ramilo

Abstract Advances in RNA interference (RNAi) and transcription studies have facilitated the application of systematic cell-based loss- or gain-of-function and cell response screening that enable genome-wide analysis of cell factors involved in viral replication and disease. Application of both experimental and computational biology approaches have led to crucial insights into virus infection, its life cycle, and host gene targets for disease intervention. A better understanding of the spatial and temporal host gene interactions during viral infection has enabled insights into mechanisms by which viral proteins co-opt host cell function, and host regulatory mechanisms that influence disease and treatment outcome. In this chapter, approaches to host gene discovery and transcriptome profiling for respiratory syncytial virus (RSV) are discussed in the context of biological relevance for disease intervention in the clinical setting and vaccine development.

R. A. Tripp (✉)
Department of Infectious Diseases, College of Veterinary Medicine, University of Georgia, Athens, GA, USA
e-mail: ratripp@uga.edu

A. Mejias · O. Ramilo
Center for Vaccines and Immunity, The Research Institute at Nationwide Children's Hospital, Columbus, OH, USA
e-mail: asuncion.mejias@nationwidechildrens.org

O. Ramilo
e-mail: octavio.ramilo@nationwidechildrens.org

A. Mejias · O. Ramilo
Department of Pediatrics, Division of Infectious Diseases, Nationwide Children's Hospital, The Ohio State University School of Medicine, Columbus, OH, USA

L. J. Anderson and B. S. Graham (eds.), *Challenges and Opportunities for Respiratory Syncytial Virus Vaccines*, Current Topics in Microbiology and Immunology 372, DOI: 10.1007/978-3-642-38919-1_10, © Springer-Verlag Berlin Heidelberg 2013

Contents

1 Introduction

Although viruses have small genomes, they have the ability to reprogram host cells to promote replication and subvert host responses. Studies evaluating how viruses exploit host genes have been instructive for virus biology, and have led to the identification of host genes that regulate or are required for virus replication. This knowledge has been central in drug discovery, drug repositioning, and discovery of new targets for vaccine development. These studies are also helping our understanding of disease pathogenesis. Historically, the development of antiviral therapeutics and vaccines has focused on targeting viral proteins with limited or short-term success because of high viral mutation rates. However, targeting host genes offers a strategy that is refractory to virus mutation and allows for potentially multiple targets to prevent viral replication. To facilitate large-scale functional genomics studies, several approaches have been developed based on microarray or microwell assays to look at inhibiting expression of host genes. Recent establishment of large libraries of RNA interference (RNAi) reagents combined with a variety of endpoint detection assays and gene pathway analyses has made it possible to do genome-wide screening for genes affecting cellular functions and virus replication in mammalian cells.

High-throughput screening (HTS) of gene-specific RNAi is an emerging means to explore biological processes related to virus infections and the associated disease. Genome-wide genetic analyses of mammalian cells was aided by advances in the application and delivery of small interfering RNA (siRNA) to suppress cellular gene expression achieved by lipid-based transfection of synthesized siRNAs (Meliopoulos et al. 2012; Krausz 2007), or by expression of short hairpin RNAs (shRNAs) that are processed to siRNAs (Krausz 2007). siRNAs are double-stranded RNA molecules that are commonly about 21 base pairs in length, and inside the cell, these molecules are unwound and introduced into the RNA-induced silencing complex (RISC) (Moss 2001; Tijsterman and Plasterk 2004). This complex, generally directed by the antisense strand of the siRNA molecule, targets complementary cellular mRNA sequences, and upon binding to RISC the target mRNAs are cleaved leading to a loss of gene function (Tijsterman and Plasterk 2004). One of the first proof-of-concept HTS study performed in mammalian cells used an RNAi-based forward genomics approach toward understanding the biology of TRAIL-induced apoptosis (Aza-Blanc et al. 2003). This study demonstrated the feasibility of large-scale screening using synthetic siRNAs targeting 510 genes encoding human kinases. With the advent of robust cell-based HTS technologies, it is easier to screen all host genes involved in the cycle of virus replication, identify the host cell pathways exploited, and identify all host genes that promote or inhibit virus infection and replication. This systems wide approach has enormous potential for the discovery of new therapeutic targets and vaccine targets.

An alternative approach to identifying host genes that respond to virus infection involves microarray-driven gene expression systems (Ziauddin and Sabatini 2001; Leroy and Raoult 2010). A variety of strategies comparable to DNA microarrays have been developed for transcriptional profiling, and yeast two-hybrid arrays have been developed for determining protein–protein interactions that facilitate analyses of gene function (Gonzalez 2012; Miller and Tang 2009). HTS approaches that use sets of cDNAs or DNA constructs for evaluating host genes encoding desired products are complementary techniques that allow for rapid screening for gene products involved in biological responses to infection. As many biological processes rely on the formation of protein complexes, the ability to investigate protein–protein interactions via two-hybrid assays can be essential for understanding cellular functions. These technologies are of pharmaceutical interest and can assist in the development and assessment of new vaccine strategies (Gobert 2010; Monsurro and Marincola 2011; Uttamchandani et al. 2009; Oyelaran and Gildersleeve 2007), although differences in post-translational processing in the yeast two-hybrid assays can limit the detection of some viral/mammalian protein interactions.

1.1 Human Host Genes in the Response to RSV Infection

The human host genes responding to respiratory syncytial virus (RSV) infection are not fully understood, and to date, there have been no comprehensive, validated RNAi-based genome-wide screens. However, developments in microarray technologies have broadened their use in virus studies to provide a view of the host gene response to infection that can be presented on a chip allowing for multiplexed experimentation (Schuurhof et al. 2010; Stark et al. 2009; Huang et al. 2008) (see chapters by E. H. Choi et al., S. Mukherjee and N.W. Lukacs, and by S.M. Varga and T.J. Braciale, this volume). To gain an understanding of the tempo of gene expression following RSV infection, cDNA microarray analysis was performed of differentially expressed genes in human type II respiratory epithelial (A549) cells infected with RSV (Martinez et al. 2007) (see chapter by R.J. Pickles, this volume). This study revealed that 85 genes were up-regulated at early times post-infection, i.e., 0–6 h pi. Among the genes most strongly up-regulated were those involved in chemotaxis and inflammation, and included genes encoding several integrins. Genes up-regulated between 6 and 12 h pi included interferon-stimulated genes, and genes in the non-canonical NF-κB pathway. At later times post-infection, genes involved in the immune response were up-regulated. These findings show a temporal relationship between RSV gene expression and the host response to replication, but as typical with microarray assays, the findings are difficult to interpret as they require validation. Similarly, studies have been performed that evaluate the immune response to RSV infection. For example, a study in infants hospitalized with RSV bronchiolitis examined the PBMC gene expression responses to RSV by microarray and quantitative real-time PCR, and reported that 30 genes had significant dysregulated expression, particularly for the interferon responsive genes (Fjaerli et al. 2006). The controls were infants from the same birth-cohort that never suffered from RSV bronchiolitis and were not RSV infected at the time of blood sampling. Similar to host gene expression in RSV-infected cells (Huang et al. 2008; Bakre et al.; van Diepen et al), these studies show that different subsets of host genes are dysregulated at different time-points in the immune response following RSV infection. Together, these findings emphasize the importance of considering the biological relevance of the cell types examined, the time-points post-RSV infection to be considered, and the validation needed to interpret the results.

Since all viruses need to exploit the host cellular machinery to replicate, it is not surprising that RSV has been shown to co-opt cellular kinases to achieve phosphorylation of their gene products needed to facilitate subcellular targeting in the infected cell (Dupuy et al. 1999; Alvisi et al. 2008). For example, the RSV M protein utilizes protein kinase CK2 for intracellular localization during replication (Alvisi et al. 2008), thus development of new antivirals directed at inhibiting CK2 may be therapeutic targets. Further, a number of nuclear and cytoplasmic binding partners of RSV M protein, for example the nuclear export receptor Crm1, have recently been identified by a variety of assays that include yeast two-hybrid screens using a human lung epithelial library, as well as by immunoprecipitation and subcellular fractionation mass spectrophotometry (Ghildyal et al. 2012; Ghildyal et al. 2005, 2009). Using these approaches a number of host genes are being identified as critical for RSV infection and replication.

1.2 Murine Host Genes Responding to RSV Infection and in Disease Pathogenesis

The BALB/c mouse is the prototypical model used to investigate the host response to RSV infection because it reproduces important features of human disease, and has provided predictive insights as to some of the mechanisms linked to protection from RSV disease (see chapter by P.J. Openshaw, this volume). Genetic differences in the response to RSV infection, and to RSV vaccine enhanced disease, have been investigated using 15 inbred and 7 first generation F_1 crossbred mice (Hussell et al. 1998). The findings based on these host genetic determinants showed that MHC is a determinant of T-cell responses that govern RSV disease severity in mice. Other systemic gene expression signatures have been examined in the intact mouse lung following acute RSV infection (Pennings et al. 2011). In this study, the host gene response was evaluated at days 1, 2, and 5 in the lung, bronchial lymph nodes, and blood following primary RSV. The results identified 53 interferon-associated and innate immunity genes common among the tissues evaluated. Findings that examined the tempo of host gene expression following infection showed a robust transcriptional response occurred at day 1 pi, but this response was reduced by day 3 pi, and the peak lung transcriptional response preceded the peak of viral replication. The genes expressed were diverse and involved in the IFN response, inflammation, chemoattraction, and antigen processing. This study again emphasizes the importance of evaluating the response to RSV infection at multiple time points, and addresses the processes and pathways induced that can be used for the selection of candidate genes for therapeutic and vaccine approaches. Unfortunately, there are limited studies in mice that have validated host genes affected by RSV replication, but a recent meta-analysis of published microarray data has shown that host genes are dysregulated in stages, i.e. at early, middle, and at late time-points post-RSV infection (Bakre et al. 2012). In this study which focused on the role of microRNAs (miRNAs) and post-transcriptional regulation of host genes responding to RSV infection, 117 host genes were found to be deregulated among all available published studies that examined RSV infection in a mammalian cell line.

As RSV infection can cause severe respiratory diseases in young and old as well as immune compromised individuals, a substantial level of research in mice has investigated host gene products that affect inflammation and disease outcome. These studies have shown that genes expressing cytokines orchestrate the pro-inflammatory response, such as IL-1beta, while others are effectors of inflammation, e.g., intracellular reactive oxygen species (ROS) (Segovia et al. 2012; Janssen et al. 2007; Tripp et al. 2005). Given that disease severity is a complex trait involving host genetic differences, evaluating those genetic differences in mouse strains has helped to identify genes that may control RSV infection, replication, and resolution of infection. Often a single endpoint assay can be used to determine if genetic differences are relevant by evaluating lung viral titers (Graham et al. 1988). It has been shown that lung viral titers vary between the AKR/J (RSV-sensitive) and C57BL/6 J (RSV-resistant) mouse strains (Stark et al. 2002; Prince et al. 1979). In this study, RSV titers in the F1 progeny were similar to those found in the resistant parent, suggesting RSV-resistance was inherited. Because RSV titers in backcross progeny were discordant with that predicted for a single gene effect, it was concluded that RSV-susceptibility was influenced by more than one gene; however, these findings have yet to be resolved by genomic analysis.

1.3 High-Throughput RNAi Assays for Host Genome Analysis

Currently, most genome-wide RNAi studies are performed using microwell-based screening platforms. Based on this format and the large amount of reagents needed, most studies analyze a single endpoint parameter such as cell death, cell morphology, virus titer, or use of a reporter gene assay (Pan et al. 2012; Meliopoulos et al. 2012; Houzet and Jeang 2011). However, multi-parametric endpoint assays and comparative analyses using different susceptible cell types perhaps infected by testing the response to more than one strain of virus is required to understand the complex nature of gene networks. For RNAi-based HTS, efficient delivery of siRNA into biologically relevant cells is required because of cell tropism. Lipid-mediated siRNA transfection is a common siRNA delivery approach, however, many cell types are not compatible with this technology, and lipid delivery reagents cause cytotoxicity, and other off-target effects that may alter gene expression profiles (Jafari et al. 2012). These features are emphasized by the lack of concordance in genome-wide RNAi-based screens identifying HIV host factors and influenza virus host factors, where the disparity has been attributed to the difference in cell lines used in each screen and in the viruses used for infection (Min and Subbarao 2010; Karlas et al. 2010; Konig et al. 2010; Bushman et al. 2009).

For screens employing genome-wide siRNA libraries it is important that the libraries are composed of 2–4 siRNAs targeting each gene such that the genes target distinct non-overlapping mRNA seed sites/sequences. Utilizing siRNA pools allows for lower concentrations of each individual siRNA to be used which decreases off-target effects (Pan et al. 2012; Houzet and Jeang 2011). However, "hits" identified from a pooled siRNA library do not identify which individual siRNAs are effective for a particular gene and validation is needed using siRNAs targeting the same gene at a different seed site. Importantly, this redundant screen using a non-overlapping siRNA for the same gene reduces the likelihood of false positives due to off-target effects (Cullen 2006). In addition, it is important to deconvolute siRNA pools in subsequent validation steps to confirm siRNAs that are efficaciously targeting the host gene.

1.4 shRNA Libraries for HTS RNAi Assays

When considering host gene discovery approaches for RSV, it may be advantageous to consider viral RNAi delivery by taking advantage of lentivirus (LV)-based shRNA libraries. LV vectors allow the delivery of shRNA constructs to a wide range of cell types, including primary and non-dividing cells, however, most shRNA libraries are pooled libraries, thus this prevents the identification of genes using individual gene-specific shRNAs. An important advantage of LV-based shRNA libraries is the permanent nature of gene suppression compared to temporary gene silencing by siRNA transfection (Tiscornia et al. 2003). The premise by which LV vectors work is that LV delivers genome-integrated shRNA that encodes a DNA sequence which is transcribed to a single-stranded RNA molecule containing sense and antisense sequences to the targeted mRNA separated by a hairpin loop (Brummelkamp et al. 2002). Cleavage of the hairpin RNA by the RISC protein, Dicer, results in short double-stranded RNA fragments analogous to siRNAs (Paddison et al. 2002).

1.5 Endpoint Assay Considerations

RSV has been shown to co-opt and dysregulate sets of host genes throughout the course of infection (Bakre et al. 2012; Tayyari and Hegele 2012; Graham 2011), thus the time of assay can affect the interpretation of the results. Similarly, the time allowed for RNAi-based host gene silencing prior to RSV infection can substantially influence the screen as many cell proteins are constitutively expressed, thus time is needed to generate a loss-of-function by RNAi. It is also important to consider that the duration of gene silencing is linked to the doubling time of a cell line because cell division dilutes the siRNA concentration (Kimpton and Emerman 1992), and cell doubling times vary among cell types. To investigate the biology of RSV replication from entry, to packaging, and release, a replication-competent virus is required. It is also important to consider when the endpoint assay is evaluated as host factors may not be essential for virus infection, but required for viral replication. To overcome these issues endpoint analysis using multiple endpoints assays is necessary, and in all assays it is critical to have a non-targeting negative control siRNA, and a positive control known to silence RSV.

2 Implications of RNAi Host Gene Discovery

What are the considerations for RNAi, host gene discovery, and vaccine development? Beyond the concerns for RNAi screening noted, it is important to emphasize that moving forward will require validation of individual host genes/factors, likely using primary cells and multiple clinical RSV isolates, to ensure that the host genes are critical for all relevant RSV strains. Additional considerations include the need to address the possibility that interactions between host genes and RSV may be specific to cellular types, e.g., type I versus type II pneumocytes, and to be cognizant that silencing genes have secondary effects that can influence the phenotype of the endpoint assay being measured. Despite this, it is possible to identify networks of host genes important to RSV replication, and through meta-analysis of existing microarray and related data, to discover and validate host genes identified as essential for RSV replication.

3 Transcriptome Profiling in the Clinical Setting: The Need and the Opportunity

Despite many years of active investigation, our understanding of the immunopathogenesis of RSV infection in humans remains incomplete and still represents a significant hurdle to advance vaccine development. One major challenge has been to translate the advances on the characterization of viral proteins and immune response mechanisms identified in mouse models into the relevant clinic context. This has been further complicated by the technical difficulties related to studying immune responses in young infants, the major target population for RSV infection.

Different classes of pathogens trigger specific pattern-recognition receptors (PRRs) that are differentially expressed on immune cells to initiate the host response (Medzhitov and Janeway 1997, 2000). We can now study the host response to specific pathogens using transcriptome analysis which has been proposed as an unbiased strategy to measure the global immune response (Fig. 1). Blood represents both a reservoir and a migration compartment for these immune cells that become educated and implement their function by circulating between central and peripheral lymphoid organs and migrating to and from the site of infection via blood. For that reason, blood leukocytes constitute an accessible source of clinically relevant information, and a comprehensive molecular phenotype of these cells can be obtained using gene expression microarrays from the relevant patient population (Chaussabel et al. 2010). Because they provide a comprehensive assessment of the immune-related cells and pathways, gene expression studies have shown to be well suited to study the host-pathogen interaction. In fact, studies have shown that different classes of pathogens induce distinct gene expression profiles that can be identified by analyses of blood leukocytes (Fig. 2) (Ramilo et al. 2007; Berry et al. 2010; Chaussabel et al. 2005, 2008; Pankla et al. 2009).

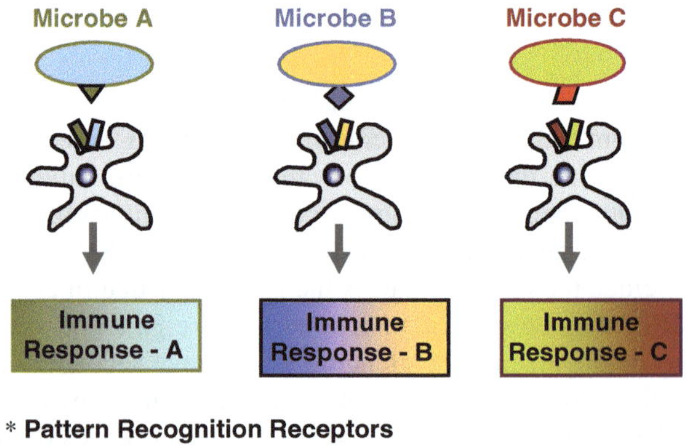

Fig. 1 Different classes of pathogens trigger specific pattern-recognition receptors (PRRs) differentially expressed on immune cells and elicit distinct immune responses

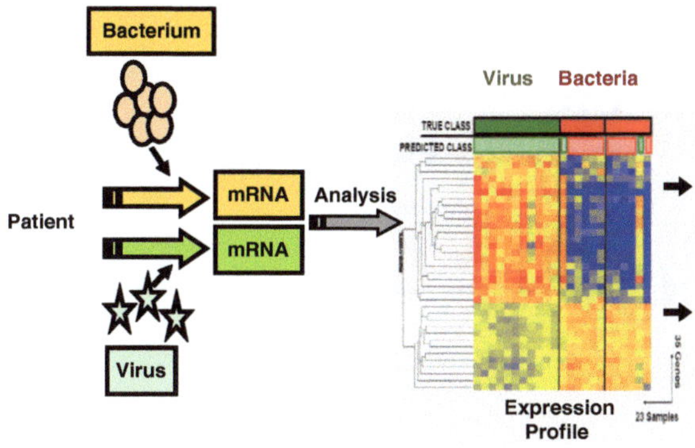

Fig. 2 Using microarray technology one can measure the differences in gene expression patterns present in blood leukocytes as induced by various types of infectious agents. The expression profile is a representative heat maps (*middle panel*) illustrating elements that corresponds to individual genes. Each column represents a sample (*patient*) and each row represents a gene transcript. Overexpressed genes are shown in red and underexpressed genes are shown in *blue*, while *yellow* represents the reference normal expression value. Modular analysis (*right panel*) illustrates the differences in expression by modules, which are groups of genes that have a common biological function (Chaussabel et al. 2008). *Red dots* represent gene modules that are overexpressed and *blue dots* gene modules that are underexpressed

4 Initial Human Studies in Acute Respiratory Infections

Global changes in transcriptional responses have been measured in the blood of patients with a variety of infectious diseases. Initial studies demonstrated that leukocytes isolated from peripheral blood of patients with acute bacterial and viral infections carry unique transcriptional signatures, which would in turn permit pathogen discrimination and patient classification (Ramilo et al. 2007). Bacterial infections induced a robust response, which was expected, as those patients were bacteremic, but children infected with influenza A infection also demonstrated a peripheral blood mononuclear cell (PBMC) signature as robust as that induced by bacteria. In addition, gene expression profiles in PBMCs from a cohort of children with a variety of acute gram-positive (*S. aureus*, *S. pneumoniae*) gram-negative bacteria (*E. coli*) and influenza A infection identified 137 classifier genes, which were applied to a population of 27 pediatric patients with pneumonia and 7 healthy children to determine whether one could differentiate patients presenting with respiratory symptoms but infected by different pathogens. Hierarchical clustering of genes and samples identified four prototypical expression profiles: healthy controls, Influenza A infection, which showed increased expression of interferon-inducible genes and was clearly different from a third profile, defining bacterial infections caused by *Staphylococcus aureus* and *Streptococus pneumoniae*, which showed over-expression of neutrophil-associated genes. Three samples belonging to the influenza A group and one from the *S. aureus* group were characterized by a

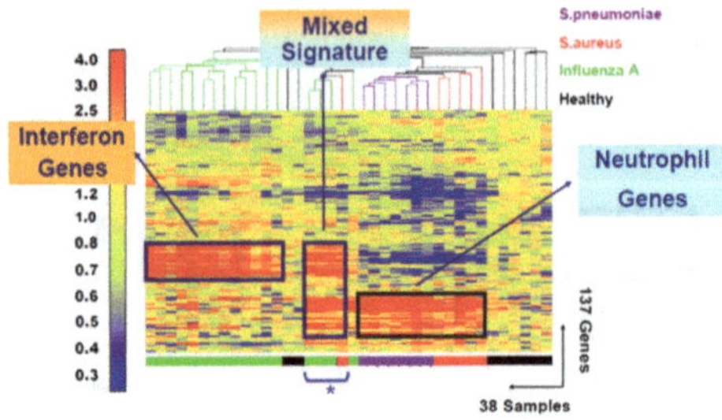

Fig. 3 Blood diagnostic biosignatures discriminate patients with pneumonia caused by viral and bacterial pathogens. Hierarchical clustering of 137 discriminative genes in 34 individuals (27 patients with pneumonia and 7 age-matched healthy controls). Each column represents an individual patient ($n = 38$); each row represents one gene ($n = 137$). Values were normalized to the median expression of each gene across all samples. *Red indicates* genes expressed at higher levels and blue lower levels than the median expression value (Ramilo et al. 2007)

fourth profile, which combined elements of the previous ones, suggesting the possibility of a co-infection caused by both a viral and a bacterial pathogen (Fig. 3) (Ramilo et al. 2007). These initial studies demonstrated that transcriptome analysis of blood leukocytes can be used to distinguish patients with acute infections caused by the most common respiratory pathogens that lead to severe pneumonia and hospitalization in children.

5 Experimental Viral Infections in Humans

A recent study provided evidence of how blood gene expression profiles can be applied to study respiratory viral infections (Zaas et al. 2009). In this study, blood gene expression profiles from healthy adult volunteers experimentally infected with human rhinovirus (HRV), RSV, or influenza A virus were analyzed. Using factor analysis they identified a 30-gene acute respiratory viral signature that clearly distinguished symptomatic infected individuals from asymptomatic subjects at the peak of the disease for each specific virus, and also from the uninfected volunteer baseline. In this "acute respiratory viral signature" the investigators identified genes (e.g., RSAD2, IFI44L, and LAMP3) that were expressed in individuals infected with any of these three viruses. They also identified genes that were specific for each viral infection, such as FCRGR1A, GBP1, and LAP3 in RSV infection, OAS2, CXCL10, and SOCS1 in individuals with HRV infection, and TNFAIP1, SEPT4, and IFI27 in cases of influenza. These observations provide evidence of the potential value of this approach to identify host response patterns that are common for a group of pathogens and those which are pathogen-specific. To validate their findings, they applied the newly identified viral signature to an independent, previously published dataset of patients with community-acquired respiratory infections (Ramilo et al. 2007). Despite the technical challenges involved in such analysis and the differences in the patient cohorts analyzed (children with naturally acquired infection vs. adults with experimental infection), their experimentally identified "acute respiratory viral signature" classified pediatric patients naturally infected with influenza A from healthy age-matched controls with 100 % accuracy (Zaas et al. 2009). This is a fundamental observation that confirms the reproducibility and potential value of blood transcriptome analysis to study host immune responses to respiratory viruses in the clinical setting. The data clearly support the hypothesis that each respiratory virus elicits a unique host immune response, that is reproducible, and that the information is easily accessible from blood immune cells. It is remarkable that blood signatures can achieve such accuracy for diagnosis of respiratory viral pathogens that are thought to be confined to the respiratory tract. Since blood specimens are usually easier to obtain than infected tissue, this approach could have practical applications in clinical medicine.

6 RSV Induces a Unique Transcriptional Profile in Peripheral Blood

Few studies have analyzed transcriptome profiles in children, the ideal target population for RSV infection. An initial study included five infants whose gene expression profiles were characterized at birth from cord blood samples (Fjaerli et al. 2007), and at the time of RSV hospitalization 3 months later (Fjaerli et al. 2006). It was shown that during acute RSV infection there was a clear activation of interferon-related genes. Activation of interferon genes was also found in the RSV signature derived from the adult volunteers challenged with RSV (Zaas et al. 2009), although there were discrepancies, which could be attributed to the differences in age. More recent studies have characterized the genome-wide transcriptional profiles in PBMCs isolated from children <2 years of age hospitalized with acute RSV ($n = 51$) and influenza A ($n = 27$) infections, and healthy matched controls ($n = 10$) (Ioannidis et al. 2012). A distinct transcriptional signature was defined in patients with acute RSV and influenza infection by using a combination of gene-expression level, statistical filters, and hierarchical clustering (Fig. 4). Statistical group comparisons using a restrictive approach (False Discovery Rate with a $p < 0.05$ followed by Bonferroni correction for multiple test corrections and a two-fold change filter) of patients with RSV infection and healthy controls yielded 110 differentially expressed transcripts, of which 70 % were overexpressed. Among the RSV-induced significantly differentially expressed genes were IFI27, defensins 1 and 3 (DEFA1, DEFA3), sialoadhesin (SN) and the interleukin-1 receptor antagonist (IL1RN). On the other hand, statistical group comparisons of patients with influenza virus and healthy controls yielded 142 differentially expressed genes, of which 64 % were overexpressed. Among the significant genes there were many interferon

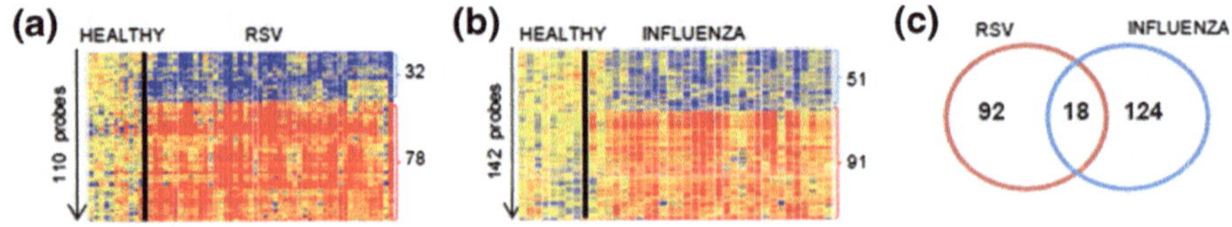

Fig. 4 Transcriptional profiles in PBMCs from patients with acute RSV and influenza virus infection. **a** Heat map of differentially expressed transcripts in patients with RSV bronchiolitis compared with healthy controls. **b** Heat map of differentially expressed transcripts in patients with acute influenza compared with healthy controls. **c** Venn diagram showing the common and virus-specific transcripts for each infection normalized to healthy controls. PBMCs isolated from patients with acute RSV bronchiolitis or acute influenza virus infection and from healthy controls. A distinct transcriptional signature was defined in patients with acute influenza or acute RSV infection by using a combination of probe expression level, statistical filters, and hierarchical clustering. Group comparisons of patients with confirmed RSV infection and healthy controls yielded 110 significant transcripts, of which 78 were overexpressed and 32 were under expressed. Statistical group comparisons of patients with influenza virus infection and healthy controls yielded 142 of which 91 were overexpressed and 51 were under expressed (Ioannidis et al. 2012)

inducible genes such as IFI27, IFI44L, IFI44, IFIT3, OAS1, OAS2, OAS3, OASL, MX1, MX2, G1P2, and DEFA1. There were 18 common interferon-inducible genes during RSV and influenza virus infection in children with a similar pattern of expression. However, the magnitude of activation was greater in influenza A infection, and specifically the activation of IFI27, which was the top common overexpressed gene in both cohorts of patients, was almost three-fold greater in children hospitalized with influenza than in those with RSV (103- vs. 38-fold change for influenza and RSV respectively). Comparison of the transcriptional profiles using additional analytical tools such as the modular analysis (Chaussabel et al. 2008), and Ingenuity pathway analysis (IPA, Ingenuity Systems), confirmed the significant differences in the magnitude of the interferon-related response between PBMCs from patients with influenza or RSV infection. Altogether, these results indicate that although the transcriptional response of PBMCs from patients with acute RSV and influenza virus infection was predominantly virus specific, it also shared a common interferon-inducible signature that was more robust in patients with influenza than in those with RSV infection. Indeed, the attenuated interferon response observed in children with RSV infection is one of the proposed mechanisms used by RSV to evade the immune system and prevent the development of an effective memory immune response (Fig. 5).

7 Clinical Application of Transcriptional Profiles

Analysis of PBMC samples requires processing in real time, which has limitations from a practical application in the clinical setting with large numbers of patients. In addition, PBMC samples do not include the majority of neutrophils, and this cell

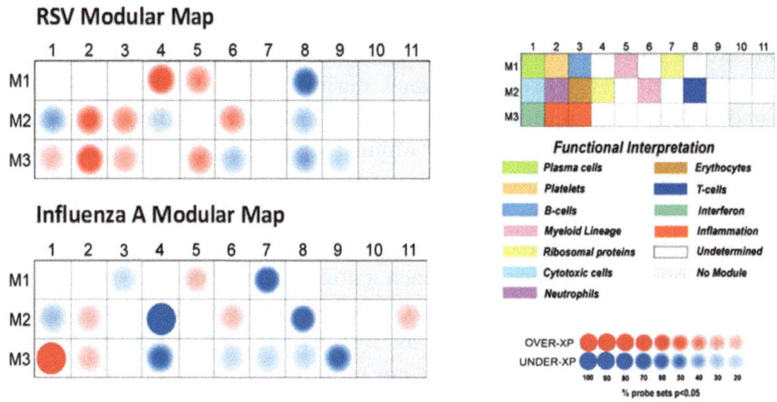

Fig. 5 Module mapping of gene expression profiles illustrates the differences between RSV and Influenza infections. Gene expression levels were compared between patients with RSV and Influenza infections on a module-by-module basis. Colored spots represent the percentage of significantly ($p < 0.05$, Mann–Whitney) over-expressed (*red*) or under-expressed (*blue*) transcripts within a module in patients with RSV and influenza infections compared with their respective matched healthy controls; spot intensity represents the magnitude of the gene expression change, blank modules demonstrate no significant differences between groups ($p > 0.05$). Information is displayed on a grid, with the coordinates corresponding to one of 28 modules with the key shown on the on the right representing the functional interpretation of modules. RSV patients display marked over expression of the neutrophil (M2.2) and myeloid lineage gene modules (M1.5 and M2.6), while influenza patients show significant over expression of the interferon module (M3.1), which is also overexpressed but much less significantly in RSV patients

population has shown to be very relevant to the pathogenesis of viral infections. For these reasons in recent years there has been a shift from PBMCs to whole blood samples as the new standard method approach to study transcriptional profiles in the clinical setting (Banchereau et al. 2012). Initial studies using this approach are being conducted in infants with RSV and other viral respiratory infections. Early analyses confirm the original observations derived from PBMCs and demonstrate that RSV induces a robust, reproducible and distinct transcriptional profile in whole blood from children naturally infected with this respiratory virus (Mejias et al. 2012).

To facilitate the application of transcriptional profiles in the clinical setting, novel analytical tools are been developed so the relationship between host transcriptional changes and the clinical outcome of the infection can be better and more objectively assessed. One of these novel tools, the Molecular Distance to Health (MDTH) is a genomic score that measures the overall transcriptional perturbation in patient's samples compared with those obtained from healthy controls (Pankla et al. 2009). The MDTH score summarizes in a single number the global perturbation of the immune profile and allows correlations between the transcriptional profiles and the clinical findings. Recent studies conducted in patients with sepsis, pulmonary tuberculosis, and *Staphylococcus aureus* infections have provided substantial evidence of the value of the MDTH score applied in the clinical setting (Berry et al. 2010; Pankla et al. 2009; Banchereau et al. 2012). In those studies, MDTH scores calculated during the acute disease significantly correlated with the magnitude and severity of the clinical findings. In addition, as patients were treated and showed clinical improvement the MDTH scores decreased and eventually showed values comparable to healthy controls.

Studies are exploring the value of using the MDTH scores in children with acute RSV and influenza infections to help stratify these patients more objectively and more importantly to correlate each patient transcriptional profile with the clinical manifestations of the respiratory disease. These observations suggest that transcriptome profiles could be used as objective tools to help identify immune pathways or cellular components that are more relevant in controlling and regulating the clinical manifestations of the disease, and as such provide relevant information to facilitate the development of RSV vaccines.

8 Conclusions

Studies of host transcription profiles and gene suppression have begun to provide insights into the host response to and regulation of RSV infection. These tools have great potential and will, no doubt, provide many new and important insights into our understanding of RSV disease. For example, recent studies demonstrate that although RSV and influenza infections induce activation of common immune pathways, the magnitude of the responses, and their patterns of activation or suppression also demonstrate viral specific responses and on-going studies suggest that this approach can identify biomarkers of the severity of RSV disease. Application of these tools to expand the characterization of the immune response to vaccines and the impact of vaccines on the host response to later infection should be of great value in the evaluation and design of future RSV vaccines.

References

Alvisi G, Rawlinson SM, Ghildyal R, Ripalti A, Jans DA (2008) Regulated nucleocytoplasmic trafficking of viral gene products: a therapeutic target? Biochim Biophys Acta 1784:213–227

Aza-Blanc P, Cooper CL, Wagner K, Batalov S, Deveraux QL, Cooke MP (2003) Identification of modulators of TRAIL-induced apoptosis via RNAi-based phenotypic screening. Mol Cell 12:627–637

Bakre A, Mitchell P, Coleman JK, Jones LP, Saavedra G, Teng M, Tompkins SM, Tripp RA. (2012) Respiratory syncytial virus modifies microRNAs regulating host genes that affect virus replication. J Gen Virol 93(11):2346–2356

Banchereau R, Jordan-Villegas A, Ardura M, Mejias A, Baldwin N, Xu H, Saye E, Rossello-Urgell J, Nguyen P, Blankenship D, Creech CB, Pascual V, Banchereau J, Chaussabel D, Ramilo O (2012) Host immune transcriptional profiles reflect the variability in clinical disease manifestations in patients with Staphylococcus aureus infections. PLoS ONE 7:e34390

Berry MP, Graham CM, McNab FW, Xu Z, Bloch SA, Oni T, Wilkinson KA, Banchereau R, Skinner J, Wilkinson RJ, Quinn C, Blankenship D, Dhawan R, Cush JJ, Mejias A, Ramilo O, Kon OM, Pascual V, Banchereau J, Chaussabel D, O'Garra A (2010) An interferon-inducible neutrophil-driven blood transcriptional signature in human tuberculosis. Nature 466:973–977

Brummelkamp TR, Bernards R, Agami R (2002) A system for stable expression of short interfering RNAs in mammalian cells. Science 296:550–553

Bushman FD, Malani N, Fernandes J, D'Orso I, Cagney G, Diamond TL, Zhou H, Hazuda DJ, Espeseth AS, Konig R, Bandyopadhyay S, Ideker T, Goff SP, Krogan NJ, Frankel AD, Young JA, Chanda SK (2009) Host cell factors in HIV replication: meta-analysis of genome-wide studies. PLoS Pathog 5:e1000437

Chaussabel D, Allman W, Mejias A, Chung W, Bennett L, Ramilo O, Pascual V, Palucka AK, Banchereau J (2005) Analysis of significance patterns identifies ubiquitous and disease-specific gene-expression signatures in patient peripheral blood leukocytes. Ann N Y Acad Sci 1062:146–154

Chaussabel D, Quinn C, Shen J, Patel P, Glaser C, Baldwin N, Stichweh D, Blankenship D, Li L, Munagala I, Bennett L, Allantaz F, Mejias A, Ardura M, Kaizer E, Monnet L, Allman W, Randall H, Johnson D, Lanier A, Punaro M, Wittkowski KM, White P, Fay J, Klintmalm G, Ramilo O, Palucka AK, Banchereau J, Pascual V (2008) A modular analysis framework for blood genomics studies: application to systemic lupus erythematosus. Immunity 29:150–164

Chaussabel D, Pascual V, Banchereau J (2010) Assessing the human immune system through blood transcriptomics. BMC Biol 8:84

Cullen BR (2006) Enhancing and confirming the specificity of RNAi experiments. Nat Methods 3:677–681

Dupuy LC, Dobson S, Bitko V, Barik S (1999) Casein kinase 2-mediated phosphorylation of respiratory syncytial virus phosphoprotein P is essential for the transcription elongation activity of the viral polymerase; phosphorylation by casein kinase 1 occurs mainly at Ser(215) and is without effect. J Virol 73:8384–8392

Fjaerli HO, Bukholm G, Krog A, Skjaeret C, Holden M, Nakstad B (2006) Whole blood gene expression in infants with respiratory syncytial virus bronchiolitis. BMC Infect Dis 6:175

Fjaerli HO, Bukholm G, Skjaeret C, Holden M, Nakstad B (2007) Cord blood gene expression in infants hospitalized with respiratory syncytial virus bronchiolitis. J Infect Dis 196:394–404

Ghildyal R, Ho A, Wagstaff KM, Dias MM, Barton CL, Jans P, Bardin P, Jans DA (2005) Nuclear import of the respiratory syncytial virus matrix protein is mediated by importin beta1 independent of importin alpha. Biochemistry 44:12887–12895

Ghildyal R, Ho A, Dias M, Soegiyono L, Bardin PG, Tran KC, Teng MN, Jans DA (2009) The respiratory syncytial virus matrix protein possesses a Crm1-mediated nuclear export mechanism. J Virol 83:5353–5362

Ghildyal R, Jans DA, Bardin PG, Mills J (2012) Protein-protein interactions in RSV assembly: potential targets for attenuating RSV strains. *Infect Disord Drug Targets* 12:103–109

Gobert GN (2010) Better understanding of anti-schistosomal strategies through microarray analysis. *Infect Disord Drug Targets* 10:251–257

Gonzalez LC (2012) Protein microarrays, biosensors, and cell-based methods for secretome-wide extracellular protein-protein interaction mapping. Methods 57(4):448–458

Graham BS (2011) Biological challenges and technological opportunities for respiratory syncytial virus vaccine development. Immunol Rev 239:149–166

Graham BS, Perkins MD, Wright PF, Karzon DT (1988) Primary respiratory syncytial virus infection in mice. J Med Virol 26:153–162

Houzet L, Jeang KT (2011) Genome-wide screening using RNA interference to study host factors in viral replication and pathogenesis. Exp Biol Med (Maywood) 236:962–967

Huang YC, Li Z, Hyseni X, Schmitt M, Devlin RB, Karoly ED, Soukup JM (2008) Identification of gene biomarkers for respiratory syncytial virus infection in a bronchial epithelial cell line. Genomic Med 2:113–125

Hussell T, Georgiou A, Sparer TE, Matthews S, Pala P, Openshaw PJ (1998) Host genetic determinants of vaccine-induced eosinophilia during respiratory syncytial virus infection. J Immunol 161:6215–6222

Ioannidis I, McNally B, Willette M, Peeples ME, Chaussabel D, Durbin JE, Ramilo O, Mejias A, Flano E (2012) Plasticity and virus specificity of the airway epithelial cell immune response during respiratory virus infection. J Virol 86:5422–5436

Jafari M, Soltani M, Naahidi S, Karunaratne DN, Chen P (2012) Nonviral approach for targeted nucleic acid delivery. Curr Med Chem 19:197–208

Janssen R, Pennings J, Hodemaekers H, Buisman A, van Oosten M, de Rond L, Ozturk K, Dormans J, Kimman T, Hoebee B (2007) Host transcription profiles upon primary respiratory syncytial virus infection. J Virol 81:5958–5967

Karlas A, Machuy N, Shin Y, Pleissner KP, Artarini A, Heuer D, Becker D, Khalil H, Ogilvie LA, Hess S, Maurer AP, Muller E, Wolff T, Rudel T, Meyer TF (2010) Genome-wide RNAi screen identifies human host factors crucial for influenza virus replication. Nature 463:818–822

Kimpton J, Emerman M (1992) Detection of replication-competent and pseudotyped human immunodeficiency virus with a sensitive cell line on the basis of activation of an integrated beta-galactosidase gene. J Virol 66:2232–2239

Konig R, Stertz S, Zhou Y, Inoue A, Hoffmann HH, Bhattacharyya S, Alamares JG, Tscherne DM, Ortigoza MB, Liang Y, Gao Q, Andrews SE, Bandyopadhyay S, De Jesus P, Tu BP, Pache L, Shih C, Orth A, Bonamy G, Miraglia L, Ideker T, Garcia-Sastre A, Young JA, Palese P, Shaw ML, Chanda SK (2010) Human host factors required for influenza virus replication. Nature 463:813–817

Krausz E (2007) High-content siRNA screening. Mol BioSyst 3:232–240

Leroy Q, Raoult D (2010) Review of microarray studies for host-intracellular pathogen interactions. *J Microbiol Methods* 81:81–95

Martinez I, Lombardia L, Garcia-Barreno B, Dominguez O, Melero JA (2007) Distinct gene subsets are induced at different time points after human respiratory syncytial virus infection of A549 cells. J Gen Virol 88:570–581

Medzhitov R, Janeway CA Jr (1997) Innate immunity: the virtues of a nonclonal system of recognition. Cell 91:295–298

Medzhitov R, Janeway C Jr (2000) Innate immune recognition: mechanisms and pathways. Immunol Rev 173:89–97

Mejias A, Dimo B, Jordan-Villegas A, Jartti T, Zu X, Suarez N, Banchereau J, Chaussabel D, Ramilo O (2012) Respiratory syncytial virus infection induces a unique gene expression profile that correlates with disease severity. In: 8th annual respiratory syncytial virus symposium, Santa Fe, NM

Meliopoulos VA, Andersen LE, Birrer KF, Simpson KJ, Lowenthal JW, Bean AG, Stambas J Stewart CR, Tompkins SM, van Beusechem VW, Fraser I, Mhlanga M, Barichievy S, Smith Q, Leake D, Karpilow J, Buck A, Jona G, Tripp RA (2012) Host gene targets for novel influenza therapies elucidated by high-throughput RNA interference screens. *FASEB* J 26:1372–1386

Miller MB, Tang YW (2009) Basic concepts of microarrays and potential applications in clinical microbiology. Clin Microbiol Rev 22:611–633

Min JY, Subbarao K (2010) Cellular targets for influenza drugs. Nat Biotechnol 28:239–240

Monsurro V, Marincola FM (2011) Microarray analysis for a comprehensive immunological-status evaluation during cancer vaccine immune monitoring. *J Biomed Biotechnol* 2011:307297

Moss EG (2001) RNA interference: it's a small RNA world. Curr Biol 11:R772–R775

Oyelaran O, Gildersleeve JC (2007) Application of carbohydrate array technology to antigen discovery and vaccine development. Expert Rev Vaccines 6:957–969

Paddison PJ, Caudy AA, Bernstein E, Hannon GJ, Conklin DS (2002) Short hairpin RNAs (shRNAs) induce sequence-specific silencing in mammalian cells. Genes Dev 16:948–958

Pan Q, van der Laan LJ, Janssen HL, Peppelenbosch MP (2012) A dynamic perspective of RNAi library development. Trends Biotechnol 30:206–215

Pankla R, Buddhisa S, Berry M, Blankenship DM, Bancroft GJ, Banchereau J, Lertmemongkolchai G, Chaussabel D (2009) Genomic transcriptional profiling identifies a candidate blood biomarker signature for the diagnosis of septicemic melioidosis. Genome Biol 10:R127

Pennings JL, Schuurhof A, Hodemaekers HM, Buisman A, de Rond LC, Widjojoatmodjo MN, Luytjes W, Kimpen JL, Bont L, Janssen R (2011) Systemic signature of the lung response to respiratory syncytial virus infection. PLoS ONE 6:e21461

Prince GA, Horswood RL, Berndt J, Suffin SC, Chanock RM (1979) Respiratory syncytial virus infection in inbred mice. Infect Immun 26:764–766

Ramilo O, Allman W, Chung W, Mejias A, Ardura M, Glaser C, Wittkowski KM, Piqueras B, Banchereau J, Palucka AK, Chaussabel D (2007) Gene expression patterns in blood leukocytes discriminate patients with acute infections. Blood 109:2066–2077

Segovia J, Sabbah A, Mgbemena V, Tsai SY, Chang TH, Berton MT, Morris IR, Allen IC, Ting JP, Bose S (2012) TLR2/MyD88/NF-kappaB pathway, reactive oxygen species, potassium efflux activates NLRP3/ASC inflammasome during respiratory syncytial virus infection. PLoS ONE 7:e29695

Stark JM, Barmada MM, Winterberg AV, Majumber N, Gibbons WJ Jr, Stark MA, Sartor MA, Medvedovic M, Kolls J, Bein K, Mailaparambil B, Krueger M, Heinzmann A, Leikauf GD, Prows DR (2009) Genomewide association analysis of respiratory syncytial virus infection in mice. *J Virol* 84:2257–2269

Stark JM, McDowell SA, Koenigsknecht V, Prows DR, Leikauf JE, Le Vine AM, Leikauf GD (2002) Genetic susceptibility to respiratory syncytial virus infection in inbred mice. J Med Virol 67:92–100

Schuurhof A, Bont L, Pennings JL, Hodemaekers HM, Wester PW, Buisman A, de Rond LC, Widjojoatmodjo MN, Luytjes W, Kimpen JL, Janssen R (2010) Gene expression differences in lungs of mice during secondary immune responses to respiratory syncytial virus infection. *J Virol* 84:9584–9594

Tayyari F, Hegele RG (2012) Identifying targets in the hunt for effective respiratory syncytial virus interventions. Expert rev respir med 6:215–222

Tijsterman M, Plasterk RH (2004) Dicers at RISC; the mechanism of RNAi. Cell 117:1–3

Tiscornia G, Singer O, Ikawa M, Verma IM (2003) A general method for gene knockdown in mice by using lentiviral vectors expressing small interfering RNA. Proc Natl Acad Sci USA 100:1844–1848

Tripp RA, Oshansky C, Alvarez R (2005) Cytokines and respiratory syncytial virus infection. Proc Am Thorac Soc 2:147–149

Uttamchandani M, Neo JL, Ong BN, Moochhala S (2009) Applications of microarrays in pathogen detection and biodefence. Trends Biotechnol 27:53–61

van Diepen A, Brand HK, Sama I, Lambooy LH, van den Heuvel LP, van der Well L, Huynen M, Osterhaus AD, Andeweg AC, Hermans PW Quantitative proteome profiling of respiratory virus-infected lung epithelial cells. *J Proteomics* 73:1680–1693

Zaas AK, Chen M, Varkey J, Veldman T, Hero AO 3rd, Lucas J, Huang Y, Turner R, Gilbert A, Lambkin-Williams R, Oien NC, Nicholson B, Kingsmore S, Carin L, Woods CW, Ginsburg GS (2009) Gene expression signatures diagnose influenza and other symptomatic respiratory viral infections in humans. Cell Host Microbe 6:207–217

Ziauddin J, Sabatini DM (2001) Microarrays of cells expressing defined cDNAs. Nature 411:107–110

Consequences of Immature and Senescent Immune Responses for Infection with Respiratory Syncytial Virus

Allison M. W. Malloy, Ann R. Falsey and Tracy J. Ruckwardt

Abstract Infants in the first 6 months of life and older adults (>65 years of age) are disproportionately burdened with respiratory syncytial virus (RSV)-associated morbidity and mortality. While other factors play a role in the risk these groups assume, shortcomings of the immune response make a substantial contribution to the predisposition to severe disease. Ineffectual antibody production with misdirected cytokine responses and excess inflammation in the airways are common to both groups. However, the mechanisms underlying these immune responses differ between infants and older adults and need to be better understood. Preventative approaches to decreasing the burden of disease are preferable to therapeutic intervention and effective vaccination strategies will need to target the strengths of the immune responses in these populations.

Contents

A. M. W. Malloy · T. J. Ruckwardt
Vaccine Research Center, NIAID, NIH, Bethesda, MD, USA

A. R. Falsey (✉)
University of Rochester School of Medicine and Dentistry, Rochester, NY, USA
e-mail: ann.falsey@viahealth.org

L. J. Anderson and B. S. Graham (eds.), *Challenges and Opportunities for Respiratory Syncytial Virus Vaccines*, Current Topics in Microbiology and Immunology 372,
DOI: 10.1007/978-3-642-38919-1_11, © Springer-Verlag Berlin Heidelberg 2013

1 Introduction: Disease Burden in Infants and the Elderly

Young age is the most significant risk factor for severe RSV disease, and infants in the first 4 months of life are at highest risk. The first 6 months of life carry a threefold higher risk of hospitalization than the second 6 months (Hall 2012), with half of all hospitalizations associated with RSV infection occurring before 6 months. Although there are other risk factors for severe disease (see chapter by C.B. Hall et al., this volume), the majority of infants hospitalized with severe RSV disease were previously healthy with no known risk factors other than age (Boyce et al. 2000). This suggests that the developing airway and the limitations of the infant immune system are responsible for severe disease in otherwise healthy infants. These factors combined with viral mechanisms of immune subversion (see chapters by S. Mukherjee and N.W. Lukacs, S.M. Varga and T.J. Braciale, and by S. Barik, this volume), unique genetic attributes of the host (see chapters by E. H. Choi et al., and by R.A. Tripp et al., this volume) and environmental factors result not only in severe disease, but may create an immunological response pattern unique to RSV because of its proclivity to infect the youngest, most vulnerable infants (see chapters by S. Mukherjee and N.W. Lukacs, S.M. Varga and T.J. Braciale, and by S. Barik, this volume).

Immunity to RSV is incomplete and reinfections occur throughout life. Generally, RSV infection in immunocompetent young adults and through mid-life causes an upper respiratory infection reflecting a primed, healthy immune response with efficient viral clearance. However, later in life, RSV infection can once again result in severe disease reminiscent of illness in infants. Aging results in perturbations of immune function and along with an accumulation of chronic medical conditions leads to severe RSV illness in older populations. In this chapter, we will review features of the immature and senescent immune response and how these polar extremes influence the manifestations of RSV infection and prospects for vaccination in these age groups. Interestingly, as the pendulum swings from early to late life several deficits in the immune response are common to both including limited access to efficient T-cell responses. The neonate has many naïve T cells but their precursor frequency is low and more stimulation is required for activation whereas the elderly have a high precursor frequency of memory cells against a few pathogen-specific epitopes, but few naïve T cells to respond to new specificities (Fig. 1). While the basis for poor antibody and cellular imune responses differ between the young and the elderly, these deficits provide a challenge that must be met for effective immunization of these two high risk populations against RSV.

2 Immunity in the Infant

Human newborns suffer from a higher frequency and severity of invasive microbial infections compared to healthy adults (Klein and Remington 2001). Globally, much of the burden disproportionately occurs in the neonatal period, or first

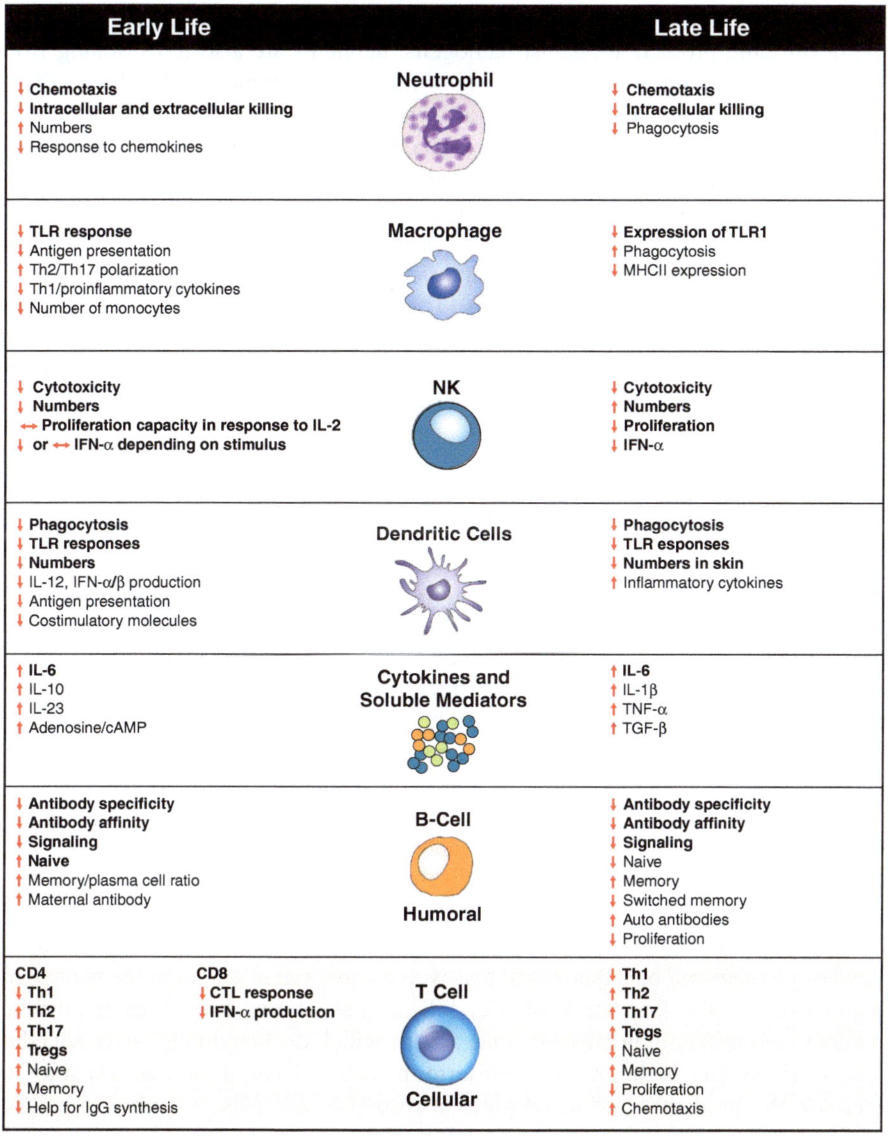

Fig. 1 Characteristics of early and late life immunity. Features common to both early and late life are shown in bold

28 days of life. In the first 5 years of life, 38 % of mortality occurs in the initial 28 days while 62 % occurs in the remaining 1,800 days, and approximately 50 % of the mortality is clearly linked to infection (Lawn et al. 2005). The susceptibility of neonates to infection is generally attributed to an "immature and naïve immune system." However, neonatal immune responses are also inherently different from

those of adults' due to critically dissimilar environmental pressures. In addition to the developing protection against pathogens, neonates are also transitioning from the fetal environment where dampening of fetal proinflammatory/T helper 1 (Th1) responses is integral in preventing an alloimmune response from the mother against the fetus which might result in fetal demise. Neonates are also transitioning from an antigenically-limited in utero environment to an outside world rife with new pathogens, some beneficial and others detrimental to their health. Through infancy, the immune response is not primed to quickly respond to harmful pathogens, but must be educated during primary exposure while allowing essential colonization by commensal pathogens. The infant immune response is therefore challenged with effectively balancing excess inflammation and protection from disease. The first year of life requires both flexibility and adaptability of the infant immune response to accommodate the diverse exposures of early life.

2.1 Neonatal Innate Immunity

Due to limited exposure to antigens in utero, the neonate must initially rely more on innate immunity which provides early non-antigen-specific pathogen protection to prevent infection. Thus, innate immune effector cells play a relatively larger role in determining how neonates respond to infections compared to antigen-experienced children and adults. Although there are multiple cells that comprise the innate immune response, neutrophils and antigen presenting cells [APCs; monocyte, macrophage, and dendritic cells (DC)] are instrumental in the neonatal response to infection.

2.1.1 Neutrophils

Neutrophils are the primary responders to pathogen-induced inflammation. In newborns, there can be a quantitative as well as qualitative defect in the neutrophil population. In the first week of life, full-term newborns have greater absolute numbers of peripheral neutrophils than adults, which decline quickly to reach adult levels. However, both pre-and full-term neonates have a limited capacity to increase the production of neutrophils in response to sepsis, resulting in neutropenia under conditions of stress (Koenig and Yoder 2004). Neonatal neutrophils also have functional impairments in chemotaxis and higher levels of circulating IL-6 demonstrated in the infant may inhibit neutrophil migration to inflammatory sites (Levy 2007). Neonatal neutrophils exhibit impairment in microbicidal mechanisms and contain reduced amounts of antimicrobial proteins and peptides (Levy 2007). Intracellular killing mediated by oxidase activity is impaired in preterm newborns in response to staphylococci, which are common neonatal pathogens, and in both pre- and full-term newborns in response to LPS (Levy 2007). Extracellular killing is also deficient due to an inability to form neutrophil extracellular traps which are lattices of extracellular DNA, chromatin, and antibacterial proteins that mediate capture and extracellular killing of microorganisms (Yost et al. 2009). The limitations in functionality found in infant neutrophils present weaknesses in this arm of the innate response which may contribute to the pathology of RSV disease.

An abundance of neutrophils in the lung is a hallmark of RSV-bronchiolitis in neonates (Everard et al. 1994). RSV acute lower respiratory tract infection has been found to be associated with production of TNF-α, IL-6, IL-8, RANTES, MIP-1α, and IFN-γ among others, and robust inflammatory responses in the airways (Collins and Graham 2008). IL-8, in particular, is a potent chemoattractant for neutrophils. Neutrophils in the airway of RSV-infected infants are highly activated, release cytokines and have delayed apoptosis which may exacerbate lung damage (Halfhide et al. 2011).

2.1.2 Antigen Presenting Cells

APC are important in phagocytosis, cytokine production and antigen presentation and are integral in directing the adaptive immune response. The capacity of the neonate to produce monocytes and the number of monocytes found in neonatal peripheral blood is equivalent to or greater than that found in adults (Weinberg et al. 1985). However, there are qualitative defects in both monocytes and other APCs. Migration to sites of inflammation and innate phagocytic function are impaired in human neonatal monocytes (Marodi 2006). DCs from cord blood have decreased TNF-α production, as well as decreased upregulation of costimulatory molecules and stimulation of T-cells after phagocytosis of necrotic cells in comparison to adult dendritic cells (Levy 2007).

Comparisons of cord blood mononuclear cells (CBMC) to adult peripheral blood mononuclear cells (PBMC) have shown differences in the way APCs respond to RSV infection. One study compared the responses of adult PBMC to CBMC following in vitro RSV infection. Live and inactivated RSV induced both innate and adaptive cytokine production by adult PBMC (IL-6 $>$ IFN-γ $>$ IL-2 $\sim$ TNF-α $>$ IL-10). Exposure of CBMC to live and inactivated RSV induced a profile of innate cytokine production (IL-6, TNF-α, IL-10, IFN-γ, and IL-12) higher than that stimulated in adult PBMC, and an absence of adaptive cytokine production consistent with a lack of prior exposure to RSV (Krishnan et al. 2003). Thornburg et al. (2010) directly compared primary DCs isolated from cord blood and adult peripheral blood following in vitro RSV infection. They found higher TGF-β1 production in RSV-stimulated cord blood DCs. Following co-culture with autologous T cells, adult RSV-infected DCs induced secretion of IFN-γ, IL-12p70, IL-2, and TNF-α, all of which were reduced by the addition of exogenous TGF-β1 to the cultures. Conversely, co-culture of autologous T cells with RSV-infected cord blood DCs stimulated production of IL-4, IL-6, IL-1β, and IL-17. These studies demonstrate that "default" neonatal immunity depends on innate responses and is deficient in adaptive responses that tend to be Th2-biased providing insight into the mechanisms underlying the reduced ability of neonates to clear RSV.

2.1.3 Cytokine/Chemokines and Soluble Mediators

Phenotypic and stimulus-specific functional immaturities in cytokine production are present at birth in both mice and humans (Levy 2007). Th2 polarizing cytokines, such as IL-2, IL-4, and IL-6 are more frequently observed during neonatal inflammatory responses while Th1 polarizing/proinflammatory cytokines, such as IFN-γ, IL-12, TNF-α, and INF-α, tend to be lower. Th2-skewing is thought to be regulated by the unique neonatal responses to pathogens via pattern recognition receptors (PRRs), of which the best studied are the Toll-like receptors (TLRs). Basal TLR expression of full-term human neonatal blood monocytes is similar to that of adults, however, there is reduced activation and functional consequence to engagement of neonatal TLRs. In general, there is an impairment of Th1-polarizing responses to TLR agonists. Monocytes from newborns have impaired TNF-α production in response to agonists of TLR1-7 (Levy 2007). Type I IFN production is also decreased in neonatal APCs. Studies have shown a decreased production of IFN-α by neonatal plasmacytoid DCs in response to polyinosinic-polycytidylic acid (polyI:C; a TLR3 agonist) and CpG (a TLR9 agonist) and impaired IFN-β production by neonatal monocyte-derived DCs in response to lipopolysaccharide (LPS), a TLR4 agonist (Levy 2007). Neonatal leukocytes exhibited impaired proinflammatory cytokine production, such as TNF-α, IL-12p70, and IFN-γ in response to LPS (Belderbos et al. 2009).

In contrast, TLR-mediated production of IL-10 (anti-inflammatory), IL-6 and IL-23 (Th17 promoting) by neonatal APCs is enhanced relative to adult cells (Levy 2007). One potential mechanism for these differences involves the high level of adenosine found in neonatal cord blood plasma resulting in a 20-fold increase in intracellular concentration of cyclic AMP (cAMP) (Levy et al. 2006). cAMP is known to decrease LPS-stimulated production of TNFα and inhibit the production of IFN-α, IFN-γ and IL-12 while preserving IL-6, IL-10 and IL-23 expression (Levy 2007). Overall, neonatal murine and human studies have demonstrated decreased Th1-polarizing/proinflammatory responses with increased production of Th2-polarizing and anti-inflammatory cytokine production. This cytokine/chemokine milieu is thought to promote anti-infective proteins and peptides within the blood and at mucosal surfaces to decrease the risk of infection during colonization, but limit proinflammatory cytokines detrimental to in utero survival (Levy 2007).

While the early-life Th1/Th2 imbalance is necessary, it may contribute to RSV disease in infants. A study comparing infants with severe RSV disease to those with mild disease and uninfected controls demonstrated increases in plasma cytokine concentrations of IL-6 and IL-8 were associated with prolonged hospitalization (Mella et al. 2013). Using CBMC, a prospective study demonstrated that infants subsequently hospitalized with RSV had higher IL-6 and IL-8 production following LPS-stimulation of CBMC. CBMC from these same infants showed lower LPS-stimulated IL-1β, IL-2, IL-4, IL-5, and IL-10 responses than infants that experienced a mild course following infection. Additional studies using cord blood found IL-12 levels were significantly lower in infants who developed RSV-bronchiolitis following infection (Blanco-Quiros et al. 1999), and that high concentrations of the proinflammatory cytokines IL-8 and TNF-α in amniotic fluid are associated with a low risk of RSV bronchiolitis in healthy term infants (Houben et al. 2012). Both prospective studies of CBMC and studies of the peripheral blood of infants with severe disease reiterate a propensity toward Th2-type responses and high production of IL-6 and IL-8.

2.2 *Neonatal Adaptive Immunity*

Clinically severe infection among healthy children is almost always seen during the primary RSV infection as an infant (Hall 2012). This first antigenic encounter with RSV occurs in the absence of an RSV-specific memory response and importantly, has the potential to direct or imprint future responses to RSV. The initiation and function of adaptive responses to RSV are covered in detail in the chapter by S.M. Vara and T.J. Braciale, this volume.

2.2.1 Neonatal CD4+ T-Cell

Neonatal CD4+ T-cell responses have been found to be Th2-skewed in murine models (Adkins et al. 2004). In human neonates, both Th1 and Th2 responses are often lower than those of adults and vaccine studies indicate a tendency toward low INF-γ production and a preference for Th2 memory responses (Wood and Siegrist 2011). However, both murine and human neonates can develop more adult-like Th1 responses in the setting of Th1-promoting adjuvants or mycobacterium bovis bacillus Calmette-Guerin vaccination (Wood and Siegrist 2011).

Detailed studies of CD4+ T-cell responses in RSV-infected human infants have been limited. Complexities of sample acquisition combined with a dearth of described epitopes and class II tetramer reagents to clearly identify RSV-specific CD4+ T cells has led to relatively unrefined descriptions of canonical cytokines in sites and samples that may contain relatively few CD4+ T cells. The ever-growing family tree of CD4+ T-cell subpopulations further complicates matters. Of importance, CD4+ T cells can be critical for optimal CD8+ T-cell function and the development of mature, high-affinity B cell responses. A paucity of RSV-specific CD4+ T-cell responses in infected infants may be one reason why RSV-specific antibody appears to lack the durability associated with other antibody responses (see chapter by S.M. Vara and T.J. Braciale, this volume).

2.2.2 CD8+ T-Cells

CD8+ T-cells require antigen stimulation, appropriate costimulatory and cytokine signals to develop into CTLs. The neonatal bias toward a Th2 pattern as well as limited or ineffectual interactions with APCs can result in deficiencies in the CTL response. Neonatal cord blood T cells have been found to be moderately less effective than adult T cells as cytotoxic effector cells (Risdon et al. 1994). Poor CTL responses were associated with a lack of protection from disease in mice following neonatal infection with murine leukemia virus (Sarzotti et al. 1996). However, both murine and human neonatal CD8+ T cells are capable of developing potent CTL activity under certain circumstances. Murine neonatal antigen-specific CTLs developed in similar frequency and with similar levels of effector function and avidity for antigen as adults following DNA vaccination (Zhang et al. 2002). In humans, potent CTL function has been detected in infants congenitally infected with cytomegalovirus or Trypanosoma cruzi (Adkins et al. 2004).

There are relatively few descriptions of T-cell responses during severe primary RSV infection, but two elegant studies of hospitalized infants have demonstrated that a systemic neutrophil response precedes what appears to be a robust CD8+ T-cell response that peaks at convalescence (Heidema et al. 2007; Lukens et al. 2010). CD8+ T-cell responses peak in the infected infant after the height of viral replication and disease expression. Thus linking CD8+ T cells to viral clearance and protection, and suggesting that in the setting of natural primary infection in human infants these benefits outweigh the negative potential for immune pathology.

2.2.3 B Cell

B cell development and antibody responses in neonates differ quantitatively and qualitatively from those generated by adults. Children younger than 18–24 months of age are particularly susceptible to encapsulated bacteria (i.e., pneumococcus, Haemophilus influenza type b and meningococcus) demonstrating poor T-cell-independent B-cell responses. However, T-cell-dependent B-cell responses are also insufficient, evident in the limited antibody response to RSV in young infants. Vaccine studies in infants have shown that the development of detectable vaccine-specific antibodies in the serum and the magnitude of those response are age-dependent, and that acceleration of vaccination schedules may fail to elicit protective antibodies (Siegrist and Aspinall 2009).

The factors that determine neonatal B cell responses and antibody production are both B cell intrinsic and environmental. Human neonatal B cells have been shown to express lower levels of the costimulatory molecules CD40, CD80, and CD86, which alters their sensitivity to signals received from other cells of the immune system (Siegrist and Aspinall 2009). In addition, neonatal splenic marginal zone B cells express lower levels of CD21, which limits their capacity to respond to polysaccharide-complement complexes in a T cell-independent fashion (Siegrist and Aspinall 2009). Reduced levels of CD21 also support the generation of memory B cells and impair the development of plasma cells resulting in a decreased immediate antibody response in deference to expanding the pool of memory B cells needed for future pathogen exposures (Siegrist and Aspinall 2009). Somatic hypermutation, which results in progressive diversification of the IgG repertoire, and affinity maturation progressively develop after the first

3 months of age reaching adult levels around 8 months of age (Williams et al. 2009; Ridings et al. 1998).

The development of the lymph node structure during the neonatal period also influences B-cell responses. Follicular dendritic cells (FDCs) reside in the germinal centers of lymph nodes, attract antigen-specific B cells, retain antigens in the form of immune complexes which stimulate B cells and provide signals that lead to somatic hypermutation and class-switching. Delayed maturation of FDCs in 7-day old mice limits B-cell responses by impairing induction and limiting the magnitude of germinal center responses (Pihlgren et al. 2003). Neonates often lack durable antibody responses. The etiology is not understood in humans, however, in mice there is a failure to establish and maintain an optimal plasma-cell pool in the bone marrow (Pihlgren et al. 2001).

Maternal antibody (matAb) passively acquired during the third trimester of pregnancy and through breast feeding provides surrogate immune protection to the neonate. IgG is passed transplacentally prior to birth and IgA, IgM, and IgG are provided through breastfeeding. The levels of circulating matAb are highest at birth, waning quickly after the first 3 months of life and are predominantly absent after 1 year. Though matAbs can be protective, they can also suppress neonatal primary antibody responses and impede effective vaccination. MatAbs can bind to vaccine antigen and prevent infant B cells from accessing immunodominant vaccine epitopes (Siegrist 2003). High titers of maternal antibody against measles have been shown to inhibit measles vaccination in infants up to 12 months of age (Gans et al. 1999).

Transplacentally transferred RSV-specific antibodies (RSV-matAbs) are predominantly of the IgG1 subclass and can be neutralizing and have a role in early protection against severe disease (Ochola et al. 2009). RSV-matAbs may account for considerably lower severe infection rates among infants under 2 months of age, and a peak of illness between 2 and 4 months of age when the waning of RSV-matAbs outpaces the ability of an infant to generate a protective immune response. Studies have shown RSV-matAbs wane in the early months of life. ELISA of RSV-specific IgG in uninfected infants estimated a half-life of around 79 days or 2.5 months (Ochola et al. 2009). Due to the rapid decline of anti-RSV antibodies even in healthy adults, there is seasonal variation in the amount of RSV-matAb transferred such that infants born right before the RSV season when mothers have the lowest antibody titers are more likely to be hospitalized (Stensballe et al. 2009). Premature infants, those at the greatest risk for severe manifestations of disease, do not acquire the same levels of RSV-matAbs, particularly those born before 28 weeks gestation (de Sierra et al. 1993).

2.3 Animal Models of Neonatal Immunity

The mouse model of RSV infection has been instrumental in studying RSV (see chapter by P.J. Openshaw, this volume) and led to the description of several CD4+ and CD8+ epitopes in both H-2^d and H-2^b mice. While neonatal mice are somewhat less permissive for infection, they offer an opportunity to compare early and late life immune responses (Cormier et al. 2010). The neonatal mouse model recapitulates some aspects of early life RSV infection in humans, including relatively weak adaptive immune responses and a Th2-bias following infection, and enhanced disease and pulmonary dysfunction result from reinfection during adulthood after primary infection during the neonatal period (Cormier et al. 2010; Culley et al. 2002). Ripple et al. demonstrated that immunomodulation with IL-4Rα antisense oligonucleotides during primary infection of neonatal mice prevents Th2-skewing of the immune response and abrogates pulmonary dysfunction following adult reinfection. In treated mice, the mechanism for protection from disease involves dampening of the expression of IL-4Rα on Th1 cells, which undergo apoptosis following exposure to IL-4 produced during neonatal infection. Facilitating survival of Th1 cells in the presence of Th2 cytokines promotes a less pathological response following infection during adulthood of neonatally-primed mice (Ripple et al. 2010).

The neonatal CB6F1 hybrid mouse model provided the first demonstration of a difference in CD8+ T-cell epitope hierarchy between neonates and adults infected with the same pathogen. Adult mice infected with RSV generate a reproducible epitope hierarchy with a strongly immunodominant K^dM2_{82-90} response. The response to a subdominant $D^bM_{187-195}$ epitope is 5–10-fold lower in infected adults. Hybrid mice infected neonatally, however, generate a codominant response to both epitopes that appears to be due, at least in part, to intrinsic CD8+ T-cell factors. Infection at or after day of life 10 revealed a dynamic and dramatic shift toward an adult-like K^dM2_{82-90}-dominated response demonstrating an age-dependent change in CD8+ T-cell response (Ruckwardt et al. 2011). Current understanding of human epitope hierarchy is limited and this unique murine model provides opportunities to determine differential CD8+ T-cell responses between neonates and adults, and understand how neonatal responses can be selectively modified to generate more adult-like immune responses.

The pre-term/newborn lamb and calf models also reflect the unique immunological and developmental consequences of RSV infection of the neonate. RSV-infected lambs demonstrate several features of human infection including apnea (Lindgren et al. 1996). Importantly, neutrophils are rapidly recruited to the lungs following RSV infection (Derscheid and Ackermann 2012). Unlike humans, lambs lack transplacentally-transferred antibody. Colostrum-deprived lambs lack any matAbs offering the ability to study intact neonatal immunity without the contribution of maternally derived components. In contrast, maternally derived bRSV antibody is transferred transplacentally and protects against severe bRSV disease (Sacco et al. 2012).

2.4 Implications for Vaccination

Immunization approaches for RSV-naïve neonates and infants must prime for more effective and safer immune responses than those generated following natural RSV infection or immunization with formalin-inactivated RSV (FI-RSV), which primed for a pathological Th2-biased response. As discussed, responses to natural infection are Th2-biased and complicated by early life limitations in the generation of adaptive immunity, particularly in the development of high affinity antibodies known to be a correlate of protection from severe disease. MatAbs and the default limitations of the early-life immune system further complicate immunization of infants under 6 months of age. At the age of 6 months, maturation of the immune system allows for somatic hypermutation and the development of high affinity antibodies, and the waning of matAbs may allow for more robust protective immune response. Although severe infection occurs about three times as often during the first 6 months of life as the second (Hall 2012; Lee et al. 2005) making this age group an important target population for immunization, up to 70 % of infants are still RSV-naïve at 6 months of age, and about 50 % of RSV-associated hospitalizations in children occur after 6 months of age. Therefore, providing effective and durable vaccine-induced immunity for children at 6 months of age may provide protection to this still-vulnerable population as well as younger siblings, parents or caregivers, and the elderly, and could have a large impact on the toll of RSV (Graham 2011).

Immunizing infants before the age of 2 months would be optimal, and children this age are capable of generating protective immunity, particularly when there is sufficient signaling from TLR, or other PRR agonists. These types of approaches may improve Th1 polarization of early life responses, provide immune "help" required to generate more effective responses and surmount the typical lackluster adaptive responses of early life. A variety of PRRs have been studied and provide varying responses during early life. In particular, TLR8 agonists have been shown to be refractory to inhibition by high adenosine concentrations found in early life and have a unique efficacy in activating human neonatal APCs (Philbin et al. 2012). B cell development and the infrastructure that supports effective antibody production are deficient in the first several months of life making truly effective antibody production prior to 4–6 months challenging. RSV-matAbs also present an obstacle to early life immunization. Delay of immunization (in the case of measles) and a series of boosting (in the case of DTaP) has resulted in effective vaccination despite the challenges presented by matAb. Alternatively, given the protective effects of RSV-matAbs, maternal immunization approaches may be a viable way to increase early protection of high-risk groups of infants until they are old enough to achieve more durable protection through active immunization.

3 The Senescent Immune Response

3.1 General Aspects of Immunosenescence

Aging is a complex process which results in a multitude of changes in immune function rendering the host more susceptible to infection, malignancy, and auto-immune disorders (Ponnappan and Ponnappan 2011). Age-related dysfunction of the innate and adaptive immune response is often referred to as "immunosenescence" and because aging is also associated with chronic inflammation, the term "inflammaging" has also been used (Franceschi et al. 2007). The result of immunosenescence is that older individuals experience more frequent and severe infections and respond less well to vaccination than younger persons (Gomez et al. 2008). Progress has been made in understanding the mechanisms of age-related changes in immune function using animal models and human studies and data clearly show that all components of the immune system are affected including; granulocytes, monocytes, and macrophages, natural killer cells, dendritic cells as well as B and T lymphocytes (Ponnappan and Ponnappan 2011). Less is known about RSV-specific immunity with aging and severity of infection is undoubtedly multifactorial with immune dysfunction only partially responsible. The role of immunosenescence in RSV disease pathogenesis is a subject of active investigation.

3.1.1 Innate Response

Aging exerts significant effects on most aspects of the innate immune system (Gomez et al. 2008). Although neutrophil numbers in the peripheral circulation do not change, there is evidence that chemotaxis, phagocytic capacity, intracellular killing efficiency, and generation of reactive oxygen intermediates are impaired in the elderly (Gomez et al. 2008; Busse and Mathur 2010). The effect of aging on macrophage is less clear with contradictory evidence regarding impaired chemotaxis and phagocytosis. Human studies have generally shown increased production of pro-inflammatory cytokines (IL-1, IL-6, and IL-8) and some defects in the TLR-induced cytokines have been observed (Gomez et al. 2008; Ershler and Keller 2000). With aging, Natural killer (NK) cells demonstrate an increase in mature phenotypes with impaired cytotoxicity and decreased IL-2 and IFN-γ production (Mariani et al. 2000; Solana and Mariani 2000). In contrast to NK cells, NK T-cell numbers are decreased and have diminished proliferation (Busse and Mathur 2010). DCs interface between innate and adaptive immune responses and demonstrate defective phagocytosis, diminished TLR expression and function and increased pro-inflammatory cytokines with advanced aging (Busse and Mathur 2010).

3.1.2 Adaptive Response

T Lymphocytes

Altered T-cell function is the most profound and consistent change observed with aging (Ponnappan and Ponnappan 2011; Busse and Mathur 2010; Haynes and Maue 2009; Maue and Haynes 2009). By 70 years of age, the thymopoietic space has decreased to approximately 10 % of the total thymus and output of naïve T cells drops to 95 % of peak levels (Flores et al. 1999). Parallel with reduced naïve T-cell output, is an increase in antigen-experienced memory cells so that the overall numbers of T cells in circulation is unchanged with aging (Busse and Mathur 2010). T-cell proliferation is reduced and T cells from elderly persons have a diminished ability to clonally expand to new antigens due to "replicative senescence" or cell cycle arrest (Effros 2000). As the memory pool expands over a lifetime of accumulated antigen experience, the ratio of memory to naïve cells increases, and T-cell repertoire diversity is decreased (Blackman and Woodland 2011). An emerging theory of immunosenescence is that chronic CMV infection stimulates CD8+ T cells over prolonged periods of subclinical infection resulting in a large pool of terminally differentiated memory T cells (Blackman and Woodland 2011). The age-related diminished CD8+ T cell repertoire leads to a reduced ability to generate an effective CD8+ T-cell response to new viral infections in older adults (Haynes and Maue 2009).

Thymic involution also leads to a decreased output of naïve CD4+ T cells. CD4+ T cells in the peripheral compartment are long lived and can compensate for decreased output however, these persistent cells accumulate intrinsic defects over time (Maue and Haynes 2009). In mouse models, CD4+ T cells from old mice have been shown to not form immunologic synapses with antigen presenting cells efficiently and have less intense T-cell receptor signaling than the young mice. In addition, after primary stimulation, CD4+ T cells from aged mice produce less IL-2 and expand poorly compared with those from young animals. These CD4+ T cells defects result in decreased B cell help with diminished antibody production and germinal center formation (Haynes and Maue 2009; Maue and Haynes 2009).

In addition to diminished T-cell functional responsiveness to new antigens, there is a shift in the cytokine profile from a predominantly Th1 (IL-2, IFN-α) type response to a Th2 (IL-4, IL-10) type response (Ponnappan and Ponnappan 2011; Sandmand et al. 2002). Th17 cells are a subgroup of T cells that are differentiated from naïve T cells in the presence of IL-6. Possibly because IL-6 expression is up regulated with aging, Th17 cells and IL-17 expression are also increased (Busse and Mathur 2010). IL-17A and IL-17F are felt to be important in airway hyper-responsiveness and may account for asthma onset in older age. Finally, regulatory T cells (Tregs) suppress several effector functions of T-cell subsets including Th1, Th2, and Th17. Tregs are an important protection against autoimmune disease but can lead to immunosuppression if dysregulated (Busse and Mathur 2010). Investigations of Tregs in older adults have yielded inconsistent results with some investigators noting similar numbers of Tregs in young and old volunteers, but with increased inhibitory function in cells from the elderly (Ponnappan and Ponnappan 2011). However, other studies show evidence of increased numbers of circulating Tregs in older adults (Ponnappan and Ponnappan 2011; Lages et al. 2008).

B Lymphocytes

The ability to generate new naïve B cells from bone marrow appears to diminish with increasing age (McKenna et al. 2001). As with T cells, the B-cell compartment is largely composed of antigen experienced B cells or CD27+ B (memory) cells (Ademokun et al. 2010). The numbers and percentages of switched memory B cells (IgD-CD27+) are also reduced in the elderly. Despite reduced numbers of naïve B lymphocytes, serum immunoglobulin levels remain relatively unchanged with age (Ponnappan and Ponnappan 2011). Antibodies from older people may have lower affinity and higher levels of circulating auto antibodies are noted in the elderly. Because of diminished T-cell help, immunity is retained to pathogens where immunity was acquired early in life, but there is a limited ability to generate an effective antibody response to new antigens (Maue and Haynes 2009). For example, protective antibodies can be found in elderly survivors of the 1918 influenza pandemic, but those same individuals would mount a poor response to vaccination with a novel antigen such as avian influenza (Ponnappan and Ponnappan 2011; Maue and Haynes 2009).

3.2 RSV Specific Immunosenescence

Immunity to RSV infection is incomplete and reinfections occur throughout life. As previously noted, the increased severity of RSV disease observed in the elderly is likely multifunctional and partially related to chronic underlying medical conditions and functional disability (Walsh et al. 2004; Falsey et al. 2005). Thus, sorting out changes in immune function specifically related to the effects of aging without the confounding effects of comorbidities can be challenging. Nonetheless, the numerous age-related changes in immune function outlined above are likely at play in RSV disease pathogenesis in older adults.

3.2.1 Humoral Immunity

As in children, the presence of neutralizing mucosal and serum antibodies is felt to be beneficial in the prevention of and amelioration of RSV illness in older adults, although precise correlates of protection are lacking (Hemming and Prince 1992). All adults have experienced multiple RSV infections during life and thus all have detectable serum binding and neutralizing antibody (Henderson et al. 1979; Walsh and Falsey 2004). Frail elderly persons attending day care who became RSV infected during a 2-year surveillance study had significantly lower serum neutralizing antibody titers than did their uninfected counterparts (Falsey and Walsh 1998). These findings were confirmed in a larger cohort of community dwelling older persons where RSV-infected subjects were noted to have significantly lower mucosal antibody, IgG to the fusion (F) and attachment (G) proteins of RSV as well as lower neutralizing antibody titers than those without RSV infection (Walsh and Falsey 2004). In the same study, low RSV-specific nasal IgA was independently associated with risk of RSV infection.

Low serum and nasal antibody levels have also been identified as risk factors for severe disease (Walsh et al. 2004; Duncan et al. 2009; Luchsinger et al. 2012).

Using logistic regression, a number of clinical variables were assessed as risk factors for hospitalization with RSV infection and the presence of chronic pulmonary disease (OR 4.0), functional disability (OR 1.7) and serum neutralizing antibody titer <10 log 2 (OR 5.9) were independently associated with risk of hospitalization (Walsh et al. 2004). A more recent study conducted in Santiago, Chile described similar findings with twofold lower serum neutralizing titers found in patients hospitalized with RSV infection compared to those hospitalized for other reasons (Luchsinger et al. 2012). Finally, nasal IgA levels were found to be significantly lower in 32 older adults hospitalized with RSV infection compared to 26 more mildly ill outpatients (Duncan et al. 2009).

The evidence to date indicates that antibody is important in the protection of older adults from infection and severe disease with RSV. However, it does not appear that there is a specific deficiency in RSV humoral immunity with aging. Agius and colleagues noted that elderly residents who were infected with RSV during a nursing home outbreak mounted vigorous RSV specific IgA and IgG responses (Agius et al. 1990). In addition, baseline serum antibody levels and rates of antibody decay have been shown to be comparable in healthy young and older adults (Falsey et al. 1999; 2006). Of note, frail elderly enrolled in a senior day care program (mean age 80 years old) had comparable neutralizing titers but significantly higher binding antibody to F and G proteins than healthy young and older adults (Falsey et al. 1999). This finding may indicate more recent RSV infection due to communal living or could represent the production of greater nonfunctional antibody with very advanced age.

Interestingly several studies have shown that older adults have a more vigorous antibody response after natural RSV infection compared to young adults (Falsey et al. 1999; Walsh and Falsey 2004). Not only was the magnitude of the response greater in the elderly, the percentage of subjects demonstrating $\geq$fourfold rises in antibody was significantly greater in the older age group (Walsh and Falsey 2004). Studies of the kinetics of antibody secreting cells (ASC) in RSV-infected adults have shown that ASCs can be detected in circulation longer in subjects with prolonged viral shedding (Lee et al. 2010). In addition, antibody responses are greater in more severely ill patients suggesting that antigen load may be a driving factor (Murata et al. 2010).

3.2.2 Cellular Immunity

Aging related changes in the RSV specific cellular immune response are presumed, however, at the present time direct evidence is sparse. Several rodent models support the concept of diminished RSV-specific T-cell function with aging as do the few human studies that have been conducted. The first investigation of RSV-specific cellular immunity was undertaken by Zhang et al. in 2001 using a BALB/c mouse model where young (2–4 months old) and aged (22–24 months old) mice were infected with RSV A2 (Zhang et al. 2002). Viral load in the nose was not different; however, a trend toward higher RSV titers in the lungs of aged mice was noted. Splenic lymphocytes from old mice exhibited significantly lower RSV specific MHC class I restricted CD8+ T lymphocyte responses as measured by chromium release assay. In addition, T cells from aged mice produced significantly less IFN-γ and slightly more IL-4 compared to young mice infected with RSV. Boukhvalova et al. using the cotton rat model examined the effect of age on viral clearance and cytokine expression (Boukhvalova et al. 2007). Although direct measures of T-cell function were not made, they noted young rats cleared virus more efficiently, shedding 2 days less than old rats. Corresponding to the difference in viral clearance, mRNA expression of cytokines was delayed in the aged animals. Peak levels of IFN-γ, IL-4, IL-10, IL-6, monocyte chemo-attractant protein 1 (MCP 1) were similar, whereas growth-regulated oncogene (GRO) mRNA transcripts were more abundant in the lungs of aged cotton rats compared to young animals. A study by Liu using an aging accelerated mouse model found old mice had significantly greater clinical illness with higher viral titers in lung and more prolonged viral shedding compared to the young (Liu and Kimura 2007). Additionally, old mice had delayed and decreased IFN-γ and increased IL-4 levels in the lung. Cytotoxic T-cell and NK-cell activity was also significantly diminished. Bronchoalveolar lavage fluid levels of RSV-specific IgG were not different but IgA levels were significantly lower in the aging accelerated mice. Although the rodent models are limited in their application to the elderly human since reinfection has not been studied, the data are overall consistent with the concept of diminished RSV CTL activity and skewing of the Th1 to Th2 response as a result of aging.

Investigations of RSV specific T-cell function in humans are few, however, several small studies show similar changes in human T-cell function with age that have been observed in the RSV rodent models. In a study of healthy young (mean age 31 years) and older (mean age 75 years) adults, IFN-γ production from PBMCs was compared (Looney et al. 2002). In both young and old, IFN-γ ELISpots were significantly greater with influenza-stimulated cells compared to RSV. Of note, significantly fewer IFN-γ ELISpots were observed in response to RSV from older subjects compared to the young, whereas no difference was noted for influenza. A Dutch study of 31 adults demonstrated elderly persons retained functional RSV-specific CD8+ memory T cells, but the frequency of these cells was significantly lower in than young subjects (de Bree et al. 2005). In addition, the RSV-specific CD8+ memory T-cell pool was considerably smaller than that of the influenza specific CD8+ memory T-cell pool. Interestingly, a study by Lee et al. found no difference in the frequencies of CD4+ IL-10, and CD4+ and CD8+ IFN-γ-secreting memory T cells specific for either RSV or influenza (Lee et al. 2005). However, the ratio of IL-10/IFN-γ in response to RSV stimulation was significantly reduced in the elderly. Lastly, age-related T-cell immune response was examined in 45 healthy volunteers, ages 20–65 years old (Cusi et al. 2010). Older subjects showed a significant increase in Treg cells and a marked decrease in long-lasting RSV-specific CD8+ memory T-cell precursors. In addition, cytokine expression was also altered in cells from elderly donors with a decrease in IFN-γ and TNF-α and increased IL-10 expression after stimulation with RSV.

In summary, RSV disease pathogenesis in older adults is likely due to a combination of factors including underlying medical conditions, physical frailty and immunosenescence. Although low serum and mucosal antibody titers are risk factors for infection and severe illness, the RSV-specific humoral immune response appears relatively intact with aging. Thus, subunit vaccine candidates to stimulate serum neutralizing antibody to RSV may be valuable in this age group. However, animal and human data also indicate that defects in cellular immunity and a shift from a Th1 to Th2 dominant immune response likely explain some of the severe disease manifestations in this population. The addition of adjuvants or alternative methods of vaccination to enhance RSV-specific cellular immunity may provide additional benefits given the cellular immune deficits associated with aging.

4 Conclusion

Our current ability to manipulate early life and late life immune responses to overcome the defects described above is limited making effective vaccination of these susceptible populations challenging. During early life, the immune response is dynamically changing in an age-dependent manner resulting in differing responses to vaccination throughout the first year of life, while people late in life experience difficulty with effective responses to new pathogens as well as to those they have previously overcome.

A lack of specific and effective antibody responses is a risk factor for severe RSV found in both of these susceptible segments of the population. Poor germinal center infrastructure and decreased B-cell help are common to both. Neonates are also challenged to create antibody responses in the setting of competing matAbs. Deficiencies in T-cell immunity may contribute to failure to efficiently clear virus for both populations. The Th2 cytokine-biased response common to both groups may also decrease effective cellular responses and misdirect inflammatory responses.

Vaccination for both populations is likely the most effective route to prevent RSV morbidity and mortality. Neonates are hampered by an inability to produce durable high-affinity antibody and some adaptive cellular responses to RSV in the first 4–6 months of life. After 6 months of age, neonates generate more robust humoral and cellular immune responses making this age group an easier target for vaccine design. For those late in life, waning immune responsiveness must be manipulated or coaxed into achieving protective RSV responses. Live-attenuated vaccines, which are still under active evaluation in infants, result in limited replication in those who have previously been exposed and therefore are not a good platform for immunizing the elderly. RSV protein vaccines may be better at generating neutralizing antibodies for those later in life, particularly as native glycoprotein structures are elucidated (see chapter by J.S. McLellan et al., and by T.G. Morrison and E.E. Walsh, this volume). Alternatively viral-like particles may provide effective vaccination that could be given periodically to maintain immunity (see chapter by T.G. Morrison and E.E. Walsh, this volume). Finally, enhancement of RSV-specific cellular immunity through adjuvants may improve viral clearance given the known deficits in cellular immunity in the aging.

Acknowledgments We thank Brenda Hartman for Fig. 1 graphics.

References

Ademokun A, Wu YC, Dunn-Walters D (2010) The ageing B cell population: composition and function. Biogerontology 11(2):125–137

Adkins B, Leclerc C, Marshall-Clarke S (2004) Neonatal adaptive immunity comes of age. Nat Rev Immunol 4(7):553–564

Agius G et al (1990) An epidemic of respiratory syncytial virus in elderly people: clinical and serological findings. J Med Virol 30(2):117–127

Belderbos ME et al (2009) Skewed pattern of toll-like receptor 4-mediated cytokine production in human neonatal blood: low LPS-induced IL-12p70 and high IL-10 persist throughout the first month of life. Clin Immunol 133(2):228–237

Blackman MA, Woodland DL (2011) The narrowing of the CD8 T cell repertoire in old age. Curr Opin Immunol 23(4):537–542

Blanco-Quiros A et al (1999) Decreased interleukin-12 levels in umbilical cord blood in children who developed acute bronchiolitis. Pediatr Pulmonol 28(3):175–180

Boukhvalova MS et al (2007) Age-related differences in pulmonary cytokine response to respiratory syncytial virus infection: modulation by anti-inflammatory and antiviral treatment. J Infect Dis 195(4):511–518

Boyce TG et al (2000) Rates of hospitalization for respiratory syncytial virus infection among children in medicaid. J Pediatr 137(6):865–870

Busse PJ, Mathur SK (2010) Age-related changes in immune function: effect on airway inflammation. J Allergy Clin Immunol 126(4):690–699; quiz 700–701

Collins PL, Graham BS (2008) Viral and host factors in human respiratory syncytial virus pathogenesis. J Virol 82(5):2040–2055

Cormier SA, You D, Honnegowda S (2010) The use of a neonatal mouse model to study respiratory syncytial virus infections. Expert Rev Anti Infect Ther 8(12):1371–1380

Culley FJ, Pollott J, Openshaw PJ (2002) Age at first viral infection determines the pattern of T cell-mediated disease during reinfection in adulthood. J Exp Med 196(10):1381–1386

Cusi MG et al (2010) Age related changes in T cell mediated immune response and effector memory to respiratory syncytial virus (RSV) in healthy subjects. Immun Ageing 7:14

de Bree GJ et al (2005) Respiratory syncytial virus-specific CD8+ memory T cell responses in elderly persons. J Infect Dis 191(10):1710–1718

de Sierra TM et al (1993) Respiratory syncytial virus-specific immunoglobulins in preterm infants. J Pediatr 122(5 Pt 1):787–791

Derscheid RJ, Ackermann MR (2012) Perinatal lamb model of respiratory syncytial virus (RSV) infection. Viruses 4(10):2359–2378

Duncan CB et al (2009) Risk factors for respiratory failure associated with respiratory syncytial virus infection in adults. J Infect Dis 200(8):1242–1246

Effros RB (2000) Long-term immunological memory against viruses. Mech Ageing Dev 121(1–3):161–171

Ershler WB, Keller ET (2000) Age-associated increased interleukin-6 gene expression, late-life diseases, and frailty. Annu Rev Med 51:245–270

Everard ML et al (1994) Analysis of cells obtained by bronchial lavage of infants with respiratory syncytial virus infection. Arch Dis Child 71(5):428–432

Falsey AR, Walsh EE (1998) Relationship of serum antibody to risk of respiratory syncytial virus infection in elderly adults. J Infect Dis 177(2):463–466
Falsey AR et al (1999) Comparison of respiratory syncytial virus humoral immunity and response to infection in young and elderly adults. J Med Virol 59(2):221–226
Falsey AR et al (2005) Respiratory syncytial virus infection in elderly and high-risk adults. N Engl J Med 352(17):1749–1759
Falsey AR, Singh HK, Walsh EE (2006) Serum antibody decay in adults following natural respiratory syncytial virus infection. J Med Virol 78(11):1493–1497
Flores KG et al (1999) Analysis of the human thymic perivascular space during aging. J Clin Invest 104(8):1031–1039
Franceschi C et al (2007) Inflammaging and anti-inflammaging: a systemic perspective on aging and longevity emerged from studies in humans. Mech Ageing Dev 128(1):92–105
Gans HA et al (1999) IL-12, IFN-gamma, and T cell proliferation to measles in immunized infants. J Immunol 162(9):5569–5575
Gomez CR et al (2008) Innate immunity and aging. Exp Gerontol 43(8):718–728
Graham BS (2011) Biological challenges and technological opportunities for respiratory syncytial virus vaccine development. Immunol Rev 239(1):149–166
Halfhide CP et al (2011) Respiratory syncytial virus binds and undergoes transcription in neutrophils from the blood and airways of infants with severe bronchiolitis. J Infect Dis 204(3):451–458
Hall CB (2012) The burgeoning burden of respiratory syncytial virus among children. Infect Disord Drug Targets 12(2):92–97
Haynes L, Maue AC (2009) Effects of aging on T cell function. Curr Opin Immunol 21(4):414–417
Heidema J et al (2007) CD8+ T cell responses in bronchoalveolar lavage fluid and peripheral blood mononuclear cells of infants with severe primary respiratory syncytial virus infections. J Immunol 179(12):8410–8417
Hemming VG, Prince GA (1992) Respiratory syncytial virus: babies and antibodies. Infect Agents Dis 1(1):24–32
Henderson FW et al (1979) Respiratory-syncytial-virus infections, reinfections and immunity. A prospective, longitudinal study in young children. N Engl J Med 300(10):530–534
Houben ML et al (2012) High concentrations of amniotic fluid proinflammatory cytokines in healthy neonates are associated with low risk of respiratory syncytial virus bronchiolitis. Pediatr Infect Dis J 31(9):931–934
Klein J, Remington J (2001) Current concepts of infections of the fetus and newborn infant. In: Remington J, Klein J (ed) Infectious diseases of the newborn and infant, 5th edn. Saunders, Philadelphia
Koenig JM, Yoder MC (2004) Neonatal neutrophils: the good, the bad, and the ugly. Clin Perinatol 31(1):39–51
Krishnan S et al (2003) Differences in participation of innate and adaptive immunity to respiratory syncytial virus in adults and neonates. J Infect Dis 188(3):433–439
Lages CS et al (2008) Functional regulatory T cells accumulate in aged hosts and promote chronic infectious disease reactivation. J Immunol 181(3):1835–1848
Lawn JE, Cousens S, Zupan J (2005) 4 million neonatal deaths: when? Where? Why? Lancet 365(9462):891–900
Lee MS, Walker RE, Mendelman PM (2005a) Medical burden of respiratory syncytial virus and parainfluenza virus type 3 infection among US children. Implications for design of vaccine trials. Hum Vaccine 1(1):6–11
Lee FE et al (2005b) The balance between influenza- and RSV-specific CD4 T cells secreting IL-10 or IFNgamma in young and healthy-elderly subjects. Mech Ageing Dev 126(11):1223–1229
Lee FE et al (2010) Circulating antibody-secreting cells during acute respiratory syncytial virus infection in adults. J Infect Dis 202(11):1659–1666

Levy O (2007) Innate immunity of the newborn: basic mechanisms and clinical correlates. Nat Rev Immunol 7(5):379–390
Levy O et al (2006) The adenosine system selectively inhibits TLR-mediated TNF-alpha production in the human newborn. J Immunol 177(3):1956–1966
Lindgren C et al (1996) Respiratory syncytial virus infection enhances the response to laryngeal chemostimulation and inhibits arousal from sleep in young lambs. Acta Paediatr 85(7):789–797
Liu B, Kimura Y (2007) Local immune response to respiratory syncytial virus infection is diminished in senescence-accelerated mice. J Gen Virol 88(Pt 9):2552–2558
Looney RJ et al (2002) Effect of aging on cytokine production in response to respiratory syncytial virus infection. J Infect Dis 185(5):682–685
Luchsinger V et al (2012) Role of neutralizing antibodies in adults with community-acquired pneumonia by respiratory syncytial virus. Clin Infect Dis 54(7):905–912
Lukens MV et al (2010) A systemic neutrophil response precedes robust CD8(+) T-cell activation during natural respiratory syncytial virus infection in infants. J Virol 84(5):2374–2383
Mariani E, Alonso C, Solana R (2000) Age-related alterations to natural killer cell function. Methods Mol Med 38:311–320
Marodi L (2006) Neonatal innate immunity to infectious agents. Infect Immun 74(4):1999–2006
Maue AC, Haynes L (2009) CD4+ T cells and immunosenescence—a mini-review. Gerontology 55(5):491–495
McKenna RW et al (2001) Immunophenotypic analysis of hematogones (B-lymphocyte precursors) in 662 consecutive bone marrow specimens by 4-color flow cytometry. Blood 98(8):2498–2507
Mella C et al (2013) Innate immune dysfunction is associated with enhanced disease severity in infants with severe respiratory syncytial virus bronchiolitis. J Infect Dis 207(4):564–573
Murata Y et al (2010) Humoral response to the central unglycosylated region of the respiratory syncytial virus attachment protein. Vaccine 28(38):6242–6246
Ochola R et al (2009) The level and duration of RSV-specific maternal IgG in infants in Kilifi Kenya. PLoS ONE 4(12):e8088
Philbin VJ et al (2012) Imidazoquinoline Toll-like receptor 8 agonists activate human newborn monocytes and dendritic cells through adenosine-refractory and caspase-1-dependent pathways. J Allergy Clin Immunol 130(1):195–204 e9
Pihlgren M et al (2001) Delayed and deficient establishment of the long-term bone marrow plasma cell pool during early life. Eur J Immunol 31(3):939–946
Pihlgren M et al (2003) Unresponsiveness to lymphoid-mediated signals at the neonatal follicular dendritic cell precursor level contributes to delayed germinal center induction and limitations of neonatal antibody responses to T-dependent antigens. J Immunol 170(6):2824–2832
Ponnappan S, Ponnappan U (2011) Aging and immune function: molecular mechanisms to interventions. Antioxid Redox Signal 14(8):1551–1585
Ridings J et al (1998) Somatic mutation of immunoglobulin V(H)6 genes in human infants. Clin Exp Immunol 114(1):33–39
Ripple MJ et al (2010) Immunomodulation with IL-4R alpha antisense oligonucleotide prevents respiratory syncytial virus-mediated pulmonary disease. J Immunol 185(8):4804–4811
Risdon G et al (1994) Proliferative and cytotoxic responses of human cord blood T lymphocytes following allogeneic stimulation. Cell Immunol 154(1):14–24
Ruckwardt TJ et al (2011) Neonatal CD8 T-cell hierarchy is distinct from adults and is influenced by intrinsic T cell properties in respiratory syncytial virus infected mice. PLoS Pathog 7(12):e1002377
Sacco RE et al (2012) Neonatal calf infection with respiratory syncytial virus: drawing parallels to the disease in human infants. Viruses 4(12):3731–3753
Sandmand M et al (2002) Is ageing associated with a shift in the balance between Type 1 and Type 2 cytokines in humans? Clin Exp Immunol 127(1):107–114
Sarzotti M, Robbins DS, Hoffman PM (1996) Induction of protective CTL responses in newborn mice by a murine retrovirus. Science 271(5256):1726–1728

Siegrist CA (2003) Mechanisms by which maternal antibodies influence infant vaccine responses: review of hypotheses and definition of main determinants. Vaccine 21(24):3406–3412
Siegrist CA, Aspinall R (2009) B-cell responses to vaccination at the extremes of age. Nat Rev Immunol 9(3):185–194
Solana R, Mariani E (2000) NK and NK/T cells in human senescence. Vaccine 18(16):1613–1620
Stensballe LG et al (2009) Seasonal variation of maternally derived respiratory syncytial virus antibodies and association with infant hospitalizations for respiratory syncytial virus. J Pediatr 154(2):296–298
Thornburg NJ, Shepherd B, Crowe JE Jr (2010) Transforming growth factor beta is a major regulator of human neonatal immune responses following respiratory syncytial virus infection. J Virol 84(24):12895–12902
Walsh EE, Falsey AR (2004a) Humoral and mucosal immunity in protection from natural respiratory syncytial virus infection in adults. J Infect Dis 190(2):373–378
Walsh EE, Falsey AR (2004b) Age related differences in humoral immune response to respiratory syncytial virus infection in adults. J Med Virol 73(2):295–299
Walsh EE, Peterson DR, Falsey AR (2004) Risk factors for severe respiratory syncytial virus infection in elderly persons. J Infect Dis 189(2):233–238
Weinberg AG et al (1985) Neonatal blood cell count in health and disease. II. Values for lymphocytes, monocytes, and eosinophils. J Pediatr 106(3):462–466
Williams JV et al (2009) The human neonatal B cell response to respiratory syncytial virus uses a biased antibody variable gene repertoire that lacks somatic mutations. Mol Immunol 47(2–3):407–414
Wood N, Siegrist CA (2011) Neonatal immunization: where do we stand? Curr Opin Infect Dis 24(3):190–195
Yost CC et al (2009) Impaired neutrophil extracellular trap (NET) formation: a novel innate immune deficiency of human neonates. Blood 113(25):6419–6427
Zhang J et al (2002a) Neonates mount robust and protective adult-like CD8(+)-T-cell responses to DNA vaccines. J Virol 76(23):11911–11919
Zhang Y et al (2002b) An aged mouse model for RSV infection and diminished CD8(+) CTL responses. Exp Biol Med (Maywood) 227(2):133–140

Part III
Vaccine Platforms and Treatment Options

Respiratory Syncytial Virus Disease: Prevention and Treatment

Helen Y. Chu and Janet A. Englund

Abstract Respiratory syncytial virus (RSV) is one of the most clinically important viruses infecting young children, the elderly, and the immunocompromised. Over the past decade, the most significant advance in the prevention of RSV disease has been the development of high-titered antibody products. Infection control is the only other strategy to prevent RSV disease. A humanized monoclonal antibody directed against the fusion (F) protein palivizumab, (Synagis®, MedImmune, Inc., Gaithersburg, MD), is given routinely on a monthly basis to premature infants and young children less than 24 months of age with underlying medical problems including prematurity, chronic lung disease, or cardiac disease to prevent RSV disease and hospitalization. Other products utilizing polyclonal or monoclonal antibodies or antibody fragments against the F protein have been developed and some already tested in patient populations. The only licensed antiviral treatment available today is ribavirin, a guanosine analogue generally administered as a small particle aerosol to immunocompromised patients with lower respiratory tract disease due to RSV. This drug has also been utilized in oral and intravenous forms, again mainly in immunocompromised patients. Promising new antiviral agents under development by multiple pharmaceutical and biotechnology companies include small molecule fusion inhibitors, attachment inhibitors, inhibitors of RNA synthesis, and small interfering RNA particles (siRNA).

H. Y. Chu
Division of Allergy and Infectious Diseases, Department of Medicine,
University of Washington, Seattle, WA, USA

J. A. Englund
Division of Infectious Diseases, Seattle Children's Hospital, and Department of Pediatrics,
University of Washington, Seattle, WA, USA

J. A. Englund (✉)
Pediatric Infectious Diseases, Seattle Children's Hospital, 4500 Sand Point Way NE
#R5441, Seattle, WA 98105, USA
e-mail: Janet.englund@seattlechildrens.org

L. J. Anderson and B. S. Graham (eds.), *Challenges and Opportunities for Respiratory Syncytial Virus Vaccines*, Current Topics in Microbiology and Immunology 372,
DOI: 10.1007/978-3-642-38919-1_12,

Contents

1 Introduction

Respiratory syncytial virus (RSV) is the most important cause of viral lower respiratory tract infections (LRTI) in children, and the second leading cause of death from respiratory viral infections in the elderly (Falsey et al. 2005; Nair et al. 2010). The majority of RSV infections occur in previously healthy children, and re-infections are common throughout a person's lifetime (Macartney et al. 2000). In most cases, RSV manifests as a self-limited upper respiratory tract infection and is treated symptomatically without requiring hospitalization or RSV-specific treatment. Populations at risk for progression to lower respiratory tract disease include infants, particularly under age 6 months, elderly adults, and persons of all ages with compromised cardiac, pulmonary or immune systems (Hall et al. 2009). The risk of progression to serious LRTI is especially high in hematopoietic stem cell transplant (HSCT) recipients, lung or heart transplant recipients, and patients with underlying pulmonary disease or severe underlying acquired or congenital immunodeficiencies. (Falsey et al. 2006; Ghosh et al. 2000). These populations could benefit the most from RSV-specific prophylaxis and/or treatment (Whimbey et al. 1995).

RSV is a single-stranded RNA paramyxovirus with a genome that encodes 11 proteins. Drug targets include the two surface glycoproteins, the flusion (F) and attachment (G) glycoproteins, as well as the viral polymerase (L) protein. The RSV G and F glycoproteins participate in viral attachment and fusion to host cells. The G glycoprotein is present on both the membrane surface, where it facilitates viral attachment to ciliated epithelial cells, and in a secreted form. In its secreted form, the G glycoprotein may serve as a decoy that increases lymphocyte migration and inflammation (Bukreyev et al. 2008). The F glycoprotein is responsible for fusion to the target cell. Upon attachment of the viral membrane, the F glycoprotein undergoes a conformational change that allows the viral and cell membrane to fuse together and allow viral entry. The F-protein exists in a pre-triggered state prior to attachment to a target cell, at which point it enters into a post-triggered conformational state that allows for membrane fusion between the target cell and virion (Costello et al. 2012). Antibody responses to the F protein are broadly reactive across RSV strains while those to G tend to be group-or strain-specific (Johnson et al. 1987). There are two major antigenic groups of RSV isolates, A and B, and multiple strains within each group.

2 Prevention

Transmission occurs through auto-inoculation of mucous membranes with secretions acquired from infected persons or fomites, as well as through aerosolized droplets (Hall and Douglas 1981) (see chapter by C.B. Hall et al., this volume). RSV is efficiently transmitted from infectious respiratory secretions. These secretions may contain virus for prolonged periods of greater than 7 days, particularly in infants experiencing their first infection or in immunocompromised adults (Hall et al. 1975). Prevention of disease acquisition can be accomplished through frequent handwashing as well as restriction of childcare attendance during RSV season (American Academy of Pediatrics Subcommittee on Diagnosis and Management of Bronchiolitis 2006). Breast milk has both neutralizing activity and RSV-specific antibodies against RSV (Toms et al. 1980), and appears to have a protective effect against hospitalization for RSV disease. One study showed that breastfeeding was associated with decreased risk of hospitalization for RSV disease among children under 3 years of age in Quebec, although a study from Argentina found that breastfeeding was only protective for female infants (Papenburg et al. 2012; Libster et al. 2009). By contrast, passive tobacco smoke exposure is associated with increased risk of RSV hospitalization (Singleton 1996).

2.1 *Infection Control Measures*

Prevention of nosocomial outbreaks of RSV is important in any setting where patients are at increased risk from RSV infection. HSCT recipients are at exceedingly high risk and infection control has been demonstrated to be best accomplished for these patients through use of a multi-faceted infection control strategy. Routine infection control precautions in HSCT wards include staff education, strict attention to good handwashing practices, use of gowns and gloves, and screening of symptomatic patients and hospital personnel with rapid diagnostic tests (Isaacs et al. 1991). Compliance is the key to effective infection control; staff education concerning the epidemiology and modes of transmission of RSV disease has been shown to be an effective intervention when coupled with other infection control measures (Macartney et al. 2000). The use of gown and glove precautions, as compared to standard precautions, was associated with a relative risk of 0.34 nosocomial infections during RSV season in an inpatient infant and toddler ward (Leclair et al. 1987). Screening for RSV on admission with subsequent assignment to a RSV-infected or RSV-uninfected cohort reduces nosocomial transmission rates in the pediatric inpatient population (Krasinski et al. 1990). Use of surgical masks and eye protection may also reduce nosocomial transmission. Other measures recommended in outbreak settings include restricting personnel caring for RSV-infected patients from caring for RSV-uninfected patients and restricting symptomatic staff and visitors from patient contact (Tablan et al. 2004). A study evaluating a multi-faceted infection control strategy in a hematoipoietic stem cell transplant unit showed that a combination of rapid diagnosis and isolation of RSV-infected patients, therapy with ribavirin, restriction of visitors and staff with respiratory symptoms from visiting or working in a bone marrow transplant unit, and use of masks and gloves was effective in reducing the incidence of RSV infection by 77% (Garcia et al. 1997). The Fred Hutchinson Cancer Research Center in Seattle, Washington, U.S.A. currently utilizes a multifaceted strategy of infection control during RSV outbreaks that includes strict hand hygiene, employee symptom screening and testing, patient testing and isolation, restriction of visitors with respiratory symptoms, and closure of schools and classes. Centers for Disease Control guidelines recommend use of standard and contact precautions in infants, young children, and immunocompromised adults diagnosed with RSV, as well as hand hygiene, cohorting or use of single patient rooms, visitor screening, and sterilization of shared surfaces (Tarrac 2008).

2.2 RSV-Specific Prophylaxis and Treatment

Prevention and treatment of RSV is hindered by the lack of effective antiviral therapy. Only prophylactic therapy with palivizumab, an RSV-specific humanized monoclonal IgG antibody product, is licensed for the prevention of RSV in high-risk infants. This is expensive and needs to be administered by monthly injections. Ribavirin is a broad antiviral agent that is approved for treatment of RSV infection in high-risk young children, and is currently utilized mainly in high-risk individuals such as hematopoietic stem cell transplant recipients, particularly allogeneic stem cell transplant recipients with severe lymphopenia. The evidence to support its efficacy for treating RSV in otherwise healthy children is limited, and its routine use is not recommended by the American Academy of Pediatrics. New prophylactic therapies in development include small interfering RNAs as well as fusion inhibitors and functional domains of neutralizing antibodies.

2.2.1 Polyclonal Antibody Prophylaxis

Historically, it was observed that infants with high titers of maternally derived RSV-specific antibody were protected from severe disease, though it was unknown whether maternal antibody directly protected against disease or served as a proxy for another mechanism of protection (Glezen et al. 1981). In the early 1990s, two trials were performed to evaluate the role of monthly infusions of intravenous immunoglobulin (IVIG) to prevent RSV disease in high-risk infants. The safety of human IVIG administered to infants with congenital heart disease was determined by administering monthly IVIG, monitoring for side effects and measuring protective antibody levels after infusion (Groothuis et al. 1991). Antibody titers of 1:100 were achieved after infusion of IVIG, though 12 of the 23 patients developed RSV disease in the following 2 seasons. Meissner et al. (1993) performed a randomized controlled trial of IVIG versus placebo in 49 infants with bronchopulmonary dysplasia or severe congenital heart disease over the course of two RSV seasons. Equal numbers of culture-proven RSV disease were documented in both groups, though the duration of hospitalization was shorter (35 versus 51 days) in the IVIG group. The results from these initial studies guided the development of RSV-specific immunoprophylaxis regimens.

High-titered RSV intravenous immunoglobulin (RSV-IVIG; RespiGam®, MedImmune, Gaithersburg, MD, USA) was developed in 1996 as the first RSV-specific immunoprophylaxis agent. It was developed from pooled serum from donors with high RSV-specific neutralizing antibody titers, with five times greater activity against RSV than standard IVIG. It also contained less fluid volume, and thus a shorter infusion time. In a randomized double-blind placebo controlled multicenter trial in the United States, RSV-IVIG administered monthly reduced the incidence and duration of hospitalization of infants with bronchopulmonary dysplasia and/or a history of prematurity by 41 and 53% respectively, as well as the frequency and duration of intensive care unit stay and mechanical ventilation (The PREVENT Study Group 1997). However, use of RSV-IVIG in children with congenital heart disease was initially associated with fluid overload, oxygen desaturation, and fever (Groothuis et al. 1993). Among children with cyanotic congenital heart disease in a follow-up study, more cyanotic events and poor post-operative outcomes occurred among the RSV-IVIG group, thereby initially excluding these children as a target population for receipt of RSV-IVIG (Simoes et al. 1998). Other concerns regarding RSV-IVIG use was the potential interference with live vaccine immunogenicity, the potential source of bloodborne pathogens, and the requirement of intravenous access for monthly administration. RSV-IVIG has now been replaced by the anti-RSV neutralizing monoclonal antibody described below and is no longer available.

The use of IVIG or RSV-IVIG therapy for RSV treatment has not been well studied in HSCT populations, and it is not routinely recommended. A study evaluating RSV-IVIG prophylaxis in a high-risk adult HSCT recipients was underpowered to show efficacy (Cortez et al 2002). Pooled IVIG as a mechanism to prevent cytomegalovirus or bacterial infections has not been shown to have an effect on mortality, and its effects specifically on prevention of RSV infection in the post-transplant setting is unknown (Cordonnier et al. 2003).

2.2.2 Monoclonal Antibody Prophylaxis

Pediatric Populations

Palivizumab (Synagis®, MedImmune Vaccines, Inc., Gaithersburg, MD) is a humanized murine monoclonal antibody targeting the fusion protein of RSV that has both neutralizing and fusion inhibitory activity against RSV. Palivizumab and subsequent monoclonal antibodies produced by MedImmune that recognize antigenic sites on the F protein are associated with broad neutralization across multiple strains.

In 1998, palivizumab was licensed for use in prophylaxis of young children under two years of age with high risk conditions including birth at or before 35 weeks gestation, chronic lung disease of prematurity, and hemodynamically significant heart disease. Palivizumab is given every 30 days as an intramuscular or intravenous injection at a dose of 15 mg/kg. Two major randomized placebo controlled trials have evaluated the efficacy of palivizumab. The first trial, the RSV-IMpact trial, focused on children two years or younger with chronic lung disease requiring medical therapy with supplemental oxygen, bronchodilators, diuretics, or corticosteroid therapy within the last six months, as well as children born at 35 weeks gestation or less who were six months of age or younger at the start of RSV season. In this study, use of palivizumab was associated with a 55% rate reduction in RSV-related hospitalization (The Impact-RSV Study Group 1998). The Cardiac Synagis Study Group performed a study showing that palivizumab prophylaxis was associated with a 43% rate reduction of RSV-related hospitalizations among children with hemodynamically significant congenital heart disease, without the side effects noted in the RSV-IVIG trials (Feltes et al. 2003). Palivizumab has not been shown to have a significant effect on mortality in clinical trials. A cohort study examining palivizumab prophylaxis in premature infants without chronic lung disease showed reduction in clinician-diagnosed recurrent wheezing in the palivizumab prophylaxis group as compared to the group that did not receive palivizumab prophylaxis (Simoes et al. 2007). A recent study showed that among infants receiving palivizumab, those with underlying medical disorders including Down's syndrome and cystic fibrosis were at higher risk for respiratory illness compared to premature infants and were at equivalent risk for RSV hospitalizations (Paes et al. 2012). This suggests that the criteria for palivizumab administration may need to be expanded to include this population.

The cost and method of administration of palivizumab are limiting factors in its use. Although various studies have reported on cost-effectiveness of this agent in various populations, one recent cohort study of infants under one year of age enrolled in Medicaid who did or did not receive palivizumab prophylaxis showed that administration of palivizumab in infants born at 32–35 weeks gestation was not a cost-effective intervention in this group. Using Medicaid reimbursement rates in 2003, the cost-per-person was $5117 USD in the prophylaxis group and $371 USD in the non-prophylaxis group, factoring in the cost of RSV prophylaxis and/ or treatment (Wegner et al. 2004). The cost of palivizumab at our institution in 2012 is $2,800 per single 100 mg vial (personal communication, R. Jain Univ. Washington). The administration of palivizumab by monthly subcutaneous injection may also be problematic in terms of adherence. Studies of home-based palivizumab administration have shown lower rates of receiving all doses in a timely fashion and increased rates of RSV-associated hospitalizations compared to children receiving palivizumab in clinic-based programs (Frogel et al. 2008).

Currently, palivizumab is administered monthly in up to five doses at 15 mg/kg. Five monthly doses are associated with protective serum antibody levels lasting greater than 20 weeks. Doses should continue to be administered even if the infant acquires RSV infection; multiple strains co-circulate during each season and it is possible that re-infection may occur with a different strain.

The American Academy of Pediatrics (AAP) recommends that premature infants born at 28 weeks gestation or earlier may receive prophylaxis during RSV season in their first year of life. Premature infants born between 29 and 32 weeks gestation may receive prophylaxis during RSV season in their first six months of life; however, once therapy is initiated, five monthly doses should be administered. Specific conditions that are eligible for administration of palivizumab in children up to 24 months include chronic lung disease as well as those with certain other high risk cardiopulmonary conditions. Though concern about the emergence of RSV resistance in recipients of palivizumab has been raised, a surveillance of RSV isolates from 458 infants did not show evidence of resistant mutants (DeVincenzo et al. 2004).

The AAP updated their recommendations in 2009 (American Academy of Pediatrics 2009), with newer guidelines recommending initiation of immunoprophylaxis based on the onset of RSV season in specific regions of the country. Other 2009 updates include recommendations that premature infants born between 32 and 35 weeks gestation who are under three months of age at the start of RSV season should receive prophylaxis if they also fulfill one of two risk criteria: attendance at childcare or presence of sibling under the age of five at home. However, these infants should receive prophylaxis only until they reach three - months of age or have received a maximum of three doses.

Hematopoietic Stem Cell Transplant Populations

The use of palivizumab for prophylaxis in the immunocompromised adult and pediatric populations is not well studied. The 2009 international HSCT guidelines recommend immunoprophylaxis in young children undergoing HSCT, though there are no controlled data to support this recommendation (Tomblyn et al. 2009). The American Academy of Pediatrics also suggests the use of palivizumab in children with severe combined immunodeficiency or children under the age of two with HSCT or lung transplantation. Palivizumab prophylaxis in adult HSCT recipients is associated with decreased progression from URTI to LRTI in small studies (Boeckh et al. 2001), though larger randomized controlled trials are needed to accurately assess its impact on disease. Palivizumab was included in infection control measures used to successfully control an RSV outbreak in the bone marrow transplant setting. Its role in controlling the outbreak was not studied.

3 Treatment

Treatment for RSV is usually reserved for patients with lower respiratory tract disease, or to prevent progression from upper respiratory to lower respiratory tract infection in a subset of high-risk individuals. Populations requiring treatment for RSV bronchiolitis or pneumonia include hospitalized infants and children and HSCT recipients, as well as lung transplant recipients. Those at particularly high risk include patients with severe lymphopenia, recipients of allogeneic stem cell transplants, and lung transplant recipients (Boeckh et al. 2005). Treatment options include supportive care with supplemental oxygen and mechanical ventilation as well as pharmacotherapy with bronchodilators, corticosteroids, ribavirin, and intravenous immunoglobulin. Studies of adult healthy volunteers and hospitalized infants have shown that RSV viral load correlates with disease severity and symptom score, and that peak viral load and symptom score occurs a median of three days after onset of infection (DeVincenzo et al. 2010; El Saleeby et al. 2011), although similar data is not yet available in immunocompromised patients. Treatments that are initiated after patients are admitted to the hospital may not be as effective, particularly if they target viral load reduction as a mechanism of action. Treatment options for RSV disease, with dosing, side effects, and contra-indications, are outlined in Table 1.

3.1 Supportive Care

Supplemental oxygen is customarily used when oxygen saturations are less than 90%. Mechanical ventilation for respiratory failure is required occasionally as well, particularly among infants as well as immunocompromised children and adults. Extracorporeal membrane oxygenation (ECMO) has also been used successfully to treat severe cases of RSV pneumonia in children (Khan et al. 1995).

Table 1 Currently available treatment modalities for RSV

Therapies	Indications	Dose	Adverse events	Contraindications/precautions	Suggested monitoring
Aerosolized ribavirin	Prevention of progression of RSV URTI to LRTI in HSCT (Boeckh et al. 2007); Treatment of RSV LRTI in HSCT (Shah and Chemaly 2011)	2 g/dose three times a week or 6 g over 12–18 h for 3–7 days, or 6 g/100 mL over 2 h 3 times daily for 6 h/day	Nausea, bronchospasm, fatigue, insomnia, anorexia	Dose adjustment for renal function, consider addition of a bronchodilator; caution use in patients with COPD or asthma, precipitation may lead to malfunction of expiratory valve in mechanically ventilated patients	Monitoring for broncho-spasm or high positive-end-expiratory pressures
Oral ribavirin	RSV LRTI in HSCT or in lung transplant (Pelaez et al. 2009)	>40 kg: 10 mg/kg as loading dose, 400 mg three times a day on day 2 and 600 mg three times a day thereafter	Hemolytic anemia, nausea	Dose adjustment for renal function, risk of teratogenicity for 6 months after therapy	Complete blood count
Intravenous ribavirin	RSV LRTI in HSCT (Lewinsohn et al. 1996); Concern for poor lung penetration with aerosolized ribavirin	< 40 kg:33 mg/kg loading dose, followed 6 h later by 16 mg/kg every 6 h for 4 days (total 16 doses), then 8 mg/kg every 8 h for 3 days (total of 9 doses)	Hemolytic anemia, leucopenia, hypocalcemia, elevated ALT	Dose adjustment for renal function, risk of teratogenicity for 6 months after therapy	Complete blood count, biochemistry, liver function
Hypertonic saline	Bronchiolitis in children (Zhang et al. 2008)	No established dose	Bronchospasm		
Palivizumab	May use in combination with ribavirin for RSV LRTI in HSCT or lung transplant (Liu et al. 2010; Chavez-Bueno et al. 2007)	15 mg/kg	Fever, rash, rarely anaphylaxis	Intramuscular injection—caution when patient has thrombocytopenia	Monitor for anaphylactic reactions
IVIG	May use in combination with ribavirin for RSV LRTI in HSCT (Shah and Chemaly 2011)	400 mg/kg	Aseptic meningitis, anaphylaxis		Close follow up for anaphylactic reaction

3.2 Pharmacotherapeutic Agents

3.2.1 Nonspecific Agents

Bronchodilators are indicated when evidence of lower airway obstruction exists, as manifested by wheezing or respiratory failure. A prospective nonrandomized trial of infants on mechanical ventilation secondary to respiratory failure from RSV demonstrated that 45% had significant improvements in lung function after administration of albuterol, though 50% had no benefit (Hammer et al. 1995). Corticosteroids have not been shown to be beneficial in treatment of RSV LRTI. Two randomized controlled trials demonstrated no benefit to nebulized or intravenous corticosteroids among hospitalized infants with RSV bronchiolitis (van Woensel and Vyas 2011). Nebulized budesonide given in the acute phase of RSV bronchiolitis did not decrease hospitalization duration, medical visits, or re-admission rates. Intravenous dexamethasone administered to children mechanically ventilated for severe RSV-associated LRTI with both mild and severe oxygenation abnormalities did not reduce duration of mechanical ventilation. There have been no randomized controlled trials examining the use of corticosteroids in adults.

Hypertonic saline has been studied as a mechanism to decrease airway edema in bronchiolitis. In a meta-analysis of four trials evaluating hypertonic saline for acute viral bronchiolitis in children up to 24 months of age, the use of hypertonic saline was associated with a significant reduction in length of hospital stay and reduction of severity score. This has not yet been reported specifically for RSV bronchiolitis (Zhang et al. 2008).

3.2.2 Licensed Antiviral Agents

Ribavirin is a broad spectrum antiviral guanosine analogue that is active against RSV and other RNA viruses through multiple mechanisms that limit viral transcription. Perhaps due to the multiple mechanisms of action, RSV resistance to ribavirin has not been documented and this drug is used routinely in the laboratory as a standard to assess antiviral activity. Ribavirin has been documented to be teratogenic in rodents, and therefore it is generally contraindicated in pregnant women. Ribavirin may be administered in aerosolized, intravenous or oral formulations but is most commonly used in aerosolized form. It is administered either continuously or intermittently using a small particle aerosol generator via face mask in a scavenging tent to prevent environmental contamination. A standard dose is 20 mg/mL (6 g reconstituted with 300 mL of sterile water) administered for 12–18 h/day for 3–7 days, although it is frequently administered in a high-dose, short duration protocol as 60 mg/mL (6 g reconstituted with 100 mL of sterile water) given for 2 h every 8 h, for a total of 6 h daily (Chemaly et al. 2012). Aerosolized ribavirin is frequently administered for longer periods of time in patients with severe lower respiratory tract disease. Side effects of aerosolized ribavirin may include cough, dyspnea, fatigue, headache, insomnia, nausea, and anorexia; pretreatment with bronchodilators may ameliorate some of the bronchospasm and bronchoconstriction associated with aerosol administration. Ribavirin has been associated with deterioration of respiratory function in some patients with chronic obstructive pulmonary disease or asthma. In mechanically ventilated patients, aerosolized administration may lead to malfunction or obstruction of the expiratory valve of ventilators, leading to high positive-end-expiratory pressures.

Intravenous ribavirin may be considered when there is concern for poor lung penetration due to consolidative pneumonia (Shah and Chemaly 2011). Side effects include hemolytic anemia when given in high doses, as well as leukopenia and hyperbilirubinemia. In a Phase I study of intravenous ribavirin for HSCT recipients with RSV pneumonia, the dose administered included a loading dose of 35 mg/kg of body weight in three divided doses, followed by a maintenance dose of 25 mg/kg/day given in three divided doses every 8 h for 6 more days. In this study, two of ten patients developed acute hemolysis requiring discontinuation of the drug, and only two of ten patients survived (Lewinsohn et al. 1996).

Oral ribavirin has decreased bioavailability compared to aerosolized ribavirin, but is more convenient to administer. This formulation has been associated with hemolytic anemia and nausea. There is concern about its penetration in the setting of graft versus host disease in HSCT recipients, and studies have shown poorer outcomes with oral versus inhaled ribavirin in HSCT patients. In one study of lung transplant recipients with RSV LRTI, use of oral ribavirin in combination with corticosteroids was associated with no complications or deaths, and cost $700 as compared to $14,000 for aerosolized ribavirin therapy (Pelaez et al. 2009).

In a meta-analysis, no statistically significant differences in mortality, duration of hospitalization or ventilation were associated with use of ribavirin in pediatric populations (Ventre and Randolph 2004). The AAP no longer recommends its routine use in the pediatric population. However, ribavirin is commonly used in the HSCT setting where RSV is associated with a mortality of up to 80% if untreated. (Kim et al. 2007). The data to support its use are largely based on single-center observational trials showing improvements in mortality among patients infected with RSV following HSCT. A prospective multicenter randomized clinical trial examining ribavirin to prevent progression from upper respiratory to lower respiratory tract disease among HSCT recipients showed that preemptive ribavirin therapy is safe and could reduce viral load over time, although the study was limited by low patient accrual (Boeckh et al. 2007). Among lung transplant recipients, receipt of ribavirin has been observed to be associated with favorable outcomes (Wendt et al. 1995).

Combination ribavirin and immunoglobulin therapy has been utilized in immunocompromised patients with severe RSV disease, including HSCT and lung transplant recipients. Cotton rat models have demonstrated decreased lung viral titers after combination treatment with IVIG and ribavirin, as compared to treatment with placebo or either agent alone. (Gruber et al. 1987). In uncontrolled trials of HSCT and lung transplant recipients with severe RSV lower respiratory tract disease, use of ribavirin and IVIG was associated with decreased mortality (Ghosh et al. 2000; Liu et al. 2010; Whimbey et al. 1995). In an uncontrolled report from MD Anderson Cancer Center, combination therapy with aerosolized ribavirin (18 h/day) and high RSV-titered IVIG (0.5 g/kg every other day) was associated with a favorable response in adult HSCT recipients with RSV pneumonia in whom therapy was initiated prior to respiratory failure (Whimbey et al. 1995). Four of 13 patients in whom combination therapy was initiated prior to respiratory failure died of respiratory disease compared with all nine patients who did not receive antiviral therapy or received ribavirin after respiratory failure. At the Dana Farber Cancer Institute, combination therapy with aerosolized ribavirin (18 h/day) and RSV-IVIG (1.5 g/kg for one dose) was similarly associated with a favorable response in two HSCT recipients with clinically severe RSV pneumonia occurring early following transplant (Small et al. 2002). In a study of 18 lung transplant recipients with RSV-associated lower respiratory tract disease, administration of intravenous ribavirin and high dose oral steroids was associated with no deaths (Glanville et al. 2005).

The use of palivizumab in combination with ribavirin for RSV LRTI has not been studied in a controlled study. A retrospective review of high-risk pediatric patients with both RSV URTI and LRTI showed that the use of combination ribavirin and palivizumab was associated with survival in 93.6 % of patients (Chavez-Bueno et al. 2007). At Stanford University, the combination of ribavirin, IVIG, palivizumab, and high dose corticosteroids for treatment of RSV URTI and LRTI in lung transplant recipients was associated with preserved lung function over time (Liu et al. 2010). Retrospective studies of combination therapy with ribavirin and palivizumab are underway at our institution.

Immunocompromised patients at our institution with RSV URTI or asymptomatic shedding are currently treated with ten days of inhaled ribavirin if they have severe lymphopenia (absolute lymphocyte count $\leq 300/mm^3$). Patients with lymphocytes $>300/mm^3$ are usually only observed closely for signs of hypoxia, respiratory distress, or clinical deterioration. For patients with RSV LRTI, ten days of inhaled ribavirin is given in addition to potential administration of 15 mg/kg of palivizumab or 400 mg/kg of IVIG, since RSV-IVIG is not routinely available. Even if the addition of RSV-specific immunoglobulins to ribavirin were more effective than ribavirin alone in reducing the progression from URTI to LRTI according to retrospective analysis, the additional high cost of RSV-specific immunoglobulins (approximately $28,000 for a 70 kg adult at our institution) is worthy of further supportive data.

The use of monoclonal antibodies for treatment in non-immunocompromised populations is not recommended. RSV-IVIG has not been shown to have an effect on reducing length of stay, intensive care unit admission, or use of mechanical ventilation (Rodriguez et al. 1997). Palivizumab has likewise been shown to be ineffective when used as treatment for RSV disease in previously healthy hospitalized children under age two years (Saez-Llorens et al. 2004).

Table 2 Therapeutic agents currently under development

Antiviral class	Name	Mechanism of action	Stage of development	Resistance mutations	Evidence	Company	Reference
Immunoglobulin	Motavizumab	Monoclonal antibody targeting the F protein	Phase I–III		Significant reductions in medically attended LRTIs for RSV when studied as a prophylactic agent; associated with injection site reactions and not approved by the FDA. No differences in outcomes in hospitalized children as a therapeutic agent.	MedImmune	Carbonell-Estrany et al. 2010, Ramilo et al. 2012, Feltes et al. 2011, Fernandez et al. 2010
	MEDI-557	Recombinant humanized monoclonal immunoglobulin G1 derived from motavizumab	Phase I		Extended half life as compared to motavizumab	MedImmune	(Criste R et al. 2012)
	mAb 131-2G	Monoclonal antibody targeting the CXC3 motif of the G glycoprotein	Preclinical	None reported	Associated with reduced lung inflammation and RSV viral titers in animal models	N/A	(Haynes et al. 2009)
	mAb 131-2A	Monoclonal antibody targeting the post-fusion F glycoprotein	Preclinical	A355 V, C313 W		N/A	(Graepel KW et al. 2012)
	R-001	Hyperimmune polyclonal antibody	Phase II	None reported	Studies ongoing evaluating ability to prevent progression from RSV URTI to LRTI in immunosuppressed individuals	ADMA Biologics	(Empey et al. 2010)

(continued)

Table 2 (continued)

Antiviral class	Name	Mechanism of action	Stage of development	Resistance mutations	Evidence	Company	Reference
siRNA	ALN-RSV01	RNA interference	Phase II	None reported	Reduces RSV infection among healthy adult volunteers and preserves lung function in lung transplant recipients with RSV URTI	Alnylam	(DeVincenzo et al. 2008; Zamora et al. 2011)
Fusion Inhibitor	BTA-9881	Fusion inhibitor	Phase I	None reported	None published	Biota Holdings	(Bonfanti and Roymans 2009)
	TMC-353121	Fusion inhibitor	Preclinical	S398L, K394R, K399I, D486 N, E487D, D489Y	Associated with reduced lung inflammation and RSV viral titers in animal models	Johnson & Johnson	(Olszewska et al. 2011)
	MDT-637	Inhalable small molecular fusion inhibitor	Phase I	None reported	Safe as inhaled therapy in healthy volunteers and asthmatics	Gilead Sciences	(MicrDoseTherapeutx.http://mdtx.com/2012/04/24/mdtx-concludes-phase1-studies/))
Other Mechanisms	MBX-300	Negatively charged lipid compound targeting G glycoprotein	Preclinical	L97P, F101L, I107T, I114T, F163P, F165L, F168S, F170S, I189T, L215P, F265L, L274P	Associated with reduced lung inflammation and RSV viral titers in animal models	Microbiotix	(Costello et al. 2012; Douglas 2004)

(continued)

Table 2 (continued)

Antiviral class	Name	Mechanism of action	Stage of development	Resistance mutations	Evidence	Company	Reference
	YM-53403	RNA synthesis inhibitor	Preclinical	Y163	Inhibits replication of RSV A and B strains in in vitro studies	Yamanouchi	(Sudo et al. 2005)
	RSV-604	Benzodiazepine targeting the nucleocapsid protein	Phase II	N105D, K107 N, I129L, L139I	Inhibits viral replication after mucosal entry in in vitro studies	Arrow	Therapeutics
(Chapman et al. 2007)							
	AS-1411	Oligonucleotide aptamer targeting nucleolin on cell surfaces	Preclinical	None reported	Reduces RSV lung viral titers in mouse models	N/A	(Mastrangelo P et al. 2012)
	AZ-27	RNA polymerase inhibitor	Preclinical	Y1631H	Inhibits viral replication in vitro in RSV subtype A and B clinical isolates	Astra-Zeneca	(Tiong-Yip CL et al. 2012)

4 Investigational Antiviral Agents

Several new classes of antiviral agents and specific antiviral therapies are under development for treatment of RSV (Empey et al. 2010), as summarized in Table 2.

4.1 Antibodies

Based on the success of palivizumab, other immunoglobulin products are currently under development. Motavizumab (MEDI-524, Numax, MedImmune Vaccines, Inc., Gaithersburg, MD) is a more potent RSV-neutralizing antibody that was developed by modification of palivizumab. It is associated with a 70-fold increase in antigen binding and a 20-fold increase in neutralization activity in the cotton rat model (Wu et al. 2007). Compared to palivizumab, motavizumab was associated with significant reductions in medically attended lower respiratory tract infections for RSV in a non-inferiority trial (Carbonell-Estrany et al. 2010). However, due to concerns over mild injection site reactions and development of anti-drug antibodies, the United States Food and Drug Administration elected not to support the drug application of motavizumab as a prophylactic agent. This product, however, could act as a potential therapeutic agent. In a recent randomized, placebo-controlled trial of motavizumab treatment in previously healthy hospitalized children under 12 months with RSV LRTI, no effect was seen in decreasing RSV viral load, ICU admission, mechanical ventilation, or duration of hospitalization (Ramilo et al. 2012). In this study, motavizumab was administered after hospitalization, and it is possible that earlier administration in the outpatient or emergency room setting would have been more effective. Currently, motavizumab is not licensed by the Food and Drug Administration. Another monoclonal product, MEDI-557 (MedImmune, Gaithersburg, MD) is a recombinant humanized monoclonal immunoglobulin G1 derived from motavizumab that has been shown in Phase I clinical trials to be safe and to have an extended half life as compared to wild type IgG1 (Criste et al. 2012).

An example of another potential monoclonal antibody under development is mAb 131-2G, which targets the G glycoprotein CX3C chemokine motif. The CX3C chemokine motif is located in a conserved portion of the G protein. Blocking of RSV G-protein binding to the CX3C receptor, CX3C-R1, reduces RSV infection, lung inflammation, and leukocyte migration (Choi et al. 2012). In mouse models, mAb 131-2G is associated with reduction of lung inflammation and RSV viral titers when administered within three days of infection (Haynes et al. 2009). Another mechanism by which the G-protein may serve as a drug target is illustrated by a study showing that shed G protein serves as a decoy for neutralizing antibodies and assists with RSV viral replication (Bukreyev et al. 2008). Therefore, using shed G protein as a specific drug target may increase the efficacy of antibody neutralization and decrease viral replication.

A hyperimmune IVIG product (RI-001, ADMA Biologics, Hackensack, NJ, USA) has recently been developed using pooled plasma donated from individuals with high RSV-specific antibody titers. Phase II clinical trials are planned in immunocompromised individuals to evaluate its effect on preventing progression from upper respiratory tract to lower respiratory tract disease (Empey et al. 2010).

Another potential target for antibody intervention is the F protein in its pre-triggered state. Recently, antibodies derived from rabbits inoculated with recombinant vaccinia virus expressing RSV F protein were shown to bind the prefusion form of the F protein, and to be responsible for the majority of the neutralizing component of a protective immune response (Magro et al. 2012). Antibodies that specifically target the prefusion F protein may be more specific drug candidates for treatment and prophylaxis trials.

Antibodies with prolonged half-lives and activity against both RSV subtypes could be anticipated to play a clinical role in the future. Interestingly, these products could also potentially be used in a fashion similar to a vaccine.

4.2 Fusion Inhibitors

Fusion inhibitors block viral entry into host cells by attaching to the F glycoprotein of RSV, and are a major target for RSV antivirals. The F glycoprotein is involved in RSV attachment and mediates membrane fusion and entry of RSV into lung epithelial cells, and is essential for establishment of intracellular infection. Fusion inhibitor candidates have been largely identified through screening of small molecules for fusion inhibitory properties.

Fusion inhibitors in development include TMC-353121 (Johnson & Johnson, New Brunswick, NJ, USA) (Olszewska et al. 2011), which works by disruption of the six-helix bundle conformation of the RSV-F protein to prevent virus-host cell fusion and syncytia formation (Bonfanti and Roymans 2009). When administered to RSV-infected mice, TMC-353121 reduced lung inflammation, as measured by chemokine and cytokine levels as well as by histopathology, and decreased RSV viral load. TMC-353121 binds to Y198, an amino acid in the central portion of the pre-triggered F-protein, and is associated with resistance mutations S398L, K394R, K399I, as well as D486 N, E487D, and D489Y (Costello et al. 2012).

Other fusion inhibitors include BTA-9881 (Biota Holdings Ltd, Notting Hill, Victoria, Australia) (ClinicalTrials 2012) and MDT-637 (MicroDose Therapeutx, Monmouth Junction, NJ) (MicroDose Therapeutx 2013). BTA-9881 is an oral fusion inhibitor that is currently in phase I clinical trials. MDT-637 is an inhalable small molecule fusion inhibitor originally developed by Viropharma Inc, acquired by Microdose in 2009, and licensed to Gilead Sciences, Inc., in 2011. This drug is administered via inhalation, and has been shown to be safe and efficacious in cotton-rat models as well as safe in adult healthy volunteers (http://mdtx.com/pipeline/proprietary-proudcts/mdt-637). Because MDT-637 was extremely effective in reducing RSV viral load both pre- and post infection in the cotton rat model, it has the potential for use in both prevention and treatment of RSV, though it is currently only being developed as a therapeutic agent. According to the company website, pharmacokinetic studies of this agent in humans has revealed low plasma levels of MDT-637 but respiratory tract levels of drug are much higher than concentrations necessary to inhibit RSV in vitro. A Phase I trial has shown it to be safe as an inhaled therapy in adult asthmatics, and Phase II trials are underway. There has been no reported resistance mutations associated with MDT-637.

4.3 RSV Antivirals with Other Targets

Another agent in development is MBX-300 (Microbiotix, Worcester, MA, USA), a negatively charged lipid compound that targets the RSV G protein as an attachment inhibitor. MBX-300 has been shown in animal models to be safe and efficacious in decreasing RSV viral load (Douglas 2004). Resistance mutations associated with MBX-300 occur across the entire G protein, and are noted in Table 2 (Costello et al. 2012). An RSV receptor antagonist in development (AS-1411) is a oligodeoxynucleotide aptamer targeting nucleolin on cell surfaces (Mastrangelo et al. 2012). Originally developed for cancer treatment, AS-1411 has been shown to reduce RSV lung viral titers in animal studies.

Another potential target is the RSV polymerase (L) protein. YM-53403 (Yamanouchi Pharmaceutical, Tokyo, Japan) is a small molecule compound that is thought to interfere with primary transcription and/or replication of the viral genomic RNA (Sudo et al. 2005). YM-53403 resistant viral strains have mutations at Y1631H, suggesting that the target of YM-53403 is the RSV L protein. RSV-604 (Arrow Therapeutics/AstraZeneca Corp., London, UK), an agent that targets the RSV polymerase, is an oral benzodiazepine that is broadly protective against both A and B subgroups (Chapman et al. 2007). This agent appears to inhibit viral replication after mucosal entry in in vitro studies and has undergone testing in phase II clinical trials. RSV-604 associated resistance mutations include N105D, K107 N, I129L, and L139I (Costello et al. 2012). AZ-27 (Astra-Zeneca, Waltham, MA) is a small molecule RSV polymerase inhibitor that has been shown in in vitro studies to have inhibition of viral replication in RSV subtype A and B clinical isolates (Tiong-Yip et al. 2012). It is associated with a Y1631H resistance mutation.

4.4 Small Interfering RNAs

Other new RSV antiviral therapies include small-interfering RNAs (siRNAs). Synthetic siRNAs are short lengths of double-stranded RNA that regulate gene expression through posttranscriptional gene silencing mechanisms (DeVincenzo

2008). RNA interference is a mechanism whereby siRNAs bind to and cleave complementary mRNA, thereby preventing expression of the genes encoded by the mRNA. Respiratory viruses are particularly suited to siRNA therapy due to the ability to target conserved viral sequences and the ability to deliver the therapy intranasally at the site of primary infection. The RSV-specific siRNAs in development target two nucleocapsid protein genes, the P and N genes, as well as a nonstructural protein gene, NS1 (Zhang et al. 2005). ALN-RSV01 (Anylam Pharmaceuticals, Cambridge, MA, USA) is a siRNA that targets the synthesis of the N protein. It is the only siRNA therapy in development that is undergoing clinical trials. It was shown in a prophylaxis trial to be associated with reduction of RSV infection among healthy adult volunteers undergoing RSV challenge (DeVincenzo et al. 2008). Among 85 adult healthy volunteers who received intranasal ALN-RSV01 two days before and three days after inoculation with RSV, subjects had a 38% reduction in RSV infection and a 95% increase in number of subjects who remained free of infection. In another therapeutic trial in 24 adult lung transplant recipients with RSV respiratory tract infection, ALN-RSV01 therapy was shown to be associated with statistically significantly lower symptom scores and reduction of bronchiolitis obliterans syndrome at 90 days (Zamora et al. 2011).

5 Conclusion

In conclusion, RSV remains a significant cause of morbidity and mortality among infants, the elderly and immunocompromised patients. RSV-specific treatment and prophylaxis options are currently limited to palivizumab and ribavirin, and treatment is largely supportive. Many new promising agents are currently in development, including small molecular fusion inhibitors, and these have the potential to make a significant impact in prevention and treatment of RSV disease among vulnerable individuals. The development of these agents needs to take into consideration multiple factors including intra-patient variability in RSV disease severity, rapidity of symptom onset, ease of administration, and challenges in clinical assessment of safety and efficacy in young infants and immunocompromised persons. Many of these newer agents have shown promise in animal models and phase I clinical trials in adult healthy volunteers but their effect on RSV disease in the proposed target populations still needs to be determined.

The timing of the administration of RSV antivirals remains an important question, as the greatest use would be in young children or elderly adults who may present with non-specific symptoms making early treatment somewhat problematic. It is possible the advent of simple rapid point-of-care diagnostics for RSV will make it possible to administer therapeutic agents closer to the onset of infection, or to be used as secondary prophylaxis in particularly high-risk settings, such as neonatal intensive care units or HSCT wards. It is particularly important to design trials that take into account the time period between infection and hospital admission. An ideal trial would incorporate the use of rapid point-of-care diagnostics for RSV to evaluate the use of a therapeutic agent at time of symptom onset in a vulnerable population, perhaps in an outpatient HSCT or pediatric clinic setting. Further, it will be important to measure outcomes using indicators of immunologic protection, such as induction of protective neutralizing antibody titers, as well as clinical indicators such as hospitalization and intensive care unit admission.

Ultimately, the prevention of RSV infection remains an important goal. The current use of oral oseltamivir as a prophylaxis against influenza could serve as a model for an RSV antiviral. However, prevention of RSV disease could also be accomplished through effective vaccines or safe, effective, and long-lasting antivirals or immunoglobulin products. Prevention of RSV disease in children would have substantial health benefit in developed and developing countries worldwide.

References

The Impact-RSV Study Group (1998) Palivizumab , a humanized respiratory syncytial virus monoclonal antibody, reduces hospitalization from respiratory syncytial virus infection in high-risk infants. Pediatrics 102:531–537

American Academy of Pediatrics Subcommittee on Diagnosis and Management of Bronchiolitis (2006) Diagnosis and management of bronchiolitis. Pediatrics 118:1774–1793

Boeckh M, Berrey MM, Bowden RA, Crawford SW, Balsley J, Corey L (2001) Phase 1 evaluation of the respiratory syncytial virus-specific monoclonal antibody palivizumab in recipients of hematopoietic stem cell transplants. J Infect Dis 184:350–354

Boeckh M, Erard V, Zerr D, Englund J (2005) Emerging viral infections after hematopoietic cell transplantation. Pediatr Transplant 9(Suppl 7):48–54

Boeckh M, Englund J, Li Y et al (2007) Randomized controlled multicenter trial of aerosolized ribavirin for respiratory syncytial virus upper respiratory tract infection in hematopoietic cell transplant recipients. Clin Infect Dis: Official pub Infect Dis Soc America 44:245–249

Bonfanti JF, Roymans D (2009) Prospects for the development of fusion inhibitors to treat human respiratory syncytial virus infection. Curr Opin Drug Discov Devel 12:479–487

Bukreyev A, Yang L, Fricke J et al (2008) The secreted form of respiratory syncytial virus g glycoprotein helps the virus evade antibody-mediated restriction of replication by acting as an antigen decoy and through effects on fc receptor-bearing leukocytes. J Virol 82:12191–12204

Carbonell-Estrany X, Simoes EA, Dagan R et al (2010) Motavizumab for prophylaxis of respiratory syncytial virus in high-risk children: a noninferiority trial. Pediatrics 125:e35–e51

Chapman J, Abbott E, Alber DG et al (2007) RSV604, a novel inhibitor of respiratory syncytial virus replication. Antimicrob Agents Chemother 51:3346–3353

Chavez-Bueno S, Mejias A, Merryman RA, Ahmad N, Jafri HS, Ramilo O (2007) Intravenous palivizumab and ribavirin combination for respiratory syncytial virus disease in high-risk pediatric patients. Pediatr Infect Dis J 26:1089–1093

Chemaly RF, Torres HA, Munsell MF, Shah DP, Rathod DB, Bodey GP, Hosing C, Saifan C, Raad II, Champlin RE (2012) An adaptive randomized trial of intermittent dosing schedule of aerosolized ribavirin in cancer patients with respiratory syncytial virus infections. J Infect Dis 206:1367–1371

Choi Y, Mason CS, Jones LP, Crabtree J, Jorquera PA, Tripp RA (2012) Antibodies to the central conserved region of respiratory syncytial virus (RSV) G protein block RSV G protein CX3C-CX3CR1 binding and cross-neutralize RSV A and B strains. Viral Immunol 25:193–203

Clinical Trials gov (2012) Safety study of oral BTA9882 to treat RSV infection: NTC0050907. Accessed June 15 2012
Cordonnier C, Chevret S, Legrand M et al (2003) Should immunoglobulin therapy be used in allogeneic stem-cell transplantation? A randomized, double-blind, dose effect, placebo-controlled, multicenter trial. Ann Intern Med 139:8–18
Cortez K, Murphy BR, Almeida KN et al (2002) Immune-globulin prophylaxis of respiratory syncytial virus infection in patients undergoing stem-cell transplantation. J Infect Dis 186:834–838
Costello HM, Ray WC, Chaiwatpongsakorn S, Peeples ME (2012) Targeting RSV with vaccines and small molecule drugs. Infect Disord Drug Targets 12:110–128
Criste RGP, Losonsky G, Jensen K, Harris B, Patel N, Robbie GJ (2012) Single-dose safety, tolerability, and pharmacokinetics of MEDI-557, a humanized monoclonal antibody with an extended half-life against Respiratory Syncytial Virus (RSV), in healthy adults. In: RSV symposium, Santa Fe, NM
DeVincenzo JP (2008) RNA interference strategies as therapy for respiratory viral infections. Pediatr Infect Dis J 27:S118–S122
DeVincenzo JP, Hall CB, Kimberlin DW et al (2004) Surveillance of clinical isolates of respiratory syncytial virus for palivizumab (Synagis)-resistant mutants. J Infect Dis 190:975–978
DeVincenzo J, Cehelsky JE, Alvarez R et al (2008) Evaluation of the safety, tolerability and pharmacokinetics of ALN-RSV01, a novel RNAi antiviral therapeutic directed against respiratory syncytial virus (RSV). Antiviral Res 77:225–231
DeVincenzo JP, Wilkinson T, Vaishnaw A et al (2010) Viral load drives disease in humans experimentally infected with respiratory syncytial virus. Am J Respir Crit Care Med 182:1305–1314
Douglas JL (2004) In search of a small-molecule inhibitor for respiratory syncytial virus. Expert Rev Anti-infect Ther 2:625–639
El Saleeby CM, Bush AJ, Harrison LM, Aitken JA, Devincenzo JP (2011) Respiratory syncytial virus load, viral dynamics, and disease severity in previously healthy naturally infected children. J Infect Dis 204:996–1002
Empey KM, Peebles RS Jr, Kolls JK (2010) Pharmacologic advances in the treatment and prevention of respiratory syncytial virus. Clin Infect Dis: Official Pub Infect Dis Soc America 50:1258–1267
Falsey AR, Hennessey PA, Formica MA, Cox C, Walsh EE (2005) Respiratory syncytial virus infection in elderly and high-risk adults. N Engl J Med 352:1749–1759
Falsey AR, Formica MA, Hennessey PA, Criddle MM, Sullender WM, Walsh EE (2006) Detection of respiratory syncytial virus in adults with chronic obstructive pulmonary disease. Am J Respir Crit Care Med 173:639–643
Feltes TF, Cabalka AK, Meissner HC et al (2003) Palivizumab prophylaxis reduces hospitalization due to respiratory syncytial virus in young children with hemodynamically significant congenital heart disease. J Pediatr 143:532–540
Feltes TF, Sondheimer HM, Tulloh RM et al (2011) A randomized controlled trial of motavizumab versus palivizumab for the prophylaxis of serious respiratory syncytial virus disease in children with hemodynamically significant congenital heart disease. Pediatr Res 70:186–191
Fernandez P, Trenholme A, Abarca K et al (2010) A phase 2, randomized, double-blind safety and pharmacokinetic assessment of respiratory syncytial virus (RSV) prophylaxis with motavizumab and palivizumab administered in the same season. BMC Pediatr 10:38
Frogel M, Nerwen C, Boron M et al (2008) Improved outcomes with home-based administration of palivizumab: results from the 2000–2004 palivizumab outcomes registry. Pediatr Infect Dis J 27:870–873
From the American Academy of Pediatrics (2009) Policy statements–modified recommendations for use of palivizumab for prevention of respiratory syncytial virus infections. Pediatrics 124:1694–1701

Garcia R, Raad I, Abi-Said D et al (1997) Nosocomial respiratory syncytial virus infections: prevention and control in bone marrow transplant patients. Infect control and Hosp Epidemiol: Official J Soc Hosp Epidemiolo America 18:412–416

Ghosh S, Champlin RE, Englund J et al (2000) Respiratory syncytial virus upper respiratory tract illnesses in adult blood and marrow transplant recipients: combination therapy with aerosolized ribavirin and intravenous immunoglobulin. Bone Marrow Transplant 25:751–755

Glanville AR, Scott AI, Morton JM et al (2005) Intravenous ribavirin is a safe and cost-effective treatment for respiratory syncytial virus infection after lung transplantation. J Heart and Lung Transplant: Official Pub Int Soc Heart Transplant 24:2114–2119

Glezen WP, Paredes A, Allison JE, Taber LH, Frank AL (1981) Risk of respiratory syncytial virus infection for infants from low-income families in relationship to age, sex, ethnic group, and maternal antibody level. J Pediatr 98:708–715

Graepel KWCM, McLellan JS, Dambach K, Modjarrad K, Kwong PK, Graham BS (2012) Investigation of an attachment-inhibiting, RSV-neutralizing antibody. In: RSV symposium, Santa Fe, NM

Groothuis JR, Levin MJ, Rodriguez W et al (1991) Use of intravenous gamma globulin to passively immunize high-risk children against respiratory syncytial virus: safety and pharmacokinetics. The RSVIG study group. Antimicrob Agents Chemother 35:1469–1473

Groothuis JR, Simoes EA, Levin MJ et al (1993) Prophylactic administration of respiratory syncytial virus immune globulin to high-risk infants and young children. The Respiratory Syncytial Virus Immune Globulin Study Group. N Engl J Med 329:1524–1530

Gruber WC, Wilson SZ, Throop BJ, Wyde PR (1987) Immunoglobulin administration and ribavirin therapy: efficacy in respiratory syncytial virus infection of the cotton rat. Pediatr Res 21:270–274

Hall CB, Douglas RG Jr (1981) Modes of transmission of respiratory syncytial virus. J Pediatr 99:100–103

Hall CB, Douglas RG Jr, Geiman JM (1975) Quantitative shedding patterns of respiratory syncytial virus in infants. J Infect Dis 132:151–156

Hall CB, Weinberg GA, Iwane MK et al (2009) The burden of respiratory syncytial virus infection in young children. N Engl J Med 360:588–598

Hammer J, Numa A, Newth CJ (1995) Albuterol responsiveness in infants with respiratory failure caused by respiratory syncytial virus infection. J Pediatr 127:485–490

Haynes LM, Caidi H, Radu GU et al (2009) Therapeutic monoclonal antibody treatment targeting respiratory syncytial virus (RSV) G protein mediates viral clearance and reduces the pathogenesis of RSV infection in BALB/c mice. J Infect Dis 200:439–447

Isaacs D, Dickson H, O'Callaghan C, Sheaves R, Winter A, Moxon ER (1991) Handwashing and cohorting in prevention of hospital acquired infections with respiratory syncytial virus. Arch Dis Child 66:227–231

Johnson PR Jr, Olmsted RA, Prince GA et al (1987) Antigenic relatedness between glycoproteins of human respiratory syncytial virus subgroups A and B: evaluation of the contributions of F and G glycoproteins to immunity. J Virol 61:3163–3166

Khan JY, Kerr SJ, Tometzki A et al (1995) Role of ECMO in the treatment of respiratory syncytial virus bronchiolitis: a collaborative report. Arch Dis Child Fetal Neonatal Ed 73:F91–F94

Kim YJ, Boeckh M, Englund JA (2007) Community respiratory virus infections in immunocompromised patients: hematopoietic stem cell and solid organ transplant recipients, and individuals with human immunodeficiency virus infection. Semin Respir Crit Care Med 28:222–242

Krasinski K, LaCouture R, Holzman RS, Waithe E, Bonk S, Hanna B (1990) Screening for respiratory syncytial virus and assignment to a cohort at admission to reduce nosocomial transmission. J Pediatr 116:894–898

Leclair JM, Freeman J, Sullivan BF, Crowley CM, Goldmann DA (1987) Prevention of nosocomial respiratory syncytial virus infections through compliance with glove and gown isolation precautions. N Engl J Med 317:329–334

Lewinsohn DM, Bowden RA, Mattson D, Crawford SW (1996) Phase I study of intravenous ribavirin treatment of respiratory syncytial virus pneumonia after marrow transplantation. Antimicrob Agents Chemother 40:2555–2557

Libster R, Hortoneda JB, Laham FR et al (2009) Breastfeeding prevents severe disease in full term female infants with acute respiratory infection. Pediatr Infect Dis J 28:131–134

Liu V, Dhillon GS, Weill D (2010) A multi-drug regimen for respiratory syncytial virus and parainfluenza virus infections in adult lung and heart-lung transplant recipients. Transplant Infect Dis: Official J Transplant Soc 12:38–44

Macartney KK, Gorelick MH, Manning ML, Hodinka RL, Bell LM (2000) Nosocomial respiratory syncytial virus infections: the cost-effectiveness and cost-benefit of infection control. Pediatrics 106:520–526

Magro M, Mas V, Chappell K et al (2012) Neutralizing antibodies against the preactive form of respiratory syncytial virus fusion protein offer unique possibilities for clinical intervention. Proc Natl Acad Sci USA 109:3089–3094

Mastrangelo P MT, Duan W, Norris M, Hegele R (2012) Targeting the RSV receptor: a proof of principle for RSV prophylaxis and therapy. In: RSV symposium, Santa Fe, NM

Meissner HC, Fulton DR, Groothuis JR et al (1993) Controlled trial to evaluate protection of high-risk infants against respiratory syncytial virus disease by using standard intravenous immune globulin. Antimicrob Agents Chemother 37:1655–1658

MicrDoseTherapeutx. (2013) microdose therapeutx announces conclusion of mdt-637 phase I single and multiple ascending dose studies in healthy volunteers for treatment of RSV. http://mdtx.com/2012/04/24/mdtx-concludes-phase1-studies/

Nair H, Nokes DJ, Gessner BD et al (2010) Global burden of acute lower respiratory infections due to respiratory syncytial virus in young children: a systematic review and meta-analysis. Lancet 375:1545–1555

Olszewska W, Ispas G, Schnoeller C et al (2011) Antiviral and lung protective activity of a novel respiratory syncytial virus fusion inhibitor in a mouse model. Eur Respir J: Official J Eur Soc Clin Respir Physiol 38:401–408

Paes B, Mitchell I, Li A, Lanctot KL (2012) A comparative study of respiratory syncytial virus (RSV) prophylaxis in premature infants within the Canadian Registry of Palivizumab (CARESS). Eur J of Clin Microbiol & Infect Dis: official Publ Eur Soc of Clin Microbiol 31:2703–2711

Papenburg J, Hamelin ME, Ouhoummane N et al (2012) Comparison of risk factors for human metapneumovirus and respiratory syncytial virus disease severity in young children. J Infect Dis 206:178–189

Pelaez A, Lyon GM, Force SD et al (2009) Efficacy of oral ribavirin in lung transplant patients with respiratory syncytial virus lower respiratory tract infection. J Heart Lung Transplant: Official pub Int Soc Heart Transplant 28:67–71

Ramilo OSJ, Wang K, Jensen K, Harris B, Losonsky G, Griffin P (2012) Motavizumab treatment of children hospitalized With rsv lower respiratory tract infection does not decrease viral load or severity of illness. In: RSV symposium, Santa Fe, NM

Rodriguez WJ, Gruber WC, Groothuis JR et al (1997) Respiratory syncytial virus immune globulin treatment of RSV lower respiratory tract infection in previously healthy children. Pediatrics 100:937–942

Saez-Llorens X, Moreno MT, Ramilo O, Sanchez PJ, Top FH Jr, Connor EM (2004) Safety and pharmacokinetics of palivizumab therapy in children hospitalized with respiratory syncytial virus infection. Pediatr Infect Dis J 23:707–712

Shah JN, Chemaly RF (2011) Management of RSV infections in adult recipients of hematopoietic stem cell transplantation. Blood 117:2755–2763

Simoes EA, Sondheimer HM, Top FH Jr et al (1998) Respiratory syncytial virus immune globulin for prophylaxis against respiratory syncytial virus disease in infants and children with congenital heart disease. The cardiac study group. J pediatr 133:492–499

Simoes EA, Groothuis JR, Carbonell-Estrany X et al (2007) Palivizumab prophylaxis, respiratory syncytial virus, and subsequent recurrent wheezing. J Pediatr 151:34–42

Singleton R (1996) Association between maternal smoking and severe respiratory syncytial virus infections and sudden infant death syndrome. Alaska Med 38:34

Small TN, Casson A, Malak SF et al (2002) Respiratory syncytial virus infection following hematopoietic stem cell transplantation. Bone Marrow Transplant 29:321–327

Sudo K, Miyazaki Y, Kojima N et al (2005) YM-53403, a unique anti-respiratory syncytial virus agent with a novel mechanism of action. Antiviral Res 65:125–131

Tablan OC, Anderson LJ, Besser R et al (2004) Guidelines for preventing health-care-associated pneumonia, 2003: recommendations of CDC and the Healthcare Infection Control Practices Advisory Committee. MMWR recommendations and reports: morbidity and mortality weekly report recommendations and reports/Centers For Disease Control vol 53, pp 1–36

Tarrac SE (2008) Application of the updated CDC isolation guidelines for health care facilities. AORN J 87:534–542, quiz 43–46

The PREVENT study group (1997) Reduction of respiratory syncytial virus hospitalization among premature infants and infants with bronchopulmonary dysplasia using respiratory syncytial virus immune globulin prophylaxis. Pediatrics 99:93–99

Tiong-Yip CLFM, Aschenbrenner L, Fan J, Johnson KD, Brown D, Xiong H, Yu Q (2012) Characterization of a novel RSV L polymerase inhibitor. In: RSV symposium, Santa Fe, NM

Tomblyn M, Chiller T, Einsele H et al (2009) Guidelines for preventing infectious complications among hematopoietic cell transplant recipients: a global perspective. Preface. Bone Marrow Transplant 44:453–455

Toms GL, Gardner PS, Pullan CR, Scott M, Taylor C (1980) Secretion of respiratory syncytial virus inhibitors and antibody in human milk throughout lactation. J Med Virol 5:351–360

van Woensel JB, Vyas H (2011) Dexamethasone in children mechanically ventilated for lower respiratory tract infection caused by respiratory syncytial virus: a randomized controlled trial. Crit Care Med 39:1779–1783

Ventre K, Randolph A (2004) Ribavirin for respiratory syncytial virus infection of the lower respiratory tract in infants and young children. Cochrane Database Syst Rev CD000181

Wegner S, Vann JJ, Liu G et al (2004) Direct cost analyses of palivizumab treatment in a cohort of at-risk children: evidence from the north carolina medicaid program. Pediatrics 114:1612–1619

Wendt CH, Fox JM, Hertz MI (1995) Paramyxovirus infection in lung transplant recipients. J Heart and Lung Transplant: Official pub Int Soc Heart Transplant 14:479–485

Whimbey E, Champlin RE, Englund JA et al (1995) Combination therapy with aerosolized ribavirin and intravenous immunoglobulin for respiratory syncytial virus disease in adult bone marrow transplant recipients. Bone Marrow Transplant 16:393–399

Wu H, Pfarr DS, Johnson S et al (2007) Development of motavizumab, an ultra-potent antibody for the prevention of respiratory syncytial virus infection in the upper and lower respiratory tract. J Mol Biol 368:652–665

Zamora MR, Budev M, Rolfe M et al (2011) RNA interference therapy in lung transplant patients infected with respiratory syncytial virus. Am J Respir Crit Care Med 183:531–538

Zhang W, Yang H, Kong X et al (2005) Inhibition of respiratory syncytial virus infection with intranasal siRNA nanoparticles targeting the viral NS1 gene. Nat Med 11:56–62

Zhang L, Mendoza-Sassi RA, Wainwright C, Klassen TP (2008) Nebulized hypertonic saline solution for acute bronchiolitis in infants. Cochrane Database Syst Rev CD006458

Live-Attenuated Respiratory Syncytial Virus Vaccines

Ruth A. Karron, Ursula J. Buchholz and Peter L. Collins

Abstract Live-attenuated respiratory syncytial virus (RSV) vaccines offer several advantages for immunization of infants and young children: (1) they do not cause vaccine-associated enhanced RSV disease; (2) they broadly stimulate innate, humoral, and cellular immunity, both systemically and locally in the respiratory tract; (3) they are delivered intranasally; and (4) they replicate in the upper respiratory tract of young infants despite the presence of passively acquired maternally derived RSV neutralizing antibody. This chapter describes early efforts to develop vaccines through the classic methods of serial cold-passage and chemical mutagenesis, and recent efforts using reverse genetics to derive attenuated derivatives of wild-type (WT) RSV and to develop parainfluenza vaccine vectors that express RSV surface glycoproteins.

Contents

R. A. Karron (✉)
Center for Immunization Research, Department of International Health, Bloomberg School of Public Health, Johns Hopkins University, Baltimore, MD, USA
e-mail: rkarron@jhsph.edu

U. J. Buchholz · P. L. Collins
Laboratory of Infectious Diseases, National Institute of Allergy and Infectious Diseases, National Institutes of Health, Bethesda, MD, USA

L. J. Anderson and B. S. Graham (eds.), *Challenges and Opportunities for Respiratory Syncytial Virus Vaccines*, Current Topics in Microbiology and Immunology 372,
DOI: 10.1007/978-3-642-38919-1_13, © Springer-Verlag Berlin Heidelberg 2013

1 Introduction

Efforts to develop live-attenuated RSV vaccines began in response to the tragic results of studies of a formalin-inactivated pediatric RSV vaccine in the 1960s. This vaccine was poorly protective and, in RSV-naïve recipients, primed for enhanced disease upon subsequent natural RSV infection (see chapter by P.L. Collins et al., this volume). In contrast, live RSV vaccine viruses do not appear to prime for enhanced RSV disease based on studies in experimental animals (Waris et al. 1997), and the general observation that community-acquired infection followed by re-infection is not associated with enhanced disease. This notion was confirmed by studies of several live-attenuated RSV vaccine candidates involving over 380 RSV-naïve infants and children, none of whom showed evidence of enhanced disease after natural infection with WT RSV (Wright et al. 2007). Live-attenuated vaccines are the only type of RSV vaccines that have been demonstrated to be safe in RSV-naïve recipients.

Live-attenuated RSV vaccines are administered intranasally, which offers three advantages: (1) replication in the upper respiratory tract and immunogenicity even in the presence of passively acquired, maternally derived serum neutralizing antibody, which usually is present in young infants (see below), (2) induction of local mucosal immunity, which is important in restricting replication of respiratory viruses, and (3) needle-free administration. In general, live vaccines broadly stimulate innate, cellula, and humoral immunity (Collins and Murphy 2005; Murphy and Collins 2002), and studies of influenza vaccines in infants and children indicate that live vaccines induce a broader, more effective, and possibly a more durable response than subunit vaccines (Ambrose et al. 2011, 2010; Belshe et al. 2007; Hoft et al. 2011).

One potential limitation to a live-attenuated RSV vaccine is that, as is typical for live-attenuated vaccines in general, a low-to-moderate level of vaccine virus replication is necessary to achieve an optimal immune response. This is probably because virus replication provides pathogen-associated molecular patterns (PAMPs) that stimulate innate immunity and antigens that stimulate adaptive immunity via MHC class I and MHC class II presentation pathways. On the other hand, the level of replication of live-attenuated RSV vaccines is generally inversely correlated with attenuation (Karron et al. 2005, 1997; Wright et al. 1982, 2000, 2006). While RSV vaccine candidates have been developed that are attenuated and highly restricted in replication (Karron et al. 2005; Wright et al. 2006) it has not yet been demonstrated that these are sufficiently immunogenic to provide effective protection against WT RSV. New strategies that capitalize on our understanding of viral replication mechanisms may help to overcome these limitations. Another limitation is the notorious instability of the virus itself, which likely will complicate vaccine production, storage and usage in resource-challenged settings.

The primary targets for a live-attenuated RSV vaccine are infants and young children. Since the peak of hospitalization for RSV disease occurs at 2–4 months of age, one proposed strategy is to immunize very early in infancy, beginning at 1–2 months of age. However, a number of factors complicate RSV vaccination of the young infant, including immunologic immaturity, the immunosuppressive effects of maternal antibodies, and the natural occurrence of events such as apnea and sudden infant death syndrome that might be perceived to be vaccine related. Therefore, an alternative strategy would be to begin immunization somewhat later, at approximately 4–6 months of age. About 40 % of hospitalizations and 75 % of outpatient visits for RSV disease occur in infants and young children over 6 months of age (Hall et al. 2009). Thus, this approach would reduce the considerable burden of pediatric RSV disease that occurs beyond early infancy (Hall et al. 2009), and might reduce transmission in the community.

Live-attenuated RSV vaccines are evaluated for replication, tolerability, and immunogenicity in Phase I studies by age de-escalation with a transition from RSV-experienced to RSV-naïve study groups. Studies in adults are typically open-label, whereas studies in children are placebo-controlled to allow for evaluation in the context of the myriad community-acquired respiratory and febrile illnesses frequently observed in young children. Each new vaccine is typically evaluated sequentially in (1) adults, presumed to be RSV-seropositive (~15 subjects); (2) RSV-seropositive children aged 12–59 months (~10 and ~5 vaccine and placebo recipients, respectively); and (3) RSV-seronegative children aged 6–24 months (~20 and ~10 vaccine and placebo recipients, respectively). If a vaccine candidate is sufficiently attenuated in RSV-seronegative children, it may also be evaluated in RSV-naïve infants aged 1–3 months (~20 and ~10 vaccine and placebo recipients, respectively). This stepwise strategy has proved to be a careful method for the evaluation of vaccine safety; it is described further in Sect. 5.

Over the past 35 years, several types of live-attenuated RSV vaccines have been described, including biological and recombinant derivatives of WT RSV, as well as replication-competent parainfluenza virus vectors expressing one or both of the RSV fusion (F) and heavily glycosylated (G) surface glycoproteins, which are the two neutralization and major protective RSV antigens. Most of these experimental vaccines have been made using RSV antigenic subgroup A viruses, usually the A2 strain. This chapter will describe past and current efforts to develop live-attenuated RSV vaccines suitable for administration to RSV-naïve infants and children. The live-attenuated pediatric RSV vaccines currently under evaluation are not being considered for use in adults because they are over-attenuated in RSV-experienced individuals (Gonzalez et al. 2000; Wright et al. 2006).

2 Live-Attenuated Vaccines: Biologically Derived RSV Strains

The first live-attenuated candidate RSV vaccines (given in Table 1) were developed either by repeated passage at low temperature to yield cold-passaged (*cp*) RSV (Friedewald et al. 1968) or by chemical mutagenesis to yield temperature-sensitive (*ts*) mutants designated *ts*-1 or *ts*-2 (Gharpure et al. 1969; Wright et al. 1982, 1976) and *ts*-1 A, B, and C (McKay et al. 1988; Pringle et al. 1993). The hypothesis underlying these efforts was that mutations which rendered the virus temperature-sensitive, i.e., less able than WT RSV to replicate at temperatures ≥37 °C, would produce vaccine viruses that would preferentially replicate in the cooler upper respiratory tract but be restricted in replication in the lower respiratory tract, thereby diminishing the risk of lower respiratory tract illness (LRI) associated with vaccine virus replication. As described below, the *cp* mutations were later mapped to five missense mutations in the nucleoprotein (N), F glycoprotein, and polymerase (L) protein genes, whereas the mutations present in the early *ts* mutants have not been directly mapped and identified. Both *cp*RSV and *ts*-1 were underattenuated in RSV-seronegative children, and isolates that had lost the *ts* phenotype were recovered from children administered the *ts*-1 vaccine (Table 1; Kim et al. 1973; Wright et al. 1976). In contrast, *ts*-2 was overattenuated and replicated poorly in RSV-seronegative children (Table 1; Wright et al. 1982). *ts*-1 A, B, and C were only administered to adult subjects. *ts*-1 A and B were insufficiently attenuated in adults; *ts*–1C was satisfactorily attenuated in adults but was not evaluated further (Table 1; Pringle et al. 1993).

In the mid-1990s, additional efforts were made to develop a live-attenuated RSV vaccine through chemical mutagenesis of *cp*RSV followed by screening for temperature sensitivity. Several RSV A2 *cpts* mutants were identified, including candidate vaccines designated as *cpts*248/955, *cpts*530/1009, and *cpts*248/404 (Table 1; Crowe et al. 1994, 1995; Firestone et al. 1996). *cpts*248/955 and *cpts*530/1009 were each sequentially evaluated in adults, RSV-seropositive children, and RSV-seronegative children. While both of these vaccines were attenuated in adults and seropositive children, they were insufficiently attenuated in seronegative children: replication of *cpts*248/955 was associated with LRI (wheezing) in one child, and the *cpts*530/1009 vaccine was transmitted from a seronegative vaccinee to a placebo recipient (Karron et al. 1997). *cpts*248/404 was highly attenuated in RSV-seropositive and -seronegative children, but was associated with nasal congestion in 1–2 month-old infants (Wright et al. 2000). Although *cpts*248/404 was insufficiently attenuated for young infants, these studies demonstrated that an intranasally administered live-attenuated vaccine could replicate in the presence of maternally derived RSV serum neutralizing antibody, an important principle for future vaccine development.

During this period, efforts were also made to develop a live-attenuated RSV B candidate vaccine. The most noteworthy candidate was RSV *cp*-52B, a live-attenuated virus derived by serial passage of WT RSV B1 virus at low

Table 1 Biologically derived live-attenuated RSV vaccines

Vaccine	Manufacturer	Population evaluated	Year(s) published (reference)	Results
*cp*RSV	NIAID, NIH	Adults, children	1968 (Crowe et al. 1995)	Attenuated in adults
			1971 (Kim et al. 1971)	Attenuated in RSV seropositive children Underattenuated in RSV seronegative children and infants
RSV *ts*-1	NIAID, NIH	Adults, children	1971 (Wright et al. 1971)	Attenuated in adults
			1973 (Kim et al. 1973)	Attenuated in RSV seropositive children
			1976 (Wright et al. 1976)	Underattenuated in RSV seronegative children and infants
RSV *ts*-2	NIAID, NIH	Adults, children	1982 (Wright et al. 1982)	Attenuated in adults[1] and RSV seropositive children[1] Overattenuated in RSV seronegative children
RSV *ts*-1 A,B,C	MRC	Adults	1990 (Watt et al. 1990)	ts-1 A, ts-1B underattenuated in adults
			1993 (Pringle et al. 1993)	ts-1 C attenuated in adults
*cpts*248/955	NIAID,NIH/ Wyeth	Adults, children	1997 (Karron et al. 1997)	Attenuated in adults and RSV seropositive children Underattenuated in RSV seronegative children[2]
*cpts*530/1009	NIAID,NIH/ Wyeth	Adults, children	1997 (Karron et al. 1997)	Attenuated in adults and RSV seropositive children Underattenuated in RSV seronegative children[2]
cp-52B (subgroup B)	NIAID,NIH/ Wyeth	Adults, children	1997 (Karron et al. 1997)	Attenuated in adults[A] and SV seropositive children[1] Overattenuated in RSV seronegative children Lacked small hydrophobic (SH) and attachment (G) proteins

(continued)

Table 1 (continued)

Vaccine	Manufacturer	Population evaluated	Year(s) published (reference)	Results
*cpts*248/404	NIAID,NIH/ Wyeth	Adults, children, young infants	2000 (Wright et al. 2000)	Attenuated in adults, RSV seropositive children, and RSV seronegative children Underattenuated in young infants[3]

cp = cold = passaged, *ts* = temperature sensitive, pfu = plaque-forming units, NIAID = National Institute of Allergy and Infectious Diseases, NIH = National Institutes of Health, MRC = Medical Research Council (United Kingdom)

[1] One cannot distinguish between attenuated and over-attenuated phenotypes in RSV seropositive subjects (adults and seropositive children) because of the very low levels of infectivity and replication observed for attenuated RSV strains in these groups

[2] Respiratory or febrile illness or otitis media was observed in all RSV seronegative children infected with *cpts*248/955, and one seronegative child developed wheezing associated with shedding of vaccine virus. Transmission of *cpts*248/955 and of *cpts*530/1009 from seronegative vaccines to placebo recipients was observed at a rate of 20–25 %

[3] Nasal congestion was observed in 1–2 month-old infants administered 10^4 or 10^5 pfu of *cpts*248/404

temperature. This virus was replication competent in vitro, but was over-attenuated in seronegative children and was subsequently found to contain a large deletion that prevented expression of the SH and G proteins (Karron et al. 1997). The over-attenuation and lack of G expression made RSV *cp*-52B unsuitable as a vaccine candidate.

Attempts to further attenuate the promising *cpts*248/404 virus by an additional round of mutagenesis and selection were unsuccessful. It became clear that the conventional methods of vaccine development such as cold-passage and chemical mutagenesis were too unpredictable and labor-intensive to be practicable for RSV.

3 Attenuated RSV Strains Produced by Reverse Genetics

In 1995, a reverse genetic system for producing complete infectious RSV strain A2 was developed, thus providing the capability of inserting pre-determined mutations into live RSV via cloned cDNA. All subsequent vaccine candidates have been produced by reverse genetics (Table 2).

Reverse genetics has a number of advantages for developing attenuated strains (Collins and Murphy 2005; Murphy and Collins 2002). Reverse genetics can directly identify and characterize attenuating mutations in existing attenuated strains (Bartlett et al. 2005; Firestone et al. 1996; Juhasz et al. 1999; Nolan et al. 2007; Whitehead et al. 1998a, b) and also can produce novel mutations not found in biologically derived strains (Bermingham and Collins 1999; Luongo et al. 2013; Teng et al. 2000; Whitehead et al. 1999). The resulting well-characterized attenuating mutations from this menu can then be assembled in desired combinations to produce live-attenuated candidate vaccines with a range of phenotypes and properties (Bartlett et al. 2005; Karron et al. 2005; Luongo et al. 2012, 2013; Nolan et al. 2007; Whitehead et al. 1999; Wright et al. 2006). However, the resulting level of attenuation cannot be predicted precisely, since the aggregate attenuating effect is not always precisely the sum of the individual mutations (Karron et al. 2005), and mutations are sometimes incompatible (Whitehead et al. 1999). The defined attenuating mutations in these viruses can readily be monitored for stability during vaccine manufacture and use, and the significance of chance mutations that may occur during virus replication can be assessed. Vaccine strains can be adjusted, such as by further attenuation in response to clinical data (Karron et al. 2005), or by updating the F and G surface antigens in response to changes in circulating stains. Amino acid substitutions can be manipulated to reduce the frequency of mutations that lead to loss of attenuation (Luongo et al. 2013, 2009). Producing virus from cDNA provides a vaccine seed with a short, well-documented passage history (Surman et al. 2007), which reduces the possibility of contamination by adventitious agents. Cloned cDNA provides a stable seed from which virus can be regenerated if necessary.

Table 2 Recombinant live-attenuated RSV vaccines

Vaccine	Manufacturer	Population evaluated	Year(s) published (reference)	Attenuating mutations[2] and results
rA2cp248/404ΔSH	NIAID, NIH/ Wyeth	Adults, children	2005 (Karron et al. 2005)	(cp + 248 + 404 + ΔSH) Attenuated in adults, RSV seropositive children, and RSV seronegative children Replication in seronegative children was indistinguishable from that of cpts248/404, suggesting insufficient attenuation for infants
rA2cp248/404/1030ΔSH	NIAID, NIH/ Wyeth	Adults, children, infants	2005 (Karron et al. 2005)	(cp + 248 + 404 + 1030 + ΔSH) Further attenuated version of the preceding virus Attenuated in adults, RSV seropositive children, RSV seronegative children, and young infants Protective against 'challenge' with a second vaccine dose Evidence of genetic reversion at 248 and 1030
rA2cpΔNS2	NIAID, NIH/ Wyeth	Adults, children	2006 (Wright et al. 2006)	(cp + ΔNS2) Attenuated in adults and RSV seropositive children, but the level of replication suggested it would be insufficiently attenuated for seronegative children
rA2cp248/404/1030ΔNS2	NIAID, NIH/ Wyeth	Adults, children	2006 (Wright et al. 2006)	(cp + 248 + 404 + ΔNS2) Further attenuated version of the preceding virus Attenuated in adults[1] and RSV seropositive children[1] Overattenuated in RSV seronegative children
rA2cp530/1009ΔNS2	NIAID, NIH/ Wyeth	Adults, children	2006 (Wright et al. 2006)	(cp + 530 + 1009 + ΔNS2) Attenuated in adults[1] and RSV seropositive children[1] Overattenuated in RSV seronegative children

(continued)

Table 2 (continued)

Vaccine	Manufacturer	Population evaluated	Year(s)published (reference)	Attenuating mutations[2] and results
MEDI-559	NIAID, NIH/ MedImmune	>5 month old seronegative children (in progress, 2012)	In progress	(cp + 248 + 404 + 1030 + ΔSH) Second version of rA2cp248/404/1030ΔSH, with a different 248 codon and 35 other silent differences ClinicalTrials.gov identifier NCT00767416
ΔM2-2	NIAID, NIH/ MedImmune	Adults, children (in progress, 2012)	In progress	(ΔM2-2) Attenuated in adults, RSV seropositive children, and RSV seronegative children ClinicalTrials.gov identifier NCT01459198
RSV ΔNS2 Δ1313 I1314L	NIAID, NIH	Children (2013)	In progress	(ΔNS2 + Δ1313 + I1314L)
RSV cps2	NIAID, NIH/ MedImmune	Study in planning (2013)	Planned	(cp + 248 + 404 + 1030 + ΔSH) Stabilized version of MEDI-559 Differs by substitutions at 5 nucleotides and 1 amino acid

[1] Note that one cannot distinguish between attenuated and over-attenuated phenotypes in seropositive subjects (adults and seropositive children) because of the very low levels of infectivity and replication observed for attenuated RSV strains in these groups

[2] Attenuating mutations:

cp: N protein (V267I), F protein (E218A and T523I), and L protein (C319Y and H1690Y), non-ts

248: L protein (Q831L); ts

404: Nt substitution in M2 GS signal (GGGGCAAATA to GGGGCAAACA); ts

1030: L protein (Y1321 N); ts; changed to Y1321 K in RSV cps2

530: L protein (F521L); ts

1009: L protein (M1169 V); ts

Δ1313 = deletion of L codon 1313; ts

ΔSH, ΔNS2, ΔM2-2 = Deletion of the indicated gene or ORF

3.1 Attenuating Point Mutations

A number of attenuating mutations useful for constructing recombinant RSV vaccine strains were identified by analysis of the previously mentioned biological *cp*RSV mutant and six of its biological *ts* derivatives, namely *cpts*248, *cpts*248/404, *cpts*248/955, *cpts*530, *cpts*530/1009, and *cpts*530/1030 (Juhasz et al. 1999; Whitehead et al. 1998a, b; 1999). Each biological virus was completely sequenced, and attenuating mutations were confirmed by introduction into WT recombinant virus and phenotypic analysis. It was useful to identify attenuating mutations in existing biological mutants known to replicate efficiently in vitro, because these mutations should be compatible with efficient growth in cell culture that is essential for vaccine manufacture. It also was desirable to identify *ts* mutations since, as noted above, these preferentially restrict virus replication at higher temperatures present in the lower respiratory tract.

Using reverse genetics, we found that the attenuation phenotype of *cp*RSV was specified by five amino acid substitutions in three proteins: N (V267I), F (E218A and T523I), and L (C319Y and H1690Y). Analysis of the other six biological *ts* viruses yielded a total of six independent mutations that each conferred a *ts* attenuation phenotype. Five of these were amino acid substitutions in the L protein (which were named based on virus number rather than sequence position): "955" (N43I), "530" (F521L), "248" (Q831L), "1009" (M1169 V), and "1030" (Y1321 N). The sixth mutation (called "404") was a single nucleotide change in the gene-start signal of the M2 gene (GGGGCAAATA to GGGGCAAACA, mRNA-sense). This latter mutation is somewhat of a surprise: while it is easy to imagine how a point mutation in a protein could render it *ts*, it remains unclear how a nucleotide substitution in a short circumscribed transcription signal confers a *ts* phenotype. Since these six *ts* mutations were found in the viral polymerase or in a gene start (GS) transcription signal, they presumably act by reducing viral RNA synthesis at restrictive temperature.

3.2 Attenuating Gene Deletions

As another means of attenuating RSV, we identified five different RSV genes or ORFs (the NS1, NS2, SH, and G genes and the M2-2 ORF) that could be deleted individually, and in some combinations, with the effect of restricting viral replication, especially in vivo. Deletion of G was not considered further as an attenuating mutation because antibodies to RSV G contribute to neutralization and protection (Connors et al. 1991; Zhang et al. 2010). The ΔNS1 and ΔM2-2 mutations each conferred a 500- to 1000-fold restriction in virus replication in seronegative chimpanzees (Teng et al. 2000). This approximates the level of attenuation thought to be necessary for a live RSV vaccine, and thus either might be an appropriate stand-alone mutation. The ΔSH and ΔNS2 mutations were associated with a 10-fold and 100-fold restriction of virus replication in chimpanzees (Whitehead et al. 1999), respectively, and thus would need to be combined with other mutations in order to make a suitably attenuated vaccine candidate. Since the NS1 and NS2 proteins are antagonists for type I interferon (IFN) production and signaling (see chapter by S. Barik, this volume), the restriction of viral replication observed for these deletion viruses presumably is due at least in part to increased IFN responses and resulting increased antiviral effect. Increased IFN responses may have an adjuvant effect that may increase vaccine immunogenicity (Valarcher et al. 2003). However, the ΔNS1 mutant also was found to cause increased apoptosis during replication in vitro, resulting in reduced viral yields (Bitko et al. 2007), and it is not clear whether the ΔNS1 virus can be propagated sufficiently well for vaccine purposes. The M2-2 protein plays a role in regulating viral RNA synthesis: deletion of M2-2 results in decreased synthesis of antigenomic and genomic RNAs, which presumably is the basis for attenuation, and in increased gene transcription, which results in increased expression of the viral proteins including the viral neutralization and protective antigens (Bermingham and Collins 1999). This might increase vaccine immunogenicity, a particularly important consideration for infant immunization. The basis of attenuation conferred by the SH deletion may involve increased tumor necrosis factor expression (Fuentes et al. 2007), but is incompletely understood. An important advantage of attenuation based on the deletion of an entire gene is that it should be highly refractory to genetic reversion or compensatory mutation.

3.3 Progressive Attenuation by Reverse Genetics Yields a Promising Candidate

As noted, the biological *cpts*248/404 virus appeared to be a promising candidate based on studies in seronegative children but was insufficiently attenuated in 1–2 month-old infants (Wright et al. 2000). Therefore, efforts were made to create a recombinant derivative that is more attenuated. The biological virus contains three attenuating mutations, namely cp (note that "cp" refers to a set of 5 amino acid substitutions in the N, F, and L proteins, as already noted), 248, and 404. The first recombinant derivative contained these three mutations combined with deletion of the SH gene, yielding rA2cp248/404ΔSH (Table 2). However, this virus was not more attenuated than its *cpts*248/404 parent in seronegative children (Karron et al. 2005), indicating that inclusion of the ΔSH mutation in this background did not confer additive attenuation. Addition of the 1030 mutation resulted in rA2cpts248/404/1030ΔSH (Fig. 1; Table 2), which had increased temperature sensitivity (i.e., rA2cpts248/404/1030ΔSH did not form plaques at 35–36 °C, whereas rA2cp248/404ΔSH did not form plaques at 38 °C) and was at least 10-fold more restricted in replication in seronegative children than *cpts*248/404 (Karron et al. 2005). The vaccine was well-tolerated in 1–2 month-old infants, was moderately immunogenic, and was protective against 'challenge' with a second dose of vaccine (Table 2; Karron et al. 2005).

Although rA2cp248/404/1030ΔSH appeared to have achieved an appropriate level of attenuation, analysis of vaccine virus isolates recovered from RSV-naïve children provided evidence of phenotypic instability: more than one-third of the isolates exhibited a 1–2 °C upward shift in temperature-sensitivity, and sequence analysis identified instances of loss of either the 248 or 1030 mutation, with loss of the latter being ~4-fold more frequent. However, these partial revertants retained four of five attenuating elements, and replication of these partial revertants was not associated with enhanced replication or clinical illness in the small numbers of children evaluated. While this degree of genetic instability might be acceptable, larger studies to more fully evaluate the potential for reversion and for transmission to susceptible contacts would be needed. A second version of this virus was made in which the codon for the 248 mutation was changed from L(CTG) to L(TTA) in an effort to increase its genetic stability, since TTA requires at least two nucleotides to revert to the WT assignment of glutamine (CAA or CAG). This virus, called MEDI-559 (Fig. 1; Table 2), contains 35 other silent nucleotide changes that are thought to be inconsequential. It presently is in Phase I/IIa studies to evaluate safety, stability, and immunogenicity in a larger number of RSV-seronegative children (ClinicalTrials.gov identifier NCT00767416).

3.4 Improvements in Genetic Stability Using Reverse Genetics

We also have worked to increase the genetic and phenotypic stability of the "248" and "1030" mutations in MEDI-559. We used reverse genetics to systematically analyze all possible amino acid assignments at the 248 and 1030 loci to identify more stable alternatives. Using this approach, each amino acid assignment is evaluated phenotypically to identify all possible attenuating versus WT-like assignments. Attenuating assignments are then examined to identify codons that differ from all possible codons for WT-like assignments by 2 or, preferably, 3 nucleotides (Luongo et al. 2012, 2009). This is based on the premise that the frequency of loss of attenuation requiring a single nucleotide change will approximate the viral mutation rate of $\sim 10^{-4}$, whereas loss of attenuation requiring two or three nucleotide changes will have rates of $\sim 10^{-8}$ and 10^{-12}, respectively, and thus will be much less frequent. For the "248" locus (Q831L in the L protein), we were unable to find an alternative attenuating amino acid or codon that would require two or three nucleotide changes. However, we found differences in stability among the available leucine codons—which was somewhat surprising—and thus identified a better choice at this position, namely L(TTG) (Luongo et al. 2009). For the "1030" locus (Y1321 N in the L protein), at which loss of attenuation was more frequent, we did identify more stable alternatives, including K(AAA), G(GGA), E(GAA), and E(GAG) (Luongo et al. 2012). Increased stability was confirmed by in vitro "stress tests", in which candidate viruses are passaged at progressively higher temperatures to favor outgrowth and detection of mutants with shifts in the *ts* phenotype. We also found that attenuation conferred by the 1030 mutation could be partly abrogated by a second-site, compensatory S1313C change in the L protein. However, this compensatory mutation could be prevented by changing the naturally occurring S(AGC) codon to S(TCA), which requires two rather than one nucleotide substitutions to change to cysteine. Though the process of genetic stabilization is complex and labor-intensive, systematic analysis using reverse genetics can provide new RSV candidates with stabilized attenuating mutations.

One such candidate is designated RSV cps2 (Table 2; Luongo et al. 2012). RSV cps2 is a modified version of MEDI-559 that contains stabilized 248 and 1030 mutations: namely, 831L[TTG] and 1321 K[AAA], respectively, with the additional assignment of 1313S[TCA] to prevent the compensatory mutation noted above. This results in a total of five nucleotide changes and one amino acid change. RSV cps2 was confirmed to be indistinguishable from MEDI-559 with regard to the level of replication in RSV-seronegative chimpanzees, and the virus is being prepared for clinical evaluation as a stabilized version of MEDI-559.

We also investigated the possibility of gaining increased genetic stability by deleting one or more codons in the L gene. We focused on the sites of the amino acid substitutions identified in the biological mutants already described: 248, 530, 955, 1009, and 1030. However, deletions at or near these sites could not be recovered. Surprisingly, we found that codon 1313, which was already mentioned as the site of a compensatory mutation, could be deleted. This yielded a virus that was substantially attenuated, was *ts*, replicated with WT-like efficiency at the permissive temperature of 32 °C, and was stable during in vitro stress tests (Luongo et al. 2013). A virus containing the codon 1313 deletion combined with deletion of the NS2 gene is currently being developed as a new RSV vaccine candidate (RSV ΔNS2 Δ1313 I1314L, see below).

3.5 Other Promising Vaccine Candidates

It may be possible to increase the immunogenicity or decrease the reactogenicity of RSV vaccine viruses by manipulation or deletion of RSV proteins involved in immune evasion (e.g., NS1 and NS2) or the regulation of RSV gene expression (e.g., M2-2). One strategy combined the NS2 deletion with various point mutations. The cp and ΔNS2 mutations were combined to create rA2cpΔNS2, which was evaluated in adults as a candidate vaccine to boost immunity in RSV-experienced individuals such as the elderly, and was also evaluated as a potential pediatric vaccine in RSV-seropositive children (Table 2). While this virus was over-attenuated in adults, it was insufficiently attenuated in RSV-seropositive children to permit evaluation in RSV-seronegative children; therefore, it would be unsatisfactory in either RSV-experienced or –naïve recipients and was not evaluated further (Wright et al. 2006).

To achieve a greater level of attenuation necessary for a pediatric vaccine, the ΔNS2 mutation was combined with 2 other sets of attenuating mutations, namely (i) the cp, 248, and 404 mutations, to yield rA2cp248/404ΔNS2 and (ii) the cp, 530, and 1009 mutations, to yield rA2cp530/1009ΔNS2 (Table 2). Each of these viruses was highly attenuated in adults and seropositive children, but was over-attenuated in seronegative children (Wright et al. 2006). More recently, the ΔNS2 mutation was combined with the newly identified Δ1313 codon deletion described above, yielding the RSVΔNS2Δ1313 virus (Luongo et al. 2013). Stress tests of this virus showed that it acquired a second-site mutation I1314T that partly abrogated the *ts* phenotype of the Δ1313 mutation (Luongo et al. 2013). However, when the native isoleucine codon was replaced with one for leucine, the resulting virus was stable during passage at 37 °C (Luongo et al. 2013). This virus replicated in seronegative chimpanzees to a level similar to that of MEDI-559 and RSV cps2, and thus it appears to be suitable for clinical evaluation as a candidate pediatric vaccine (Table 2).

A third type of RSV vaccine virus that is currently being tested clinically contains a deletion of most of the M2-2 ORF (Table 2). As noted, the ΔM2-2 mutant exhibits reduced RNA replication and increased gene transcription and antigen production, which may provide for increased immunogenicity despite highly restricted replication. This virus was highly attenuated in adults and seropositive children, and presently (January 2013) is being studied in seronegative children.

3.6 Summary of Current Live-Attenuated RSV Vaccine Strains

Four promising candidate viruses representing several different attenuation strategies are presently being evaluated clinically or being prepared for clinical evaluation. MEDI-559 (currently being evaluated in seronegative children, 2013) and RSV cps2 (being prepared for evaluation in seronegative children in 2013) are highly *ts* and are attenuated mainly by point mutations affecting RNA synthesis. The mechanisms of action of the various mutations that affect RNA synthesis are generally not known, although the 530 and 1009 mutations were studied in a mini-replicon system and appeared to affect both transcription and RNA replication, and the latter mutation appeared to increase the frequency of read through transcription (Juhasz et al. 1999). RSV MEDIΔM2-2 (currently being evaluated in seronegative children, 2013) is not *ts* and is attenuated by deletion of the regulatory M2-2 protein. RSVΔNS2Δ1313I1314L (currently being evaluated in seropositive children, 2013) is moderately *ts* and combines two highly stable deletion mutations involving the NS2 IFN antagonist protein and codon 1313 in the L polymerase protein.

3.7 Other Efforts to Develop a Recombinant Live-Attenuated RSV vaccine

Bovine RSV (BRSV) was considered as a potential vaccine against human RSV because the two viruses are related antigenically, and BRSV was thought to be attenuated in primates. However, BRSV was over-attenuated and poorly immunogenic in seronegative chimpanzees (Buchholz et al. 2000). In an attempt to increase its levels of replication and immunogenicity, a recombinant version of BRSV was modified in which the G and F genes were replaced by their human RSV counterparts. However, this chimeric virus remained over-attenuated and insufficiently immunogenic in chimpanzees (Buchholz et al. 2000). Various other combinations of bovine and human RSV genes also have been evaluated (our unpublished data). A chimeric virus consisting of the human RSV backbone in which the P gene was swapped with that of BRSV yielded a promising level of attenuation in chimpanzees (unpublished data), but this virus has yet to be further evaluated.

The RSV L protein was subjected to scanning mutagenesis in which charged residues were changed to alanine residues (Tang et al. 2002). However, this did not yield particularly promising mutations, e.g., mutations that did not restrict virus replication under permissive conditions in vitro (necessary for vaccine manufacture) but were substantially attenuating in vivo.

It may be desirable to include a subgroup B (RSV-B) component in an RSV vaccine, since cross-protection between the two antigenic subgroups is incomplete. One strategy is to express the RSV-B G protein from an added gene in the subgroup A backbone, making a virus that is bivalent with respect to G, the more divergent of the two major RSV protective antigens F and G (Jin et al. 1998). A second strategy is to replace the F and G genes of recombinant RSV-A with their counterparts from RSV-B (Whitehead et al. 1999). Since F and G are the major protective antigens, this produces an AB chimeric virus that is predominantly a vaccine against RSV-B and would need to be paired with a separate RSV-A vaccine. These two approaches have the advantage of being able to directly use attenuated RSV-A backbones in developing an RSV-B vaccine: thus, when a suitably attenuated RSV-A backbone is identified, the glycoprotein addition or glycoprotein swap would expeditiously provide an RSV-B component. As a third approach, a reverse genetics system was developed for RSV-B that can be used to prepare attenuated RSV-B strains, such as by incorporating attenuated mutations identified for RSV-A. This approach has the advantage of providing the complete set of RSV-B antigens, rather than only G and F.

Continuing pre-clinical work on vaccine development should also consider additional objectives, such as improved growth in cell culture, improved immunogenicity, and reduced reactogenicity. A modest increase in growth in cell culture (accompanied by modestly decreased, rather than increased, replication in vivo) was observed with deletion of the SH gene. It may be that further reduction in genome size, such as by the deletion of non-essential sequence, would increase growth in cell culture, improving manufacture. We have already noted that deletion of NS1, NS2, or M2-2 may provide increased immunogenicity. As another strategy, the F and G genes were placed in promoter-proximal positions, which increased the level of expression of these protective antigens due to the transcriptional gradient (Krempl et al. 2002). This yielded a modest increase in immunogenicity in mice (Krempl et al. 2002). The level of expression of the F and G proteins also can be increased using versions of their genes that had been resynthesized using codons optimal for translation, although our initial efforts have yielded only modest increases in expression (unpublished data). It also may be desirable to ablate the fractalkine motif in the G protein (Harcourt et al. 2006), or to eliminate expression of the secreted form of G (Bukreyev et al. 2008), since both of these features may contribute to reactogenicity and may interfere with the host immune response.

4 Parainfluenza Viruses as RSV Vaccine Vectors

Recombinant PIVs are being evaluated as vaccine vectors for expressing the RSV F and G glycoproteins, the two RSV neutralization antigens and major protective antigens. Using reverse genetics, the ORF for RSV F or G is placed under the control of a set of PIV gene-start and gene-end transcription signals and is inserted as an extra gene into a suitably attenuated PIV genome and recovered as replication-competent virus. Priority is given to the RSV F protein because it is much more highly conserved among RSV strains than is the G protein, and because it was the more immunogenic and protective of the two antigens in studies in rodents (Connors et al. 1991). The clinical success of the RSV-F-specific monoclonal antibody palivizumab also supports the importance of the F protein as antigen.

The vectors that are being evaluated are HPIV1, HPIV2, HPIV3, and Sendai virus (SeV), the last being a murine relative of HPIV1 that shares considerable antigenic relatedness. HPIV1, 2, and 3 are significant pediatric pathogens, and thus the use of these viruses or related viruses as vectors creates a single vaccine virus capable of inducing immunity against two human pathogens: the PIV vector and RSV. HPIV3 is particularly appropriate as a vector for RSV immunization because both RSV and HPIV3 frequently infect young infants, and immunization against the two viruses should begin early in life. Also, the burden of disease associated with HPIV3 is substantially greater than with HPIV1 or 2, which more frequently cause illness in the preschool years. A bivalent vaccine against RSV and HPIV3 could provide greater protection against viral LRI than a monovalent vaccine against RSV, and a bivalent vaccine based on a single virus can be developed more rapidly than a vaccine consisting of a mixture of two viruses.

A PIV-vectored RSV vaccine has a number of advantages for development, manufacture, and use compared to a live-attenuated RSV strain, specifically: (i) the PIVs replicate more efficiently in cell culture compared to RSV; (ii) RSV grown in cell culture mostly consists of large filamentous particles that are difficult to process, compared to the smaller, spherical particles of the PIVs; and (iii) the PIVs have substantially greater physical stability than RSV. The greater stability, in particular, would facilitate distribution and use in resource-limited settings. Also, it has been speculated that RSV may substantially suppress or skew the host immune response, which provides another possible advantage of PIV vectors. Another potential advantage is that, since the RSV antigens are not involved in replication of the PIV vectors, versions of F or G can be expressed that have been engineered for improved immunogenicity but may have lost functionality, such as versions of RSV-F that have been stabilized in "post-fusion" or "pre-fusion" conformations (Magro et al. 2012; McLellan et al. 2011). Conversely, PIV vectors expressing RSV F and/or G protein have the disadvantage that only part of the RSV proteome is represented: while the presence of F and G provides the viral neutralization antigens, the absence of the other proteins would mean the absence of a substantial portion of the epitopes for CD4+ and CD8+ T cells.

Initial studies with HPIV3 showed that the vector could accommodate three foreign genes totaling 7.5 kb (a nearly 50 % increase in genome length) with only a minimal reduction in replication in cell culture (Skiadopoulos et al. 2000, 2002). In hamsters, the level of attenuation was positively correlated with the aggregate size of the foreign insert. Thus, the RSV insert confers attenuation in vivo, which must be taken into account when introducing attenuating mutations into the vector backbone. Furthermore, the level of attenuation conferred by the insert appeared to increase when in the presence of other attenuating mutations. The level of expression of a foreign insert was affected by its position in the genome: promoter-proximal inserts were expressed with greater efficiency, reflecting the viral transcriptional gradient. Initial studies indicated that the foreign genes appeared to be surprisingly stable: deletions were not observed, although chance point mutations were sometimes detected.

The first PIV vector to be developed for clinical evaluation was based on a chimeric virus called B/HPIV3 (Fig. 1, lower panel) that had been developed as an HPIV3 vaccine (Schmidt et al. 2002). This virus is based on BPIV3, which is attenuated in primates due to a natural host range restriction. In B/HPIV3, the BPIV3 F and HN genes were replaced by their counterparts in HPIV3. Thus, B/HPIV3 combines the host range restriction of BPIV3 with the viral neutralization and major protective antigens of HPIV3. The determinants of the host range restriction of BPIV3 have been shown to be polygenic, and presumably involve the cumulative effect of multiple sequence differences. This likely accounts for the observed stability of the attenuation phenotype. An initial construct expressing the RSV G and F proteins together from promoter-proximal inserts appeared to be promising based on studies in rodents and African green monkeys (Schmidt et al. 2002). Subsequently, single-insert constructs were evaluated to reduce the incidence of chance mutations incorporated in the RSV genes during replication. Chance mutations can accumulate in the added RSV genes because these are not involved in replication of the PIV vector, and thus there is no selective pressure against mutations. For example, the construct MEDI-534 (Fig. 1) was made in which the B/HPIV3 vector expresses the RSV F protein from an insert added

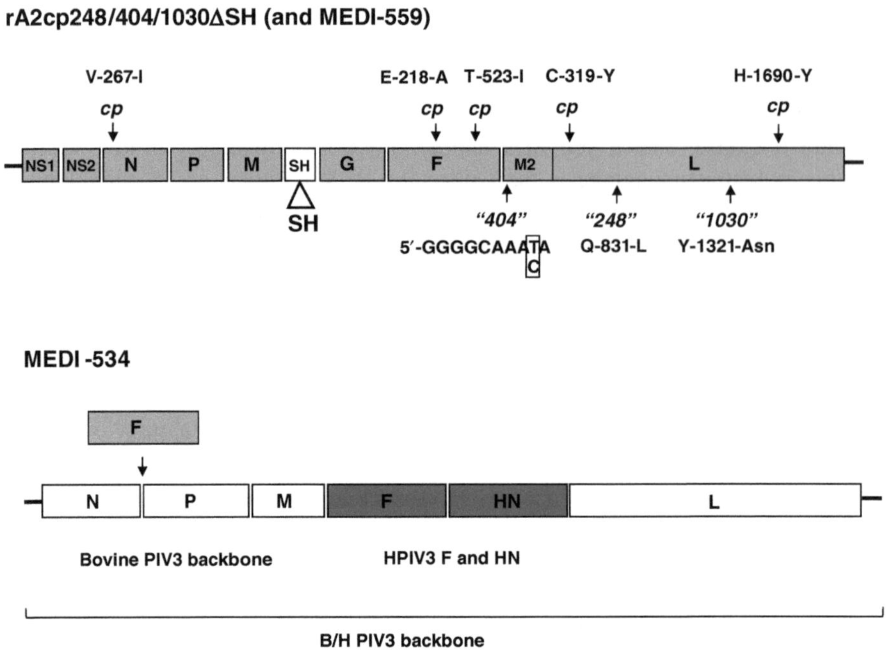

Fig. 1 Two types of live RSV candidate vaccines made by reverse genetics that have been evaluated clinically. The rA2cp248/404/1030ΔSH virus (*upper panel*) represents the strategy of developing an attenuated derivative of WT RSV. This mutant contains five cold-passaged (cp) mutations in the N, F, and L proteins, two amino acid substitutions (248 and 1030) in L, and one nucleotide substitution in the M2 GS signal (404) that independently confer temperature-sensitivity, and deletion of the SH gene. This virus was previously shown to be well-tolerated and immunogenic in infants and young children (Karron et al. 2005). MEDI-559 is another version of this virus that differs by a number of silent mutations and is phenotypically indistinguishable from rA2cp248/404/1030ΔSH. As noted in the text, MEDI-559 is currently being evaluated in a phase I-II clinical trial. The MEDI-534 virus (*lower panel*) represents the strategy of using a parainfluenza virus vector to express RSV antigen, in this case the RSV F protein, to provide a bivalent vaccine against RSV and HPIV3. MEDI-534 is a chimeric virus in which the F and HN genes from BPIV3 have been replaced by their counterparts from HPIV3 and the coding sequence for the RSV F gene was placed under the control of HPIV3 transcription signals and inserted as an added gene between the N and P genes. This virus also has been evaluated in clinical studies (Bernstein et al. 2012; Gomez et al. 2009; Tang RS 2012; Tang et al. 2008)

between the N and P genes (Gomez et al. 2009; Tang et al. 2008). In seropositive children, this virus was well-tolerated and shedding was not detected (Gomez et al. 2009). In children seronegative to both RSV and HPIV3, MEDI-534 replicated in the majority of recipients, was well-tolerated, and induced a serum antibody response to RSV and HPIV3 in 50–55 % and 100 % of recipients, respectively (Bernstein et al. 2012). The percentage of recipients with a serum antibody response to RSV was substantially less than in previous studies in similar populations with live-attenuated RSV strains. Recent evaluation of MEDI-534 virus shed by seronegative children indicated that some isolates had genetic changes in the noncoding sequences of RSV F and/or in the RSV F ORF, which could lead to decreased expression of RSV F (Tang RS et al. 2012). Additional work is needed to improve the genetic stability and immunogenicity of this vaccine candidate.

HPIV1 and HPIV2 also are being developed as vectors for RSV F. The burden of RSV disease in children aged between 1 and 5 years highlights the need for re-immunization after the first year of life (Hall et al. 2009), and an HPIV1- or HPIV2-vectored RSV vaccine might be used to boost immunity in children previously immunized with an RSV vaccine. The use of an HPIV1- or HPIV2-vectored vaccine may be preferable to re-immunization with a live-attenuated RSV strain, because RSV-specific immunity from prior immunization would restrict replication of a live-attenuated RSV strain, but not that of an HPIV1 or HPIV2 vector. This has been confirmed in studies in rodents, in which secondary immunization of RSV-immune animals with an HPIV1-RSV-F virus was more immunogenic than re-immunization with a live-attenuated RSV strain (unpublished data). In the case of HPIV1 and HPIV2 vectors, it will be necessary to develop suitably attenuated backbones. A number of attenuating mutations for HPIV1 and HPIV2 have been characterized (Bartlett et al. 2005; Nolan et al. 2007) and attenuated versions of HPIV1 and HPIV2 are presently in clinical studies in seropositive and seronegative children (ClinicalTrials.gov identifiers NCT00641017 and NCT01139437, respectively).

Sendai virus (SeV), the murine relative of HPIV1, also is being evaluated as a vaccine vector for human use. Immunization of experimental animals with SeV was protective against challenge with HPIV1, reflecting the antigenic relatedness of the two viruses. SeV may be attenuated in humans due to a host range restriction, although this was not evident in studies in non-human primates (Skiadopoulos et al. 2002). Shedding of SeV was not detected in HPIV1-seropositive human adults, which is consistent with attenuation (Slobod et al. 2004). SeV is presently being evaluated for safety and immunogenicity in children and toddlers >12 months of age (Clinicaltrials.gov identifier NCT00186927). In preclinical studies, SeV has been engineered to express a number of glycoproteins from other paramyxoviruses including the RSV F and G proteins. SeV expressing the RSV F or G protein was immunogenic and protective against RSV challenge in cotton rats, and SeV expressing RSV F was immunogenic and protective in African green monkeys (Jones et al. 2012). It is anticipated that an SeV-RSV F construct will be evaluated in clinical trials once the safety of the parent SeV in young children has been established.

5 Lessons Learned from the Evaluation of Live-Attenuated RSV Vaccines

As biologically derived and recombinant live-attenuated RSV vaccines have been evaluated over the past two decades, certain principles have emerged which can be used to guide future development and evaluation of live-attenuated vaccine candidates:

1. *Substantial replication in adults or RSV-seropositive children is a marker of insufficient attenuation for RSV-seronegative children and RSV-naïve infants.* Initial evaluation of a 10^5 $TCID_{50}$ dose of *cpts*248/955 in RSV-seropositive children demonstrated that this virus replicated in 38 % of vaccinees for a mean duration of 5 days and at a mean peak titer of $10^{2.7}$ $TCID_{50}$/mL. This vaccine appeared to be well-tolerated. However, in RSV-seronegative children, *cpts*248/955 was minimally restricted in replication (mean peak titer, $10^{4.4}$ $TCID_{50}$/mL; mean duration of replication, 9 days), and was associated with fever or lower respiratory tract illness (Table 1; Karron et al. 1997). Thus, prolonged, substantial replication of a live-attenuated RSV vaccine candidate in RSV-experienced populations, even if well-tolerated, is an indication that the vaccine may be insufficiently attenuated for RSV-naïve populations. In contrast, vaccine candidates with a greater level of attenuation, such as *cpts*248/404 and *cpts*248/404/1030/ΔSH, could only be recovered sporadically or not at all from nasal washes of RSV-seropositive children, and were infectious, well-tolerated, and immunogenic in seronegative children (Table 2; Karron et al. 2005; Wright et al. 2000). Thus, new vaccine candidates are typically evaluated sequentially in adults and RSV-seropositive children, and minimal or no virus shedding in these RSV-experienced groups is a necessary signal that the vaccine can proceed to evaluation in seronegative children. A corollary of this principle is that a live-attenuated RSV vaccine that is satisfactorily attenuated for RSV-naïve children will be over-attenuated for adults and older children, underscoring the need for alternative vaccine strategies for these populations.
2. *In children over 6 months of age, a serum RSV neutralizing antibody titer of ≤1:40, as measured in a classic complement-enhanced 60 % plaque reduction neutralization assay, is a useful determinant of susceptibility to infection with highly attenuated live RSV vaccine viruses.* While this titer was initially chosen empirically, based upon the range of antibody titers measured in young children, it has proved to be a reliable predictor of infectivity with highly attenuated RSV vaccine candidates such as *cpts*248/404 (Wright et al. 2000), rA2cp28/404/ΔSH, rA2cp28/404/1030/ΔSH (Karron et al. 2005; Wright et al. 2000) and rRSV MEDI ΔM2-2 (unpublished observations). We therefore define RSV "seropositive" as children >6 months of age with neutralizing antibody titers >1:40, and "seronegative" as those with neutralizing antibody titers ≤1:40. If other assays that measure RSV neutralizing activity are used in future clinical trials, cut-off values will need to be established.

Table 3 Replication of live-attenuated RSV vaccines is comparable in infants and young children

Vaccine	Dose (log_{10} PFU/mL)	Population evaluated	No. of subjects	Mean peak titer shed (log_{10} PFU/mL)
*cpts*248/404[1]	4.0	Seronegative children 6–24 months old	11	2.4
	4.0	Infants 1–2 months old	7	4.2
	5.0	Seronegative children 6–24 months old	38	4.2
	5.0	Infants 1–2 months old	17	4.0
rA2cp248/404/1030ΔSH[2]	4.3	Seronegative children 6–24 months old	13	2.6
	4.3	Infants 1–2 months old	16	2.4
	5.3	Seronegative children 6–24 months old	8	2.5
	5.3	Infants 1–2 months old	16	3.5

1. Ref. (Wright et al. 2000)
2. Ref. (Karron et al. 2005)

3. *Maternally derived RSV neutralizing antibody does not interfere with replication of live-attenuated RSV vaccines in the upper respiratory tract of young infants.* *cpts*248/404 and 248/404/1030/ΔSH have both been evaluated in 1–2 month-old infants, who were presumed to be RSV-naïve but have residual maternally derived antibody. For each virus, comparison of peak viral titers in nasal wash specimens demonstrated levels of vaccine virus replication in young infants that were equal to or greater than those observed in RSV-seronegative 6–24 month-old children, indicating a lack of restriction of replication by maternal antibody (Table 3; Karron et al. 2005; Wright et al. 2000). This finding is particularly important as vaccination strategies that incorporate both maternal and infant immunization are contemplated.
4. *Restriction of a second dose 'challenge' with intranasally administered vaccine virus demonstrates that young infants can develop protective immunity even in the absence of a measurable antibody response.* As noted above, intranasally delivered respiratory viruses replicate in the upper respiratory tracts of young infants even in the presence of maternal antibody. However, passive antibody has been shown to suppress humoral responses in these young infants (Murphy et al. 1986), which may occur through epitope masking or other immunologic mechanisms. Nonetheless, immunity capable of restricting viral replication can be induced in these situations, and this is best measured by administering a second 'challenge' dose of the vaccine virus. As an example, replication of the second dose of the investigational vaccine virus rA2cp28/404/1030/ΔSH was restricted approximately 100-fold compared to replication of the first dose in 1–3 month-old infants (Karron et al. 2005). This finding underscores the utility of a 'challenge' dose for assessment of induction of immunity, but also highlights the need for development of additional assays for measurement of immune responses in young infants.

6 Conclusions: The Future of Live-Attenuated RSV Vaccine Development

The live-attenuated approach continues to offer a number of advantages for development of an RSV vaccine for RSV-naive infants and young children. The most important advantage is the lack of disease enhancement associated with live RSV vaccines. Also, as previously noted, live-attenuated respiratory virus vaccines generally induce broad and durable immunity in infants and young children. In addition, the administration of a live-attenuated RSV vaccine should prime a young infant for development of a protective and non-pathologic immune response, and therefore could be safely followed by administration of a non-replicating RSV vaccine. The availability of reverse genetics and a panel of characterized attenuating mutations has led to the systematic development of a number of live-attenuated RSV vaccine candidates, including rA2cp248/404/1030ΔSH, the first vaccine candidate that appeared to be well-tolerated and moderately immunogenic in phase I trials in infants as young as 1 month of age (Karron et al. 2005).

One potential limitation of live-attenuated RSV vaccines is that the level of viral replication is generally linked to the magnitude of the antibody response, and it might therefore be difficult to achieve a balance between the level of viral replication required for safety and the level required for induction of protective immunity. However, improved understanding of RSV gene function has allowed for the development of new recombinant vaccine candidates that may de-link replication and immune response, either by enhancing antigen production while limiting viral replication (ΔM2-2 deletion), or by bolstering the innate immune response (ΔNS2 deletion). Clinical evaluation of these promising vaccine candidates is underway.

Acknowledgments PLC and UJB were supported by the Intramural Research Program of NIAID, NIH. RAK was supported by NIAID contract HHSN272200900010C.

References

Ambrose CS, Wu X, Belshe RB (2010) The efficacy of live attenuated and inactivated influenza vaccines in children as a function of time postvaccination. Pediatr Infect Dis J 29:806–811

Ambrose CS, Levin MJ, Belshe RB (2011) The relative efficacy of trivalent live attenuated and inactivated influenza vaccines in children and adults. Influenza Respir Viruses 5:67–75

Bartlett EJ, Amaro-Carambot E, Surman SR, Newman JT, Collins PL, Murphy BR, Skiadopoulos MH (2005) Human parainfluenza virus type I (HPIV1) vaccine candidates designed by reverse genetics are attenuated and efficacious in African green monkeys. Vaccine 23:4631–4646

Belshe RB, Edwards KM, Vesikari T, Black SV, Walker RE, Hultquist M, Kemble G, Connor EM (2007) Live attenuated versus inactivated influenza vaccine in infants and young children. N Engl J Med 356:685–696

Bermingham A, Collins PL (1999) The M2–2 protein of human respiratory syncytial virus is a regulatory factor involved in the balance between RNA replication and transcription. Proc Natl Acad Sci U S A 96:11259–11264

Bernstein DI, Malkin E, Abughali N, Falloon J, Yi T, Dubovsky F (2012) Phase 1 study of the safety and immunogenicity of a live, attenuated respiratory syncytial virus and parainfluenza virus type 3 vaccine in seronegative children. Pediatr Infect Dis J 31:109–114

Bitko V, Shulyayeva O, Mazumder B, Musiyenko A, Ramaswamy M, Look DC, Barik S (2007) Nonstructural proteins of respiratory syncytial virus suppress premature apoptosis by an NF-kappaB-dependent, interferon-independent mechanism and facilitate virus growth. J Virol 81:1786–1795

Buchholz UJ, Granzow H, Schuldt K, Whitehead SS, Murphy BR, Collins PL (2000) Chimeric bovine respiratory syncytial virus with glycoprotein gene substitutions from human respiratory syncytial virus (HRSV): Effects on host range and evaluation as a live-attenuated HRSV vaccine. J Virol 74:1187–1199

Bukreyev A, Yang L, Fricke J, Cheng L, Ward JM, Murphy BR, Collins PL (2008) The secreted form of respiratory syncytial virus G glycoprotein helps the virus evade antibody-mediated restriction of replication by acting as an antigen decoy and through effects on Fc receptor-bearing leukocytes. J Virol 82:12191–12204

Collins PL, Murphy BR (2005) New generation live vaccines against human respiratory syncytial virus designed by reverse genetics. Proc Am Thorac Soc 2:166–173

Connors M, Collins PL, Firestone CY, Murphy BR (1991) Respiratory syncytial virus (RSV) F, G, M2 (22 K), and N proteins each induce resistance to RSV challenge, but resistance induced by M2 and N proteins is relatively short-lived. J Virol 65:1634–1637

Crowe JE Jr, Bui PT, London WT, Davis AR, Hung PP, Chanock RM, Murphy BR (1994) Satisfactorily attenuated and protective mutants derived from a partially attenuated cold-passaged respiratory syncytial virus mutant by introduction of additional attenuating mutations during chemical mutagenesis. Vaccine 12:691–699

Crowe JE Jr, Bui PT, Siber GR, Elkins WR, Chanock RM, Murphy BR (1995) Cold-passaged, temperature-sensitive mutants of human respiratory syncytial virus (RSV) are highly attenuated, immunogenic, and protective in seronegative chimpanzees, even when RSV antibodies are infused shortly before immunization. Vaccine 13:847–855

Firestone CY, Whitehead SS, Collins PL, Murphy BR, Crowe JE Jr (1996) Nucleotide sequence analysis of the respiratory syncytial virus subgroup A cold-passaged (cp) temperature sensitive (ts) cpts-248/404 live attenuated virus vaccine candidate. Virology 225:419–422

Friedewald WT, Forsyth BR, Smith CB, Gharpure MA, Chanock RM (1968) Low-temperature-grown RS virus in adult volunteers. JAMA 203:690–694

Fuentes S, Tran KC, Luthra P, Teng MN, He B (2007) Function of the respiratory syncytial virus small hydrophobic protein. J Virol 81:8361–8366

Gharpure MA, Wright PF, Chanock RM (1969) Temperature-sensitive mutants of respiratory syncytial virus. J Virol 3:414–421

Gomez M, Mufson MA, Dubovsky F, Knightly C, Zeng W, Losonsky G (2009) Phase-I study MEDI-534, of a live, attenuated intranasal vaccine against respiratory syncytial virus and parainfluenza-3 virus in seropositive children. Pediatr Infect Dis J 28:655–658

Gonzalez IM, Karron RA, Eichelberger M, Walsh EE, Delagarza VW, Bennett R, Chanock RM, Murphy BR, Clements-Mann ML, Falsey AR (2000) Evaluation of the live attenuated cpts 248/404 RSV vaccine in combination with a subunit RSV vaccine (PFP-2) in healthy young and older adults. Vaccine 18:1763–1772

Hall CB, Weinberg GA, Iwane MK, Blumkin AK, Edwards KM, Staat MA, Auinger P, Griffin MR, Poehling KA, Erdman D, Grijalva CG, Zhu Y, Szilagyi P (2009) The burden of respiratory syncytial virus infection in young children. N Engl J Med 360:588–598

Harcourt J, Alvarez R, Jones LP, Henderson C, Anderson LJ, Tripp RA (2006) Respiratory syncytial virus G protein and G protein CX3C motif adversely affect CX3CR1+ T cell responses. J Immunol 176:1600–1608

Hoft DF, Babusis E, Worku S, Spencer CT, Lottenbach K, Truscott SM, Abate G, Sakala IG, Edwards KM, Creech CB, Gerber MA, Bernstein DI, Newman F, Graham I, Anderson EL, Belshe RB (2011) Live and inactivated influenza vaccines induce similar humoral responses, but only live vaccines induce diverse T-cell responses in young children. J Infect Dis 204:845–853

Jin H, Clarke D, Zhou HZ, Cheng X, Coelingh K, Bryant M, Li S (1998) Recombinant human respiratory syncytial virus (RSV) from cDNA and construction of subgroup A and B chimeric RSV. Virology 251:206–214

Jones BG, Sealy RE, Rudraraju R, Traina-Dorge VL, Finneyfrock B, Cook A, Takimoto T, Portner A, Hurwitz JL (2012) Sendai virus-based RSV vaccine protects African green monkeys from RSV infection. Vaccine 30:959–968

Juhasz K, Whitehead SS, Boulanger CA, Firestone CY, Collins PL, Murphy BR (1999) The two amino acid substitutions in the L protein of cpts530/1009, a live-attenuated respiratory syncytial virus candidate vaccine, are independent temperature-sensitive and attenuation mutations. Vaccine 17:1416–1424

Karron RA, Buonagurio DA, Georgiu AF, Whitehead SS, Adamus JE, Clements-Mann ML, Harris DO, Randolph VB, Udem SA, Murphy BR, Sidhu MS (1997a) Respiratory syncytial virus (RSV) SH and G proteins are not essential for viral replication in vitro: clinical evaluation and molecular characterization of a cold-passaged, attenuated RSV subgroup B mutant. Proc Natl Acad Sci U S A 94:13961–13966

Karron RA, Wright PF, Crowe JE Jr, Clements ML, Thompson J, Makhene M, Casey R, Murphy BR (1997b) Evaluation of two live, cold-passaged, temperature-sensitive respiratory syncytial virus (RSV) vaccines in chimpanzees, adults, infants and children. J Infect Dis 176:1428–1436

Karron RA, Wright PF, Belshe RB, Thumar B, Casey R, Newman F, Polack FP, Randolph VB, Deatly A, Hackell J, Gruber W, Murphy BR, Collins PL (2005) Identification of a recombinant live attenuated respiratory syncytial virus vaccine candidate that is highly attenuated in infants. J Infect Dis 191:1093–1104

Kim HW, Arrobio JO, Pyles G, Brandt CD, Camargo E, Chanock RM, Parrott RH (1971) Clinical and immunological response of infants and children to administration of low-temperature adapted respiratory syncytial virus. Pediatrics 48:745–755

Kim HW, Arrobio JO, Brandt CD, Wright P, Hodes D, Chanock RM, Parrott RH (1973) Safety and antigenicity of temperature sensitive (TS) mutant respiratory syncytial virus (RSV) in infants and children. Pediatrics 52:56–63

Krempl C, Murphy BR, Collins PL (2002) Recombinant respiratory syncytial virus with the G and F genes shifted to the promoter-proximal positions. J Virol 76:11931–11942

Luongo C, Yang L, Winter CC, Spann KM, Murphy BR, Collins PL, Buchholz UJ (2009) Codon stabilization analysis of the "248" temperature sensitive mutation for increased phenotypic stability of respiratory syncytial virus vaccine candidates. Vaccine 27:5667–5676

Luongo C, Winter CC, Collins PL, Buchholz UJ (2012) Increased genetic and phenotypic stability of a promising live-attenuated respiratory syncytial virus vaccine candidate by reverse genetics. J Virol 86:10792–10804

Luongo C, Winter CC, Collins PL, Buchholz UJ (2013) Respiratory syncytial virus modified by deletions of the NS2 gene and amino acid S1313 of the L polymerase protein is a temperature-sensitive, live-attenuated vaccine candidate that is phenotypically stable a physiologic temperature. J Virol 87: 1985–1996

Magro M, Mas V, Chappell K, Vazquez M, Cano O, Luque D, Terron MC, Melero JA, Palomo C (2012) Neutralizing antibodies against the preactive form of respiratory syncytial virus fusion protein offer unique possibilities for clinical intervention. Proc Natl Acad Sci USA 109:3089–3094

McKay E, Higgins P, Tyrrell D, Pringle C (1988) Immunogenicity and pathogenicity of temperature-sensitive modified respiratory syncytial virus in adult volunteers. J Med Virol 25:411–421

McLellan JS, Yang Y, Graham BS, Kwong PD (2011) Structure of the respiratory syncytial virus fusion glycoprotein in the post-fusion conformation reveals preservation of neutralizing epitopes. J Virol 85:7788–7796

Murphy BR, Collins PL (2002) Live-attenuated virus vaccines for respiratory syncytial and parainfluenza viruses: applications of reverse genetics. J Clin Invest 110:21–27

Murphy BR, Graham BS, Prince GA, Walsh EE, Chanock RM, Karzon DT, Wright PF (1986) Serum and nasal-wash immunoglobulin G and A antibody response of infants and children to respiratory syncytial virus F and G glycoproteins following primary infection. J Clin Microbiol 23:1009–1014

Nolan SM, Skiadopoulos MH, Bradley K, Kim OS, Bier S, Amaro-Carambot E, Surman SR, Davis S, St Claire M, Elkins R, Collins PL, Murphy BR, Schaap-Nutt A (2007) Recombinant human parainfluenza virus type 2 vaccine candidates containing a 3′ genomic promoter mutation and L polymerase mutations are attenuated and protective in non-human primates. Vaccine 25:6409–6422

Pringle CR, Filipiuk AH, Robinson BS, Watt PJ, Higgins P, Tyrrell DA (1993) Immunogenicity and pathogenicity of a triple temperature-sensitive modified respiratory syncytial virus in adult volunteers. Vaccine 11:473–478

Schmidt AC, Wenzke DR, McAuliffe JM, St. Clair M, Elkins WR, Murphy BR, Collins PL (2002) Mucosal immunization of rhesus monkeys against respiratory syncytial virus subgroups A and B and human parainfluenza virus type 3 using a live cDNA-derived vaccine based on a host range-attenuated bovine parainfluenza virus type 3 vector backbone. J Virol 76:1089–1099

Skiadopoulos MH, Surman SR, Durbin AP, Collins PL, Murphy BR (2000) Long nucleotide insertions between the HN and L protein coding regions of human parainfluenza virus type 3 yield viruses with temperature-sensitive and attenuation phenotypes. Virology 272:225–234

Skiadopoulos MH, Surman SR, Riggs JM, Elkins WR, St Claire M, Nishio M, Garcin D, Kolakofsky D, Collins PL, Murphy BR (2002a) Sendai virus, a murine parainfluenza virus type 1, replicates to a level similar to human PIV1 in the upper and lower respiratory tract of African green monkeys and chimpanzees. Virology 297:153–160

Skiadopoulos MH, Surman SR, Riggs JM, Orvell C, Collins PL, Murphy BR (2002b) Evaluation of the replication and immunogenicity of recombinant human parainfluenza virus type 3 vectors expressing up to three foreign glycoproteins. Virology 297:136–152

Slobod KS, Shenep JL, Lujan-Zilbermann J, Allison K, Brown B, Scroggs RA, Portner A, Coleclough C, Hurwitz JL (2004) Safety and immunogenicity of intranasal murine parainfluenza virus type 1 (Sendai virus) in healthy human adults. Vaccine 22:3182–3186

Surman SR, Collins PL, Murphy BR, Skiadopoulos MH (2007) An improved method for the recovery of recombinant paramyxovirus vaccine candidates suitable for use in human clinical trials. J Virol Methods 141:30–33

Tang RS, M. E, Stillman E, Nelson C, Yang C, Song E, Liang B, Shambaugh C, Zuo F, Liem A, Balagot C, Wang K, Galinski M, Esser M (2012) Implication of genetic changes observed in Phase I evaluation of MEDI-534, a live attenuated chimeric bovine human parainfluenza type 3 vectored RSV vaccine. Poster #38, RSV 2012, Santa FE, MN

Tang RS, Nguyen N, Zhou H, Jin H (2002) Clustered charge-to-alanine mutagenesis of human respiratory syncytial virus L polymerase generates temperature-sensitive viruses. Virology 302:207–216

Tang RS, Spaete RR, Thompson MW, MacPhail M, Guzzetta JM, Ryan PC, Reisinger K, Chandler P, Hilty M, Walker RE, Gomez MM, Losonsky GA (2008) Development of a PIV-vectored RSV vaccine: preclinical evaluation of safety, toxicity, and enhanced disease and initial clinical testing in healthy adults. Vaccine 26:6373–6382

Teng MN, Whitehead SS, Bermingham A, Clair MS, Elkins WR, Murphy BR, Collins PL (2000) Recombinant respiratory syncytial virus that does not express the NS1 or M2–2 protein is highly attenuated and immunogenic in chimpanzees. J Virol 74:9317–9321

Valarcher JF, Furze J, Wyld S, Cook R, Conzelmann KK, Taylor G (2003) Role of alpha/beta interferons in the attenuation and immunogenicity of recombinant bovine respiratory syncytial viruses lacking NS proteins. J Virol 77:8426–8439

Waris ME, Tsou C, Erdman DD, Day DB, Anderson LJ (1997) Priming with live respiratory syncytial virus (RSV) prevents the enhanced pulmonary inflammatory response seen after RSV challenge in BALB/c mice immunized with formalin-inactivated RSV. J Virol 71:6935–6939

Watt PJ, Robinson BS, Pringle CR, Tyrrell DA (1990) Determinants of susceptibility to challenge and the antibody response of adult volunteers given experimental respiratory syncytial virus vaccines. Vaccine 8:231–236

Whitehead SS, Firestone CY, Collins PL, Murphy BR (1998a) A single nucleotide substitution in the transcription start signal of the M2 gene of respiratory syncytial virus vaccine candidate cpts248/404 is the major determinant of the temperature-sensitive and attenuation phenotypes. Virology 247:232–239

Whitehead SS, Juhasz K, Firestone CY, Collins PL, Murphy BR (1998b) Recombinant respiratory syncytial virus (RSV) bearing a set of mutations from cold-passaged RSV is attenuated in chimpanzees. J Virol 72:4467–4471

Whitehead SS, Bukreyev A, Teng MN, Firestone CY, St Claire M, Elkins WR, Collins PL, Murphy BR (1999a) Recombinant respiratory syncytial virus bearing a deletion of either the NS2 or SH gene is attenuated in chimpanzees. J Virol 73:3438–3442

Whitehead SS, Firestone CY, Karron RA, Crowe JE Jr, Elkins WR, Collins PL, Murphy BR (1999b) Addition of a missense mutation present in the L gene of respiratory syncytial virus (RSV) cpts530/1030 to RSV vaccine candidate cpts248/404 increases its attenuation and temperature sensitivity. J Virol 73:871–877

Whitehead SS, Hill MG, Firestone CY, St Claire M, Elkins WR, Murphy BR, Collins PL (1999c) Replacement of the F and G proteins of respiratory syncytial virus (RSV) subgroup A with those of subgroup B generates chimeric live attenuated RSV subgroup B vaccine candidates. J Virol 73:9773–9780

Wright PF, Mills Jv, Chanock RM (1971) Evaluation of a temperature-sensitive mutant of respiratory syncytial virus in adults. J Infect Dis 124:505–511

Wright PF, Shinozaki T, Fleet W, Sell SH, Thompson J, Karzon DT (1976) Evaluation of a live, attenuated respiratory syncytial virus vaccine in infants. J Pediatr 88:931–936

Wright PF, Belshe RB, Kim HW, Van Voris LP, Chanock RM (1982) Administration of a highly attenuated, live respiratory syncytial virus vaccine to adults and children. Infect Immun 37:397–400

Wright PF, Karron RA, Belshe RB, Thompson J, Crowe JE Jr, Boyce TG, Halburnt LL, Reed GW, Whitehead SS, Anderson EL, Wittek AE, Casey R, Eichelberger M, Thumar B, Randolph VB, Udem SA, Chanock RM, Murphy BR (2000) Evaluation of a live, cold-passaged, temperature-sensitive, respiratory syncytial virus vaccine candidate in infancy. J Infect Dis 182:1331–1342

Wright PF, Karron RA, Madhi SA, Treanor JJ, King JC, O'Shea A, Ikizler MR, Zhu Y, Collins PL, Cutland C, Randolph VB, Deatly AM, Hackell JG, Gruber WC, Murphy BR (2006) The interferon antagonist NS2 protein of respiratory syncytial virus is an important virulence determinant for humans. J Infect Dis 193:573–581

Wright PF, Karron RA, Belshe RB, Shi JR, Randolph VB, Collins PL, O'Shea AF, Gruber WC, Murphy BR (2007) The absence of enhanced disease with wild type respiratory syncytial virus infection occurring after receipt of live, attenuated, respiratory syncytial virus vaccines. Vaccine 25:7372–7378

Zhang W, Choi Y, Haynes LM, Harcourt JL, Anderson LJ, Jones LP, Tripp RA (2010) Vaccination to induce antibodies blocking the CX3C-CX3CR1 interaction of respiratory syncytial virus G protein reduces pulmonary inflammation and virus replication in mice. J Virol 84:1148–1157

Subunit and Virus-Like Particle Vaccine Approaches for Respiratory Syncytial Virus

Trudy G. Morrison and Edward E. Walsh

Abstract Despite its impact on global health, there is no vaccine available for the prevention of respiratory syncytial virus (RSV) infection. Failure to develop a licensed vaccine is not due to lack of effort, as numerous vaccine candidates have been characterized in preclinical and clinical studies spanning five decades. The vaccine candidates thus far explored can be generally divided into four categories: (1) whole inactivated virus, (2) replication competent, attenuated virus including recombinant viruses, (3) gene-based vectors, and (4) subunit and particulate forms of RSV antigens. The first clinically tested RSV vaccine candidate was a formalin-inactivated purified virus preparation administered to infants and children in the late 1960s. Due to the disastrous outcome of these trials and results of animal models investigating the mechanisms involved, there have been no further studies with inactivated RSV vaccines. Rather, efforts have focused on development of other approaches. In this chapter, we review the history and status of purified proteins, peptides, virus-like particles, virosomes, and nanoparticles and discuss their future potential.

Contents

T. G. Morrison (✉)
Department of Microbiology and Physiological Systems, University of Massachusetts Medical School, Worcester, MA 01655, USA
e-mail: trudy.morrison@umassmed.edu

E. E. Walsh
Division of Infectious Diseases, University of Rochester School of Medicine and Dentistry, Rochester General Hospital, Rochester, NY 14621, USA
e-mail: Edward.Walsh@rochestergeneral.org

L. J. Anderson and B. S. Graham (eds.), *Challenges and Opportunities for Respiratory Syncytial Virus Vaccines*, Current Topics in Microbiology and Immunology 372,
DOI: 10.1007/978-3-642-38919-1_14,

1 Target Populations for RSV Vaccines

Human RSV infection affects different populations with variable severity, and each can be considered as a potential vaccine target group. RSV is the single most important cause of acute viral lower respiratory tract disease in infants and young children worldwide (Karron 2008), with an estimated 33.8 million RSV cases and 199,000 deaths worldwide in 2005 alone (Nair et al. 2010). The incidence and severity of RSV is greatest in the first 6 months of life, with hospitalization rates of 17/1000 infants in the USA. However, children 2–5 years of age, another target group, comprise a substantial proportion of RSV illness burden, and likely provide the reservoir for infection of newborns. The epidemiology and impact of RSV in the elderly, adults with underlying cardiopulmonary disease and hematologic stem cell and solid organ transplant recipients have also been described (Walsh et al. 1999). Estimates of RSV attributable deaths in adults in the USA range from 10,000 to 17,000 annually, with hospitalizations ranging from 14,000 to 177,000 (Falsey et al. 2005; Falsey and Walsh 2000; Han et al. 1999; Thompson et al. 2003; Zhou et al. 2012). Another vaccine target population is pregnant women because of the potential protective role of maternal antibodies for very young infants. Because each of these populations has notable differences in the properties and robustness of their immune responses and immunological history, it will be likely necessary to tailor RSV vaccines to each specific population.

2 Considerations for Types of RSV Vaccine Candidates

Live-attenuated virus vaccines are often considered, in general, the most effective due to their stimulation of broad innate and adaptive T and B cell immunity. However, they may cause disease in immunocompromised persons, an expanding population due to increased use of immune modulators for inflammatory diseases and cancer, the HIV pandemic, and organ transplantation. Recombinant live viruses, currently being developed for RSV (reviewed in the chapter by Karron 2008, this volume), likely similarly pose dangers to immunocompromised populations. Live virus vaccines in infants also raise concerns, particularly the ones that require intranasal inoculation such as those based on other respiratory viruses (reviewed in (Graham)). In addition, perhaps due to the immunological immaturity of young infants, some live-attenuated RSV vaccines may be incompletely attenuated in this population. Thus, infectious virus vaccine candidates, discussed in the chapter by Karron 2008 in this volume, may not always be appropriate for these populations.

In contrast, inactivated or protein subunit vaccines usually do not pose an immediate safety risk and are often the vaccine of choice for infants and immunocompromised populations. However, using classical methods for inactivated vaccine preparation, a formalin-inactivated preparation of purified virus (FI-RSV) not only failed to protect infants from infection, but also unexpectedly resulted in enhanced respiratory disease (ERD) upon subsequent infection with RSV (reviewed in Collins and Crowe 2007; Openshaw and Tregoning 2005). Eighty percent of infected infants vaccinated, when less than 6 months of age, were hospitalized and two died. In contrast, older RSV-experienced infants receiving this vaccine did not manifest ERD. The mechanisms responsible for this unusual response to a classical approach to vaccine preparation are not completely understood even after decades of research using animal models (see chapters by M.S. Boukhvalova and J.C.G. Blanco, P.J. Openshaw, R.J. Pickles and by G. Taylor, this volume). However, this experience has significantly affected the development of all inactivated or subunit RSV vaccine candidates and has heightened concerns about their safety, especially in sero-negative infants.

3 Challenges for Subunit and Particulate RSV Vaccine Development

Three interrelated problems have significantly and uniquely impeded RSV vaccine development, each of which must be overcome. First and foremost is safety, as discussed above. The ERD observed after FI-RSV vaccination was associated with poor neutralizing responses in infants, as well as in experimental animals, perhaps related, in part, to elimination of protective epitopes by formalin treatment (reviewed in Collins and Crowe 2007; Collins and Graham 2007). In addition, it has been reported that, in mice, FI-RSV immunization as well as immunization with purified F protein or UV-inactivated RSV, resulted in unbalanced Th2-biased cytokine responses and low affinity, poorly neutralizing antibodies (Delgado et al. 2009). Upon RSV challenge, immunized animals had weak CD8 + T cell responses, pronounced lung eosinophilia, and significant lung inflammation (reviewed in Collins and Crowe 2007; Collins and Graham 2007). It was proposed that inactivated viruses or purified proteins fail to adequately stimulate innate immune responses necessary for effective adaptive responses including affinity maturation necessary for effective neutralizing antibodies. This failure, for reasons that are still unclear, results in the immunopathology observed upon subsequent live virus infection (Delgado et al. 2009). In any event, the absence of the types of immune responses associated with enhanced disease is now considered a benchmark for development of a successful RSV vaccine.

A second critical issue is predicting and determining efficacy. Some vaccine candidates have proved to be highly protective in animal models, but failed to have equivalent protection in humans (reviewed in Power 2008). In infants, this problem may be related to immunological immaturity or the inhibitory effect of maternal antibodies. In other populations, the problem may be related to preexisting antibody or non-responsiveness of aging or a compromised immune system. It is also likely that there are undefined differences in stimulation of anti-RSV responses and in the character of these responses in animal model systems and in humans. The lack of an animal model that is directly translatable to human protection, coupled with the lack of clearly defined immune correlates of protection in different human populations, makes efficacy in human populations difficult to predict for new vaccine candidates.

A third but related problem is an understanding of requirements for long-lived and memory responses to RSV. One of the hallmarks of RSV is the observation that humans experience repeated infection caused by the same virus serogroup multiple times over several years or even within the same season (reviewed in Hall 2001; Power 2008). The reasons for the failure of RSV infection to protect against subsequent infection are not clear but the inadequate immune response to RSV natural infection illustrates a major problem to be overcome in RSV vaccine development. As noted by Pulendran and Ahmed (2011), an optimally effective RSV vaccine must stimulate immune responses that are more protective and more durable than those resulting from natural infection (Pulendran and Ahmed Pulendran and Ahmed 2011). Indeed, many RSV vaccine candidates have failed to stimulate long-term protective responses (discussed in Hall 2001; Power 2008), illustrating the lack of knowledge of immune mechanisms required to generate long-term, protective anti-RSV immune responses in humans.

Another issue that also should be addressed in any vaccine candidate is the existence of two major serotypes of RSV, designated group A and B, each of which has 5–6 genotypes (reviewed in Collins and Crowe 2007; Collins and Graham 2007; and chapters by C.B. Hall et al. and by E.H. Choi et al. this volume). The G proteins of different RSV strains demonstrate significant antigenic diversity while the F proteins are virtually identical. Thus, inclusion of F protein in any vaccine candidate is usually considered necessary in order to cover both serotypes.

4 Subunit and Particulate Vaccine Candidates Overview

The primary goal of immunization with vaccines is the stimulation of protective antibody responses. Indeed, the only currently effective prophylaxis for RSV disease is a humanized monoclonal antibody specific for RSV F protein (Cardenas et al. 2005). While too expensive for use in general populations, this reagent clearly demonstrates that serum antibodies specific to the RSV F protein, while not preventing infection, can reduce severity and underscores the importance of humoral immune responses to this virus and, in particular, the importance of antibodies to the F protein. The protective role of G protein antibodies is less clear although recent studies have suggested that antibodies specific to the conserved central domain of the G protein are protective in animal models and prevent ERD stimulated by FI-RSV in those systems (Simard et al. 1995; Zhang et al. 2010).

Protein subunit and particulate vaccine candidates for RSV fall into three categories, intact purified F protein or both F and G proteins, peptide fragments of G protein, and particles of various types containing F and/or G proteins (summarized in Table 1). All candidates have been characterized in preclinical studies in model systems, usually in mice and/or in cotton rats, and a few in primate models. Many of the earlier candidates have also been tested in phase I or II human

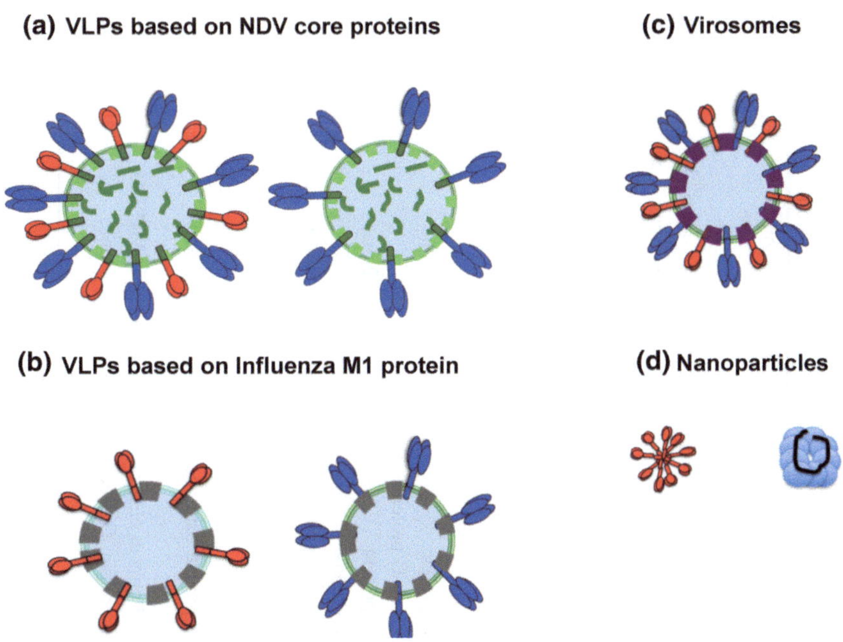

Fig. 1 Particulate RSV vaccine candidates figure illustrates four types of particulate RSV vaccine candidates described in detail. VLP, virus-like particles. *Red*, RSV F protein sequences; *blue*, RSV G protein sequences; *green*, NDV sequences; *purple*, RSV M sequences; *gray*, influenza M1 protein; *light blue*, RSV N protein; *black line*, *E. coli* RNA

Table 1 Summary of subunit and particle RSV vaccine candidates

	Antigen	Adjuvant	Preclinical trials	Clinical trials
PFP-1, PFP-2, PFP-3	F protein purified from infected VERO cells	Alum	Mice, cotton rats	Seropositive children, elderly, healthy adults pregnant women
F+G+M	purified virion proteins	+/−alum	Cotton rats	Elderly
G protein peptides		Various in animal models	Cotton rats, mice macaques	Healthy adults, elderly
BBG2Na	G protein sequences			
Other Peptides				
F-G fusion protein	Fused ectodomains of F and G proteins	Various in animal models	Cotton rats, mice	None reported
VLPs				
NDV based	G protein F and G protein	None	Mice	None
Influenza based	G protein F protein	None	Mice	None
Virosomes	F+G+M protein from virions	None MPL P3CSK4	Mice, cotton rats	None reported
Nanoparticles				
F protein rosettes	F protein from insect cells	Alum	Not reported	Yes
NP + M2	NP protein oligomers + CTL Epitope	LT9R192G	Mice	None reported
DNA	F protein, G protein, G peptide sequences	None	Mice, macaques	None reported

clinical trials although for various reasons have failed to move forward to large-scale phase III efficacy trials. These candidates will be described with an emphasis on the reasons for their failure to progress and their potential for future development (Fig. 1).

5 Subunit Vaccine Candidates

5.1 Purified Protein Vaccine Candidates

Following the failed FI-RSV trials, the next vaccine candidates to be developed were purified F protein (PFP) isolated from RSV infected VERO cells. Three versions, PFP-1, PFP-2, and PFP-3 (Table 1), were developed sequentially, each with increasing purity. PFP-1 was isolated by affinity chromatography using a monoclonal antibody and contained up to 10 % contaminating G protein. PFP-2 and 3 were purified by ion exchange chromatography and were 98–99 % pure. The F proteins were combined with either aluminum hydroxide (PFP-1, PFP-2) or aluminum phosphate (PFP-3) as adjuvant. Characterization of the conformation of the F protein in these preparations is not reported, thus it is unknown if the antigen is composed of pre-fusion or post fusion mature forms or unfolded, immature forms or a mixture of these forms. It is unknown if the protein was aggregated into rosettes as often the case in purified preparations of paramyxovirus F proteins (Calder et al. 2000).

Limited preclinical studies in animal models are described for these early candidates. Some reports suggest that they may prime for ERD upon subsequent RSV challenge (Delgado et al. 2009; Murphy et al. 1990) although this conclusion has been challenged (Hildreth et al. 1993). There have been, however, a number of clinical trials of these vaccine candidates in various human populations, primarily seropositive children and older adults. For example, placebo-controlled trials of PFP-1 in seropositive children (and two inadvertently included sero-negative children) (Belshe et al. 1993; Paradiso et al. 1994; Piedra et al. 1995; Tristram et al. 1993) and of PFP-2 in RSV experienced children with bronchopulmonary dysplasia (Groothuis et al. 1998) showed generally fourfold or greater increases in serum neutralizing antibody titers with few adverse reactions. Notably, PFP-1, but not PFP-2 or 3, also induced significant anti-G responses. Although there was a trend toward a reduction of RSV infection in vaccine recipients during the subsequent winter, the small study size precludes firm conclusions (Falsey and Walsh 1996). However, a meta-analysis of six double-blind randomized control trials with PFP was reported in 2002 (Simoes et al. 2001). The authors concluded that the vaccinees had a statistically significant reduction in the incidence of all RSV infections, but not in more serious lower respiratory tract infection. Despite these findings, the authors of this meta-analysis questioned the validity of this conclusion due to possible bias in reporting of negative studies. PFP-2 was also evaluated in pregnant women with the goal of enhancing transmission of RSV-specific IgG to their newborns (Munoz et al. 2003). Immunized mothers had only modest increases in anti-RSV serum titers, as did their newborns. Again, subject numbers were too small to assess efficacy. A much larger phase II clinical trial of PFP-3 in older RSV seropositive children with cystic fibrosis showed a robust increase in serum antibody titers and neutralizing antibody titers but no significant reduction in the incidence of RSV infection (Piedra et al. 2003). Finally, PFP-2 was shown to be immunogenic and safe in two phase I trials in healthy and frail elderly adults (Falsey and Walsh 1996, 1997). Importantly, none of the above studies found evidence of ERD although no formal studies in sero-negative children have been reported. Nevertheless, the lingering specter of the FI-RSV experience and the inconsistent results from animal models of ERD with these vaccine candidates have inhibited their study in newborn infants. The lack of an indication of efficacy of these vaccines in other human populations has also discouraged their further development.

Subsequently, candidates containing both F and G proteins as well as M protein have been characterized in two different studies (Falsey et al. 2008; Langley et al. 2009) of elderly populations. This protein mix was co-purified from RSV grown in VERO cells. It has since been reported by others that RSV G protein made in VERO cells lacks the carboxyl terminal sequences and is, thus, a truncated version of the authentic G protein (Kwilas et al. 2009). The effect of this deletion on immune responses is unknown. Furthermore, the conformation of the F protein was uncharacterized. This vaccine formulation was reported by Murphy and colleagues not to induce ERD upon RSV challenge in cotton rats (Murphy et al. 1989), in contrast to their results with PFP. In phase I and II clinical trials, the protein mix was formulated with and without alum, with both studies reporting better F, G and neutralizing antibody responses without alum. One possible interpretation of this result is that alum alters the conformation of the proteins, thus interfering with stimulation of protective neutralizing antibodies, a consideration in evaluating all RSV vaccine candidates with alum. While these studies were specifically designed to test safety and immunogenicity and were not powered to test efficacy, there was little indication that the vaccine protected from subsequent RSV infection. However, there was the suggestion that risk of RSV infection was lowered in those with the highest serum antibody. At this time, there is no indication that the F, G, M protein mixture has been further developed as a vaccine in either adults or infants.

5.2 Peptide Vaccine Candidates

Other subunit vaccine candidates have been formulations that contain various peptide sequences primarily from the G protein. BBG2Na is a peptide with sequences from amino acid 130-230 of the G protein fused to the albumin-binding domain of the Streptococcus G protein, which served as carrier. The protein was made in *E. coli* and formulated in alum. In mice, this antigen resulted in antibody responses, and immunization of cotton rats with this antigen protected against lower respiratory tract infection but not upper respiratory tract infection. Studies of responses in healthy, seropositive humans (18–45 years) to different doses after intramuscular inoculation were also reported by Power et al. (2001). At the highest dose, the vaccine candidate showed no evidence of adverse reactions and stimulated >4 fold increases in serum antibody titers and >2 fold increases in neutralization titers. Thus the vaccine appeared to be safe and immunogenic in RSV sero-positive adults. As reported by Murata (2009), trials in the elderly also stimulated antibody responses but their rapid declined diminished interest in further development of this candidate for elderly populations. Furthermore, subsequent studies in rhesus macaques showed evidence of Th2 skewed responses as determined by IL13 levels and eosinophils in lungs of challenged animals (de Waal et al. 2004) and protective efficacy was also unclear. In addition, clinical phase III trials of BBG2Na were halted due to adverse events in a small number of individuals (cited in Nguyen et al. 2012). More recently, it was reported that these adverse reactions were due to the Streptococcus component (BB component), and work has explored the use of diphtheria toxin fragments as a carrier (Nguyen et al. 2012). These results have stimulated renewed interest in sequences from this region (amino acids 130–230) of the G protein as a vaccine or a component of a vaccine candidate.

It has been shown that passive transfer of anti-G protein antibodies directed to the central cysteine rich region of the ectodomain sequence (amino acids 174–187) is protective in mice (for example Miao et al. 2009; Trudel et al. 1991). The G protein ectodomain includes a motif analogous to a motif responsible for the binding of the CX3C chemokine fractalkine to its receptor (Tripp et al. 2001). Fractalkine is involved in migration of immune cells, mainly CD8+ and NK T cells, to sites of inflammation. It has been suggested that the soluble form of G protein acts as a fractalkine antagonist and is involved in immune evasion by the virus, ultimately resulting in disease (Tripp 2004). Tripp, Anderson, and colleagues have shown that antibodies specific to this CX3C sequence motif block the binding of G protein to the CX3C receptor, and that immunization with a peptide encoding the CX3C motif protected mice from RSV challenge and decreased pulmonary inflammation (Zhang et al. 2010). These studies have recently been extended showing some cross reaction of antibodies raised to this region to RSV-B G protein in mice (Choi et al. 2012) indicating that inclusion of this region of the G protein in vaccine candidates may increase protective responses to both RSV A and B.

Other studies in mice and cotton rats utilized an antigen made in insect cells that was a fusion of most of the ectodomain sequences from both the F and G proteins, designated F-G. Other details of the structure of this fusion protein are not available. Studies in cotton rats of the F-G antigen showed low levels of neutralizing antibodies, only partial protection from challenge, and possible evidence of ERD (Connors et al. 1992). A modified version of F-G linked to cholera toxin subunit B was also studied, and when given intranasally or via intramuscular injection, induced serum antibody responses, but only intranasal inoculation resulted in protection (Oien et al. 1993; Oien et al. 1994). No human clinical trials have been reported.

6 Particulate Vaccine Candidates

6.1 Virus-Like Particles (VLPs) as Vaccines for RSV

6.1.1 General Properties of VLPs

VLPs are particles with sizes similar to authentic virus and contain repeating protein arrays that mimic those of infectious viruses (Jennings and Bachmann 2008) and account, in part, for the potent immunogenicity of viruses (Jennings and Bachmann 2008). VLPs are released from cells expressing viral structural proteins. The viral surface glycoproteins are folded and inserted into the VLP membranes typical of infectious virus, thus antigenicity of VLPs is likely very similar to live virus. No inactivation is required, thus important epitopes are retained and new ones are not likely generated. Since VLPs are assembled without a genome, VLPs cannot replicate and spread from cell to cell typical of an infectious virus. Nor is there is any chance of reversion to virulence.

Immune responses to VLPs are usually quite robust. Not only do they stimulate neutralizing antibody, but, because of their particulate nature, VLPs are taken up and processed for presentation by both MHC class II and class I, by cross presentation pathways, resulting in broad range of T cell responses (reviewed in Jennings and Bachmann 2008) at least in murine systems. There is evidence that VLPs are also potent stimulators of innate responses since no adjuvant is required for robust immune responses in most systems. Indeed, VLPs are described as "self adjuvanting" (Grgacic and Anderson 2006; Ludwig and Wagner 2007). All these responses usually translate into vaccine candidates that provide good protection of experimental animals from challenge by live, virulent viruses (reviewed in Jennings and Bachmann 2008; Kang et al. 2009). The memory responses to VLPs are less well characterized.

Not all virus systems yield VLPs at levels sufficient for their use as immunogens. Indeed, VLP release from cells expressing the structural proteins of RSV has been reported to be very inefficient (McGinnes et al. 2010; Teng and Collins 1998). To overcome this problem, two different approaches, summarized below, have been reported both of which take advantage of the very efficient release of VLPs formed with core proteins of other viruses.

6.1.2 Newcastle Disease Virus-Based VLPs

One type of RSV VLP vaccine candidate is based on Newcastle disease virus (NDV) VLPs. Expression of the structural proteins of NDV, another paramyxovirus, results in extremely efficient release of VLPs and these VLPs (ND VLPs) (Pantua et al. 2006) stimulated, without adjuvant, robust, neutralizing anti-NDV antibodies and CD8+ and CD4+ T cell immune responses in mice that were comparable to responses to virus (McGinnes et al. 2010). These VLPs have been used as a platform to construct particles containing ectodomains of the RSV glycoprotein ectodomains (McGinnes et al. 2011; Murawski et al. 2010). Efficient incorporation of the RSV glycoproteins into ND VLPs was achieved by constructing chimera protein genes composed of the sequences encoding the ectodomains of G or F proteins fused to the sequences encoding the transmembrane (TM) and cytoplasmic (CT) domains of the NDV HN or F glycoprotein, respectively. Upon expression in cells of the chimera protein genes along with the NDV M (membrane or matrix) protein and NP (nucleocapsid protein) genes, the NDV domains in the chimera proteins specifically interact with the NDV NP and M protein resulting in very efficient incorporation of the chimera protein into VLPs. Using this approach, VLPs containing the RSV G protein ectodomain or VLPs containing ectodomains of both the RSV G and F proteins have been generated (McGinnes et al. 2011; Murawski et al. 2010).

The VLPs containing the RSV G (Murawski et al. 2010) or both the G and F proteins (McGinnes et al. 2011) demonstrated striking effectiveness as a vaccine for RSV in mice. A single intramuscular immunization of BALB/c mice with either VLP stimulated, without adjuvant, antibody levels that were comparable to or higher than responses to infectious RSV delivered by intranasal inoculation to mimic natural infection (McGinnes et al. 2011). The ratios of IgG subtypes during infection or immunization have been used as one indicator of Th1 or Th2 biased immune responses (for example Delgado et al. 2009). VLP immunization resulted in anti-G protein and anti-F protein IgG2a/IgG1 ratios that indicated a predominant Th1 immune response. The neutralizing antibody titers after immunization with VLPs containing only the G protein were relatively weak (Murawski et al. 2010). However, after a single immunization with VLPs containing both glycoproteins, the neutralizing antibody titers were robust (McGinnes et al. 2011). Immunization with either VLP completely protected mice from virus replication upon live virus challenge. Neither VLP stimulated ERD after RSV challenge as determined by inflammation around blood vessels, airways, and in interstitial spaces (McGinnes et al. 2011; Murawski et al. 2010).

Assessment of long-term immune responses to VLPs containing both G and F protein in mice have also shown that these particles can stimulate durable serum neutralizing antibody levels and long-lived memory responses, properties important for an effective RSV vaccine (Schmidt et al. 2012). BALB/c mice immunized with a single dose of VLPs, without adjuvant, generated stable neutralizing antibody titers that lasted for 14 months whereas those of RSV immunized animals declined significantly by 3 months. This finding was reinforced by detection of significant levels of long-lived, bone marrow associated anti-F protein antibody secreting cells in VLP immunized mice while none were detected in mice immunized with an RSV infection. In addition, VLPs stimulated memory responses while RSV infection did not, as determined by adoptive transfer of splenic B cells from VLP immunized mice into immunodeficient *Rag-/-* mice. After transfer, the recipient mice had significant levels of anti-F and anti-G protein serum IgG antibody responses that were protective upon RSV challenge. In contrast, transfer of splenic B cells from RSV immunized mice produced no detectable serum antibody in the recipients nor could these mice inhibit RSV replication upon virus challenge. Thus these VLPs promise to be more effective than natural infection, a requirement for a successful vaccine.

6.1.3 Influenza-Based RSV VLPs

Expression of baculovirus encoded influenza M1 (matrix) protein in SF9 insect cells results in efficient release of VLPs (Latham and Galarza 2001). Taking advantage of this observation, Quan et al. (2011) produced influenza M1 based VLPs containing an RSV glycoprotein by co-expressing influenza M1 and either the RSV G or F proteins. The RSV glycoproteins were presumably passively incorporated into M1 containing particles released from the insect cells. In mice, these particles stimulated anti-F or anti-G antibody responses, both of which were neutralizing. Antibodies, particularly after both a prime and boost immunization were predominantly IgG2a suggesting Th1 biased immune responses. Furthermore, immunization with either of these particles provided protection from RSV challenge. Surprisingly, immunization with G protein containing VLPs showed higher neutralization titers and marginally better protection from challenge than the F protein containing VLPs. Whether this difference between the two VLPs is related to antigen dose is unclear since the amounts of F or G proteins in these particles were not reported. The safety of these VLPs, as determined by analysis of markers of ERD upon RSV challenge of immunized animals was not assessed. In addition, the authors have not reported construction of VLPs containing both RSV glycoproteins. Clinical trials of these VLPs as RSV vaccine candidates have not been reported.

6.2 Virosomes

Another type of particulate RSV vaccine candidate recently described is a virosome (Kamphuis et al. 2012; Stegmann et al. 2010). Virosomes are defined as phospholipid vesicles containing viral glycoproteins. The recently described RSV virosomes contain both RSV F and G proteins and were formed by reconstituting solublized virus envelopes with various combinations of phosphatidyl choline, phosphatidyl serine, cholesterol, and sphingomyelin. The particles formed were relatively homogeneous with a diameter from 70–130 nm. These particles have also been reconstituted to include the adjuvants MPL, a TLR-4 agonist, or P3CSK4, a TLR-2 agonist. The particles stimulated robust anti-RSV antibody responses and neutralizing antibody responses. Without inclusion of either adjuvant, the particles stimulated predominantly a Th2 response as determined by IgG2a/IgG1 ratios, and by IFNγ and IL5 levels. However, inclusion of either adjuvant shifted the response to a more balanced one, but MPL was more effective in promoting this shift.

Particles with adjuvants included stimulated protective responses as determined by virus lung titers after challenge. Importantly, the adjuvanted particles showed no evidence of ERD upon virus challenge in either mice or cotton rats as assessed by cytokine secretion and lung histology. Thus these virosomes show promise as potential vaccine candidates. No clinical trials have been reported. In addition, the conformation of the F protein in the particles has not been characterized.

6.3 Nanoparticles

Another RSV vaccine candidate is described as a nanoparticle (Patent application #WO 2010/077717A1). This particle is composed entirely of a mutant version of the RSV F protein, which was synthesized in baculovirus-infected insect cells and purified by column chromatography. These particles are described as rosettes of F protein trimers that are 20–40 nm in diameter. The conformation of the F protein has not been reported. How this form of F protein differs from previously characterized purified protein vaccine candidates (PFP-1, 2, and 3) is unclear. Results of preclinical trials in either mice or cotton rats have not been reported, and, importantly, nor have safety studies in mice or cotton rats. However, phase 1 clinical trials of the material have recently been conducted (clinical trials.gov, NCT 01290419).

7 Considerations for Future Development of Subunit and Particulate RSV Vaccines

7.1 Overview

There are a number of lessons from past preclinical and, particularly, clinical trials of subunit vaccine candidates. In addition, results of recent studies of RSV protein structure and function and the immunology of RSV infections and vaccine candidates suggest important modifications to future RSV vaccine candidates, modifications that could significantly impact their safety and efficacy and, ultimately, their approval for use in human populations. Studies of potential improvements should minimally consider F protein conformation, G protein conformation, adjuvant incorporation, and supramolecular structures containing repeating arrays of antigens. Because murine or cotton rat immune responses are likely different than human responses, there must be careful assessment of the human immune response, perhaps informed by studies in humanized murine systems.

7.2 Role of Conformation of F Protein in Effective Antibody Stimulation

The paramyxovirus F protein is folded into a metastable conformation and upon fusion activation refolds into the post fusion conformation, which is structurally very different from the pre-fusion form (see chapter by McLellan et al. (2011) this volume; Yin et al. 2005; Yin et al. 2006). Given current models of paramyxovirus fusion, it is logical to assume that optimally neutralizing antibodies should bind to pre-fusion F protein in order to block virus entry, and that an effective vaccine candidate should contain the pre-fusion form of the protein. Recent structural studies of the RSV F protein demonstrate that at least two neutralizing monoclonal antibody binding sites, including the Palivizumab epitope, are accessible on the post fusion form of the protein, suggesting that post fusion forms should stimulate at least a subset of the potential protective neutralizing antibodies (McLellan et al. 2011; Swanson et al. 2011), at least in mice. However, it has also been reported that a significant proportion of the neutralizing antibodies in human immune sera do not bind to the post fusion form of the protein (Magro et al. 2012) raising the possibility that there are human neutralizing antibody binding sites on the pre-fusion form not present on the post fusion form. Thus, it remains to be determined which of the two forms of the F protein is the optimal antigen for inclusion in human vaccine candidates. Conformational intermediates between the pre and post fusion forms should also be considered. Presumably, the PFP vaccine candidates tested in the past were primarily in the post fusion form and stimulated only a subset of neutralizing antibodies, a possibility that could reduce their protective effect, as noted in clinical trials. Formulation of future vaccine candidates should focus on inclusion of the pre-fusion or a conformational intermediate form of F protein.

7.3 G Protein Conformation

Recent studies have indicated that antibodies to the central region of the G protein ectodomain have a significant role in protective responses to RSV (Zhang et al 2010). There is virtually no information about the conformation of the G protein or any conformational changes that may take place during virus entry. A more detailed understanding of the conformation of this protein and any changes that occur during virus attachment and entry could suggest more directed approaches for the inclusion of a form of the G protein that will stimulate effective antibodies to this central region. Perhaps future vaccine candidates should present only peptides from this region of the G protein rather than the intact protein.

7.4 Adjuvants

Stimulation of innate immunity is necessary for affinity maturation of antibodies and long-term T and B cell responses (reviewed in Bessa et al. 2010; Guay et al. 2007; Lanzavecchia and Sallusto 2007; Pasare and Medzhitov 2005). Adjuvants stimulate innate immunity and are often included in vaccines to enhance adaptive immune responses. Different adjuvants stimulate through different innate immune response pathways with different outcomes on adaptive immunity. The only licensed adjuvant that has been tested with RSV vaccine candidates is alum. Indeed, until just recently, alum is the only adjuvant licensed for use in vaccines in the US. While the pathways stimulated by alum are only recently described, precise mechanisms responsible for enhancement of antibody responses by alum are still not clearly defined (reviewed in Lambrecht et al. 2009). It is clear, however, that alum stimulates primarily Th2 responses (Lambrecht et al. 2009). Thus, alum is not the appropriate adjuvant for subunit RSV vaccine candidates, especially for use in young infants, since predominantly Th2 responses are associated with ERD. In addition, many subunit vaccine candidates discussed above are less likely to induce T cell responses, particularly Th1 responses, and optimal adjuvants will be those that increase stimulation of Th1 responses. Indeed, over the past decade, subunit or inactivated RSV vaccines combined with many different adjuvants, particularly TLR agonists, have shown significantly improved adaptive immune responses in animal models. For example, in a head-to-head comparison of alum and TLR agonists, it was recently reported that inclusion of TLR 4, 3, and 7 agonists with UV-inactivated RSV promoted antibody affinity maturation and protective responses without ERD while inclusion of alum did not (Delgado et al. 2009).

Recently one TLR4 agonist, MPL, has been approved for use in the HPV vaccine. However, a number of other TLR agonists are in phase I-III clinical trials (Steinhagen et al. 2011). Furthermore, recent reports clearly show that combinations of TLR agonists synergize to enhance adaptive immunity (Kasturi et al. 2011; Napolitani et al. 2005; Querec et al. 2006) to a noninfectious, particulate influenza vaccine candidate suggesting that inclusion of several different adjuvants in vaccine candidates may be optimal. Another adjuvant included in influenza vaccines used in Europe is MF59, an oil in water emulsion that improves immunogenicity. However, the mechanism of action of this adjuvant is not entirely understood (Lambrecht et al. 2009). Testing of these adjuvants and combinations of adjuvants with RSV vaccine candidates, including older ones such as the PFP, will likely significantly improve efficacy and yield formulations with acceptable properties in at least certain target populations. Inclusion of these new adjuvants with newer particulate RSV vaccine candidates will potentially further enhance the effectiveness of these candidates.

7.5 Supramolecular Structure of the Antigens

It is increasingly recognized that immunization with particulate antigens often results in more robust antibody responses, in more efficient maturation of high affinity antibodies, and in enhanced development of T cell responses compared to responses to purified proteins. Indeed, there are two particulate vaccines licensed for use, the vaccine for the human papilloma virus (HPV) and the vaccine for hepatitis B virus (HBV). The HPV vaccine is a virus-like particle (VLP) composed of the major capsid structural protein and is assembled into a structure very similar to the virus particle. The HBV vaccine is the envelope protein embedded in a lipid vesicle resembling, to some extend, the virus particle. Preclinical studies of particulate vaccine candidates for RSV, described above, show promise, some of which do not require addition of adjuvant.

Potentially other types of particles may also be developed for future candidates. Indeed, particles, described as microparticles, have recently been reported as a potential vaccine candidate for RSV. These 10-nm polyphosphazene microparticles were composed of poly[di(sodium carboxylatoethylphenoxyl)] phosphazene and contained the truncated secreted form of F protein as well as TLR agonists (Garlapati et al. 2012). Nanoparticles composed of the influenza HA protein as well as TLR agonists encapsulated in 300-nm particles composed of poly(D,L-lactic-co-glycolic acid) have been reported as potent influenza vaccine candidates (Kasturi et al. 2011). These could potentially be developed for RSV vaccine candidates.

Another type of nanoparticle reported contains the RSV N protein and a carboxyl terminal fragment of the P protein (Roux et al. 2008) and is an alternative approach to RSV vaccines. N protein sequence includes human specific RSV T cell epitopes. It has been reported that CD8 T cells are important for protection from vaccine induced eosinophilia (Hussell et al. 1997, 1998) thus the goal of these studies was to explore the potential role of N protein induced T cell responses in protective responses. The N protein, expressed in *E. coli*, formed 15 nM particles composed of a ring of 10–11 N proteins associated with 70 nucleotide fragments of *E. coli* RNA. In adult mice, intranasal immunization with this particle resulted in N protein specific antibody, CD4 and CD8 T cell responses, and protection from RSV replication upon challenge as determined by quantitative PCR of NP sequences in lungs (Roux et al. 2008). However, in neonatal mice (Remot et al. 2012), while there was some protection, there was evidence for ERD and a Th2 biased immune response upon RSV challenge. Inclusion of the adjuvant CpG only partially eliminated ERD. Thus, the efficacy and safety of this type of vaccine candidate must be further optimized and assessed in preclinical trials.

Thus, future endeavors that focus on defining the most appropriate form of particulate antigen for RSV vaccines are warranted. A consideration is that many of the early vaccine candidates, notably PFPs, were purified proteins mixed with alum. This combination does result in particulate multivalent structures (Tritto et al. 2009) yet these particles were not very effective vaccines in humans. It is likely that the protein antigens in alum are not organized into repeating arrays typical of a virus. Thus, the organization of the viral antigens in any particulate vaccine candidate may need to be carefully considered.

7.6 Human Immune Responses to Vaccine Candidates

A major difficulty in developing an effective RSV vaccine has been the failure to translate the positive results obtained in rodent models to humans (Power 2008). It is possible that RSV vaccine candidate stimulation of innate and B cell immunity in humans is different than in model animal systems (Lanzavecchia and Sallusto 2009). Definition of immune responses to vaccine candidates in humans will be a key to the development of an effective human vaccine. The recent development of humanized murine systems for vaccine testing (Schmidt et al. 2008; Shultz et al. 2007; Zhang et al. 2007) should facilitate these studies. Not only will formulations most favorable for human innate immunity and B cell stimulation be identified but also optimal routes, schedules, and doses of immunization can be tested.

8 Conclusions

Existing data suggest that subunit vaccines, despite primarily inducing antibody without T cell responses, are safe and relatively immunogenic in adults of all ages, although efficacy will need to be demonstrated and may require enhanced immunogenicity. Similarly, this approach should also be safe and effective in the RSV-experienced older child. However, subunit vaccines will need to reliably induce balanced Th1/Th2 responses as well as neutralizing antibodies and T cells if they are to be successfully developed for use in very young RSV-naïve infants, the primary group in need of a vaccine. VLPs may be particularly successful for this group as they can induce T cell immunity.

Given the history and current status of subunit and particulate vaccine candidates summarized here as well as considerations described above, one may predict that an optimal RSV vaccine candidate will include the pre-fusion form of the F protein and a conformation of G protein or G protein fragment that exposes the central conserved region of the protein. These proteins should be assembled in a virus-sized particle in an organized array typical of an infectious virus. In addition, a combination of adjuvants targeting different innate immunity pathways will likely increase the affinity maturation of antibodies as well as induction of long-lived antibody secreting bone marrow plasma cells and memory B cells. Testing of new candidates should include neonatal rodents, humanized murine systems, as well as different human target populations. Approval of any of subunit or particulate vaccine candidates for use in humans will depend upon optimization of efficacy and safety in these model systems. Approval for vaccines targeted to adults and sero-positive children will be far less problematic than those targeted for sero-negative infants.

References

Belshe RB, Anderson EL, Walsh EE (1993) Immunogenicity of purified F glycoprotein of respiratory syncytial virus: clinical and immune responses to subsequent natural infection in children. J Infect Dis 168:1024–1029

Bessa J, Kopf M, Bachmann MF (2010) Cutting edge: IL-21 and TLR signaling regulate germinal center responses in a B cell-intrinsic manner. J Immunol 184:4615–4619

Calder LJ, Gonzales-Reyes L, Garcia-Barreno B, Wharton SA, Skehel JJ, Wiley DC, Melero JA (2000) Electron microscopy of the human respiratory syncytial virus fusion protein and complexes that it forms with monoclonal antibodies. Virology 271:122–131

Cardenas S, Auais A, Piedimonte G (2005) Palivizumab in the prophylaxis of respiratory syncytial virus infection. Expert Rev Anti Infect Ther 3:719–726

Choi Y, Mason CS, Jones LP, Crabtree J, Jorquera PA, Tripp RA (2012) Antibodies to the central conserved region of respiratory syncytial virus (RSV) G protein block RSV G protein CX3C-CX3CR1 binding and cross-neutralize RSV A and B strains. Viral Immunol 25:193–203

Collins PL, Crowe JE (2007) Respiratory syncytial virus and metapneumovirus, 5th edn. LippincottWilliams and Wilkins, Philadelphia

Collins PL, Graham BS (2007) Viral and host factors in human respiratory syncytial virus pathogenesis. J Virol 82:2040–2055

Connors M, Collins PL, Firestone C-Y, Sotnikov AV, Waitze A, Davis AR, Hung PP, Chanock RM, Murphy BR (1992) Cotton rats previously immunized with a chimeric RSV FG glycoprotein develop enhanced pulmonary pathology when infected with RSV, a phenomenon not encountered following immunization with vaccinia RSV recombinants or RSV. Vaccine 10:475–484

de Waal L, Power UF, Yuksel S, van Amerongen G, Nguyen TN, Niesters HGM, de Swart RL, Osterhaus ADME (2004) Evaluation of BBG2Na in infant macaques: specific immune responses after vaccination and RSV challenge. Vaccine 22:915–922

Delgado MF, Coviello S, Monsalvo AC, Melendi GA, Hernandez JZ, Batalle JP, Diaz L, Trento A, Chang H-Y, Mitzner W, Ravetch J, Melero JA, Irusta PM, Polack FP (2009) Lack of antibody affinity maturation due to poor toll-like receptor stimulation leads to enhanced respiratory syncytial virus disease. Nat Med 15:34–41

Falsey AR, Hennessey PA, Formica MA, Cox C, Walsh EE (2005) Respiratory syncytial virus infection in elderly and high-risk adults. N Engl J Med 352:1749–1759

Falsey AR, Walsh EE (1996) Safety and immunogenicity of a respiratory syncytial virus subunit vaccine (PFP-2) in ambulatory adults over age 60. Vaccine 14:1214–1218

Falsey AR, Walsh EE (1997) Safety and immunogenicity of a respiratory syncytial virus subunit vaccine (PFP-2) in the institutionalized elderly. Vaccine 15:1130–1132

Falsey AR, Walsh EE (2000) Respiratory syncytial virus infection in adults. Clin Microbiol Rev 13:371–384

Falsey AR, Walsh EE, Capellan J, Gravenstein S, Zambon M, Yau E, Gorse GJ, Edelman R, Hayden FG, McElhaney JE, Neuzil KM, Nichol KL, Simoes EAF, Wright PF, Sales VM-P (2008) Comparison of the safety and immunogenicity of 2 respiratory syncytial virus (RSV) vaccines–nonadjuvanted vaccine or vaccine adjuvanted with alum–given concomitantly with influenza vaccine to high-risk elderly individuals. J Infect Dis 198:1317–1326

Garlapati S, Garg R, Brownlie R, Latimer L, Simko E, Hancock REW, Babiuk LA, Gerdts V, Potter A, van Drunen Littel-van den Hurk S (2012) Enhanced immune responses and protection by vaccination with respiratory syncytial virus fusion protein formulated with CpG oligodeoxynucleotide and innate defense regulator peptide in polyphosphazene microparticles. Vaccine 30:5206–5214

Graham BS (2012) Biological challenges and technological opportunities for respiratory syncytial virus vaccine development. Immunol Rev 239:149–166

Grgacic EVL, Anderson DA (2006) Virus-like particles: passport to immune recognition. Methods 40:60–65

Groothuis JR, King SJ, Hogerman DA, Paradiso PR, Simoes EAF (1998) Safety and immunogenicity of a purified F protein respiratory syncytial virus (PFP-2) vaccine in seropositive children with bronchopulmonary dysplasia. J Infect Dis 177:467–469

Guay HM, Andreyeva TA, Garcea RL, Welsh RM, Szomolanyi-Tsuda E (2007) MyD88 is required for the formation of long-term humoral immunity to virus infection. J Immunol 178:5124–5131

Hall CB (2001) Respiratory syncytial virus and parainfluenza virus. N Engl J Med 344:1917–1928

Han LL, Alexander JP, Anderson LJ (1999) Respiratory syncytial virus pneumonia among the elderly: an assessment of disease burden. J Infect Dis 179:25–30

Hildreth SW, Baggs RR, Brownstein DG, Castleman WL, Paradiso PR (1993) Lack of detectable enhanced pulmonary histopathology in cotton rats immunized with purified F glycoprotein of respiratory syncytial virus (RSV) when challenged at (3–6) months after immunization. Vaccine 11:615–618

Hussell T, Baldwin CJ, O'Garra A, Openshaw PJM (1997) CD8 + T cells control Th2-driven pathology during pulmonary respiratory syncytial virus infection. Eur J Immunol 27:3341–3349

Hussell T, Georgiou A, Sparer TE, Matthews S, Pala P, Openshaw PJM (1998) Host genetic determinants of vaccine-induced eosinophilia during respiratory syncytial virus infection. J Immunol 161:6215–6222

Jennings GT, Bachmann MF (2008) The coming of age of virus-like particle vaccines. J Biol Chem 389:521–536

Kamphuis T, Meijerhof T, Stegmann T, Lederhofer J, Wilschut J, de Haan A (2012) Immunogenicity and protective capacity of a virosomal respiratory syncytial virus vaccine adjuvanted with monophosphoryl lipid a in mice. PLoS ONE 7:e36812

Kang S-M, Song J-M, Quan F-S, Compans RW (2009) Influenza vaccines based on virus-like particles. Virus Res 143:140–146

Karron RA (2008) Respiratory syncytial virus and parainfluenza virus vaccines, 5th edn. Saunders-Elsevier, Philadelphia

Kasturi SP, Skountzou I, Albrecht RA, Koutsonanos D, Hua T, Nakaya HI, Ravindran R, Stewart S, Alam M, Kwissa M, Villinger F, Murthy N, Steel J, Jacob J, Hogan RJ, Garcia-Sastre A, Compans R, Pulendran B (2011) Programming the magnitude and persistence of antibody responses with innate immunity. Nature 470:543–547

Kwilas S, Liesman RM, Zhang L, Walsh E, Pickles RJ, Peeples ME (2009) Respiratory syncytial virus grown in vero cells contains a truncated attachment protein that alters its infectivity and dependence on glycosaminoglycans. J Virol 83:10710–10718

Lambrecht BN, Kool M, Willart MAM, Hammad H (2009) Mechanism of action of clinically approved adjuvants. Curr Opin Immunol 21:23–29

Langley JM, Sales V, McGeer A, Guasparini R, Predy G, Meekison W, Li M, Capellan J, Wang E (2009) A dose-ranging study of a subunit respiratory syncytial virus subtype A vaccine with and without aluminum phosphate adjuvantation in adults greater than or equal to 65 years of age. Vaccine 27:5913–5919

Lanzavecchia A, Sallusto F (2007) Toll-like receptors and innate immunity in B-cell activation and antibody responses. Curr Opin Immunol 19:268–274

Lanzavecchia A, Sallusto F (2009) Human B cell memory. Curr Opin Immunol 21:298–304

Latham T, Galarza JM (2001) Formation of wild type and chimeric influenza virus-like particles following simultaneous expression of only four structural proteins. J Virol 75:6154–6165

Ludwig C, Wagner R (2007) Virus-like particles–universal molecular toolboxes. Curr Opin Biotechnol 18:537–545

Magro M, Mas V, Chappell K, Vazquez M, Cano O, Luque D, Terron MC, Melero JA, Palomo C (2012) Neutralizing antibodies against the preactive form of respiratory syncytial virus fusion protein offer unique possibilities for clinical intervention. Proc Natl Acad Sci 109:3089–3094

McGinnes LW, Gravel KA, Finberg RW, Kurt-Jones EA, Massare MJ, Smith G, Schmidt MR, Morrison TG (2011) Assembly and immunological properties of Newcastle disease virus-like particles containing the respiratory syncytial virus F and G proteins. J Virol 85:366–377

McGinnes LW, Pantua H, Laliberte JP, Gravel KA, Jain S, Morrison TG (2010) Assembly and biological and immunological properties of Newcastle disease virus-like particles. J Virol 84:4513–4523

McLellan JS, Yang Y, Graham BS, Kwong PD (2011) Structure of respiratory syncytial virus fusion glycoprotein in the postfusion conformation reveals preservation of neutralizing epitopes. J Virol 85:7788–7796

Miao C, Radu GU, Caidi H, Tripp RA, Anderson LJ, Haynes LM (2009) Treatment with respiratory syncytial virus G glycoprotein monoclonal antibody or $F(ab')_2$ components mediated reduced pulmonary inflammation in mice. J Gen Virol 90:1119–1123

Munoz FM, Piedra PA, Glezen WP (2003) Safety and immunogenicity of respiratory syncytial virus purified fusion protein-2 vaccine in pregnant women. Vaccine 21:3465–3467

Murata Y (2009) Respiratory syncytial virus vaccine development. Clin Lab Med 29:725–739

Murawski MR, McGinnes LW, Finberg RW, Kurt-Jones EA, Massare M, Smith G, Heaton PM, Fraire A, Morrison TG (2010) Newcastle disease virus-like particles containing respiratory syncytial virus G protein induced protection in BALB/c mice with no evidence of immunopathology. J Virol 84:1110–1123

Murphy BR, Sotnikov A, Paradiso PR, Hildreth SW, Jenson AB, Baggs RB, Lawrence L, Zubak JJ, Chanock RM, Beeler JA, Prince GA (1989) Immunization of cotton rats with the fusion (F) and large (G) glycoproteins of respiratory syncytial virus (RSV) protects against RSV challenge without potentiating RSV disease. Vaccine 7:533–540

Murphy BR, Sotnikov AV, Lawrence LA, Banks SM, Prince GA (1990) Enhanced pulmonary histopathology is observed in cotton rats immunized with formalin-inactivated respiratory syncytial virus (RSV) or purified F glycoprotein and challenged with RSV 3–6 months after immunization. Vaccine 8:497–502

Nair H, Nokes DJ, Gessner BD, Dherani M, Madhi SA, Singleton RJ, O'Brien KL, Roca A, Wright PF, Bruce N, Chandran A, Theodoratou E, Sutanto A, Sedyaningsih ER, Ngama M, Munywoki PK, Kartasasmita C, Simoes EAF, Rudan I, Weber MW, Campbell H (2010) Global burden of acute lower respiratory infections due to respiratory syncytial virus in young children: a systematic review and meta-analysis. The Lancet 375:1545–1555

Napolitani G, Rinaldi A, Bertoni F, Sallusto F, Lanzavecchia A (2005) Selected Toll-like receptor agonist combinations synergistically trigger a T helper type 1-polarizing program in dendritic cells. Nat Immunol 6:769–776

Nguyen TN, Power UF, Robert A, J-Fo Haeuw, Helffer K, Perez A, Asin M-A, Corvaia N, Libon C (2012) The respiratory syncytial virus G protein conserved domain induces a persistent and protective antibody response in rodents. PLoS ONE 7:e34331

Oien NL, Brideau RJ, Thomsen DR, Homa FL, Wathen MW (1993) Vaccination with a heterologous respiratory syncytial virus chimeric FG glycoprotein demonstrates significant subgroup cross-reactivity. Vaccine 11:1040–1048
Oien NL, Brideau RJ, Walsh EE, Wathen MW (1994) Induction of local and systemic immunity against human respiratory syncytial virus using a chimeric FG glycoprotein and cholera toxin B subunit. Vaccine 12:731–735
Openshaw PJ, Tregoning JS (2005) Immune responses and disease enhancement during respiratory syncytial virus infection. Clin Microbiol Rev 18:541–555
Pantua HD, McGinnes LW, Peeples ME, Morrison TG (2006) Requirements for the assembly and release of Newcastle disease virus-like particles. J Virol 80:11062–11073
Paradiso PRP, Hildreth SWD, Hogerman DA, Speelman DJ, Lewin EBM, Oren JM, Smith DHM (1994) Safety and immunogenicity of a subunit respiratory syncytial virus vacine in children 24 to 48 months old. Pediatr Infect Dis J 13:792–798
Pasare C, Medzhitov R (2005) Control of B-cell responses by Toll-like receptors. Nature 438:364–368
Piedra PA, Cron SG, Jewell A, Hamblett N, McBride R, Palacio MA, Ginsberg R, Oermann CM, Hiatt PW (2003) Immunogenicity of a new purified fusion protein vaccine to respiratory syncytial virus: a multi-center trial in children with cystic fibrosis. Vaccine 21:2448–2460
Piedra PA, Glezen WP, Kasel JA, Welliver RC, Jewel AM, Rayford Y, Hogerman DA, Hildreth SW, Paradiso PR (1995) Safety and immunogenicity of the PFP vaccine against respiratory syncytial virus (RSV): the Western blot assay aids in distinguishing immune responses of the PFP vaccine from RSV infection. Vaccine 13:1095–1101
Power UF (2008) Respiratory syncytial virus (RSV) vaccines–two steps back for one leap forward. J Clin Virol 41:38–44
Power UF, Nguyen TN, Rietveld E, de Swart RL, Groen J, Osterhaus ADME, dc Groot R, Corvaia N, Beck A, Bouveret-le-Cam N, Bonnefoy J-Y (2001) Safety and immunogenicity of a novel recombinant subunit respiratory syncytial virus vaccine (BBG2Na) in healthy young adults. J Infect Dis 184:1456–1460
Pulendran B, Ahmed R (2011) Immunological mechanisms of vaccination. Nat Immunol 12:509–517
Quan F-S, Kim Y, Lee S, Yi H, Kang S-M, Bozja J, Moore ML, Compans RW (2011) Virus-like particle vaccine induces protection against respiratory syncytial virus infection in mice. J Infect Dis 204:987–995
Querec T, Bennouna S, Alkan S, Laouar Y, Gorden K, Flavell R, Akira S, Ahmed R, Pulendran B (2006) Yellow fever vaccine YF-17D activates multiple dendritic cell subsets via TLR2, 7, 8, and 9 to stimulate polyvalent immunity. J Exp Med 203:413–424
Remot A, Roux X, Dubuquoy C, Fix J, Bouet S, Moudjou M, Eleouet J-F, Riffault S, Petit-Camurdan A (2012) Nucleoprotein nanostructures combined with adjuvants adapted to the neonatal immune context: a candidate mucosal RSV vaccine. PLoS ONE 7:e37722
Roux X, Dubuquoy C, Durand G, Tran-Tolla T-L, Castagne N, Bernard J, Petit-Camurdan A, Eleouet J-F, Riffault S (2008) Sub-nucleocapsid nanoparticles: a nasal vaccine against respiratory syncytial virus. PLoS ONE 3:e1766
Schmidt MR, Appel MC, Giassi LJ, Greiner DL, Shultz LD, Woodland RT (2008) Human BLyS facilitates engraftment of human PBL derived B cells in immunodeficient mice. PLoS ONE 3:e3192
Schmidt MR, McGinnes LW, Kenward SA, Willems KN, Woodland RT, Morrison TG (2012) Long term and memory immune responses in mice against Newcastle disease virus-like particles containing respiratory syncytial virus glycoprotein ectodomains. J Virol 86:11654–11662
Shultz LD, Ishikawa F, Greiner DL (2007) Humanized mice in translational biomedical research. Nat Rev Immunol 7:118–130
Simard C, Nadon F, Seguin C, Trudel M (1995) Evidence that the amino acid region 124–203 of glycoprotein G from the respiratory synctial virus (RSV) constitutes a major part of the

polypeptide domain that is involved in the protection against RSV infection. Antiviral Res 28:303–315
Simoes EAF, Tan DHS, Ohlsson A, Sales V, Wang EEL (2001) Respiratory syncytial virus vaccine: a systematic overview with emphasis on respiratory syncytial virus subunit vaccines. Vaccine 20:954–960
Stegmann T, Kamphuis T, Meijerhof T, Goud E, de Haan A, Wilschut J (2010) Lipopeptide-adjuvanted respiratory syncytial virus virosomes: a safe and immunogenic non-replicating vaccine formulation. Vaccine 28:5543–5550
Steinhagen F, Kinjo T, Bode C, Klinman DM (2011) TLR-based immune adjuvants. Vaccine 29:3341–3355
Swanson KA, Settembre EC, Shaw CA, Dey AK, Rappuoli R, Mandl CW, Dormitzer PR, Carfi A (2011) Structural basis for immunization with postfusion respiratory syncytial virus fusion F glycoprotein (RSV F) to elicit high neutralizing antibody titers. Proc Natl Acad Sci USA 108:9619–9624
Teng MN, Collins PL (1998) Identification of respiratory syncytial virus proteins required for formation and passage of helper dependent infectious particles. J Virol 72:5707–5716
Thompson W, Shay DK, Weintraub E et al (2003) Mortality associated with influenza and respiratory syncytial virus in the United States. JAMA 289:179–186
Tripp RA (2004) Pathogenesis of respiratory syncytial virus infection. Viral Immunol 17:165–181
Tripp RA, Jones LP, Haynes LM, Zheng H, Murphy PM, Anderson LJ (2001) CX3C chemokine mimicry by respiratory syncytial virus G glycoprotein. Nat Immunol 2:732–738
Tristram DA, Welliver RC, Mohar CK, Hogerman DA, Hildreth SW, Paradiso P (1993) Immunogenicity and safety of respiratory syncytial virus subunit vaccine in seropositive children 18–36 months old. J Infect Dis 167:191–195
Tritto E, Mosca F, De Gregorio E (2009) Mechanism of action of licensed vaccine adjuvants. Vaccine 27:3331–3334
Trudel M, Nadon F, Seguin C, Binz H (1991) Protection of BALB/c mice from respiratory syncytial virus infection by immunization with a synthetic peptide derived from the G glycoprotein. Virology 185:749–757
Walsh EE, Falsey AR, Hennessey PA (1999) Respiratory syncytial and other virus infections in persons with chronic cardiopulmonary disease. Am J Respir Crit Care Med 160:791–795
Yin H-S, Paterson RG, Wen X, Lamb RA, Jardetzky TS (2005) Structure of the uncleaved ectodomain of the paramyxovirus (hPIV3) fusion protein. Proc Natl Acad Sci U S A 102:9288–9293
Yin H-S, Wen X, Paterson RG, Lamb RA, Jardetzky TS (2006) Structure of the parainfluenza virus 5 F protein in its metastable, prefusion conformation. Nature 439:38–44
Zhang L, Kovalev GI, Su L (2007) HIV-1 infection and pathogenesis in a novel humanized mouse model. Blood 109:2978–2981
Zhang W, Choi Y, Haynes LM, Harcourt JL, Anderson LJ, Jones LP, Tripp RA (2010) Vaccination to induce antibodies blocking the CX3C-CX3CR1 interaction of respiratory syncytial virus g protein reduces pulmonary inflammation and virus replication in mice. J Virol 84:1148–1157
Zhou H, Thompson WW, Viboud CG, Ringholz CM, Cheng P-Y, Steiner C, Abedi GR, Anderson LJ, Brammer L, Shay DK (2012) Hospitalizations Associated With Influenza and Respiratory Syncytial Virus in the United States, 1993–2008. Clin Infect Dis 54:1427–1436

Gene-Based Vaccine Approaches for Respiratory Syncytial Virus

Rebecca J. Loomis and Philip R. Johnson

Abstract A respiratory syncytial virus (RSV) vaccine has remained elusive for decades, largely due to the failure of a formalin-inactivated RSV vaccine in the 1960s that resulted in enhanced disease upon RSV exposure in the immunized individuals. Vaccine development has also been hindered by the incomplete immunity conferred by natural infection allowing for re-infection at any time, and the immature immune system and circulating maternal antibodies present in the neonate, the primary target for a vaccine. This chapter will review the use of gene delivery, both nonviral and viral, as a potential vaccine approach for human RSV. Many of these gene-based vaccines vectors elicit protective immune responses in animal models. None of the RSV gene-based platforms have progressed into clinical trials, mostly due to uncertainty regarding the direct translation of animal model results to humans and the hesitancy to invest in costly clinical trials with the potential for unclear and complicated immune responses. The continued development of RSV vaccine gene-based approaches is warranted because of their inherent flexibility with regard to composition and administration. It is likely that multiple candidate vaccines will reach human testing in the next few years.

Contents

R. J. Loomis · P. R. Johnson (✉)
Children's Hospital of Philadelphia, Philadelphia, PA, USA
e-mail: johnsonphi@email.chop.edu

P. R. Johnson
University of Pennsylvania, Philadelphia, PA, USA

L. J. Anderson and B. S. Graham (eds.), *Challenges and Opportunities for Respiratory Syncytial Virus Vaccines*, Current Topics in Microbiology and Immunology 372,
DOI: 10.1007/978-3-642-38919-1_15,

1 Introduction

Most licensed vaccines for humans were developed without the use or benefit of recombinant DNA (rDNA) technology. In recent years, a few vaccines have been produced on a commercial scale using rDNA methods (e.g., hepatitis B and human papilloma virus). However, no vaccines approved for humans have exploited the decades old concept of injecting genes directly in vivo (reviewed in Ferraro et al. 2011; Kutzler and Weiner 2008), wherein the vaccine immunogen of choice is synthesized endogenously (as opposed to exogenously). In this scheme, the vaccine gene of interest is delivered to the prospective vaccinee by one of two general methods: (i) direct injection of the rDNA, usually as a bacterial plasmid or (ii) the rDNA, as a vector, within a virus or bacterium. Regardless of the delivery method, once inside the host cell, the gene directs the production of mRNA, and subsequently, the protein immunogen of interest.

The purpose of this chapter is to review the use of gene delivery as a potential vaccine approach for human respiratory syncytial virus (RSV). The well-known failure of a formalin-inactivated vaccine in the 1960s due to poor induction of virus-neutralizing antibodies and an exaggerated Th2 response (reviewed in Collins and Melero 2011; Graham 2011) has haunted the search for an RSV vaccine for almost 50 years. Subsequent attempts to develop live-attenuated vaccine viruses and subunit protein immunogens have fallen short for a variety of reasons (reviewed in Blanco et al. 2010; Chang 2011; Graham 2011; Murata 2009; Schickli et al. 2009). Gene-based RSV vaccines are viable candidates worthy of continued development and testing in humans.

2 DNA Vaccines

Many "non-viral" methods have been proposed and tested as a means to introduce genes into mammalian cells in vitro. Over 20 years ago, Wolff and colleagues (Jiao et al. 1992; Wolff et al. 1990) showed that naked plasmid DNA carrying a eukaryotic gene, injected directly into animal muscle, resulted in endogenous protein biosynthesis. Importantly, there was no evidence of anti-DNA or anti-nuclear antibodies (or other unanticipated side effects) arising in injected animals. These seminal studies provided the basis for using direct DNA injection as a method to deliver genes of interest for vaccine purposes.

The potential advantages of using DNA as an immunizing agent include: DNA is easy to manufacture and purify and it can be produced rapidly and efficiently on a commercial scale, the gene of vaccine interest (transgene) can be modified almost at will; transgene size is usually not a constraint, the DNA itself is not immunogenic, host immune responses are focused on the transgene product rather than the gene carrier, and the endogenously synthesized protein is subject to host translational and post-translational modifications, thereby presumably making the product more "authentic". Endogenous synthesis of the vaccine immunogen engenders both humoral and cellular responses, stimulating natural infection by a pathogen.

While the naked DNA platform continues to hold tremendous promise, translation of the technology into humans has not proceeded smoothly. In general, expression levels of the transgene following DNA injection in humans have been low, leading to less than desirable immune responses. Nonetheless, early on, many groups recognized the potential of using plasmid DNAs to generate specific immune responses (reviewed in Kutzler and Weiner 2008). A variety of pathogens were targeted including hepatitis B virus, HIV, influenza viruses, and rabies virus, to name a few. Since these early reports, numerous enhancements have been made to DNA vaccines including modifications to transcriptional elements in the plasmid backbone, codon-optimized gene sequences, incorporation of adjuvant moieties, and better delivery methods. Advantages and disadvantages of using DNA vaccines are summarized in Table 1.

Table 1 Nonviral vector vaccines advantages and disadvantages

Nonviral gene-based vectors	Advantages	Disadvantages
DNA immunization	Stable and easy to manipulate	Poor transgene transduction/expression
	Accommodates large transgene size	May require special formulations and delivery approaches
	DNA not immunogenic	
	Elicits robust humoral and cellular immune response	
RNA immunization	Does not need to enter the nucleus to initiate transcription	Potential obstacles in manufacturing, delivery, and stability
	Self-amplifying replicons	
	RNA is not immunogenic	
	Can elicit both humoral and cellular immune responses	
Bacteria-mediated transfer of DNA plasmid	Inherent immune stimulatory activity	Potentially lack transgene post-translational modifications
	Elicits both mucosal and systemic immunity	Pre-existing immunity
		Safety issues
	Noninvasive route of immunization possible	
	Some bacterial species are nonpathogenic	

The attractive features of DNA vaccines notwithstanding, relatively few studies have been published using this approach for an RSV vaccine. Early work in a mouse model showed that plasmids carrying the RSV F gene were comparable to live RSV infection at inducing a strong humoral (neutralizing) response that was protective from challenge infection and inducing a balanced lung cytokine expression after challenge (Li et al. 1998). The RSV fusion (F) glycoprotein is a known target of neutralizing antibodies in humans and is widely believed to be an essential component of an effective RSV vaccine (Olmsted et al. 1986; Pemberton et al. 1987). One unexpected finding was that the DNA vaccine caused a switch from a previously established Th2 profile to a Th1 profile. This later finding suggested that a DNA vaccine might be an important component in a "prime-boost" RSV immunization strategy (Li et al. 1998).

In a similar fashion, studies of plasmids that carried the RSV G gene were carried out (Li et al. 2000). The RSV attachment G glycoprotein is also a target of human neutralizing antibodies, but is less well conserved across RSV subgroups when compared to the F glycoprotein (reviewed in Chang 2011; Collins and Melero 2011; Murata 2009; Schickli et al. 2009). The RSV G DNA vaccine generated significant virus neutralizing antibodies and conferred protection in the lower respiratory tract against RSV challenge. The vaccines induced a balanced Th1/Th2 anti-G response regardless of route of administration or form of the antigen encoded, similar to responses observed after intranasal live RSV infection. Unlike RSV F DNA, immunization with RSV G DNA did not elicit detectable virus-specific cytotoxic T cell responses.

Before the advent of codon-optimization, RSV F expression after in vitro transient transfection of plasmid DNA into cultured cells was always lower than expected, especially when compared to many other viral and nonviral genes driven by strong constitutive promoters. Codon-optimization of the RSV F gene alleviated this problem (Ternette et al. 2007). When compared to nonoptimized sequences, only animals immunized with optimized expression plasmids carrying the F ectodomain were afforded significant reduction in RSV viral loads after challenge (Ternette et al. 2007).

The consensus from the aforementioned studies is that RSV DNA vaccines that carry either the F or G glycoproteins can be engineered to provide at least partial protection in the mouse model. The use of codon-optimized genes appears to be important since RSV replication occurs in the cytoplasm and has never adapted to nuclear environment. Future improvements in DNA vaccine strategies for humans will likely be required to move DNA-based RSV vaccines further down the path to human clinical trials (Ferraro et al. 2011).

3 Bacterial Vector Vaccines

The use of live bacterial vaccine vectors to deliver vaccine antigens mucosally may prove a successful strategy at inducing protective immunity against RSV infection. The vaccine antigen can be made as a bacterial protein, either surface-expressed or secreted, or the bacteria can mediate the transfer of the plasmid DNA into the mammalian cells (aka bactofection) where the infected mammalian cells generate the vaccine antigen. Advantages of using bacteria include the inherent immunostimulatory activity of the bacterial carrier, early induction of innate immune responses and the potential to deliver DNA by both oral and mucosal routes. However, gene products expressed by bacteria do not exhibit post-translational modifications typical of the mammalian host and the resulting effects on immune responses are not well understood. Additionally, bacterial vectors may pose potential safety risks and pre-existing immunity may blunt efficacy (Hurwitz 2011). A variety of bacterial strains have been developed as gene transfer vectors (reviewed in Palffy et al. 2006), but relatively few have been tested as candidate RSV vaccines. Advantages and disadvantages to bacterial vector vaccines are summarized in Table 1.

Staphylococcal species (primarily food-grade organisms) have been developed as gene delivery vectors and several are nonpathogenic in mice when administered orally or subcutaneously. A *S. carnosus* vector, enhanced by the introduction of the cholera toxin B subunit peptide into chimeric surface proteins, was used to express three different peptides derived from the RSV G protein (residues 130–230) on the bacterial surface (Cano et al. 2000). Intranasal immunization of mice with the live, recombinant staphylococci elicited significant anti-RSV G peptide and anti-virus serum IgG responses with balanced IgG1/IgG2a responses. After intranasal RSV challenge, about half of the mice were protected.

The commensal bacterium *Streptococcus gordonii* was engineered to express on its surface a conserved domain (residues 130–230) of the RSV G glycoprotein and then tested as a systemic and mucosal vaccine vector (Falcone et al. 2006). Mice immunized subcutaneously had RSV-specific serum IgG partially controlled RSV replication in the lungs after challenge. Intranasal immunization of mice stimulated the production of RSV-specific IgA in both serum and bronchoalveolar fluid. When challenged, the vaccinated mice had lower virus titers in the lungs relative to controls (Falcone et al. 2006).

Oral administration is an attractive feature for attenuated strains of Salmonella. Mice immunized orally with Salmonella carrying the DNA plasmid encoding the RSV F gene had significantly higher serum anti-RSV IgG and bronchoalveolar lavage secretory IgA when compared with controls (Xie et al. 2007). After challenge with live RSV, immunized mice were able to significantly reduce RSV replication in lungs and reduce pulmonary inflammation.

In summary, bacterial vectors display a number of attractive features including the capacity to induce local and systemic immune responses. Multiple issues remain to be addressed, however, before such vectors can be considered as viable human RSV vaccine vectors. As noted for DNA vectors, general improvements in the vector technology are underway and time will tell if bacterial vectors can be adapted for routine human use.

4 RNA Vaccines

Two decades ago, Wolff et al. showed that intramuscular injection of mRNA in mice led to local production of an encoded reporter protein (Wolff et al. 1990). Follow-up studies showed general utility of RNA vaccines to elicit immune responses to a variety of gene targets (reviewed in Pascolo 2004) and their versatility in animal models. Long-standing issues, such as the inherent instability of mRNA and the slow development of manufacturing processes that can yield not only a stable formulation but also a cost-effective product, have made RNA vaccines unfeasible for vaccine development.

Recently, Novartis Vaccines and Diagnostics showed that a synthetic lipid nanoparticle formulation of self-amplifying RNA encoding RSV F efficiently expressed the antigen, elicited an immunogenic response in vivo and protected against RSV challenge in cotton rats (Geall et al. 2012). Their self-amplifying RNA vector is based on the alphavirus genome, containing the genes necessary for replication but lacking the structural genes required for infectious particle production. Following immunization, the RNA molecule is replicated and amplified in the cytoplasm of transfected cells, thereby eliminating concerns of potential genomic integration and cell transformation that are seen with DNA and viral vectors (Smerdou and Liljestrom 1999). This novel RNA vaccine technology could yield a new generation of vaccines that are easily produced, versatile, and elicit potent immune responses that could revolutionize vaccine development.

5 Viral Vector Vaccines

Shortly after the advent of rDNA technology, recombinant viral genomes were rapidly exploited as gene carriers. One of the first applications of this concept was in the vaccine arena. Vaccinia virus, with its long history as an attenuated live smallpox vaccine, was among the first viruses to be engineered as a vaccine vector (Piccini and Paoletti 1986). Over the years, recombinant vaccinia viruses carrying genes from almost every conceivable human pathogen have been made. In addition to vaccinia, many other viral vector systems (e.g., alphaviruses, herpesviruses, adenovirus viruses, adeno-associated viruses, and rhabdoviruses, to name a few) have emerged (reviewed in Brave et al. 2007; Graham 2011; Liniger et al. 2007; Schickli et al. 2009; Small and Ertl 2011).

Viral vectors have many attractive features. Most of the vector genomes are easy to construct and virus stocks are usually straightforward to produce. Transgene products are generally expressed at high levels in vivo and broad immune responses are induced including antigen-specific T cells and pathogen-specific antibodies. These characteristics notwithstanding, viral vectors are not a panacea. Pre-existing immunity to the vector can block transduction, concerns over vector pathogenicity are always present, and in some cases large-scale manufacturing is challenging. Advantages and disadvantages to all viral vector vaccines are summarized in Table 2.

5.1 Poxvirus Vectors

As noted above, of all the viral vectors, vaccinia (and related poxviruses) has been the most popular. The vaccinia genome can accommodate large inserts and the resulting recombinant vectors (rVV) are usually stable and easy to propagate. rVV has a broad host cell range and while pre-existing immunity was a concern many years ago, most people have not been vaccinated against smallpox since 1972, making this less of a concern for young adults.

Early work with vaccinia demonstrated its utility as a vector with insertion of the HSV thymidine kinase gene into the genome of infectious vaccinia virus (Panicali and Paoletti 1982). The coding capacity of foreign DNA in poxviruses was tested by constructing a recombinant that contained 24,700 bp of bacteriophage λ DNA inserted into the vaccinia virus (Smith and Moss 1983). The recombinant was stable, infectious and replicated in vitro. Subsequently, rabbits were immunized with rVV containing the influenza hemagglutinin (HA) gene, which led to production of antibodies reactive to an authentic influenza HA capable of neutralizing influenza virus infection (Panicali et al. 1983).

Building on this early work, Olmsted et al. constructed a cDNA clone containing the mRNA coding sequence of the RSV F gene inserted into the thymidine kinase gene of vaccinia virus under the control of a vaccinia virus promoter (Olmsted et al. 1986). Cotton rats vaccinated with the rVV-RSV F construct developed high-titered RSV F-specific antibodies that provided almost complete resistance to RSV challenge in the lower respiratory tract. When compared with rVV-RSV G, the F construct provided superior protection, but both transgenes engendered balanced immune responses including RSV-specific antibody and T cells.

Previous rodent studies provided the foundation for Collins et al. to evaluate the immunogenicity and protective efficacy of rVV-RSV F and G proteins in chimpanzees (Collins et al. 1990). A single intradermal immunization induced only moderate levels of RSV F- and G-specific serum antibodies that did not efficiently neutralize RSV infectivity in vitro. Upon challenge, there was a marginal reduction in the magnitude and duration of RSV shedding that did not modify clinical disease. Interestingly, RSV challenge did induce unusually strong secondary antibody responses (Collins et al. 1990; Crowe et al. 1993), and this observation supports the premise that immunizing with vectors expressing selected RSV immunogens has the capacity to prime for better responses than natural virus infection.

Concerns with the safety of standard strains of vaccinia virus stalled further development of rVV-RSV vaccines. The prospects for use of rVV vectors was improved with the adaption of highly attenuated, host-restricted derivatives of

Table 2 Viral vector vaccines advantages and disadvantages

Viral gene-based vectors	Advantages	Disadvantages
Poxvirus	Stable, easy to propagate system	Low to moderate RSV-specific responses
	Highly attenuated, replication-defective vaccinia backbone	Poor protective efficacy in chimpanzees
	Inexpensive to manufacture	
	Safe to administer in humans	
	High expression levels of transgene	
	Transgene exhibits authentic post-translational modifications	
	Elicits strong and balanced immune responses	
	Potential to boost pre-existing immunity	
Alphavirus	Replication-defective	Anti-VEE responses potentially hinder transgene immune responses
	Very efficient particle (VRP) infection relative to DNA transfection	Large-scale manufacturing may be difficult due to use of cell-lines and capacity for production
	High levels of transgene expression (cytoplasmic replication)	Safety concerns when scaling up?
	Sequential immunization possible	Translation of immune responses in mice to primates?
	Strong mucosal responses	
	Induction of humoral and cellular immune responses	
	Limited pre-existing immunity	
Adenovirus	Stable and easy to manipulate system	Pre-existing immunity
		Size restrictions for transgene insert
	Robust manufacturing capacity	Host-restriction for Ad serotypes difficult for study in animal models
	Proper post-translational modification of transgene	Innate inflammatory immune response varies among serotypes
	Induction of balanced T cell responses	
	Potential use in prime-boost strategy	
	Multiple serotypes available	
Adeno-associated Virus (AAV)	Replication-defective (genome contains no endogenous AAV genes)	Pre-existing immunity
		Size restrictions for transgene insert
	Stable long-term transgene expression	Poor translation of immune responses in animal models to humans?
	Systems for large-scale production have been developed	
	Excellent safety profile in humans	
	Elicits strong immune responses	

vaccinia virus with replication-defects in human cells such as modified vaccinia virus Ankara (MVA) (Olszewska et al. 2004; Wyatt et al. 1999) and the engineering of powerful promoters to enhance recombinant gene expression (Moss 1996; Paoletti 1996). Compared to replication-competent strains of vaccinia virus, MVA provides similar levels of recombinant gene expression even in nonpermissive cells, induces equal or better immune responses in animals (de Waal et al. 2004) and has an excellent safety profile in humans (tested during the smallpox eradication campaign) (Mayr et al. 1978).

The immunogenicity of recombinant MVA expressing RSV F and G glycoproteins individually and in combination was evaluated to determine if these recombinants were capable of boosting pre-existing immunity to RSV in mice (Wyatt et al. 1999). Intranasal and intramuscular immunizations of mice induced higher RSV-antibody titers than those achieved by RSV infection and greatly restricted the replication of RSV challenge virus in both the upper and lower respiratory tracts. Interestingly, the antibody levels induced by intranasal immunization with RSV could be boosted by the double recombinant MVA (MVA-RSV F/G) suggesting that RSV seropositive persons at-risk of serious RSV disease might benefit from this type of immunization. Olszewska et al. performed a detailed analysis of T cell and pathogenic responses induced by MVA-RSV F and MVA-RSV G in mice (Olszewska et al. 2004). MVA-RSV F and G vaccinated animals lost weight after challenge infection, but there was no evidence of vaccine-enhanced RSV disease. The weight loss (illness) was attributed to immune dysregulation resulting from the large inoculation and antigen load given directly into the mouse lung rather than uncontrolled viral replication (Olszewska et al. 2004).

Building on the previous studies, de Waal et al. evaluated the safety and immunogenicity of a recombinant MVA vector expressing RSV F and G proteins in infant macaques (de Waal et al. 2004). RSV seronegative infant macaques were vaccinated twice with a combination of the two single recombinants (RSV F and RSV G) and were challenged intratracheally with a macaque-passaged RSV strain 4 months after the second vaccination. Prior to challenge, vaccinated macaques had vaccinia virus-specific IgG responses and detectable RSV-specific serum antibodies. After challenge, the RSV-specific IgG and vaccinia antibody responses were both rapidly boosted but protection was not observed.

In summary, rVV-RSV vaccines have shown high levels of transgene expression and primed for strong antibody responses, but protection in nonhuman primates has been lacking. Likewise, the inability of vaccinia MVA-RSV constructs to protect vaccinated infant macaques (which are semi-permissive to begin with) is concerning and underscores the necessity to look beyond mice and cotton rats when attempting to find an RSV vaccine for development in humans. An important observation was that animals previously infected with RSV did experience a boost in serum RSV antibodies following immunization with vaccinia MVA-RSV F/G suggesting that vaccination could afford additional protection in "primed" populations such as young children, the elderly and other immunocompromised individuals at-risk for severe RSV infection.

5.2 Alphavirus Vectors

Alphavirus vectors contain self-replicating RNA molecules that retain the gene required for replication and transcription of the viral RNA (the nonstructural proteins), but a heterologous gene replaces the genes encoding the viral structural proteins (a replicon). Replicons can be efficiently packaged when the structural proteins are provided *in trans* by co-transfection of cells with defective helper genomes. Particle (VRPs) infection is inherently more efficient than DNA transfection in vivo and the infection cycle takes place entirely in the cytoplasm, avoiding potential problems of RNA splicing, or integration of DNA into host cell chromosomes. Sindbis, Semliki Forest and Venezuelan Equine Encephalitis (VEE) virus-based vectors have all individually been shown to be effective in inducing both humoral and cellular immune responses at levels sufficient to protect against lethal challenge with the infectious agent corresponding to the replicon-expressed antigen in animals (reviewed in Graham 2011; Schlesinger and Dubensky 1999). Immunization with VEE VRPs has been shown to induce mucosal immune responses after nonmucosal inoculation in animals (Mok et al. 2007) and to confer protection to the primary mucosal target tissue, a key determinant in induction of protection in the upper respiratory tract (Greer et al. 2007).

In general, alphavirus-based vector vaccines offer several key advantages. They are replication-defective (single round of infection only), express high levels of the heterologous genes, and the self-limiting nature of the replicon particle infection minimizes immune responses to the vector, allowing multiple immunizations with the same vector.

Pushko et al. developed a replicon vaccine system using an attenuated strain of VEE wherein the VEE structural proteins were replaced with influenza virus hemagglutinin (HA) or Lassa virus nucleocapsid (N) gene (Pushko et al. 1997). Subcutaneous immunization of Balb/C mice with either of these VRPs-induced antibody responses directed against the expressed protein. After two inoculations of HA-VRP, complete protection against intranasal challenge with influenza was observed. Additionally, sequential immunization of mice with two inoculations of N-VRP prior to two inoculations of HA-VRP induced an immune response to both HA and N equal to immunization with either VRP construct alone. Protection against influenza challenge was unaffected by previous N-VRP immunization suggesting that despite an immune response to the VEE vector, reactivity to the heterologous proteins in a sequential immunization scheme is possible (i.e., the same individual could be sequentially immunized against multiple pathogens).

VEE replicons encoding RSV Fa or RSV Ga or RSV Gb proteins of the A or B subgroups of RSV were engineered to evaluate immunogenicity in mice and rhesus macaques (Elliott et al. 2007). Intramuscular immunization with a formulation composed of equal amounts of each replicon particle (3vRSV) generated neutralizing antibodies against A and B strains of RSV in both BALB/c mice and rhesus macaques (Elliott et al. 2007). In mice, the 3vRSV replicon vaccine-induced protective immune responses after a secondary immunization and following RSV challenge, there was a significant reduction of infectious virus in the lungs. In macaques, two weeks after a secondary immunization, significant neutralizing antibody titers were measured against both RSV A and B subgroups that were maintained over the course of the study. The presence of anti-VEE responses did not appear to hinder the enhancement of anti-RSV neutralization titers following secondary immunization with 3vRSV vaccine.

Mok et al. evaluated the immunogenicity and efficacy of RSV F or RSV G VEE VRPs against RSV challenge in mice and cotton rats (Mok et al. 2007). Intranasal administration of VRPs induced RSV (F or G)-specific virus-neutralizing antibodies and mucosal IgA antibodies in both the upper and lower respiratory tract. In animals vaccinated with RSV F VRPs, challenge virus replication was reduced below the level of detection in both the upper and lower respiratory tract, while in those vaccinated with RSV G VRPs, challenge virus was detected in the upper but not the lower respiratory tract. This study found that a single immunization was sufficient to elicit effective neutralizing antibodies against RSV in vivo.

Very recently, Novartis Vaccines and Diagnostics demonstrated that a self-amplifying RNA (based on the alphavirus genome) encoding RSV F formulated with a synthetic lipid nanoparticle elicited a protective immune response in cotton rats (also refer to RNA Vaccine section above) (Geall et al. 2012).

Considered together, these data suggest that recombinant alphaviruses hold great promise as vaccine vectors, and in this case, for an RSV vaccine. They elicit strong humoral responses that are protective and it appears that pre-existing immunity may not be an issue.

5.3 Adenovirus Vectors

Recombinant adenovirus (rAd) vectors have a long history as expression vectors in mammalian cells and as gene transfer vehicles in animals and humans (Graham 1990). They are easy to construct, grow, are generally stable, and can be designed as either replication-competent or replication-defective. Pre-existing immunity against adenoviruses (particularly Ad5) has been shown to reduce vaccine efficacy (reviewed in Graham 2011) but can sometimes be overcome with larger amounts of vector or using alternate routes of administration (Swenson et al. 2008; Xiang et al. 2003). Adenovirus vectors have another disadvantage compared to some other vectors in that the transgene insert size is limited (<800 amino acids) making the expression of one or two average-sized genes in a single vector nearly impossible. Despite these challenges, rAd vectors have been widely used as vaccine vehicles because of their high antigen expression and rapid induction of immune responses, and can easily accommodate the 574 amino acids of the F glycoprotein.

One early study examined the efficacy of human adenoviruses as vaccines by studying the immunity induced by a recombinant adenovirus vector, AdgB2 (McDermott et al. 1989) which had previously been shown to express high levels of HSV gB in human and murine cells (Johnson et al. 1988). Mice inoculated with

AdgB2 produced antibodies specific for gB that neutralized HSV in the presence of complement and were protected from a lethal challenge with HSV. These findings demonstrate the utility of rAd as a vaccine vector.

The earliest studies of replication-competent rAd vectors encoding RSV F and RSV G genes were done in dog and ferret models because cotton rats and hamsters support limited replication of human adenoviruses (Hjorth et al. 1988; Pacini et al. 1984). rAd vectors (representing serotypes 4, 5, and 7) containing RSV F, RSV G or both were made and tested in a dog model (Hsu et al. 1992). Intratracheal immunization with Ad7-RSV F induced moderate titers of RSV-neutralizing antibodies. Following a booster immunization with Ad4-RSV F, the dogs developed high titers of RSV-specific antibody. Several different sequential dose regimens (prime-boost) were evaluated, and all completely protected the lungs of dogs from RSV infection. In a second study (Hsu et al. 1994), these same adenoviruses were evaluated in a ferret model. Ad4-RSV and Ad5-RSV recombinants, when administered intranasally to ferrets, induced stronger RSV-antibody responses than Ad7-RSV recombinants. Following intranasal RSV challenge, most of the vaccination regimens protected ferrets from RSV infection in a dose-dependent manner.

Yu et al. designed a replication-defective vaccine (rAd/RSV3 × G) that expressed the soluble core domain of the G glycoprotein (residues 130–230) engineered by codon-optimization and tandem repetition (×3) for higher-level expression (Yu et al. 2008). A single intranasal immunization in mice with rAd/RSV3 × G provided significant protection against RSV challenge that lasted more than 10 weeks. Intranasal immunization, but not intramuscular or oral administration, of rAd/RSV3 × G induced strong mucosal IgA responses. Fu et al. constructed Ad-RSV F and evaluated its potential as an RSV vaccine in the murine model (Fu et al. 2009). Intranasal immunization with Ad-RSV F generated systemic IgG responses, secretory IgA in the lungs and balanced Th1/Th2 responses. After challenge, intranasal immunization with Ad-RSV F reduced RSV lung viral titer against RSV infection, but was inferior to immunization with Ad-RSV G, perhaps due to poor expression of full-length RSV F.

The role of a codon-optimized open reading frame for both full-length RSV F and its soluble form in a replication-deficient adenoviral serotype 5 vector on immunogenicity and efficacy of the adenovirus vector vaccines was determined (Kohlmann et al. 2009). BALB/c mice were immunized twice with Ad-RSV F-FL or Ad-RSV F-sol and challenged intranasally with RSV. Side-by-side comparison of Ad-RSV F-FL and Ad-RSV F-sol showed that both induced similar levels of RSV F-specific IgG1 and IgG2a antibodies, but neutralizing antibody titers were slightly higher using the soluble form of RSV F. Both vaccines reduced viral load after challenge.

rAd vectors are viable vaccine delivery vehicles, but challenges remain before they are suitable for widespread human use. One major challenge is the prevalence of pre-existing immunity, specifically with rAd5. Replication-defective human serotype 5 Ad vectors (rAd5) with HIV-1 antigens (clade B HIV-1 Gag, Pol, Nef), in the STEP clinical trial, not only showed lack of efficacy, but also appeared to slightly increase rates of HIV-1 acquisition in uncircumcised individuals with pre-existing neutralizing antibodies to AdHu5 (Buchbinder et al. 2008; McElrath et al. 2008). Some newer vectors based on less prevalent serotypes derived from non-human primates (chimpanzees or gorillas) might overcome this obstacle. Another potential issue with some adenovirus serotypes (e.g., rAd35 and rAd28) is the innate inflammatory immune response elicited following adenovirus vector administration that could reduce vector transduction efficiency and limit the duration of transgene expression, thereby reducing T cell immunity. In contrast, Type C adenovectors [rAd5, chimp Ad3, or a gorilla-derived adenovirus vector (GC46)] induce more limited innate inflammation, enhance antigen expression and T cell immunity, and may have less reactogenicity (personal communication from Robert A. Seder, NIAID, VRC).

5.4 Adeno-Associated Virus Vectors

Recombinant adeno-associated viruses (rAAV) have emerged as a platform for genetic vaccines. Wild-type AAV was originally engineered as a transduction vector almost 30 years ago (Samulski et al. 1982). The vector genome contains only a small fraction of the wild-type AAV genome, namely the 145 nucleotide inverted terminal repeats that contain signals for replication and packaging; no endogenous AAV genes are included. The recombinant genome can be packaged into a variety of AAV capsids (often called serotypes), some of which convey unique targeting properties to the vector (Rabinowitz et al. 2002; Wu et al. 2006).

Like plasmid DNA, rAAV vectors can be used to deliver "self-genes" [e.g., alpha-1-antitrypsin (Johnson et al. 2005)] or "foreign genes" designed to elicit immune responses (a vaccine). Early experiments suggested that transgenes expressed from rAAV vectors could elicit immune responses (Clark et al. 1997) and rAAV vectors have been used successfully as vaccine carriers to induce transgene-specific immune responses in various animal models (Brockstedt et al. 1999; Johnson et al. 2005; Xin et al. 2001). Xin et al. demonstrated that Balb/C mice immunized intranasally with an AAV vector expressing the influenza virus hemagglutinin (HA) gene showed protective immunity against homologous influenza virus challenge (Xin et al. 2001). In a macaque model of AIDS, Johnson et al. showed that after a single intramuscular dose, rAAV/SIV vaccines elicited SIV-specific T cells and antibodies in macaques and these animals were able to significantly restrict replication of a live, virulent SIV challenge (Johnson et al. 2005). In a dose-escalation Phase I study evaluating safety and immunogenicity, rAAV/HIV Gag was weakly immunogenic at the highest dose given with responses of moderate magnitude to the Gag epitope (Mehendale et al. 2008). In a follow-up Phase II study, a higher dose of rAAV/HIV Gag was given for intramuscular immunization, but fewer than half of the recipients had any T cell response to HIV Gag (Vardas et al. 2010). While the absolute responses were disappointing, these trials showed that rAAV vectors were safe in healthy human volunteers and that it was feasible to use rAAV vectors as vaccine vehicles.

In unpublished work, our group has shown that rAAV vectors carrying the codon-optimized RSV F gene (either secreted or anchored) can elicit neutralizing antibodies in rabbits. Since the rabbit is not a challenge model, in the future, similar experiments in the well-characterized BALB/c mouse model of RSV infection will be performed. Another potential use for rAAV vectors in RSV prophylaxis is to deliver genes that represent antibodies that neutralize RSV (e.g., palivizumab). This approach was tested in the macaque model of AIDS where long-lived serum neutralizing activity and complete protection against intravenous challenge with a virulent SIV was shown (Johnson et al. 2009). The antibody gene transfer concept could offer a viable alternative strategy for RSV since several RSV F neutralizing epitopes have been identified and it is known that immunoprophylaxis with palivizumab can prevent severe RSV infection in high-risk infants (reviewed in Collins and Melero 2011; Graham 2011; Hurwitz 2011).

6 Conclusion

The RSV vaccine field has come a long way since the failed RSV vaccine trials of the 1960s. Vaccine development, as a whole, is moving forward largely due to advances in technology and improved immunological methods. As technology and manipulation capabilities continue to become more advanced and we gain more knowledge about RSV pathogenesis and the host immune response to RSV, the chances of developing a viable human vaccine increase.

Today, there are many strategies worthy of pursuit, many of which induce protective immune responses in animal models. The vaccine entities are generally easy to produce and most appear safe in humans. More work is necessary to extend the knowledge from animal models in applying it to humans in clinical trials to develop an efficacious RSV vaccine. None of the RSV gene-based platforms have gone into human trials, mostly due to uncertainty about whether protection demonstrated in animal studies will translate to humans, and consequent cost considerations. Natural RSV infection yields incomplete immunity, meaning that an individual can be re-infected with RSV any time. In the neonate, the primary target for an RSV vaccine, an immature immune system and circulating maternal antibodies present additional barriers.

The continued development of gene-based approaches to an RSV vaccine is warranted because of their inherent flexibility with regard to composition and administration, and potential for eliciting immunity that exceeds the protection afforded by natural infection. It is likely that multiple candidate vaccines will reach human testing in the next few years.

References

Blanco JC, Boukhvalova MS, Shirey KA, Prince GA, Vogel SN (2010) New insights for development of a safe and protective RSV vaccine. Hum vaccines 6:482–492

Brave A, Ljungberg K, Wahren B, Liu MA (2007) Vaccine delivery methods using viral vectors. Mol Pharm 4:18–32. doi:10.1021/mp060098+

Brockstedt DG, Podsakoff GM, Fong L, Kurtzman G, Mueller-Ruchholtz W, Engleman EG (1999) Induction of immunity to antigens expressed by recombinant adeno-associated virus depends on the route of administration. Clin Immunol 92:67–75. doi:10.1006/clim.1999.4724

Buchbinder SP, Mehrotra DV, Duerr A, Fitzgerald DW, Mogg R, Li D, Gilbert PB, Lama JR, Marmor M, Del Rio C, McElrath MJ, Casimiro DR, Gottesdiener KM, Chodakewitz JA, Corey L, Robertson MN (2008) Efficacy assessment of a cell-mediated immunity HIV-1 vaccine (the Step Study): a double-blind, randomised, placebo-controlled, test-of-concept trial. Lancet 372:1881–1893. doi:10.1016/S0140-6736(08)61591-3

Cano F, Plotnicky-Gilquin H, Nguyen TN, Liljeqvist S, Samuelson P, Bonnefoy J, Stahl S, Robert A (2000) Partial protection to respiratory syncytial virus (RSV) elicited in mice by intranasal immunization using live staphylococci with surface-displayed RSV-peptides. Vaccine 18:2743–2752

Chang J (2011) Current progress on development of respiratory syncytial virus vaccine. BMB reports 44:232–237. doi:DOI10.5483/BMBRep.2011.44.4.232

Clark KR, Sferra TJ, Johnson PR (1997) Recombinant adeno-associated viral vectors mediate long-term transgene expression in muscle. Hum Gene Ther 8:659–669. doi:10.1089/hum.1997.8.6-659

Collins PL, Melero JA (2011) Progress in understanding and controlling respiratory syncytial virus: still crazy after all these years. Virus Res 162:80–99. doi:10.1016/j.virusres.2011.09.020

Collins PL, Purcell RH, London WT, Lawrence LA, Chanock RM, Murphy BR (1990) Evaluation in chimpanzees of vaccinia virus recombinants that express the surface glycoproteins of human respiratory syncytial virus. Vaccine 8:164–168

Crowe JE Jr, Collins PL, London WT, Chanock RM, Murphy BR (1993) A comparison in chimpanzees of the immunogenicity and efficacy of live attenuated respiratory syncytial virus (RSV) temperature-sensitive mutant vaccines and vaccinia virus recombinants that express the surface glycoproteins of RSV. Vaccine 11:1395–1404

de Waal L, Wyatt LS, Yuksel S, van Amerongen G, Moss B, Niesters HG, Osterhaus AD, de Swart RL (2004) Vaccination of infant macaques with a recombinant modified vaccinia virus Ankara expressing the respiratory syncytial virus F and G genes does not predispose for immunopathology. Vaccine 22:923–926. doi:10.1016/j.vaccine.2003.10.010

Elliott MB, Chen T, Terio NB, Chong SY, Abdullah R, Luckay A, Egan MA, Boutilier LA, Melville K, Lerch RA, Long D, Eldridge JH, Parks CL, Udem SA, Hancock GE (2007) Alphavirus replicon particles encoding the fusion or attachment glycoproteins of respiratory syncytial virus elicit protective immune responses in BALB/c mice and functional serum antibodies in rhesus macaques. Vaccine 25:7132–7144. doi:10.1016/j.vaccine.2007.07.065

Falcone V, Mihm D, Neumann-Haefelin D, Costa C, Nguyen T, Pozzi G, Ricci S (2006) Systemic and mucosal immunity to respiratory syncytial virus induced by recombinant Streptococcus gordonii surface-displaying a domain of viral glycoprotein G. FEMS Immunol Med Microbiol 48:116–122. doi:10.1111/j.1574-695X.2006.00130.x

Ferraro B, Morrow MP, Hutnick NA, Shin TH, Lucke CE, Weiner DB (2011) Clinical applications of DNA vaccines: current progress. Clin Infect Dis: an official publication of the Infectious Diseases Society of America 53:296–302. doi:10.1093/cid/cir334

Fu Y, He J, Zheng X, Wu Q, Zhang M, Wang X, Wang Y, Xie C, Tang Q, Wei W, Wang M, Song J, Qu J, Zhang Y, Hong T (2009) Intranasal immunization with a replication-deficient adenoviral vector expressing the fusion glycoprotein of respiratory syncytial virus elicits

protective immunity in BALB/c mice. Biochem Biophys Res Commun 381:528–532. doi:10.1016/j.bbrc.2009.02.075
Geall AJ, Verma A, Otten GR, Shaw CA, Hekele A, Banerjee K, Cu Y, Beard CW, Brito LA, Krucker T, O'Hagan DT, Singh M, Mason PW, Valiante NM, Dormitzer PR, Barnett SW, Rappuoli R, Ulmer JB, Mandl CW (2012) Nonviral delivery of self-amplifying RNA vaccines. Proc Natl Acad Sci USA 109:14604–14609. doi:10.1073/pnas.1209367109
Graham BS (2011) Biological challenges and technological opportunities for respiratory syncytial virus vaccine development. Immunol Rev 239:149–166. doi:10.1111/j.1600-065X.2010.00972.x
Graham FL (1990) Adenoviruses as expression vectors and recombinant vaccines. Trends Biotechnol 8:85–87
Greer CE, Zhou F, Legg HS, Tang Z, Perri S, Sloan BA, Megede JZ, Uematsu Y, Vajdy M, Polo JM (2007) A chimeric alphavirus RNA replicon gene-based vaccine for human parainfluenza virus type 3 induces protective immunity against intranasal virus challenge. Vaccine 25:481–489. doi:10.1016/j.vaccine.2006.07.048
Hjorth RN, Bonde GM, Pierzchala WA, Vernon SK, Wiener FP, Levner MH, Lubeck MD, Hung PP (1988) A new hamster model for adenoviral vaccination. Arch Virol 100:279–283
Hsu KH, Lubeck MD, Bhat BM, Bhat RA, Kostek B, Selling BH, Mizutani S, Davis AR, Hung PP (1994) Efficacy of adenovirus-vectored respiratory syncytial virus vaccines in a new ferret model. Vaccine 12:607–612
Hsu KH, Lubeck MD, Davis AR, Bhat RA, Selling BH, Bhat BM, Mizutani S, Murphy BR, Collins PL, Chanock RM et al (1992) Immunogenicity of recombinant adenovirus-respiratory syncytial virus vaccines with adenovirus types 4, 5, and 7 vectors in dogs and a chimpanzee. J Infect Dis 166:769–775
Hurwitz JL (2011) Respiratory syncytial virus vaccine development. Expert Rev Vaccines 10:1415–1433. doi:10.1586/erv.11.120
Jiao S, Williams P, Berg RK, Hodgeman BA, Liu L, Repetto G, Wolff JA (1992) Direct gene transfer into nonhuman primate myofibers in vivo. Hum Gene Ther 3:21–33. doi:10.1089/hum.1992.3.1-21
Johnson DC, Ghosh-Choudhury G, Smiley JR, Fallis L, Graham FL (1988) Abundant expression of herpes simplex virus glycoprotein gB using an adenovirus vector. Virology 164:1–14
Johnson PR, Schnepp BC, Connell MJ, Rohne D, Robinson S, Krivulka GR, Lord CI, Zinn R, Montefiori DC, Letvin NL, Clark KR (2005) Novel adeno-associated virus vector vaccine restricts replication of simian immunodeficiency virus in macaques. J Virol 79:955–965. doi:10.1128/JVI.79.2.955-965.2005
Johnson PR, Schnepp BC, Zhang J, Connell MJ, Greene SM, Yuste E, Desrosiers RC, Clark KR (2009) Vector-mediated gene transfer engenders long-lived neutralizing activity and protection against SIV infection in monkeys. Nat Med 15:901–906. doi:10.1038/nm.1967
Kohlmann R, Schwannecke S, Tippler B, Ternette N, Temchura VV, Tenbusch M, Uberla K, Grunwald T (2009) Protective efficacy and immunogenicity of an adenoviral vector vaccine encoding the codon-optimized F protein of respiratory syncytial virus. J Virol 83:12601–12610. doi:10.1128/JVI.01036-09
Kutzler MA, Weiner DB (2008) DNA vaccines: ready for prime time? Nature reviews. Genetics 9:776–788. doi:10.1038/nrg2432
Li X, Sambhara S, Li CX, Ettorre L, Switzer I, Cates G, James O, Parrington M, Oomen R, Du RP, Klein M (2000) Plasmid DNA encoding the respiratory syncytial virus G protein is a promising vaccine candidate. Virology 269:54–65. doi:10.1006/viro.2000.0186
Li X, Sambhara S, Li CX, Ewasyshyn M, Parrington M, Caterini J, James O, Cates G, Du RP, Klein M (1998) Protection against respiratory syncytial virus infection by DNA immunization. J Exp Med 188:681–688
Liniger M, Zuniga A, Naim HY (2007) Use of viral vectors for the development of vaccines. Expert Rev Vaccines 6:255–266. doi:10.1586/14760584.6.2.255
Mayr A, Stickl H, Muller HK, Danner K, Singer H (1978) [The smallpox vaccination strain MVA: marker, genetic structure, experience gained with the parenteral vaccination and

behavior in organisms with a debilitated defence mechanism (author's transl)]. Zentralblatt fur Bakteriologie, Parasitenkunde, Infektionskrankheiten und Hygiene. Erste Abteilung Originale. Reihe B: Hygiene, Betriebshygiene, praventive Medizin 167:375–390

McDermott MR, Graham FL, Hanke T, Johnson DC (1989) Protection of mice against lethal challenge with herpes simplex virus by vaccination with an adenovirus vector expressing HSV glycoprotein B. Virology 169:244–247

McElrath MJ, De Rosa SC, Moodie Z, Dubey S, Kierstead L, Janes H, Defawe OD, Carter DK, Hural J, Akondy R, Buchbinder SP, Robertson MN, Mehrotra DV, Self SG, Corey L, Shiver JW, Casimiro DR (2008) HIV-1 vaccine-induced immunity in the test-of-concept step study: a case-cohort analysis. Lancet 372:1894–1905. doi:10.1016/S0140-6736(08)61592-5

Mehendale S, van Lunzen J, Clumeck N, Rockstroh J, Vets E, Johnson PR, Anklesaria P, Barin B, Boaz M, Kochhar S, Lehrman J, Schmidt C, Peeters M, Schwarze-Zander C, Kabamba K, Glaunsinger T, Sahay S, Thakar M, Paranjape R, Gilmour J, Excler JL, Fast P, Heald AE (2008) A phase 1 study to evaluate the safety and immunogenicity of a recombinant HIV type 1 subtype C adeno-associated virus vaccine. AIDS Res Hum Retroviruses 24:873–880. doi:10.1089/aid.2008.0292

Mok H, Lee S, Utley TJ, Shepherd BE, Polosukhin VV, Collier ML, Davis NL, Johnston RE, Crowe JE Jr (2007) Venezuelan equine encephalitis virus replicon particles encoding respiratory syncytial virus surface glycoproteins induce protective mucosal responses in mice and cotton rats. J Virol 81:13710–13722. doi:10.1128/JVI.01351-07

Moss B (1996) Genetically engineered poxviruses for recombinant gene expression, vaccination, and safety. Proc Natl Acad Sci USA 93:11341–11348

Murata Y (2009) Respiratory syncytial virus vaccine development. Clin Lab Med 29:725–739. doi:10.1016/j.cll.2009.07.004

Olmsted RA, Elango N, Prince GA, Murphy BR, Johnson PR, Moss B, Chanock RM, Collins PL (1986) Expression of the F glycoprotein of respiratory syncytial virus by a recombinant vaccinia virus: comparison of the individual contributions of the F and G glycoproteins to host immunity. Proc Natl Acad Sci USA 83:7462–7466

Olszewska W, Suezer Y, Sutter G, Openshaw PJ (2004) Protective and disease-enhancing immune responses induced by recombinant modified vaccinia Ankara (MVA) expressing respiratory syncytial virus proteins. Vaccine 23:215–221. doi:10.1016/j.vaccine.2004.05.015

Pacini DL, Dubovi EJ, Clyde WA Jr (1984) A new animal model for human respiratory tract disease due to adenovirus. J Infect Dis 150:92–97

Palffy R, Gardlik R, Hodosy J, Behuliak M, Resko P, Radvansky J, Celec P (2006) Bacteria in gene therapy: bactofection versus alternative gene therapy. Gene Ther 13:101–105. doi:10.1038/sj.gt.3302635

Panicali D, Davis SW, Weinberg RL, Paoletti E (1983) Construction of live vaccines by using genetically engineered poxviruses: biological activity of recombinant vaccinia virus expressing influenza virus hemagglutinin. Proc Natl Acad Sci USA 80:5364–5368

Panicali D, Paoletti E (1982) Construction of poxviruses as cloning vectors: insertion of the thymidine kinase gene from herpes simplex virus into the DNA of infectious vaccinia virus. Proc Natl Acad Sci USA 79:4927–4931

Paoletti E (1996) Applications of pox virus vectors to vaccination: an update. Proc Natl Acad Sci USA 93:11349–11353

Pascolo S (2004) Messenger RNA-based vaccines. Expert Opin Biol Ther 4:1285–1294. doi:10.1517/14712598.4.8.1285

Pemberton RM, Cannon MJ, Openshaw PJ, Ball LA, Wertz GW, Askonas BA (1987) Cytotoxic T cell specificity for respiratory syncytial virus proteins: fusion protein is an important target antigen. J Gen Virol 68(Pt 8):2177–2182

Piccini A, Paoletti E (1986) The use of vaccinia virus for the construction of recombinant vaccines. BioEssays: News Rev Mole Cell Dev Biol 5:248–252. doi:10.1002/bies.950050604

Pushko P, Parker M, Ludwig GV, Davis NL, Johnston RE, Smith JF (1997) Replicon-helper systems from attenuated Venezuelan equine encephalitis virus: expression of heterologous

genes in vitro and immunization against heterologous pathogens in vivo. Virology 239:389–401. doi:10.1006/viro.1997.8878
Rabinowitz JE, Rolling F, Li C, Conrath H, Xiao W, Xiao X, Samulski RJ (2002) Cross-packaging of a single adeno-associated virus (AAV) type 2 vector genome into multiple AAV serotypes enables transduction with broad specificity. J Virol 76:791–801
Samulski RJ, Berns KI, Tan M, Muzyczka N (1982) Cloning of adeno-associated virus into pBR322: rescue of intact virus from the recombinant plasmid in human cells. Proc Natl Acad Sci USA 79:2077–2081
Schickli JH, Dubovsky F, Tang RS (2009) Challenges in developing a pediatric RSV vaccine. Hum Vaccines 5:582–591
Schlesinger S, Dubensky TW (1999) Alphavirus vectors for gene expression and vaccines. Curr Opin Biotechnol 10:434–439
Small JC, Ertl HC (2011) Viruses—from pathogens to vaccine carriers. Curr Opin Virol 1:241–245. doi:10.1016/j.coviro.2011.07.009
Smerdou C, Liljestrom P (1999) Non-viral amplification systems for gene transfer: vectors based on alphaviruses. Curr Opin Mol Ther 1:244–251
Smith GL, Moss B (1983) Infectious poxvirus vectors have capacity for at least 25 000 base pairs of foreign DNA. Gene 25:21–28
Swenson DL, Wang D, Luo M, Warfield KL, Woraratanadharm J, Holman DH, Dong JY, Pratt WD (2008) Vaccine to confer to nonhuman primates complete protection against multistrain Ebola and Marburg virus infections. Clin Vaccine Immunol CVI 15:460–467. doi:10.1128/CVI.00431-07
Ternette N, Tippler B, Uberla K, Grunwald T (2007) Immunogenicity and efficacy of codon optimized DNA vaccines encoding the F-protein of respiratory syncytial virus. Vaccine 25:7271–7279. doi:10.1016/j.vaccine.2007.07.025
Vardas E, Kaleebu P, Bekker LG, Hoosen A, Chomba E, Johnson PR, Anklesaria P, Birungi J, Barin B, Boaz M, Cox J, Lehrman J, Stevens G, Gilmour J, Tarragona T, Hayes P, Lowenbein S, Kizito E, Fast P, Heald AE, Schmidt C (2010) A phase 2 study to evaluate the safety and immunogenicity of a recombinant HIV type 1 vaccine based on adeno-associated virus. AIDS Res Hum Retroviruses 26:933–942. doi:10.1089/aid.2009.0242
Wolff JA, Malone RW, Williams P, Chong W, Acsadi G, Jani A, Felgner PL (1990) Direct gene transfer into mouse muscle in vivo. Science 247:1465–1468
Wu Z, Asokan A, Samulski RJ (2006) Adeno-associated virus serotypes: vector toolkit for human gene therapy. Mole Ther J Am Soc Gene Ther 14:316–327. doi:10.1016/j.ymthe.2006.05.009
Wyatt LS, Whitehead SS, Venanzi KA, Murphy BR, Moss B (1999) Priming and boosting immunity to respiratory syncytial virus by recombinant replication-defective vaccinia virus MVA. Vaccine 18:392–397
Xiang ZQ, Gao GP, Reyes-Sandoval A, Li Y, Wilson JM, Ertl HC (2003) Oral vaccination of mice with adenoviral vectors is not impaired by preexisting immunity to the vaccine carrier. J Virol 77:10780–10789
Xie C, He JS, Zhang M, Xue SL, Wu Q, Ding XD, Song W, Yuan Y, Li DL, Zheng XX, Lu YY, Shang Z (2007) Oral respiratory syncytial virus (RSV) DNA vaccine expressing RSV F protein delivered by attenuated *Salmonella typhimurium*. Hum Gene Ther 18:746–752. doi:10.1089/hum.2007.053
Xin KQ, Urabe M, Yang J, Nomiyama K, Mizukami H, Hamajima K, Nomiyama H, Saito T, Imai M, Monahan J, Okuda K, Ozawa K (2001) A novel recombinant adeno-associated virus vaccine induces a long-term humoral immune response to human immunodeficiency virus. Hum Gene Ther 12:1047–1061. doi:10.1089/104303401750214276
Yu JR, Kim S, Lee JB, Chang J (2008) Single intranasal immunization with recombinant adenovirus-based vaccine induces protective immunity against respiratory syncytial virus infection. J Virol 82:2350–2357. doi:10.1128/JVI.02372-07

Part IV
Animal and Tissue Culture Models of Infection

Bovine Model of Respiratory Syncytial Virus Infection

Geraldine Taylor

Abstract Bovine respiratory syncytial virus (BRSV), which is an important cause of respiratory disease in young calves, is genetically and antigenically closely related to human (H)RSV. The epidemiology and pathogenesis of infection with these viruses are similar. The viruses are host-specific and infection produces a spectrum of disease ranging from subclinical to severe bronchiolitis and pneumonia, with the peak incidence of severe disease in individuals less than 6 months of age. BRSV infection in calves reproduces many of the clinical signs associated with HRSV in infants, including fever, rhinorrhoea, coughing, harsh breath sounds and rapid breathing. Although BRSV vaccines have been commercially available for decades, there is a need for greater efficacy. The development of effective BRSV and HRSV vaccines face similar challenges, such as the need to vaccinate at an early age in the presence of maternal antibodies, the failure of natural infection to prevent reinfection, and a history of vaccine-augmented disease. Neutralising monoclonal antibodies (mAbs) to the fusion (F) protein of HRSV, which can protect infants from severe HRSV disease, recognise the F protein of BRSV, and vice versa. Furthermore, bovine and human $CD8^+$ T-cells, which are known to be important in recovery from RSV infection, recognise similar proteins that are conserved between HRSV and BRSV. Therefore, not only can the bovine model of RSV be used to evaluate vaccine concepts, it can also be used as part of the preclinical assessment of certain HRSV candidate vaccines.

Contents

G. Taylor (✉)
Compton Laboratory, The Pirbright Institute, Ash Road, Pirbright, Woking, Surrey GU24 ONF, UK
e-mail: geraldine.taylor@pirbright.ac.uk

L. J. Anderson and B. S. Graham (eds.), *Challenges and Opportunities for Respiratory Syncytial Virus Vaccines*, Current Topics in Microbiology and Immunology 372,
DOI: 10.1007/978-3-642-38919-1_16, © Springer-Verlag Berlin Heidelberg 2013

1 Introduction

Bovine respiratory syncytial virus (BRSV) is the most important viral cause of respiratory disease in young calves worldwide, resulting in significant economic losses in the farming industry. BRSV is genetically and antigenically closely related to human (H)RSV, and infections with these viruses have a similar epidemiology (Van der Poel et al. 1994) and pathogenesis (Valarcher and Taylor 2007). Thus, antibodies in bovine serum cross-react with HRSV, and there is a high degree of amino acid identity between HRSV and BRSV proteins, with >80 % amino acid identity in all but three of the viral proteins (Table 1). Outbreaks of HRSV and BRSV respiratory disease have a seasonal periodicity in countries with a temperate climate, peaking in the autumn and winter months. Similar to HRSV infections, the majority of cattle become infected with BRSV during the first year of life, and the severity of disease is variable, ranging from subclinical to fatal, with a peak incidence of severe disease in calves between 1 and 3 months of age, when maternal antibodies are present. Man and cattle can be reinfected with RSV, although reinfections with BRSV are usually subclinical or

Table 1 Relationship between BRSV and HRSV proteins

Protein	Protein name	Size		% AA identity
		BRSV	HRSV	
NS1	Non-structural protein 1	136	139	68
NS2	Non-structural protein 2	124	124	84
N	Nucleoprotein	391	391	93
P	Phosphoprotein	241	241	81
M	Matrix protein	256	256	89
SH	Small hydrophobic protein	73	64	38
G	Attachment protein	257/263	298	30
F	Fusion protein	574	574	81
M2-1	Matrix protein	186	192	80
L	Polymerase	2161	2165	84

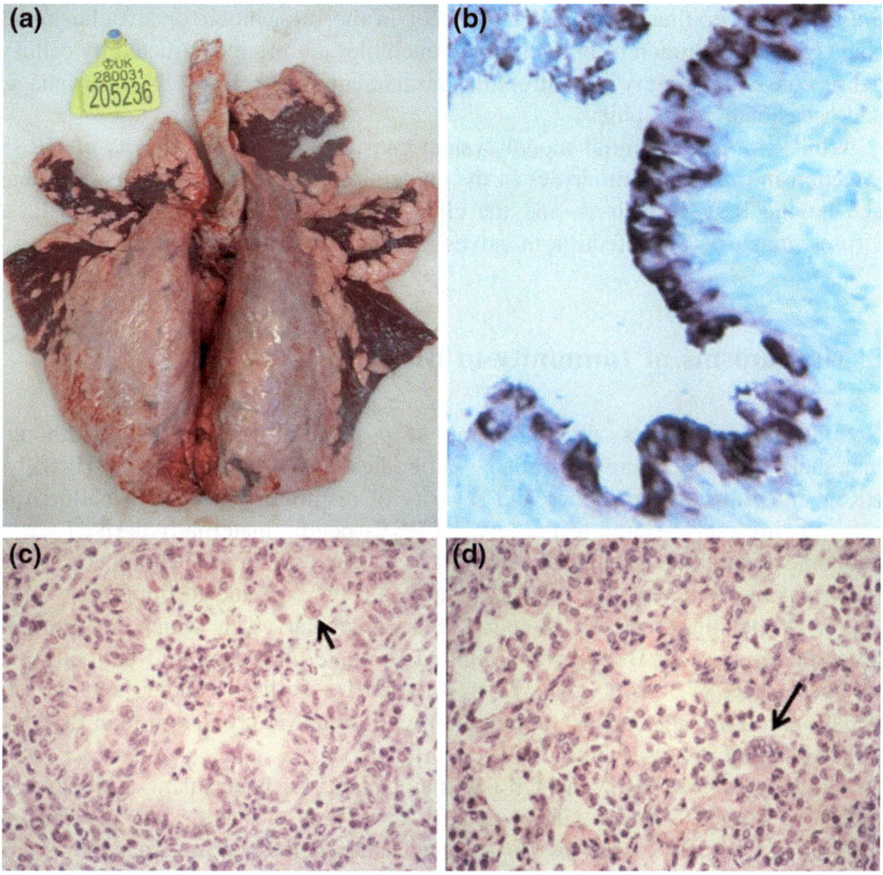

Fig. 1 Pulmonary pathology in calves following experimental BRSV challenge. **a** Pneumonic consolidation in a BRSV-seronegative, 4-week-old SPF calf infected with in vivo-passaged BRSV by aerosol; **b** in situ hybridisation showing the distribution of BRSV G mRNA in ciliated airway epithelial cells, 6 days after BRSV infection; **c** bronchiolitis, 6 days after infection, characterised by hypertrophy of bronchiolar epithelium, syncytia (*arrow*), bronchiolar exudate containing neutrophils, macrophages and desquamated epithelial cells, and peribronchiolar lymphoid infiltration; **d** alveolitis, 6 days after infection, characterised by hypertrophy of alveolar epithelium, desquamation of cells and syncytia (*arrow*)

result in mild disease (Van der Poel et al. 1994). However, primary BRSV infections in older cattle can cause disease (Elvander 1996).

Both HRSV and BRSV primarily infect ciliated airway (Fig. 1b) and alveolar epithelial cells and induce a similar pattern of pathology characterised by a proliferative and exudative bronchiolitis and, in the most severe cases, interstitial pneumonia (Viuff et al. 2002; Johnson et al. 2007; Welliver et al. 2007) (Fig. 1). Epithelial necrosis and apoptosis, with phagocytosis of apoptotic cells by neighbouring epithelial cells can be seen and occasional multinuclear syncytia may be present in the epithelium, alveolar walls or in the bronchiolar or alveolar lumen (Fig. 1c, d). The lumen of bronchi and bronchioles are often obstructed by cellular debris consisting mainly of neutrophils, desquamated epithelial cells, macrophages and occasional eosinophils.

Whilst no single animal model system can fully reflect all aspects of HRSV infection in man, the similarities in the epidemiology and pathogenesis of human and bovine RSV infections, and the close antigenic relationship between these viruses, make BRSV infection in calves a good animal model for HRSV.

2 Mechanisms of Immunity to BRSV

Neutralising antibodies are important in protection against RSV. Although maternally derived serum neutralising antibodies do not protect against HRSV or BRSV infection, the incidence and severity of respiratory disease in children or calves is inversely related to the level of RSV-specific maternally derived serum antibodies (Kimman and Westenbrink 1990). Mucosal immunity appears to be important in protection against RSV and would be expected to limit virus replication at the initial site of infection. Studies in calves in which the primary mucosal response to BRSV was inhibited by maternally acquired antibody and studies of BRSV vaccine efficacy, suggest that the ability to mount a rapid secondary mucosal IgA response may be more important in protection than the level of pre-existing mucosal antibodies.

The F protein of HRSV, which is a major target for neutralising antibodies, is highly conserved between HRSV and BRSV strains (Table 1), and neutralising monoclonal antibodies (mAbs) to antigenic sites corresponding to amino acids 255–275 and 417–438 of the F protein neutralise both HRSV and BRSV, and are cross-protective (Taylor et al. 1992; Thomas et al. 1998).

The G protein is also a target for neutralising antibodies. Although the HRSV and BRSV G proteins share the same structural features, they are antigenically distinct. The G protein is the most variable of the HRSV proteins, with up to 45 % difference in amino acid identity between HRSV isolates. In contrast, there is only 15 % difference in amino acid identity between different isolates of BRSV (Furze et al. 1997). There are at least two major antigenic subgroups of BRSV and polyclonal sera from BRSV-infected calves or calves vaccinated with a recombinant vaccinia virus (rVV) expressing the BRSV G protein recognise the G protein only from viruses in the homologous subgroup (Furze et al. 1994, 1997).

Although neutralising antibodies are important in resistance to RSV, $CD8^+$ T-cells play an important role in clearance of virus. In contrast to HRSV-infected mice, delayed clearance of BRSV from the upper and lower respiratory tract of calves (Taylor et al. 1995) depleted of $CD8^+$ T-cells is associated with more severe pulmonary pathology. Similarly, HRSV replication is prolonged and disease is severe in individuals immunocompromised by chemotherapy or as a result of a primary immunodeficiency disorder (Hall et al. 1986). Clearance of BRSV is associated with an influx of $CD8^+$ T-cells into the lungs (McInnes et al. 1999; Antonis et al. 2006) and induction of BRSV-specific, MHC class I-restricted cytotoxic T-cell responses in peripheral blood and lungs, 7–10 days after BRSV infection (Gaddum et al. 1996; West et al. 2000). Although there appears to be a paucity of pulmonary $CD8^+$ T-cells in the lungs of infants that have died from HRSV infection (Welliver et al. 2007), clearance of HRSV is associated with a $CD8^+$ T-cell response, which peaks in the peripheral blood of infants 6–10 days after admittance to hospital with severe HRSV infection (Lukens et al. 2010). RSV-specific human and bovine $CD8^+$ T-cells recognise the N, F, M and M2 proteins, which are highly conserved between HRSV and BRSV (Table 1), and bovine $CD8^+$ T-cells also recognise the BRSV P and G proteins (Gaddum et al. 2003; Antonis et al. 2006; Cherrie et al. 1992; Taylor et al. unpublished observations).

3 Experimental BRSV Infection

Studies to evaluate BRSV vaccine efficacy depend upon a reproducible model of BRSV infection. However, attempts to induce clinical signs of respiratory disease in calves experimentally infected with BRSV have yielded inconsistent results. A variety of different protocols, including administration of high titres of virus by the intranasal (i.n.) and intratracheal (i.t.) routes, aerosol, or a combination of aerosol and i.t. inoculation, have induced disease ranging from clinically unapparent to severe (Belknap et al. 1995; West et al. 1999). Failure to induce clinical signs of disease may have arisen because of the use of low virulent isolates of BRSV, or because the virus had been attenuated by passage in tissue culture. The use of low tissue culture passaged virus or virus passaged in specific-pathogen-free (SPF) calves appears to be important for the induction of disease similar to that seen in naturally occurring outbreaks. Aerosol (<0.5 μm particles) challenge of calves with $\sim 10^4$ pfu of BRSV in bronchoalveolar lavage (BAL) fluid from SPF calves, free from other bovine respiratory pathogens, reproducibly induces severe signs of respiratory disease. Virus can be isolated from the nasopharynx from day 2 to 3 post-infection and reaches peak titres from day 5 to 6. The severity of clinical disease and pulmonary pathology in calves inoculated i.n. and i.t. with BRSV in BAL fluid is variable, and lung lesions are mainly confined to the cranial and middle lobes. In contrast, clinical signs of disease and lung lesions are consistently induced by aerosol challenge, and lesions are present in all lobes of the lung (Fig. 1a). Whilst aerosol administration of in vivo-passaged BRSV provides a model for evaluating the ability of a vaccine to protect against clinical signs of disease, a challenge that delivers virus particles deep into the lungs may not mimic the course of naturally occurring BRSV infection, which starts in the upper respiratory tract several days before virus replication occurs in the lungs.

Studies of vaccine-augmented RSV infections in rodent models have been compromised by the contribution of immune responses to non-viral proteins present in the vaccine, and which were also present in the challenge virus (Ostler and Ehl 2002). Thus, vaccinated mice become sensitised to non-viral components, such as foetal calf serum, present in the vaccine and when they are challenged with RSV, the recall response to the same non-viral components in the virus preparation plays a role in the inflammatory response. Therefore, an advantage of in vivo-passaged BRSV as a challenge in vaccine efficacy studies is that it is unlikely to contain the same immunogenic non-viral proteins that are present in the vaccine to which calves may have been sensitised as a result of vaccination.

3.1 Clinical Signs of BRSV Disease and Pathophysiology

Clinical signs of disease, gross pneumonic consolidation and analysis of pulmonary function can be used to determine the severity of BRSV infection in calves. Following experimental BRSV infection, calves may develop a range of clinical signs with varying severity of fever, tachypnea, dyspnea, lung sounds, coughing, depression, anorexia, and ocular and nasal discharge. These parameters can be evaluated daily and a weighted value allocated for each observation, varying, for example, from 1 (mild) to 3 (severe) to give a clinical disease score.

Gross pneumonic consolidation of the cranial, medial and accessory lobes may be induced by experimental BRSV infection, which can extend to the caudal lobes in severely affected calves. Interstitial emphysema and formation of subpleural and parenchymal bullae may be seen, especially in the caudal lobes (Viuff et al. 2002). The extent of gross pneumonic consolidation can be scored as a percentage of the total lung area (Valarcher et al. 2003), or given a score from 0 to 5 (Viuff et al. 2002), where a score of 0 is given to lungs completely free of lesions, and a score of 5 given to lungs in which most of the cranial, medial and accessory lobes, and at least a third of the caudal lobes is consolidated.

A variety of methods for analysis of pulmonary function can also be applied to cattle. Analysis of pulmonary resistance and dynamic lung compliance is possible in conscious cattle of all ages using an oesophageal balloon catheter and a pneumotachograph. Ventilatory parameters in 2–7 month-old calves are comparable to those of humans. Therefore, lung function techniques designed for humans that do not require active cooperation are applicable to calves. Different forced oscillation techniques have been validated in conscious calves in order to assess changes in respiratory mechanics (Reinhold et al. 1998). Radiography or measurements of functional residual capacity can be used to evaluate emphysema, and impaired gas exchange can be evaluated by arterial blood gas analyses or pulse oximetry in calves. Depending upon the severity of disease, experimental BRSV infection has been reported to reduce PaO_2, increase minute ventilation, and decrease dynamic lung compliance (Gershwin et al. 1998).

3.2 Pulmonary Inflammatory Response

Airway inflammation can be assessed using bronchoscopy to obtain BAL fluid at different times post-infection and at post-mortem by infusing the whole or parts of the lung with phosphate-buffered saline. The BAL fluid can be used for cytological, immunological or virological studies. However, because of the high degree of lung lobulation, BAL samples are not necessarily representative of the whole lung. Nevertheless, phenotypic analysis of BAL cells can provide an indication of the pulmonary inflammatory response in calves. Following experimental BRSV infection, levels of TNFα and IFNγ in BAL fluid increase with a peak around day 6 (Antonis et al. 2010). This corresponds with peak virus titres in the lungs and with a marked increase in the number of neutrophils in BAL fluid (Taylor et al. 1989), similar to that seen in infants with HRSV bronchiolitis (McNamara et al. 2003). In SPF or gnotobiotic calves, the proportion of neutrophils in BAL can be used to determine the effect of vaccination on the pulmonary inflammatory response induced following BRSV challenge. BAL can also be used to analyse the kinetics, magnitude and antigenic specificity of BRSV-specific T-cells (Antonis et al. 2006).

Microscopic lung lesions of bronchiolitis and alveolitis can be scored as absent, mild, moderate or severe. However, only small areas of the bovine lung can be sampled at post-mortem examination. Therefore, lung sections can provide information on the nature of the histopathological changes but are not necessarily representative of the whole lung.

4 BRSV in Calves as a Model for HRSV Vaccine Development

As a natural bovine pathogen, BRSV engages with the innate and acquired immune responses of cattle in an evolutionarily meaningful fashion, and lessons learnt from an evaluation of BRSV vaccines in calves can inform studies on the development of HRSV vaccines. Although BRSV vaccines have been commercially available since the 1970, there have been reports of vaccine-exacerbated disease, poor efficacy and poor duration of immunity, highlighting the need for the development of more effective BRSV vaccines. BRSV vaccine development faces the same challenges as vaccine development for HRSV. Thus, protective immune responses need to be induced within the first few months of life in individuals with maternally derived antibodies; vaccines should not prime for augmented disease; and, ideally, vaccines should induce more durable protection than that induced by natural infection. Calves acquire maternal antibodies through ingestion of colostrum during the first day of life and regulating the amount of colostrum given to new born calves provides a means of evaluating vaccine efficacy in animals with different levels of maternal antibodies. Furthermore, studies on the ability of maternal vaccination to boost protection against BRSV provided by passively transferred antibodies could be undertaken in young calves. In addition to the insight that can be provided by studies on BRSV vaccine development in calves, certain HRSV vaccine candidates composed of conserved RSV proteins such as the F, a target for neutralising antibodies, and internal viral antigens such as the N, M and M2 proteins, targets for $CD8^+$ T-cell responses, can be evaluated for their ability to protect against BRSV infection in calves.

4.1 Inactivated BRSV Vaccines

Despite conflicting descriptions of efficacy and reports of enhanced respiratory disease in calves vaccinated with inactivated BRSV, such vaccines have been commercially available for many years. In contrast to live virus, inactivated BRSV vaccines induce low ratios of virus-neutralising antibodies compared to BRSV-specific IgG in calves, as detected by ELISA (West and Ellis 1997). Furthermore, a commercial inactivated BRSV vaccine primes for a Th2-biased immune response, BRSV-specific serum IgE and pulmonary eosinophilia in mice following BRSV challenge (Oumouna et al. 2005). Nevertheless, vaccination of BRSV-seronegative calves with an inactivated BRSV vaccine has induced partial protection against clinical signs of disease, development of pulmonary consolidation and nasopharyngeal excretion of virus following experimental BRSV challenge (Ellis et al. 2001). Complete protection against BRSV challenge has been induced in calves given an inactivated vaccine formulated in saponin, which induced serum neutralising antibodies similar to those induced by parenteral live BRSV vaccines (Ellis et al. 2005). A vaccine composed of gluteraldehyde-fixed, BRSV-infected bovine nasal mucosal cells given subcutaneously (s.c.) to calves with low levels of maternal antibodies also induced neutralising antibodies and complete protection in 11 out of 12 animals (Stott et al. 1984). This vaccine, formulated with Quil A, was effective in calves with high levels of maternal antibody at the time of vaccination and, in field trials, reduced mortality due to respiratory disease by $\sim$70 % and significantly reduced the proportion of calves treated for respiratory disease (Howard et al. 1987). BRSV-ISCOMs, which are open cage-like complexes, typically with a diameter of about 40 nm, formed from cholesterol, lipid, solubilised purified BRSV and saponins, were also highly effective at protecting young calves with maternal antibodies against BRSV challenge (Hagglund et al. 2004). Protection induced by BRSV-ISCOMs was significantly greater than that induced by a commercial inactivated BRSV vaccine and correlated with a rapid mucosal antibody and T-cell response in peripheral blood following challenge. ISCOM-based vaccines have been shown to be highly effective in inducing antibody, T helper and cytotoxic T lymphocyte responses in a number of animal species, including non-human primates (Sjolander et al. 2001).

In contrast to the protective efficacy of inactivated BRSV vaccines described above, some field studies have reported a lack of protection, or even disease enhancement in calves vaccinated with a commercial β-propiolactone-inactivated

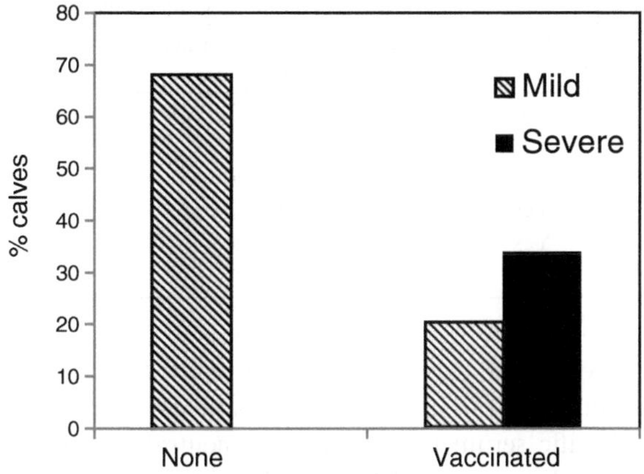

Fig. 2 Vaccine-enhanced BRSV disease in calves. Incidence of severe BRSV respiratory disease in calves (~7 months old) that had been vaccinated 3–4 months previously with a commercial inactivated BRSV vaccine, compared with that in unvaccinated calves (~2.5 months old). Mild respiratory disease was characterised by tachypnoea, coughing, nasal discharge and fever; severe disease was characterised by ptyalism, breathing with open mouth, dyspnoea, abnormal breath sounds on auscultation, cyanotic ocular mucosa and anorexia [adapted from Schreiber et al. (2000)]

BRSV vaccine, formulated with saponin and aluminium hydroxychloride, which was withdrawn from the European market (Schreiber et al. 2000; Larsen et al. 2001). Thus, in one study, nearly all calves in two herds developed BRSV respiratory disease, 2 months after vaccination, with a death rate of 1.6–8 % (Larsen et al. 2001). In another study, severe disease was seen in 34 % of vaccinated calves, all but one of which died, whereas only mild disease was seen in non-vaccinated calves, none of whom died (Fig. 2) (Schreiber et al. 2000). At the time of the BRSV outbreak, 3–4 months after vaccination, the mean age of the vaccinated calves was ~7 months and that of the non-vaccinated calves was 2.5 months. BRSV antigen was isolated from the lungs of the vaccinated calves that died and lung lesions were characterised by an interstitial pneumonia, with bronchitis and bronchiolitis obliterans, pneumocyte type II hyperplasia with syncytial cell formation, and an infiltration of neutrophils and eosinophils. These features of altered age distribution for severe disease and the pattern of lung pathology were similar to those seen in the two FI-HRSV vaccinated infants that died following natural HRSV infection (Kim et al. 1969). Furthermore, there was a correlation between the number of doses of vaccine and the incidence of severe disease, with severe disease in 42 % of calves that had been vaccinated three times, compared with 18 % of calves that had been vaccinated on only one or two occasions.

Attempts to experimentally reproduce vaccine-augmented BRSV disease have yielded inconsistent and conflicting results, and none have fully reproduced the vaccine-enhanced disease seen in infants vaccinated with FI-HRSV (Kim et al.

Table 2 Experimental models of FI-BRSV vaccine-augmented disease

Clinical disease	Pulmonary eosinophilia	Protection against infection	Reference
Enhanced	No	No	(Gershwin et al. 1998)
Not enhanced	No	Not known	(Kalina et al. 2004)
Earlier onset	Yes	No	(West et al. 1999)
Earlier onset and enhanced	Yes	Yes	(Antonis et al. 2003)

1969) or that reported in calves vaccinated with β-propiolactone-inactivated BRSV (Schreiber et al. 2000). In experimental studies, vaccination with FI-BRSV induced BRSV-specific serum IgG, little or no neutralising antibody, and a variable degree of protection against BRSV replication (Table 2). The vaccine-augmented disease described by Gershwin et al. (1998) was associated with virus replication in the lungs and a more extensive proliferative alveolitis in vaccinated compared with control calves, but was not associated with pulmonary eosinophilia. In contrast, pulmonary eosinophilia was associated with vaccine-enhanced BRSV disease in studies by other groups, both of which described an earlier onset of clinical signs of disease, but less extensive pulmonary pathology in the majority of vaccinated calves, 8–9 days after challenge (West et al. 1999; Antonis et al. 2003). In these models of FI-BRSV vaccine-enhanced disease, severe disease was associated with induction of BRSV-specific IgE (Antonis et al. 2003; Kalina et al. 2004) and a type I hypersensitivity and Arthus reaction to an intradermal inoculation of BRSV (West et al. 1999).

The failure of experimental studies in calves vaccinated with FI-BRSV to fully reproduce the severe disease seen in infants vaccinated with FI-HRSV and the inconsistent reports of the efficacy of inactivated BRSV vaccines may be related to differences in the method of BRSV inactivation, dose of vaccine (Kalina et al. 2004), adjuvant, interval between vaccination and challenge, age of the calves, breed of calves, and/or the level of maternal antibodies present at the time of vaccination. However, differences in the outcome of BRSV infection in calves vaccinated with inactivated BRSV vaccines were unlikely to have been due to breed differences since a similar breed of cattle (Holstein) was used in at least three of the studies summarised in Table 2. Further studies in calves addressing these variables could help to shed light on the mechanisms of RSV-vaccine exacerbated disease.

4.2 Live-Attenuated BRSV Vaccines

The advantage of live-attenuated RSV vaccines is that they induce neutralising antibodies and prime $CD8^+$ T-cells, which are important mediators of protection. Moreover, the presence of $CD8^+$ T-cells producing IFNγ has been associated with diminished Th2 responses and pulmonary eosinophilia in small animal models of RSV, which have been associated with vaccine-exacerbated RSV disease. Live-attenuated BRSV vaccines, produced by serial passage in cell culture, have been commercially available since the 1970s and are given either intramuscularly (i.m.) or i.n. to calves. Experimental studies have shown that one or two doses of 5×10^5 p.f.u. of a commercial live BRSV given i.m. to BRSV-seronegative calves reduced nasopharyngeal excretion of BRSV following a mild BRSV challenge (Stott et al. 1984), and reduced clinical disease and pulmonary pathology following a severe BRSV challenge (West et al. 2000). However, results of studies with this vaccine in young seropositive calves have shown variable efficacy, and there is a report suggesting that i.m. vaccination with a live BRSV vaccine during the course of a natural infection may enhance the severity of disease (Kimman et al. 1989a). Nevertheless, there is evidence that i.m. vaccination with live BRSV can reduce the impact of respiratory disease in the field (Bohlender 1984). In contrast, trials of a live HRSV vaccine ($\sim 10^4$ $TCID_{50}$) given i.m. to children showed no evidence of protection against HRSV disease (Belshe et al. 1982). The failure of live HRSV given i.m. to protect may have been related to the dose of vaccine virus, which was $\sim$50-fold less than that used in calves, and the levels of maternally acquired antibodies. Although maternal antibodies severely inhibit the antibody response to parenterally administered live virus in both children and calves (Belshe et al. 1982; Kimman et al. 1989b), there is evidence that this mode of vaccination can prime for a rapid secondary immune response following BRSV challenge (West et al. 2000). Protection correlated with a rapid BRSV-specific cytotoxic T-cell, serum and mucosal antibody responses following challenge (West et al. 2000).

Since natural RSV infections do not prime for more severe disease following reinfection, much effort has been invested in the development of live-attenuated mucosal RSV vaccines, which should induce immunity in the respiratory tract. A mucosal live-attenuated RSV vaccine needs to replicate efficiently in order to generate a protective immune response without causing disease, thereby combining safety with a high level of immunogenicity. Following i.n. vaccination of calves with a live-attenuated BRSV commercial vaccine (10^5 $TCID_{50}$), produced by serial passage in cell culture, the vaccine virus is excreted for 4–11 days (Vangeel et al. 2007) and virus can be detected by RT-PCR for up to 20 days after vaccination (Timsit et al. 2009), but does not induce clinical signs of disease. However, since other studies have demonstrated that tissue-culture adapted BRSV reverts to virulence following in vivo passage in calves, it is possible that the vaccine virus could spread to non-vaccinated susceptible animals, revert to virulence and cause disease. The efficacy of i.n. vaccination with live-attenuated BRSV has been evaluated as a single component vaccine and as a component of a multivalent vaccine. A single i.n. vaccination of young, BRSV-seronegative calves with a commercial live-attenuated BRSV vaccine, reduced nasopharyngeal excretion of BRSV as early as 10 days after vaccination (Vangeel et al. 2007), and reduced, but did not prevent, the development of clinical signs of disease and pneumonic consolidation, when challenged 3 weeks after vaccination (Ellis et al.

2007; Xue et al. 2010). However, maternally derived antibodies suppressed the BRSV-specific serum antibody response in calves vaccinated i.n. with the live-attenuated BRSV vaccine, induced only partial protection following challenge, 9 weeks after vaccination (Vangeel et al. 2007), and failed to protect against either BRSV infection or respiratory disease when challenged 4.5 months after vaccination (Ellis et al. 2010). The poor efficacy of i.n. vaccination seen in the latter study may be related to the inhibitory effects of maternally derived antibodies, the duration of immunity induced by intranasal vaccination with live-attenuated BRSV and/or the severity of the BRSV challenge. These observations in calves suggest that it may not be possible for an attenuated RSV vaccine to induce better immunity than infection with wild-type HRSV.

The ability to genetically manipulate the genome of RS viruses has provided the opportunity to identify attenuating mutations and generate genetically stable attenuated vaccine candidates that combine safety with a high level of immunogenicity. The calf provides a valuable model for studying the virulence of genetically engineered BRSV in a natural host. However, calves may not be a good model to evaluate the ability of live-attenuated mutant BRSV to induce nasal congestion, which is a particular problem with live-attenuated HRSV vaccine candidates in infants. Deletion of the G, NS1, NS2 or SH genes results in BRSV with an attenuated phenotype in calves, and mucosal vaccination with any of these mutant viruses can protect against challenge with wild-type BRSV (Schmidt et al. 2002; Valarcher et al. 2003; Taylor et al. unpublished observations). However, it remains to be determined if any of these attenuated viruses are able to induce greater and more durable protective immunity than live, classically attenuated or even wild-type BRSV in young calves with maternal antibodies.

4.3 Subunit and DNA Vaccines

Whereas a number of subunit HRSV vaccine candidates have been evaluated in small animal models and man, there have been few studies on subunit BRSV vaccine candidates in calves. A peptide corresponding to amino acid region 174–187 of the BRSV G glycoprotein, containing a Cys_{186} to Ser substitution, coupled to keyhole limpet hemocyanin (KLH), completely protected mice against a homologous strain of BRSV, and although it did not protect calves against BRSV infection, it appeared to reduce the extent of pneumonic consolidation (Bastien et al. 1997).

DNA vaccination results in the intracellular expression of the encoded antigen which is then processed and presented through both MHC class I and II pathways to induce antibodies and prime $CD8^+$ T-cells, in a similar way to virus infection. DNA vaccines are effective against a range of pathogens, especially in mice. However, they have been less successful in man and livestock species. Parenteral vaccination of young BRSV-seronegative calves with high doses of a DNA vaccine encoding the BRSV F gene induced low levels of neutralising antibodies and partial protection against BRSV challenge (Taylor et al. 2005). Similar results were seen in calves vaccinated with mixture DNA plasmids encoding the BRSV N, a target for bovine $CD8^+$ T-cells, and F protein (Boxus et al. 2007). However, protective efficacy was improved by boosting with a commercial, inactivated BRSV, Quil A-adjuvanted vaccine (Boxus et al. 2007). Although there is evidence that DNA vaccines can circumvent the inhibitory effects of maternal antibodies, immunity induced by DNA vaccines, with or without boosting with inert antigens, is slow to develop and may not be rapid enough to provide protective immunity in individuals 2–3 months of age, which have the greatest burden of severe disease.

4.4 Virus-Vectored Vaccines

Foreign proteins expressed by replication-competent or replication-defective virus vaccine vectors are also processed and presented to the immune system in the same way as live virus. HRSV antigens have been expressed in a wide range of vaccine vectors and evaluated in small animal and non-human primate models, and some of these are currently in clinical trials. However, few have been evaluated in calves. Recombinant vaccinia viruses (rVV) expressing BRSV F or G proteins, but not the N or M2 proteins, induced neutralising antibodies in calves without maternally derived antibodies, and all, apart from that expressing M2, induced partial protection against BRSV infection, 6 weeks after vaccination (Taylor et al. 1997). In contrast to the enhanced lung pathology seen following HRSV challenge in mice vaccinated with rVV expressing HRSV proteins, there was a reduction in pulmonary pathology in calves vaccinated with rVV expressing BRSV F, G or N proteins. Mucosal (i.n. and i.t.) vaccination of young, BRSV-seronegative calves with a recombinant, attenuated bovine herpesvirus (BHV-1) expressing the BRSV G protein was more effective at inducing protection against BRSV than rVV-G (Taylor et al. 1998). However, expression of the BRSV G protein increased the virulence of BHV-1 for the respiratory tract of calves, suggesting that replication competent virus vectors expressing the HRSV G protein should be used with caution (Taylor et al. 1998).

Replication defective-virus vectors have a good safety profile and modified vaccinia virus Ankara (MVA) has been used extensively in the human population, with no adverse effects. MVA replicates in avian cells, in vitro, but its replication is restricted in mammalian cells. MVA expressing HRSV F and/or G proteins induced protective immunity in mice, but failed to confer protection in non-human primates (Wyatt et al. 1999; de Waal et al. 2004). As seen in non-human primates vaccinated with MVA, vaccination of BRSV-seronegative calves on two occasions with MVA expressing BRSV F and/or G proteins induced only low levels of neutralising antibodies, but these were rapidly boosted following BRSV challenge. Although vaccination reduced lung viral load and protected calves against the development of lung pathology, there was only slight protection against clinical signs of disease following BRSV challenge (Antonis et al. 2007). Thus, parenteral vaccination with MVA expressing F and/or G proteins was only partially effective in calves. However, it may be more effective as part of a heterologous prime/boost vaccination strategy.

4.5 Evaluation of HRSV Vaccine Candidates Against BRSV Infection in Calves

Although HRSV and BRSV display a highly restricted host range in the field, limited replication of HRSV has been demonstrated in the respiratory tract of experimentally infected gnotobiotic calves (Thomas et al. 1984). However, it is likely that replication of live-attenuated HRSV mutants in the bovine respiratory tract will be too restricted to induce protective immunity. Nevertheless, a temperature-sensitive mutant of HRSV (ts-1) given i.m. to calves reduced nasopharyngeal excretion of BRSV to the same extent as i.m. vaccination with live BRSV (Stott et al. 1984).

The high degree of amino acid sequence conservation between all but three of the HRSV and BRSV proteins provides the opportunity to determine the ability of some HRSV vaccine vectors to induce cross-protective immunity against BRSV in calves. Thus, scarification of BRSV-seronegative calves with rVV expressing the HRSV F protein induced BRSV-specific serum neutralising antibodies and protected the lungs against BRSV infection (Taylor et al. 1997). Similarly, vaccination of BRSV-seronegative calves with HRSV N protein produced as ring-shaped nanoparticles, formulated with adjuvant, induced partial protection against BRSV challenge (Riffault et al. 2010). In addition, strong protection against BRSV infection has been induced in calves primed i.n. with a chimpanzee adenovirus, which is not recognised by pre-existing immunity in man or cattle, expressing a string of HRSV proteins, F, N and M2-1, and boosted i.m. with MVA expressing the same HRSV proteins (Taylor et al. manuscript in preparation).

5 Conclusions

Ideally, an HRSV vaccine should induce a rapid, protective antibody and T-cell response, not prime for enhanced disease, and be able to boost immune responses in those whose immunity has declined. Many current vaccine efforts therefore seek to induce an effective RSV-specific memory T-cell response in addition to a neutralising antibody response. Whilst the extensively studied rodent models of HRSV have provided valuable information on the types of immune responses that can contribute to protection and vaccine-exacerbated disease, they are only semi-permissive for HRSV replication, requiring high doses of virus to infect, and to induce lung pathology and clinical signs of disease. Furthermore, rodent models do not necessarily predict vaccine efficacy and safety for human neonates. BRSV infection of calves mimics human disease more closely than experimental infection of unnatural laboratory hosts. Studies on BRSV in calves therefore provide a valuable model not only to evaluate HRSV vaccine concepts but could also be part of the preclinical assessment of HRSV vaccine candidates which contain proteins that are conserved between HRSV and BRSV. Calves can be infected with low doses of in vivo-passaged virus and the outcome of infection can be influenced by the method of administration and the dose of virus. Although aerosol administration with low doses of BRSV can induce clinical signs of disease in the majority of calves, it does not recapitulate the normal progression of RSV infection, which starts with initial virus local replication in the nose, followed by spread to the lower respiratory tract. The induction of a rapid secondary mucosal immune response as a result of an initial period of virus replication in the nose in vaccinated animals would be expected to restrict spread of the virus to the lower airways. However, induction of a rapid secondary immune response as a result of the introduction of aerosolised virus deep into the lower airways may result in inflammation and, as a result, a more rapid onset and possibly faster resolution of disease than in animals undergoing a primary infection, as has been seen in some studies evaluating BRSV vaccine efficacy. Therefore, whilst the experimental reproduction of clinical signs of disease in calves is a valuable tool in vaccine efficacy studies, interpretation of the results should take into account the differences between the course of a natural infection and that induced by an aerosol challenge. Evaluation of the ability of vaccines to reduce the incidence of respiratory disease following natural exposure to BRSV could also be undertaken in field trials in which controls are age-matched on the same farm. Although naturally occurring outbreaks of respiratory disease in calves are often associated with a variety of different microorganisms, BRSV is considered to be the major contributor to disease, and protection against disease has been shown in calves vaccinated with a BRSV vaccine alone (Howard et al. 1987). Whilst the housing and handling of calves requires specialised facilities and expertise, the availability of a wide range of commercially available reagents to analyse immune responses in calves, tools to analyse airway physiology, and similarities in the pathology of BRSV in calves and HRSV in infants, makes the calf an ideal model for HRSV infection. Studies on BRSV in the calf, which closely matches that of HRSV in infants, will facilitate the efficient translation of results from basic research to successful clinical application.

References

Antonis AF, Claassen EA, Hensen EJ, de Groot RJ, de Groot-Mijnes JD, Schrijver RS, van der Most RG (2006) Kinetics of antiviral CD8 T cell responses during primary and post-vaccination secondary bovine respiratory syncytial virus infection. Vaccine 24(10):1551–1561. doi:10.1016/j.vaccine.2005.10.005

Antonis AF, de Jong MC, van der Poel WH, van der Most RG, Stockhofe-Zurwieden N, Kimman T, Schrijver RS (2010) Age-dependent differences in the pathogenesis of bovine respiratory syncytial virus infections related to the development of natural immunocompetence. J Gen Virol 91(Pt 10):2497–2506. doi:10.1099/vir.0.020842-0

Antonis AF, Schrijver RS, Daus F, Steverink PJ, Stockhofe N, Hensen EJ, Langedijk JP, van der Most RG (2003) Vaccine-induced immunopathology during bovine respiratory syncytial virus infection: exploring the parameters of pathogenesis. J Virol 77(22):12067–12073

Antonis AF, van der Most RG, Suezer Y, Stockhofe-Zurwieden N, Daus F, Sutter G, Schrijver RS (2007) Vaccination with recombinant modified vaccinia virus Ankara expressing bovine respiratory syncytial virus (bRSV) proteins protects calves against RSV challenge. Vaccine 25(25):4818–4827. doi:10.1016/j.vaccine.2007.04.002

Bastien N, Taylor G, Thomas LH, Wyld SG, Simard C, Trudel M (1997) Immunization with a peptide derived from the G glycoprotein of bovine respiratory syncytial virus (BRSV) reduces the incidence of BRSV-associated pneumonia in the natural host. Vaccine 15(12–13):1385–1390

Belknap EB, Ciszewski DK, Baker JC (1995) Experimental respiratory syncytial virus infection in calves and lambs. J Vet Diagn Invest 7(2):285–298

Belshe RB, Van Voris LP, Mufson MA (1982) Parenteral administration of live respiratory syncytial virus vaccine: results of a field trial. J Infect Dis 145(3):311–319

Bohlender RE (1984) Field trials of a bovine respiratory syncytial virus vaccine. Mod Vet Pract 65(8):606–609

Boxus M, Tignon M, Roels S, Toussaint JF, Walravens K, Benoit MA, Coppe P, Letesson JJ, Letellier C, Kerkhofs P (2007) DNA immunization with plasmids encoding fusion and nucleocapsid proteins of bovine respiratory syncytial virus induces a strong cell-mediated immunity and protects calves against challenge. J Virol 81(13):6879–6889. doi:10.1128/JVI.00502-07

Cherrie AH, Anderson K, Wertz GW, Openshaw PJ (1992) Human cytotoxic T cells stimulated by antigen on dendritic cells recognize the N, SH, F, M, 22 K, and 1b proteins of respiratory syncytial virus. J Virol 66(4):2102–2110

de Waal L, Wyatt LS, Yuksel S, van Amerongen G, Moss B, Niesters HG, Osterhaus AD, de Swart RL (2004) Vaccination of infant macaques with a recombinant modified vaccinia virus Ankara expressing the respiratory syncytial virus F and G genes does not predispose for immunopathology. Vaccine 22(8):923–926. doi:10.1016/j.vaccine.2003.10.010

Ellis J, Gow S, West K, Waldner C, Rhodes C, Mutwiri G, Rosenberg H (2007) Response of calves to challenge exposure with virulent bovine respiratory syncytial virus following intranasal administration of vaccines formulated for parenteral administration. J Am Vet Med Assoc 230(2):233–243. doi:10.2460/javma.230.2.233

Ellis J, West K, Konoby C, Leard T, Gallo G, Conlon J, Fitzgerald N (2001) Efficacy of an inactivated respiratory syncytial virus vaccine in calves. J Am Vet Med Assoc 218(12):1973–1980

Ellis JA, Gow SP, Goji N (2010) Response to experimentally induced infection with bovine respiratory syncytial virus following intranasal vaccination of seropositive and seronegative calves. J Am Vet Med Assoc 236(9):991–999. doi:10.2460/javma.236.9.991

Ellis JA, West KH, Waldner C, Rhodes C (2005) Efficacy of a saponin-adjuvanted inactivated respiratory syncytial virus vaccine in calves. Can Vet J 46(2):155–162

Elvander M (1996) Severe respiratory disease in dairy cows caused by infection with bovine respiratory syncytial virus. Vet Rec 138(5):101–105

Furze J, Wertz G, Lerch R, Taylor G (1994) Antigenic heterogeneity of the attachment protein of bovine respiratory syncytial virus. J Gen Virol 75(Pt 2):363–370

Furze JM, Roberts SR, Wertz GW, Taylor G (1997) Antigenically distinct G glycoproteins of BRSV strains share a high degree of genetic homogeneity. Virology 231(1):48–58. doi:10.1006/viro.1997.8490

Gaddum RM, Cook RS, Furze JM, Ellis SA, Taylor G (2003) Recognition of bovine respiratory syncytial virus proteins by bovine $CD8^+$ T lymphocytes. Immunology 108(2):220–229

Gaddum RM, Cook RS, Thomas LH, Taylor G (1996) Primary cytotoxic T-cell responses to bovine respiratory syncytial virus in calves. Immunology 88(3):421–427

Gershwin LJ, Schelegle ES, Gunther RA, Anderson ML, Woolums AR, Larochelle DR, Boyle GA, Friebertshauser KE, Singer RS (1998) A bovine model of vaccine enhanced respiratory syncytial virus pathophysiology. Vaccine 16(11–12):1225–1236

Hagglund S, Hu KF, Larsen LE, Hakhverdyan M, Valarcher JF, Taylor G, Morein B, Belak S, Alenius S (2004) Bovine respiratory syncytial virus ISCOMs–protection in the presence of maternal antibodies. Vaccine 23(5):646–655. doi:10.1016/j.vaccine.2004.07.006

Hall CB, Powell KR, MacDonald NE, Gala CL, Menegus ME, Suffin SC, Cohen HJ (1986) Respiratory syncytial viral infection in children with compromised immune function. N Engl J Med 315(2):77–81. doi:10.1056/NEJM198607103150201

Howard CJ, Stott EJ, Thomas LH, Gourlay RN, Taylor G (1987) Protection against respiratory disease in calves induced by vaccines containing respiratory syncytial virus, parainfluenza type 3 virus, Mycoplasma bovis and M dispar. Vet Rec 121(16):372–376

Johnson JE, Gonzales RA, Olson SJ, Wright PF, Graham BS (2007) The histopathology of fatal untreated human respiratory syncytial virus infection. Mod Pathol 20(1):108–119. doi:10.1038/modpathol.3800725

Kalina WV, Woolums AR, Berghaus RD, Gershwin LJ (2004) Formalin-inactivated bovine RSV vaccine enhances a Th2 mediated immune response in infected cattle. Vaccine 22(11–12):1465–1474. doi:10.1016/j.vaccine.2003.10.024

Kim HW, Canchola JG, Brandt CD, Pyles G, Chanock RM, Jensen K, Parrott RH (1969) Respiratory syncytial virus disease in infants despite prior administration of antigenic inactivated vaccine. Am J Epidemiol 89(4):422–434

Kimman TG, Sol J, Westenbrink F, Straver PJ (1989a) A severe outbreak of respiratory tract disease associated with bovine respiratory syncytial virus probably enhanced by vaccination with modified live vaccine. Vet Q 11(4):250–253. doi:10.1080/01652176.1989.9694231

Kimman TG, Westenbrink F (1990) Immunity to human and bovine respiratory syncytial virus. Arch Virol 112(1–2):1–25

Kimman TG, Westenbrink F, Straver PJ (1989b) Priming for local and systemic antibody memory responses to bovine respiratory syncytial virus: effect of amount of virus, virus replication, route of administration and maternal antibodies. Vet Immunol Immunopathol 22(2):145–160

Larsen LE, Tegtmeier C, Pedersen E (2001) Bovine respiratory syncytial virus (BRSV) pneumonia in beef calf herds despite vaccination. Acta Vet Scand 42(1):113–121

Lukens MV, van de Pol AC, Coenjaerts FE, Jansen NJ, Kamp VM, Kimpen JL, Rossen JW, Ulfman LH, Tacke CE, Viveen MC, Koenderman L, Wolfs TF, van Bleek GM (2010) A systemic neutrophil response precedes robust CD8(+) T-cell activation during natural respiratory syncytial virus infection in infants. J Virol 84(5):2374–2383. doi:10.1128/JVI.01807-09

McInnes E, Sopp P, Howard CJ, Taylor G (1999) Phenotypic analysis of local cellular responses in calves infected with bovine respiratory syncytial virus. Immunology 96(3):396–403

McNamara PS, Ritson P, Selby A, Hart CA, Smyth RL (2003) Bronchoalveolar lavage cellularity in infants with severe respiratory syncytial virus bronchiolitis. Arch Dis Child 88(10):922–926

Ostler T, Ehl S (2002) A cautionary note on experimental artefacts induced by fetal calf serum in a viral model of pulmonary eosinophilia. J Immunol Methods 268(2):211–218

Oumouna M, Mapletoft JW, Karvonen BC, Babiuk LA, van Drunen Littel-van den Hurk S (2005) Formulation with CpG oligodeoxynucleotides prevents induction of pulmonary immunopathology following priming with formalin-inactivated or commercial killed bovine respiratory syncytial virus vaccine. J Virol 79(4):2024–2032. doi:10.1128/JVI.79.4.2024-2032.2005

Reinhold P, Smith HJ, Close R, Genicot B, Lekeux P (1998) Validation of impulse oscillometry in Friesian and Blue Belgian calves with respect to changes in extrathoracic upper airway resistance. Res Vet Sci 65(2):93–101

Riffault S, Meyer G, Deplanche M, Dubuquoy C, Durand G, Soulestin M, Castagne N, Bernard J, Bernardet P, Dubosclard V, Bernex F, Petit-Camurdan A, Deville S, Schwartz-Cornil I, Eleouet JF (2010) A new subunit vaccine based on nucleoprotein nanoparticles confers partial clinical and virological protection in calves against bovine respiratory syncytial virus. Vaccine 28(21):3722–3734. doi:10.1016/j.vaccine.2010.03.008

Schmidt U, Beyer J, Polster U, Gershwin LJ, Buchholz UJ (2002) Mucosal immunization with live recombinant bovine respiratory syncytial virus (BRSV) and recombinant BRSV lacking the envelope glycoprotein G protects against challenge with wild-type BRSV. J Virol 76(23):12355–12359

Schreiber P, Matheise JP, Dessy F, Heimann M, Letesson JJ, Coppe P, Collard A (2000) High mortality rate associated with bovine respiratory syncytial virus (BRSV) infection in Belgian white blue calves previously vaccinated with an inactivated BRSV vaccine. J Vet Med B Infect Dis Vet Public Health 47(7):535–550

Sjolander A, Drane D, Maraskovsky E, Scheerlinck JP, Suhrbier A, Tennent J, Pearse M (2001) Immune responses to ISCOM formulations in animal and primate models. Vaccine 19(17–19):2661–2665

Stott EJ, Thomas LH, Taylor G, Collins AP, Jebbett J, Crouch S (1984) A comparison of three vaccines against respiratory syncytial virus in calves. J Hyg (Lond) 93(2):251–261

Taylor G, Bruce C, Barbet AF, Wyld SG, Thomas LH (2005) DNA vaccination against respiratory syncytial virus in young calves. Vaccine 23(10):1242–1250. doi:10.1016/j.vaccine.2004.09.005

Taylor G, Rijsewijk FA, Thomas LH, Wyld SG, Gaddum RM, Cook RS, Morrison WI, Hensen E, van Oirschot JT, Keil G (1998) Resistance to bovine respiratory syncytial virus (BRSV) induced in calves by a recombinant bovine herpesvirus-1 expressing the attachment glycoprotein of BRSV. J Gen Virol 79(Pt 7):1759–1767

Taylor G, Stott EJ, Furze J, Ford J, Sopp P (1992) Protective epitopes on the fusion protein of respiratory syncytial virus recognized by murine and bovine monoclonal antibodies. J Gen Virol 73(Pt 9):2217–2223

Taylor G, Thomas LH, Furze JM, Cook RS, Wyld SG, Lerch R, Hardy R, Wertz GW (1997) Recombinant vaccinia viruses expressing the F, G or N, but not the M2, protein of bovine respiratory syncytial virus (BRSV) induce resistance to BRSV challenge in the calf and protect against the development of pneumonic lesions. J Gen Virol 78(Pt 12):3195–3206

Taylor G, Thomas LH, Stott EJ (1989) Effect of vaccination on cell populations in lung washes from calves after infection with respiratory syncytial virus. Res Vet Sci 47(2):231–235

Taylor G, Thomas LH, Wyld SG, Furze J, Sopp P, Howard CJ (1995) Role of T-lymphocyte subsets in recovery from respiratory syncytial virus infection in calves. J Virol 69(11):6658–6664

Thomas LH, Cook RS, Wyld SG, Furze JM, Taylor G (1998) Passive protection of gnotobiotic calves using monoclonal antibodies directed at different epitopes on the fusion protein of bovine respiratory syncytial virus. J Infect Dis 177(4):874–880

Thomas LH, Stott EJ, Collins AP, Crouch S, Jebbett J (1984) Infection of gnotobiotic calves with a bovine and human isolate of respiratory syncytial virus. Modification of the response by dexamethasone. Arch Virol 79(1–2):67–77

Timsit E, Le Drean E, Maingourd C, Belloc C, Guatteo R, Bareille N, Seegers H, Douart A, Sellal E, Assie S (2009) Detection by real-time RT-PCR of a bovine respiratory syncytial virus vaccine in calves vaccinated intranasally. Vet Rec 165(8):230–233

Valarcher JF, Furze J, Wyld S, Cook R, Conzelmann KK, Taylor G (2003) Role of alpha/beta interferons in the attenuation and immunogenicity of recombinant bovine respiratory syncytial viruses lacking NS proteins. J Virol 77(15):8426–8439

Valarcher JF, Taylor G (2007) Bovine respiratory syncytial virus infection. Vet Res 38(2):153–180. doi:10.1051/vetres:2006053

Van der Poel WH, Brand A, Kramps JA, Van Oirschot JT (1994) Respiratory syncytial virus infections in human beings and in cattle. J Infect 29(2):215–228

Vangeel I, Antonis AF, Fluess M, Riegler L, Peters AR, Harmeyer SS (2007) Efficacy of a modified live intranasal bovine respiratory syncytial virus vaccine in 3-week-old calves experimentally challenged with BRSV. Vet J 174(3):627–635. doi:10.1016/j.tvjl.2006.10.013

Viuff B, Tjornehoj K, Larsen LE, Rontved CM, Uttenthal A, Ronsholt L, Alexandersen S (2002) Replication and clearance of respiratory syncytial virus: apoptosis is an important pathway of virus clearance after experimental infection with bovine respiratory syncytial virus. Am J Pathol 161(6):2195–2207. doi:10.1016/S0002-9440(10)64496-3

Welliver TP, Garofalo RP, Hosakote Y, Hintz KH, Avendano L, Sanchez K, Velozo L, Jafri H, Chavez-Bueno S, Ogra PL, McKinney L, Reed JL, Welliver RC Sr (2007) Severe human lower respiratory tract illness caused by respiratory syncytial virus and influenza virus is characterized by the absence of pulmonary cytotoxic lymphocyte responses. J Infect Dis 195(8):1126–1136. doi:10.1086/512615

West K, Ellis J (1997) Functional analysis of antibody responses of feedlot cattle to bovine respiratory syncytial virus following vaccination with mixed vaccines. Can J Vet Res 61(1):28–33

West K, Petrie L, Haines DM, Konoby C, Clark EG, Martin K, Ellis JA (1999) The effect of formalin-inactivated vaccine on respiratory disease associated with bovine respiratory syncytial virus infection in calves. Vaccine 17(7–8):809–820

West K, Petrie L, Konoby C, Haines DM, Cortese V, Ellis JA (2000) The efficacy of modified-live bovine respiratory syncytial virus vaccines in experimentally infected calves. Vaccine 18(9–10):907–919

Wyatt LS, Whitehead SS, Venanzi KA, Murphy BR, Moss B (1999) Priming and boosting immunity to respiratory syncytial virus by recombinant replication-defective vaccinia virus MVA. Vaccine 18(5–6):392–397

Xue W, Ellis J, Mattick D, Smith L, Brady R, Trigo E (2010) Immunogenicity of a modified-live virus vaccine against bovine viral diarrhea virus types 1 and 2, infectious bovine rhinotracheitis virus, bovine parainfluenza-3 virus, and bovine respiratory syncytial virus when administered intranasally in young calves. Vaccine 28(22):3784–3792. doi:10.1016/j.vaccine.2010.03.043

The Cotton Rat *Sigmodon Hispidus* Model of Respiratory Syncytial Virus Infection

Marina S. Boukhvalova and Jorge C. G. Blanco

Abstract The cotton rat *Sigmodon hispidus* is a New World rodent that has become an important model of respiratory syncytial virus (RSV) infection. This small animal is relatively permissive to RSV and can be infected throughout life. It recapitulates the pathology associated with the FI-RSV vaccine-enhanced disease, the phenomenon of maternally transmitted immunity and the ability of passive immunity to suppress efficacy of RSV vaccines. Different highly susceptible human cohort scenarios have been modeled in the cotton rat, including RSV disease in infants, elderly, and immunosuppressed individuals. The cotton rat has accurately predicted efficacy and dose of antibody immunoprophylaxis, and the lack of efficacy of antibody immunotherapy for disease treatment. With the recent development of molecular reagents and tools for the model, the cotton rat is an important model of RSV infection to consider for vaccine and drug testing, and will continue to advance our understanding of RSV disease pathogenesis.

Contents

M. S. Boukhvalova (✉) · J. C. G. Blanco
Sigmovir Biosystems, Inc, 9610 Medical Center Drive, Suite 100, Rockville, MD 20850, USA
e-mail: m.boukhvalova@sigmovir.com

L. J. Anderson and B. S. Graham (eds.), *Challenges and Opportunities for Respiratory Syncytial Virus Vaccines*, Current Topics in Microbiology and Immunology 372,
DOI: 10.1007/978-3-642-38919-1_17,

1 Introduction

Over the years, the cotton rat emerged as an important model of a wide array of human infectious diseases, in particular those of the respiratory tract caused by respiratory syncytial virus (RSV), parainfluenza, metapneumovirus, and influenza viruses (Boukhvalova et al. 2009; Niewiesk and Prince 2002). The surprising susceptibility of this New World rodent to respiratory infectious pathogens and the ability to mimic human disease may be associated with important similarities in the innate immune mechanisms (e.g., Mx system preservation (Pletneva et al. 2006, 2008)), or similarities in the pathogen receptor repertoire (Blanco et al. 2013). The first report of the potential to use the cotton rat as the model of RSV infection came from the Soviet Union in 1970s (Dreizin et al. 1971). The cotton rat model of RSV infection has since been expanded through the contribution of several laboratories in the USA, primarily those of Dr. Robert M. Chanock at NIH, and later that of Dr. Gregory A. Prince at Virion Systems, Inc. Several advantages of cotton rats over existing models of RSV infection include the facts that cotton rats are more permissive to RSV infection than most other animals and remain susceptible to RSV infection throughout life. For example, ferrets can be effectively infected only during infancy (Prince and Porter 1976) and compared to the most common laboratory mouse strains, cotton rats are 50–1,000 fold more permissive for RSV replication (Prince et al. 1979, 1978). The cotton rat has been used to model maternally transmitted passive immunity and antibody-mediated immunosuppression of RSV vaccine efficacy. The phenomenon of FI-RSV has been successfully recapitulated in the cotton rat, and the disease in highly susceptible cohorts, such as infants, elderly, and immunosuppressed has been modeled in this animal. The relative permissiveness to infection and its track record in preclinical evaluation of vaccines, antibodies, and antiviral therapeutics has made the cotton rat an attractive model system for the pharmaceutical industry. Although the available reagents are rapidly increasing, the mouse model has more often been used for RSV-related research in part because of availability of more immunologic reagents. This review will cover the stepping stones of the cotton rat RSV model development that span the last four decades and will describe features of the model (Fig. 1) important for defining disease mechanisms and for identifying effective

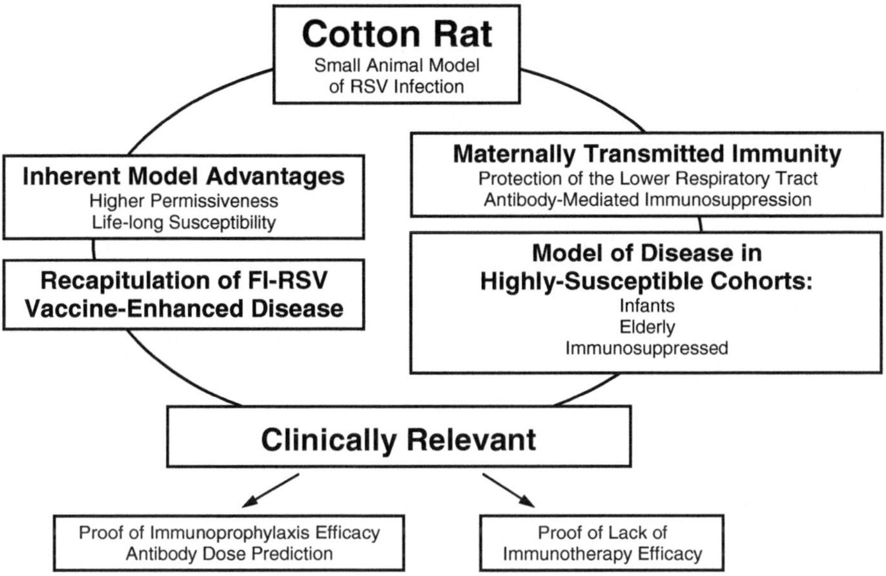

Fig 1 Diagram of the main features of the cotton rat *S. hispidus* model of RSV infection

intervention strategies. For other animal model systems see chapters by G. Taylor and by P.J. Openshaw, this volume and, and for information on human pathogenesis see chapters by P.L. Collins et al., and by M.T. Lotz et al., this volume.

2 General Features of RSV Infection in the Cotton Rat *S. hispidus*

Intranasal RSV infection of 4- to 8-week-old cotton rats (age bracket most frequently used for experimentation) with a range of viral inocula of 10^4 to 10^6 pfu results in active viral replication in lungs and nose, and weaker replication in trachea. RSV replication in the lungs lasts for 5–6 days, with viral titers peaking on day 4 post-infection and becoming undetectable by day 7. Viral replication in the nose is more prolonged than in the lungs, with RSV clearance from that site occurring by day 9 postinfection. Microscopic examination of lung tissue post-RSV infection reveals mild bronchitis/bronchiolitis (inflammatory cells, primarily lymphocytes, surrounding the small airways) which peaks on day 6 postinfection (Prince et al. 1986). Moderate desquamative exudative rhinitis is detected in the nose (Prince et al. 1978). Viral replication and histopathology are proportional to the infectious dose. Doses above 10^6 pfu cause, in addition to peribronchiolitis, interstitial pneumonitis, and alveolitis (inflammatory cells within alveolar walls and spaces, respectively) (Prince et al. 1978). No apparent signs of disease, such as nasal discharge, changes in alertness, weight, or general physical activity have been detected in RSV-infected cotton rats. A rise in serum neutralizing activity first becomes apparent by day 6 postinfection, and is established in 100 % of infected animals by day 9 postinfection (Prince et al. 1978). Reinfection of cotton rats with RSV results in no detectable virus production, and yet pulmonary inflammation is augmented early after reinfection (Prince et al. 1999). Molecular studies indicate that even though no virus is detected during secondary RSV infection in cotton rats, abortive viral replication takes place in the lungs and is accompanied by upregulation of type I interferon response and expression of interferon-induced Mx genes (Boukhvalova et al. 2007; Pletneva et al. 2008). A variety of cytokines, chemokines, interferons, and enzymes regulating the arachidonic acid pathway (e.g., cyclooxygenase 2, COX-2; arachidonate 5-lipoxygenase 5, ALOX 5) are elevated by RSV infection in cotton rats (Blanco et al. 2002; Richardson et al. 2005), manuscript in preparation]. Infection is accompanied by induction of alternatively activated macrophages (M2) in the lungs of RSV-infected animals (Shirey et al. 2010). Recent studies suggest that overexpression of interferon or COX-2 enzyme may be detrimental for RSV disease (Richardson et al. 2005; Boukhvalova et al. 2010), whereas the induction of differentiation of macrophages to the alternative lineage may reduce the signs of disease (Shirey et al. 2010).

3 Cotton Rat as a Model of RSV Immunoprophylaxis

As the model of RSV, the cotton rat owes its real recognition to the role it played in the development of two antibody formulations for preventing severe RSV disease in human infants: RespiGam® and Synagis® (MedImmune Inc., MD, USA). Clinical trials of RSVIG (RespiGam®, a purified pooled IgG preparation from plasma donors) in premature infants and infants with bronchopulmonary dysplasia were consented by FDA after preclinical data obtained solely in cotton rats was provided (Siber et al. 1994; Prince et al. 1985). Subsequent studies in cotton rats of an improved new formulation of prophylactic RSV treatment, a monoclonal antibody against the F glycoprotein of RSV (palivizumab, Synagis®) accurately predicted the positive outcome of subsequent clinical trials, leading to the licensure of Synagis®.

The fact that the cotton rat predicted success of RSV antibody immunoprophylaxis in infants was important because it contradicted the once prevailing dogma that anti-RSV antibodies in humans were detrimental. Early serologic studies in cotton rats demonstrated that infant cotton rats who acquired RSV antibodies from immune mothers were resistant to RSV infection even when nursed on nonimmune mothers (Prince et al. 1983). Resistance to infection was also conferred by intraperitoneal inoculation of convalescent sera into naïve cotton rats without induction of immunopathology upon challenge (Prince et al. 1983, 1985). Serum neutralizing antibody titers above 1:100, as determined by the 60 % plaque reduction assay, rendered some pulmonary protection, with titer of 1:380

conferring sterilizing immunity in the lungs. This was consistent with the results of extended studies of human infants which demonstrated that infants under 2 months of age with high levels of maternally acquired antibodies (1:400) were less susceptible to RSV disease, while RSV bronchiolitis and pneumonia was seen in infants with 1:100–1:200 serum antibody titers (Parrott et al. 1973). The direct correlation of antibody levels required for protection of human infants and cotton rats attested to the ability of the cotton rat model to be used not only to confirm the efficacy of antibody immunoprophylaxis, but also to predict the dose of the antibody required to achieve protective serum titer in human infants. The 15 mg/kg Synagis® dose was suggested based on the cotton rat studies. This results in trough levels of about 40 μg/ml of serum and is associated with about 50 % reduction in hospitalization of children at high risk of severe disease. The model was also reliable at predicting that RSV nasal protection is incomplete in human infants following antibody immunoprophylaxis (Siber et al. 1994).

4 Cotton Rat as a Model of RSV Immunotherapy

Once anti-RSV antibodies were demonstrated to be an effective immunoprophylaxis tool, an effort was initiated to validate them in a therapeutic setting. Cotton rat studies indicated that therapeutically administered antibodies were effective in protecting against viral replication in the lungs and nose with the dose response pattern similar to that of immunoprophylaxis (Prince et al. 1985). Despite the antiviral effect, therapeutic administration of antibodies showed no improvement in lung histology in cotton rats (Prince et al. 2000). Clinical trials of a therapeutic RSV treatment with a variety of antibody formulations including Synagis®, RespiGam®, and Generic IgG (Sandoz) also demonstrated that while being safe, they afforded no therapeutic benefit (Malley et al. 1998; Rodriguez et al. 1997). The latter findings closely correlated with the absence of pathology-reducing effect in therapeutically treated cotton rats. Additional studies in the model revealed that to achieve disease reduction, therapeutic administration of antibodies needs to be accompanied by anti-inflammatory treatment (Prince et al. 2000).

5 Vaccine-Enhanced Disease in the Cotton Rat Model

5.1 Enhanced Pulmonary Pathology with Alveolitis as the Primary Marker

Several preparations of formalin-inactivated RSV (FI-RSV) have been tested in the cotton rat model, including the original Lot 100 vaccine and newly prepared formulations of formalin-inactivated virus with or without addition of alum (Prince et al. 1986, 2001). In all cases animals vaccinated with FI-RSV and subsequently challenged with RSV developed a more severe disease, characterized by increased pulmonary histopathology, compared to primary infected or reinfected animals. Changes in lung pathology were primarily associated with increase in pulmonary infiltration of neutrophils and a dramatic rise in alveolitis. Alveolitis is currently considered the primary marker of vaccine-enhanced disease in the cotton rat model, whereas bronchiolar histopathology is a normal component of the immune-mediated resolution of RSV infection. A rise in lymphocyte infiltration was also seen in RSV-challenged FI-RSV-vaccinated animals, but was less pronounced than that of polymorphonuclear leukocytes (Prince et al. 1986). The effect on pathology was FI-RSV dose-related, with maximum pathology seen in animals vaccinated with an intermediate dose of the vaccine (Prince et al. 2001). Predominance of neutrophils among cells infiltrating lungs of cotton rats vaccinated with FI-RSV corresponded to the pathology findings from Lot 100 trials, with the majority of granulocytes described in the two autopsy reports of the human fatalities identified as neutrophils, with some mention of eosinophils.

5.2 *Reflection of the Timing of Onset of Vaccine-Enhanced Disease*

The cotton rat model has accurately recapitulated the timing of the vaccine-enhanced disease onset as described in humans. RSV vaccine-enhanced disease was manifested only during the first winter following vaccination with Lot 100 vaccine, but not thereafter. The phenomenon was explored in cotton rats by extending the time of RSV challenge after FI-RSV vaccination from 42 days to 6 months. While still visible, pulmonary pathology of animals infected 6 months after vaccination was reduced compared to animals challenged 42 days after vaccination. These results are consistent with those in the reports of the Lot 100 vaccine trials (Prince et al. 2001; Murphy et al. 1990).

5.3 *Imbalance Between Total and Neutralizing Antibody Response to FI-RSV Vaccination*

The serum antibody response of infants and children immunized with FI-RSV vaccine in the 1960's were characterized by high titers of antibodies against the fusion protein but lower levels of neutralizing antibodies compared to individuals of similar age with natural RSV infection (Murphy et al. 1986). An examination of the immunogenicity of the FI-RSV vaccine in the cotton rat showed results that parallel those observed in human serology of FI-RSV vaccine trial. Animals vaccinated with FI-RSV also developed comparable levels of total antibodies to the G and F glycoproteins to those attained in animals with pulmonary RSV infection. However, as in the human trials, vaccinated cotton rats developed levels of neutralizing antibodies that were only 1/20–1/30 of those in animals infected with RSV by intranasal or intramuscular inoculation (Prince et al. 1986).

5.4 Molecular Mechanisms of Vaccine-Enhanced Disease

The mechanism of vaccine-enhanced disease has long been associated with an aberrant activation of the Th2 arm of the immune system. With the advent of the development of cotton rat reagents for monitoring Th1 and Th2 responses, it became possible to evaluate contribution of the Th2 response in FI-RSV vaccine-enhanced disease (Boukhvalova et al. 2006). Vaccine-enhanced disease in cotton rats was also accompanied by the augmentation of the Th2 cytokine response (Boukhvalova et al. 2006). However, levels of several Th1 cytokines and chemokines were also increased. Inclusion of MPL, an adjuvant that reduces pathology of vaccine-enhanced disease by signaling through Toll-like receptor (TLR) 4, inhibited FI-RSV-induced augmentation of Th2, Th1, and chemotactic cytokine responses (Boukhvalova et al. 2006; Prince et al. 2001). The latter effect was limited to the first 2 days postinfection, suggesting that exacerbated early release of pro-inflammatory cytokines, Th1 cytokines, and chemokines contribute to enhanced pulmonary inflammation in FI-RSV-vaccinated animals. Additional adjuvants (e.g., CpG, poly ICLC) that signal through other TLRs were tested in cotton rats by admixing them with RSV vaccine formulations (Prince et al. 2003), [data not published]. Although they render increased protection against RSV, for some adjuvants, such as CpG the effect can be accompanied by enhanced pulmonary pathology (Prince et al. 2003).

6 Model of Maternally Transmitted Immunity

6.1 Protection of the Lower Respiratory Tract

Maternally transmitted immunity has been studied in weanling cotton rats born to immune mothers infected with RSV several weeks before parturition. Almost complete protection of the lungs and partial protection of the nose was seen in animals born to and suckled on immune mothers (Prince et al. 1983). Passive immunity acquired from mothers protected the lower respiratory tract of cotton rats from RSV infection for 4 weeks after birth (a.b). Foster-nursing experiments were set up to evaluate the role of colostrum versus transplacental immunity transfer in conferring RSV resistance. Both mechanisms of transfer were involved, with colostrums significantly contributing to protection of the lower respiratory tract in a mechanism generally correlated with serum neutralizing antibodies (Prince et al. 1983).

6.2 Antibody-Mediated Immunosuppression

Immunosuppression by maternally derived serum antibodies presents one of the most challenging obstacles to the development of RSV vaccine for human infants. Immunization with live RSV administered intramuscularly, once a promising vaccine approach, was found ineffective in children with detectable maternal antibodies against RSV. The phenomenon of maternal antibody-mediated immunosuppression has been reproduced in cotton rats by two different approaches: by parenteral live RSV immunization of infant animals born to RSV-immune mothers (Prince et al. 1979) and by immunization of adult animals pretreated with anti-RSV serum (Prince et al. 1982). Animals born to RSV-immune mothers and vaccinated at 20 days a.b were refractory to vaccination whereas animals vaccinated 37 days a.b generated protective immunity (Prince et al. 1979). Inoculation of anti-RSV antiserum into naïve cotton rats was sufficient to suppress immunogenicity of parenteral live immunization indicating that serum antibodies alone are sufficient for immunosuppression (Prince et al. 1982).

7 Cotton Rat as a Model of RSV Disease in Highly Susceptible Cohorts

RSV disease is particularly severe in infants, elderly, and immunosuppressed individuals. All three cohorts have been modeled in the cotton rat RSV infection model (see chapter by A.M.W Malloy et al., this volume).

7.1 Infants

Infant cotton rats (3-day-old) display higher viral RSV replication in the upper respiratory tract than 4-week-old animals. Over one $\log_{10}$ pfu more virus is recovered from noses of infant cotton rats than in adults and viral clearance is substantially delayed (Prince et al. 1978). This is different from mice in which viral titers in lung and nose are lower in neonates than in adults. Furthermore, infant cotton rats do not mount a neutralizing antibody response to RSV infection as efficiently as adult animals. It takes twice as long for a 3-day-old animal to reach the same level of antibody response as it takes for 4-week-old animals. This is similar to a diminished antibody response of human infants to RSV infection (Murphy et al. 1986), although the latter may also be due to an immunosuppression mediated by maternal antibodies in addition to an immunological immaturity.

7.2 Elderly

Severity of RSV disease is increased in the elderly (Falsey 2005). Responses of cotton rats to RSV also differ by the age of infected animals. Cotton rats display signs of immunosenescence from 6 months of age, with age-proportional delay in viral clearance seen in 6–16 months old *S. hispidus* (Curtis et al. 2002). The delay is accompanied by skewed pulmonary expression of cytokines and over expression of chemokines in animals over 9 months of age (Boukhvalova et al. 2007). The altered cytokine and chemokine responses in aged cotton rats might be associated with an age-related defect in innate immune response to RSV.

7.3 Immunosuppressed

The cotton rat has also been used to model RSV disease in immunosuppressed humans, where it is a cause of significant morbidity and mortality. Immunosuppression has been reproduced in cyclophosphamide-treated cotton rats and was found to have an effect on RSV pathogenesis (Ottolini et al. 1999). Delayed RSV clearance is the primary marker of RSV disease in immunosuppressed cotton rats. The model has been used to address effectiveness of RSVIG treatment under conditions of immunosuppression. Both prophylactic and therapeutic administration of a single-dose of RSVIG diminished viral replication in RSV-infected animals. However, viral rebound was seen after a week of the antibody treatment and continuous suppression of viral presence required multiple RSVIG treatments (Ottolini et al. 1999).

8 Model Status

There are still caveats in using the cotton rat model of RSV. (1) Although more permissive than other small animal models, the infection still requires a relatively large viral inoculum and does not appear to spread from upper to lower airway as it does in the natural host. (2) While assessment of lung pathology has been useful, the cotton rat does not exhibit other overt signs of illness. (3) While reagents are improving, the cotton rat is still relatively limited on what can be done with detailed analysis of T cell responses, B cell phenotypes, and antigen presenting cell subsets.

The work to further develop the model is ongoing. Up to date, more than 50 different selected cotton rat genes, with additional partial clones and sequences of more than 200 have been cloned (Blanco et al. 2004) (Table 1). All completed cDNA sequences are available at the GenBank site. R&D Systems, Inc. has used these sequences to develop reagents for the cotton rat and made many of these

Table 1 Current State of cotton rat reagents

Cytok/Growth factors:
IFN-γ. (A, B, C, D)
IFN-α. (A,B)
IFNβ
IL-1 α. (A, B, C)
IL-1 β. (A, B)
IL-2. (A, B, C)
IL-4. (A, B, C, D)
IL-5
IL-6. (A, B, C)
IL-9.
IL-10. (A, B, C)
IL-12p40.
IL-12p35.
IL-13.
IL-18.
IL-15
IL-17
IL-23
GM-CSF
G-CSF
M-CSF
TNFα. (A, B, C,)
TNFβ.
TGFβ1
TGFβ2
VEGF

Chemokines:
MIP-1α. (A, B, C)
MIP-1β. (A, B, C)
RANTES. (A, B)
IP-10. (A, B)
GRO/IL-8. (A, C)
MIP-2/IL-8. (A, C).
MCP-1/JE. (A,C)
Eotaxin (SCYA-11)
MCP-5
MDC (SCYA-22)
TARC (SCYA-17)

Cell surface/Receptor molecules:
CCR5
CCR3
CXCR4
CX3CR
CD3
CD4 (C)
CD8a (C)
CD11a and b
CD14
CD16
CD18
CD25
CD28
CD36 (Scavenger Receptor)
CD45/B220
CD62L (L-selectin)
CD74(MHC II)
CD80
CD83(HB15)
CD86(B7-2)
CD132
CD161
CD200
CD206 (Mannose Receptor)
IL-4Ra
IL-23 receptor
Calcium binding atopy-related autoantigen
Neutral solute channel aquaphorin 9 (Aqp9)
Prostaglandin E receptor
TLR1/TLR2/TLR3/TLR4/TLR5/TLR7
Ly-6
MHC I
MHC II A
MHC II E
β-2 microglobulin

Enzymes:
4-amino butyrate aminotransferase
Cyclooxygenase 1 (COX-1)
Cyclooxygenase 2 (COX-2) (B)
Cytochrome C oxidase subunit III
Farnesyl transferase
Glutamine synthetase
3-hydroxyisobutyrate dehydrogenase
i-nitric Oxide synthetase (i-NOS)
Macrophage metalloelastase
NADH dehydrogenase subunits 2 and 3
Sphingolipid hydrolase glycoprotein
Superoxide dismutase, mitochondrial (SOD)
Ubiquitin-conjugating enzyme 7 (Ubce7)
5-Lipoxygenase (5-LO)
Arginase I (Arg1)

Ribosomal proteins:
Elongation factor 1-a
Ribosomal proteins L13, L30, L31, S16

Housekeeping genes:
β-actin
GAPDH
18S rRNA

Ribonuclear proteins:
Heterogeneous nuclear ribonucleoprotein A/B (Hnrpab)
Small nuclear RNA auxiliary factor (RNU2)

Structural proteins:
Alpha I type III collagen
Alpha globin chain
Anexin A2
Beta gamma crystalline-like protein
Calreticulin
Cofilin 1
Ficolin A
Granulin (Grn)
Gravin (PRKA anchor protein)
Legumain
Pro-alpha-1 (V) collagen
Vimentin
Nectin-1

Other proteins:
Adrenomedullin
BASF
Cell cycle checkpoint protein (CHFR)
Complement protein C4
Dystroglycan
FIZZ1
Hsp70
Icos-L
IgE heavy chain
IRF-2
IRF-8 (ICSBP)
Mx1/Mx2/Mx3 (B)
Protein kinase C
Protein tyrosin phosphatase
RAB7 (Ras oncogene family member)
Ras suppressor protein 1
Serine protease inhibitor 3
Surfactant protein C
Thrombospondin
Thymopoietin

The letters in Table 1 refer to the status of reagents and assays available. *A*: recombinant protein; *B*: polyclonal antibodies; *C*: monoclonal antibodies; *D*: ELISA

reagents available to the scientific community. In addition, our laboratory has established the techniques to develop transgenic animals from cotton rats adding to the tools available for research with this model (Blanco et al. 2011).

References

Blanco JC, Richardson JY, Darnell ME et al (2002) Cytokine and chemokine gene expression after primary and secondary respiratory syncytial virus infection in cotton rats. J Infect Dis 185(12):1780–1785

Blanco JC, Pletneva L, Boukhvalova M, Richardson JY, Harris KA, Prince GA (2004) The cotton rat: an underutilized animal model for human infectious diseases can now be exploited using specific reagents to cytokines, chemokines, and interferons. J Interferon Cytokine Res 24(1):21–28

Blanco J et al (2011) The quest for a small animal model for HIV infection and disease. In: Chang TL (ed) *HIV-host interactions*, vol. Chapter 14 INTECH Open Access Publisher, p 349

Blanco JC, Plentneva LM, Wan H et al (2013) Receptor characterization and susceptibility of cotton rats to avian and 2009 pandemic influenza virus strains. J Virol 87(4):2036–2045

Boukhvalova MS, Prince GA, Soroush L, Harrigan DC, Vogel SN, Blanco JC (2006) The TLR4 agonist, monophosphoryl lipid A, attenuates the cytokine storm associated with respiratory syncytial virus vaccine-enhanced disease. Vaccine 24(23):5027–5035

Boukhvalova MS, Prince GA, Blanco JC (2007a) Respiratory syncytial virus infects and abortively replicates in the lungs in spite of preexisting immunity. J Virol 81(17):9443–9450
Boukhvalova MS, Prince GA, Blanco JC (2009) The cotton rat model of respiratory viral infections. Biologicals 37(3):152–159
Boukhvalova MS, Sotomayor TB, Point RC, Pletneva LM, Prince GA, Blanco JC (2010) Activation of interferon response through toll-like receptor 3 impacts viral pathogenesis and pulmonary toll-like receptor expression during respiratory syncytial virus and influenza infections in the cotton rat Sigmodon hispidus model. J Interferon Cytokine Res 30(4):229–242
Boukhvalova MS, Yim KC, Kuhn KH et al (2007b) Age-related differences in pulmonary cytokine response to respiratory syncytial virus infection: modulation by anti-inflammatory and antiviral treatment. J Infect Dis 195(4):511–518
Curtis SJ, Ottolini MG, Porter DD, Prince GA (2002) Age-dependent replication of respiratory syncytial virus in the cotton rat. Exp Biol Med (Maywood) 227(9):799–802
Dreizin RS, Vyshnevetskaia LO, Bagdamian EE, Iankevich OD, Tarasova LB (1971) Experimental RS virus infection of cotton rats. a viral and immunofluorescent study. Vopr Virusol 16(6):670–676
Falsey AR (2005) Respiratory syncytial virus infection in elderly and high-risk adults. Exp Lung Res 31(Suppl 1):77
Malley R, DeVincenzo J, Ramilo O et al (1998) Reduction of respiratory syncytial virus (RSV) in tracheal aspirates in intubated infants by use of humanized monoclonal antibody to RSV F protein. J Infect Dis 178(6):1555–1561
Murphy BR, Sotnikov AV, Lawrence LA, Banks SM, Prince GA (1990) Enhanced pulmonary histopathology is observed in cotton rats immunized with formalin-inactivated respiratory syncytial virus (RSV) or purified F glycoprotein and challenged with RSV 3–6 months after immunization. Vaccine 8(5):497–502
Murphy BR, Graham BS, Prince GA et al (1986b) Serum and nasal-wash immunoglobulin G and A antibody response of infants and children to respiratory syncytial virus F and G glycoproteins following primary infection. J Clin Microbiol 23(6):1009–1014
Murphy BR, Prince GA, Walsh EE et al (1986a) Dissociation between serum neutralizing and glycoprotein antibody responses of infants and children who received inactivated respiratory syncytial virus vaccine. J Clin Microbiol 24(2):197–202
Niewiesk S, Prince G (2002) Diversifying animal models: the use of hispid cotton rats (Sigmodon hispidus) in infectious diseases. Lab Anim 36(4):357–372
Ottolini MG, Porter DD, Hemming VG, Zimmerman MN, Schwab NM, Prince GA (1999) Effectiveness of RSVIG prophylaxis and therapy of respiratory syncytial virus in an immunosuppressed animal model. Bone Marrow Transplant 24(1):41–45
Parrott RH, Kim HW, Arrobio JO et al (1973) Epidemiology of respiratory syncytial virus infection in Washington, D.C. II. Infection and disease with respect to age, immunologic status, race and sex. Am J Epidemiol 98(4):289–300
Pletneva LM, Haller O, Porter DD, Prince GA, Blanco JC (2006) Interferon-inducible Mx gene expression in cotton rats: cloning, characterization, and expression during influenza viral infection. J Interferon Cytokine Res 26(12):914–921
Pletneva LM, Haller O, Porter DD, Prince GA, Blanco JC (2008) Induction of type I interferons and interferon-inducible Mx genes during respiratory syncytial virus infection and reinfection in cotton rats. J Gen Virol 89(Pt 1):261–270
Prince GA, Denamur F, Deschamps M et al (2001b) Monophosphoryl lipid A adjuvant reverses a principal histologic parameter of formalin-inactivated respiratory syncytial virus vaccine-induced disease. Vaccine 19(15–16):2048–2054
Prince GA, Mond JJ, Porter DD, Yim KC, Lan SJ, Klinman DM (2003) Immunoprotective activity and safety of a respiratory syncytial virus vaccine: mucosal delivery of fusion glycoprotein with a CpG oligodeoxynucleotide adjuvant. J Virol 77(24):13156–13160
Prince GA, Potash L, Horswood RL et al (1979b) Intramuscular inoculation of live respiratory syncytial virus induces immunity in cotton rats. Infect Immun 23(3):723–728

Prince GA, Horswood RL, Camargo E, Suffin SC, Chanock RM (1982) Parenteral immunization with live respiratory syncytial virus is blocked in seropositive cotton rats. Infect Immun 37(3):1074–1078

Prince GA, Porter DD (1976) The pathogenesis of respiratory syncytial virus infection in infant ferrets. Am J Pathol 82(2):339–352

Prince GA, Horswood RL, Berndt J, Suffin SC, Chanock RM (1979a) Respiratory syncytial virus infection in inbred mice. Infect Immun 26(2):764–766

Prince GA, Jenson AB, Horswood RL, Camargo E, Chanock RM (1978) The pathogenesis of respiratory syncytial virus infection in cotton rats. Am J Pathol 93(3):771–791

Prince GA, Jenson AB, Hemming VG et al (1986) Enhancement of respiratory syncytial virus pulmonary pathology in cotton rats by prior intramuscular inoculation of formalin-inactiva ted virus. J Virol 57(3):721–728

Prince GA, Prieels JP, Slaoui M, Porter DD (1999) Pulmonary lesions in primary respiratory syncytial virus infection, reinfection, and vaccine-enhanced disease in the cotton rat (Sigmodon hispidus). Lab Invest 79(11):1385–1392

Prince GA, Hemming VG, Horswood RL, Chanock RM (1985a) Immunoprophylaxis and immunotherapy of respiratory syncytial virus infection in the cotton rat. Virus Res 3(3):193–206

Prince GA, Horswood RL, Camargo E, Koenig D, Chanock RM (1983) Mechanisms of immunity to respiratory syncytial virus in cotton rats. Infect Immun 42(1):81–87

Prince GA, Horswood RL, Chanock RM (1985b) Quantitative aspects of passive immunity to respiratory syncytial virus infection in infant cotton rats. J Virol 55(3):517–520

Prince GA, Curtis SJ, Yim KC, Porter DD (2001a) Vaccine-enhanced respiratory syncytial virus disease in cotton rats following immunization with Lot 100 or a newly prepared reference vaccine. J Gen Virol 82(Pt 12):2881–2888

Prince G, Mathews A, Curtis S, Porter DD (2000) Treatment of respiratory syncytial virus bronchiolitis and pneumonia in a cotton rat model with systemically administered monoclonal antibody (palivizumab) and glucocorticosteroid. J Infect Dis 182:1326–1330

Richardson JY, Ottolini MG, Pletneva L et al (2005) Respiratory syncytial virus (RSV) infection induces cyclooxygenase 2: a potential target for RSV therapy. J Immunol 174(7):4356–4364

Rodriguez WJ, Gruber WC, Welliver RC et al (1997) Respiratory syncytial virus (RSV) immune globulin intravenous therapy for RSV lower respiratory tract infection in infants and young children at high risk for severe RSV infections: respiratory syncytial virus immune globulin study group. Pediatrics 99(3):454–461

Shirey KA, Pletneva LM, Puche AC et al (2010) Control of RSV-induced lung injury by alternatively activated macrophages is IL-4R alpha-, TLR4-, and IFN-beta-dependent. Mucosal Immunol 3(3):291–300

Siber GR, Leombruno D, Leszczynski J et al (1994) Comparison of antibody concentrations and protective activity of respiratory syncytial virus immune globulin and conventional immune globulin. J Infect Dis 169(6):1368–1373

The Mouse Model of Respiratory Syncytial Virus Disease

Peter J. Openshaw

Abstract The laboratory mouse is the species of choice for most immunological studies, ranging from simple vaccine testing to the intricate dissection of fundamental immunopathogenic mechanisms. Although not fully mouse adapted, some strains of respiratory syncytial virus (RSV) replicate in the murine respiratory tract and induce specific T and B cell responses. Passive transfer of neutralising antibody is protective and assist in viral clearance. In addition, many of RSV's complex behaviours are recapitulated in the mouse (including enhancement of disease by vaccination and delayed effects of neonatal infection). However, human studies remain essential to confirm or refute predictions from animal models.

Contents

P. J. Openshaw (✉)
National Heart and Lung Institute, Norfolk Place, St Mary's Campus, Imperial College London, London W2 1PG, UK
e-mail: p.openshaw@imperial.ac.uk

L. J. Anderson and B. S. Graham (eds.), *Challenges and Opportunities for Respiratory Syncytial Virus Vaccines*, Current Topics in Microbiology and Immunology 372,
DOI: 10.1007/978-3-642-38919-1_18,

1 Introduction

Respiratory syncytial virus (RSV) provides interesting puzzles for the immunologist. Antibody protects, but only partially. T-cells are important, since children with defects in T cell immunity suffer from prolonged and atypical infection. Most remarkably, boosting immunity with formalin inactivated vaccines enhances disease during subsequent natural infection (Graham 2011), and children who recover from severe bronchiolitis are at high risk of recurrent wheeze and asthma in later childhood (Sigurs et al. 2010). These complexities make RSV an intriguing virus for an immunologist, potentially providing insights not only into RSV disease but into the pathogenesis of asthma and other inflammatory disorders.

Over the past 30 years, the mouse has become the species of choice for most immunological experiments. Most human genes have murine equivalents and the behaviour of human and murine immune systems are broadly very similar. The availability of inbred mice, knockouts, transgenics and a multitude of specific immunological reagents allows studies of extraordinary intricacy and sophistication to be performed. The rapid breeding cycle allows age-dependant effects to be studied with relative ease. The relatively low cost allows sampling of relevant tissues at multiple time-points in dynamic studies and keen attention to animal husbandry provides clean mice, unperturbed by environmental variations.

Some of the advantages of murine studies are also weaknesses. Humans are not the same as laboratory mice: they are outbred and genetically diverse, have suffered multiple past infections and live in complex and unclean environments. Although murine studies produce dazzling insights, interpretation is difficult, complex, and subtle. Human studies should be informed by findings in animals but we should not over-interpret animal data and assume that the same applies in man (see chapters by S. Mukherjee and N.W. Lukacs, S.M. Varga and T.J. Braciale, M.S. Boukhvalova and J.C.G. Blanco, G. Taylor, and by R.J. Pickles, this volume).

2 Historical Perspectives

Gregory Prince, working under Dr. Robert Chanock, published the first detailed account of RSV infection of mice in 1979. They took 20 inbred strains of neonatal mouse and infected them with the human Long strain of RSV, harvesting lungs and nasal tissue on day 4. This showed a wide variation in susceptibility of different stains, with DBA2N the most susceptible; and C3H towards the bottom of the rank. Common strains (BALB/c and C57Bl/6) showed intermediate susceptibility. The virus loads in the most susceptible mice were approximately 100 fold higher than those in the least susceptible (Fig. 1) (Prince et al. 1979). Follow-up studies show that each mouse strain has its own way of responding to RSV infection, but most laboratories have focussed on the BALB/c mouse.

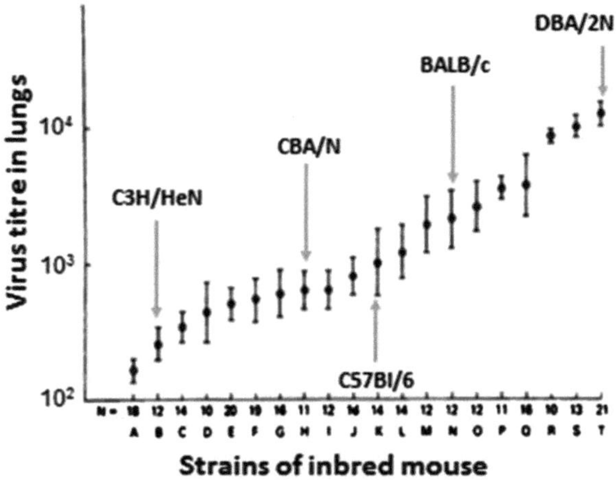

Fig. 1 RSV infection of 20 strains of inbred mice. Inoculation of virus (*Long strain*) was performed at 3 days of age. After 4 days, infant mice were sacrificed; the lungs and nose were homogenized and titrated on HEp-2 cell monolayers. The viral load is shown on the ordinate and the strains of mouse on the abscissa (with the number of infants in each group). Examples of common strains are labelled (Figure adapted with permission from Prince et al. 1979)

Subsequent studies showed that much of the virus administered intranasally does not infect, but is rapidly cleared from the respiratory tract (presumably by cilliary action). Very little virus can be recovered at 12–24 h, but thereafter viral load climbs progressively and peaks on days 4–5. Thereafter, viral loads decline with little recoverable virus being detected beyond day 10 (Fig. 2) (Taylor et al. 1984b). Viral replication is accompanied by mononuclear cell infiltration in the perivascular and peribronchial regions that resembles bronchiolitis in human infants (Johnson et al. 2007). However, unless Type I IFN pathways are interrupted it is difficult to demonstrate infection of the murine bronchiolar epithelium (Bhoj et al. 2008). In mice, the predominant infected cell is the Type I alveolar pneumocyte.

Although several investigators tried, it proved impossible to adapt human strains of RSV to mice and attempts at viral purification have usually been disappointing because of loss of infective virus. Most laboratories use crude viral suspensions containing cell debris and tissue culture medium to infect mice, relying on normal mechanical clearance to remove the adventitious debris and proteins from the respiratory tract. Using selected plaque-picked virus, the severity of disease and the viral load in the lungs shows incremental progression as the inoculated dose is increased (Pribul et al. 2008).

However, there are important differences between mouse and man, particularly that it is well recognised that aged mice (older than 15 weeks) are more susceptible to RSV replication than younger or neonatal mice (Graham et al. 1988). This would appear to contrast with the situation in human infants where the greatest risk of bronchiolitis is in children under the age of 6 months. Children over the age of 2

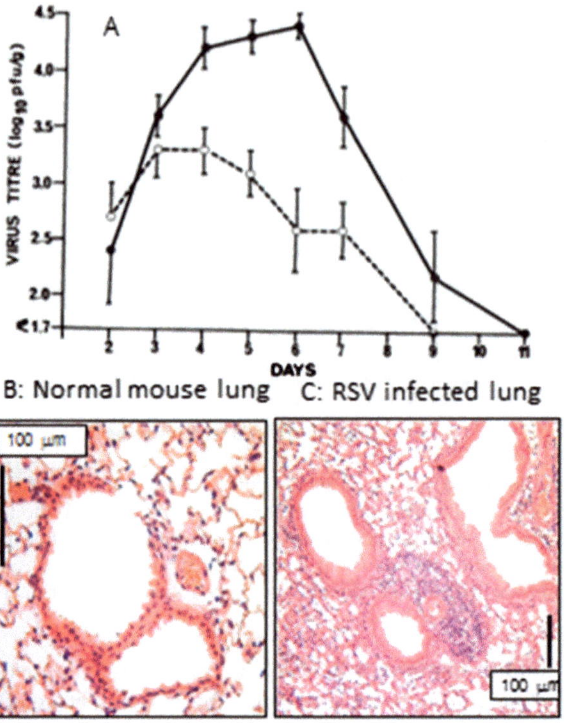

Fig. 2 RSV infection causes viral replication and histological changes in the lungs of mice. BALB/c mice infected with human A2 strain of RSV show peak viral load in nose (*dotted line, open circle*) on days 3–4 and the lungs (*solid line, filled circles*) on days 4–6 (panel **A** adapted with permission from Taylor et al. 1984b). Histological appearances of normal **B** and RSV infected mouse lung **C** are shown at day 6 after inoculation (bronchial lavage has removed the exudate within the airway; colour images kindly provided by Dr Geraldine Taylor, UK). In this case, virus-induced pathology is predominantly a perivascular accumulation of lymphocytes, but appearances depend on the stage of infection

are very unlikely to suffer severe disease even if they have hitherto escaped RSV infection. However, we do not know what would happen if a naïve population of old adults were exposed to RSV for the first time. Infection is so universal that virtually no one above the age of 3 suffers primary infection, and it is possible that adult humans infected for the first time could suffer severe disease.

During the 1980s, a number of important techniques and methods were established that facilitated the use of mice in the study of RSV disease. The first was the introduction of weight loss as an index of disease severity. This is a very convenient, cheap and reproducible objective measure, often used in place of subjective rating of disease severity by an experienced observer. However, weight loss is not a familiar feature of RSV disease in children, and does not simply reflect the severity of lung disease which might be better reflected by lung function or oxygenation measurements.

The second technique that has proved very useful is recovery of bronchiolar alveolar lavage fluid and cells. This allows enumeration of inflammatory cells and mediators and has considerable advantages of speed and cost over the detailed analysis of pulmonary histological changes, although histological clearly has an important place in evaluating disease.

Third, viral load measurement by plaque assay requires skilled, dedicated technical support and rapid handling of infected tissue that is ideally processed fresh (much of the virus is lost in freezing). The use of quantitative PCR to has become routine in many laboratories, providing a good dynamic range and circumventing the problems of the plaque assay.

All of these refinements and technical developments allowed the mouse to be used to study the evolution of immune responses over time, while the reagents available to mouse immunologists have enabled the fine dissection of immune responses, generating complex insights into the immunopathogenesis of viral lung disease.

3 Pathogenic Insights from Studies in Mice

Stemming from our interest in inflammatory lung disease, many of our studies used recombinant virus to modify the response to RSV challenge. Soon after the RSV mouse model was established in Dr Brigitte (Ita) Askonas's laboratory in London, CD8 cytotoxic T-cell lines and clones were made that recognised RSV proteins. These were injected intravenously into irradiated or normal RSV infected mice with the aim of demonstrating that T-cells were able to clear RSV from the lungs. This they did, but they also caused greatly enhanced disease in mice undergoing RSV infection (Cannon et al. 1988). When we examined the lungs of these mice, there was indeed a detectable increase in the number of recovered CD8 T-cells. However, the more obvious and numerically dominant cells present in the lungs and airways were neutrophils, reflecting the pattern of bronchial cellular exudate seen in children with severe bronchiolitis (Wang et al. 1998).

Another intriguing phenomenon became evident while we were attempting to study vaccine-enhanced disease using recombinant vaccinia viruses. We found that recombinant vaccinia virus (rVV, kindly provided by Dr Gail Wertz, Peter Collins and colleagues) expressing RSV proteins induce specific immune responses to different viral proteins, and that this protein-specific immunity produces contrasting patterns of enhanced lung disease after RSV challenge. Most remarkably, cutaneous infection with rVV expressing the attachment protein G induces very strong Th2 responses and lung eosinophilia during subsequent intranasal RSV infection (Fig. 3); (Openshaw et al. 1992). This effect depends on CD4 cells making IL-4 and IL-13. The genetic background of the host has powerful effects: about half of all mouse strains sensitised to rVV-G show lung eosinophilia after RSV challenge. In addition, strains that do not develop lung eosinophilia in this situation (e.g. C57BL/6) do develop eosinophilia if CD8 T-cell responses are

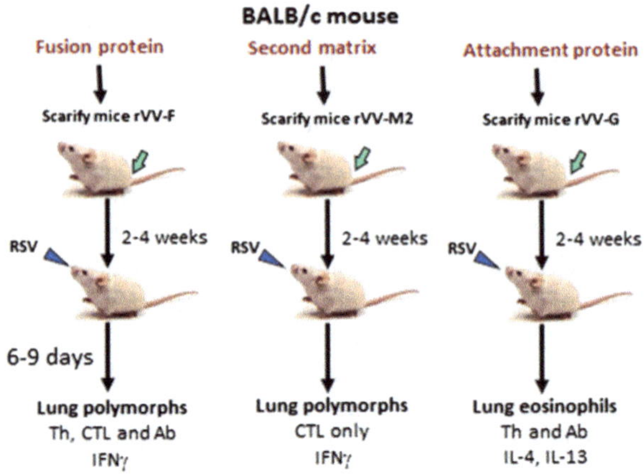

Fig. 3 The effect of sensitisation of mice to individual viral proteins is expressed by vaccinia. Mice scarified on the skin with recombinant vaccinia viruses (rVV) expressing the Fusion protein, the M2 (second matrix) or major surface glycoprotein G show contrasting patterns of immunity. The F protein- specific immunity comprises Th1, CD8 T-cell and antibody; the M2 induces CD8 T-cells and G a Th2 response. Upon challenge with RSV intranasally, this translates into specific types of lung pathology (illustrating the work of Openshaw et al. 1992 and Alwan et al. 1994)

impaired (Hussell and Openshaw 1997). These experiments provide insights into eosinophilic lung diseases, and were among the first studies to show that the concept of Th1 versus Th2 (first developed in studies of mouse T-cell clones at the DNAX Research Institute (Mosmann et al. 1986)) actually were applicable in vivo, and explained subtypes (or endophenotypes) of disease in viral infections.

In contrast, rVV-M2 induces almost exclusively CD8 T-cell responses and lung neutrophilia after RSV challenge, reminiscent of the effects of CD8 T-cell transfer described by Cannon et al. (1988). The fusion protein F (the protein usually selected for vaccine development) induces antibody, CD4 and CD8 T-cell responses. All of these responses are only partially protective against secondary infection, and can be associated with enhanced disease as measured by weight loss (Alwan et al. 1993, 1994). It is important to note that the rVV is given peripherally by cutaneous scarification, and that the first point of contact between primed T-cells and RSV itself is in the lungs.

A second avenue of fruitful collaboration with virologists has been the testing of genetically engineered recombinant RSVs (rRSV) that express murine cytokines (created by Peter Collins, Alex Bukreyev, and Ursula Bucholtz at NIH, USA). These viruses create an intense cytokine-rich environment in the respiratory tract after infection and produce remarkable changes in pathogenicity. For example, rRSV-IFNγ activates macrophages and NK cells, inhibiting Th2 CD4 T-cell responses during subsequent RSV challenge. In contrast, rRSV-IL-4 boosts Th2 cells, eosinophils and IgE after RSV challenge. The environment created by infection with these viruses is very long-lasting. If neonatal mice are infected with these rRSVs, the immune responses to adult re-challenge are altered (Harker et al.

2010) suggesting that the cytokine environment present during infantile bronchiolitis can have very profound effects even into adulthood. Indeed, the Th2 programming by rRSV-IL-4 is so profound that subsequent infection with an irrelevant pathogen (e.g. influenza) also induces an eosinophilic lung reaction, presumably because of resident Th2 cells that react non-specifically to future viral encounters (Harker et al. 2007). Although these studies are aimed primarily at obtaining fundamental insights into pathogenesis, the ultimate goal is to find a highly immunogenic and non-pathogenic combination of virus and cytokine that might form the basis of a future vaccine.

The nature of the inflammatory response to RSV re-challenge is also profoundly altered by the age at which RSV is first encountered: mice first infected as neonates develop enhanced granulocytic lung disease after re-challenge as adults (Culley et al. 2002). This effect can be attenuated by modifying the immune response to primary infection in the neonate using the TLR9 ligand CpG, or with recombinant RSV expressing IFNγ (Harker et al. 2010; Culley et al. 2002). This suggests that modifying or delaying the first encounter with RSV could be effective in avoiding the long term respiratory sequelae that are associated with severe neonatal RSV disease.

Clearly, none of these studies could have been performed in man or in less controlled conditions than an animal facility with high levels of containment. The mouse is the most suitable species in which to study novel and potentially pathogenic recombinant viruses of this sort but has the drawback of not being a natural host for RSV infection.

4 Mimicking the Complexity of RSV's Natural Behaviour

Many of RSV's complex behaviours in man are reflected in mouse studies. For example, it is well known that protection against reinfection after natural exposure is only partial, and that adult care-givers (all of whom have anti-RSV antibody and have been repeatedly infected in the past) are quite likely to develop an RSV upper respiratory illness if they are in close contact with RSV infected infants (Hall et al. 1976). This is also the case in mice, in which secondary infection causes transient early viral replication which rapidly aborts, but can be associated with disease (Graham et al. 1991). In addition, it was shown as early as 1983 that mice could be used to demonstrate the protective effects of monoclonal antibodies in vivo (Taylor et al. 1984a), a discovery which ultimately led to the humanisation of a mouse monoclonal antibody the viral fusion glycoprotein F and to the successful marketing of palivizumab (Synagis™). This advance came initially from studies in the mouse; although the cotton rat played an important part in justifying human trials, this is arguably one the greatest translational benefits to have come from study of the mouse.

The second immunological puzzle that has been virtually solved by studies in mice is that of the formalin inactivated vaccine augmentation of RSV disease (Graham et al. 1986). The phenomenon has been reproduced in a remarkably wide range of animal studies suggesting that there are common underlying pathways. It appears that disease enhancement is associated with poor quality non-neutralising antibody, which nonetheless binds very well in ELISA assays. Studies in Dr. Brian Murphy's laboratory demonstrated that depletion of CD4 T-cells from mice that had been vaccinated with formalin inactivated RSV virtually abolishes enhanced disease (Connors et al. 1994).

It is possible that different components of disease enhancement may be caused by antibody deposition and by cellular immune responses (Delgado et al. 2009; Polack et al. 2002). From our own work, it seems that the action of formaldehyde on RSV antigens creates carbonyl groups that are sensed by the immune system and processed to generate Th2 responses (Moghaddam et al. 2006). The stimulation of Toll-like receptors may be altered by formalin treatment, leading to a lack of normal antibody affinity maturation (Delgado et al. 2009). In addition, it is possible that enhancement of disease is in part caused by an absence of sufficient regulatory T-cell responses (Loebbermann et al. 2013).

5 Advantages and Limitations of the Mouse Model of Murine Studies in Vaccine Development

There are many recent examples of promising vaccine candidates that have been tested in mice. In addition to classical subunit vaccines delivered with appropriate adjuvants, there is considerable interest in induction of mucosal immunity using virus-like particles (VLP) or bacteria-like particles (BLP). For example, a nano-emulsion-adjuvanted, inactivated mucosal RSV vaccine induces durable, RSV-specific humoral responses, both systemically and locally. Mice given this vaccine are protected against subsequent live viral challenge with a skewing of the immune response towards Th1/Th17 responses, reduced mucin production and without evidence of Th2 mediated immunopotentiation (Lindell et al. 2011).

To give a second example, reconstituted RSV viral envelopes (virosomes) with monophosphoryl lipid A (MPLA) adjuvant given intramuscularly induces neutralising RSV-specific IgG2a, at levels similar to those induced by infection with live RSV. This novel vaccine induces high levels of IFNγ and low levels of IL-5 in lungs of immunised, RSV challenged mice, indicating a Th1-skewed response. Vaccinated mice were protected from live RSV challenge, clearing the inoculated virus without showing signs of lung pathology (Kamphuis et al. 2012). It is to be hoped that new vaccines based on novel, modern adjuvants and routes of delivery will result in protective and non-pathogenic vaccines for RSV disease.

It is important to recognise that vaccines that appear promising in mice are not necessarily successful in man. BBG2Na is a fragment of the RSV G protein (residues 130–230) fused to the albumin-binding domain of streptococcal protein G, produced in bacteria and formulated in an alum-based adjuvant. In small and large animals, BBG2Na has elicited a good immune response in mice (Goetsch et al. 2000) and initial studies of BBG2Na in RSV seropositive humans were also promising (Power et al. 2001). However, clinical trials were terminated because of the development of rashes and renal complications in some volunteers that were not predicted by animal testing.

In terms of guiding human testing of vaccines, a vaccine candidate that induces poorly neutralising antibody (even of high titre in ELISA), strong IL-4/5/13 responses, enhanced disease severity and lung eosinophilia after RSV infectious challenge in rodents would not be a safe candidate for testing in seronegative human neonates. Indeed, it seems advisable that any vaccine so reminiscent of the notorious 'Lot 100' formalin-inactivated vaccines should not be administered to human volunteers of any age. In contrast, rodent studies showing a lack of Th2 responses and the absence of enhanced disease after RSV challenge, with strong neutralising antibody (with or without demonstrable Th1 responses) would help make the case for human testing, but could not guarantee that the vaccine will be safe and effective in man.

6 Concluding Remarks

Mice are often an ideal species in which to gain insights into complex biological interactions, particularly through experimental studies of immune responses. They allow controlled interventions to reveal genetic and environmental effects, illustrate underlying principals, provide a roadmap and generate hypotheses that guide and motivate experimental studies in man.

Although they tell us a great deal about protective and pathogenic mechanisms, what actually happens in man may be substantially different from that which is predicted from studies in mice. Effects that easily reproducible in mice may never be seen in man, and rare events in human studies can be impossible to anticipate from even the most extensive animal studies.

There are many important principals that have emerged from studies in animals that are essential to our understanding of human disease. We now know that a disease which appears unitary on clinical grounds may in fact arise from diverse pathogenic pathways; particularly with acute infections, it is vital to study disease over time and to examine cells from relevant tissues. Indeed, sampling cells the peripheral compartment (i.e. blood) when the disease is localised to the lung may be entirely misleading: cells may be depleted from the periphery when they are recruited to the lung, and may only recirculate once the acute disease has passed. Most importantly, counting the number of cells present at the site of disease may be an inconstant indicator of which cells are actually causing the responses (Cannon et al. 1988).

Although animal studies show what might happen, they do not necessarily reflect what actually happens in man. An intelligent dialogue between those who perform animal experiments and those observing events following natural infection is essential in our continued mission to prevent and treat RSV disease.

Acknowledgments and dedication I thank the Wellcome Trust (UK) for providing generous financial support for over the past 25 years, and all past and present laboratory members for their inestimable input and commitment. I would like apologise to all whose work has been necessary to omit from this brief review, which I dedicate to Dr Brigitte Askonas.

References

Alwan WH, Kozlowska WJ, Openshaw PJM (1994) Distinct types of lung disease caused by functional subsets of antiviral T cells. J Exp Med 179:81–89

Alwan WH, Record FM, Openshaw PJM (1993) Phenotypic and functional characterization of T cell lines specific for individual respiratory syncytial virus proteins. J Immunol 150:5211–5218

Bhoj VG, Sun Q, Bhoj EJ, Somers C, Chen X, Torres JP, Mejias A, Gomez AM, Jafri H, Ramilo O, Chen ZJ (2008) MAVS and MyD88 are essential for innate immunity but not cytotoxic T lymphocyte response against respiratory syncytial virus. Proc Natl Acad Sci USA 105:14046–14051

Cannon MJ, Openshaw PJM, Askonas BA (1988) Cytotoxic T cells clear virus but augment lung pathology in mice infected with respiratory syncytial virus. J Exp Med 168:1163–1168

Connors M, Giese NA, Kulkarni AB, Firestone C-Y, Morse HC III, Murphy BR (1994) Enhanced pulmonary histopathology induced by respiratory syncytial virus (RSV) challenge of formalin-inactivated RSV-immunized BALB/c mice is abrogated by depletion of interleukin-4 (IL-4) and IL-10. J Virol 68:5321–5325

Culley FJ, Pollott J, Openshaw PJ (2002) Age at first viral infection determines the pattern of T cell-mediated disease during reinfection in adulthood. J Exp Med 196:1381–1386

Delgado MF, Coviello S, Monsalvo AC, Melendi GA, Hernandez JZ, Batalle JP, Diaz L, Trento A, Chang HY, Mitzner W, Ravetch J, Melero JA, Irusta PM, Polack FP (2009) Lack of antibody affinity maturation due to poor toll-like receptor stimulation leads to enhanced respiratory syncytial virus disease. Nat Med 15:34–41

Goetsch L, Plotnicky-Gilquin H, Champion T, Beck A, Corvaia N, Stahl S, Bonnefoy JY, Nguyen TN, Power UF (2000) Influence of administration dose and route on the immunogenicity and protective efficacy of BBG2Na, a recombinant respiratory syncytial virus subunit vaccine candidate. Vaccine 18:2735–2742

Graham BS (2011) Biological challenges and technological opportunities for respiratory syncytial virus vaccine development. Immunol Rev 239:149–166

Graham BS, Bunton LA, Wright PF, Karzon DT (1991) Reinfection of mice with respiratory syncytial virus. J Med Virol 34:7–13

Graham BS, Perkins MD, Wright PF, Karzon DT (1988) Primary respiratory syncytial virus infection in mice. J Med Virol 26:153–162

Graham BS, Wright PF, Karzon DT (1986) Enhanced illness in BALB/c mice following nasal infection with respiratory syncytial virus (RSV) after intramuscular priming with live or formalin inactivated RSV. ICAA Abs 78:110

Hall CB, Geiman JM, Biggar R, Kotok D, Hogan PM, Douglas RG (1976) Respiratory syncytial virus infection within families. N Engl J Med 294:414–419

Harker J, Bukreyev A, Collins PL, Wang B, Openshaw PJ, Tregoning JS (2007) Virally delivered cytokines alter the immune response to future lung infections. J Virol 81:13105–13111

Harker JA, Lee DC, Yamaguchi Y, Wang B, Bukreyev A, Collins PL, Tregoning JS, Openshaw PJ (2010) Delivery of cytokines by recombinant virus in early life alters the immune response to adult lung infection. J Virol 84:5294–5302

Hussell T, Openshaw PJM (1997) Genetic influences on the induction of eosinophilic pathology during viral infection. Immunol Lett 56:305–306

Johnson JE, Gonzales RA, Olson SJ, Wright PF, Graham BS (2007) The histopathology of fatal untreated human respiratory syncytial virus infection. Mod Pathol 20:108–119

Kamphuis T, Meijerhof T, Stegmann T, Lederhofer J, Wilschut J, de HA (2012) Immunogenicity and protective capacity of a virosomal respiratory syncytial virus vaccine adjuvanted with monophosphoryl lipid A in mice. PLoS ONE 7:e36812

Lindell DM, Morris SB, White MP, Kallal LE, Lundy PK, Hamouda T, Baker JR Jr, Lukacs NW (2011) A novel inactivated intranasal respiratory syncytial virus vaccine promotes viral clearance without Th2 associated vaccine-enhanced disease. PLoS ONE 6:e21823

Loebbermann J, Thornton H, Johansson C, Openshaw PJM (2013) Defective immunoregulation in vaccine-augmented viral lung disease, restored by selective chemoattraction of regulatory T cells. Proc Natl Acad Sci USA 110(8):2987–2992

Moghaddam A, Olszewska W, Wang B, Tregoning JS, Helson R, Sattentau QJ, Openshaw PJ (2006) A potential molecular mechanism for hypersensitivity caused by formalin-inactivated vaccines. Nat Med 12:905–907

Mosmann TR, Cherwinski HM, Bond MW, Giedlin MA, Coffman RL (1986) Two types of murine helper T cell clone. I. Definition according to profiles of lymphokine activities and secreted proteins. J Immunol 136:2348–2357

Openshaw PJM, Clarke SL, Record FM (1992) Pulmonary eosinophilic response to respiratory syncytial virus infection in mice sensitized to the major surface glycoprotein G. Int Immunol 4:493–500

Polack FP, Teng MN, Collins L, Prince GA, Exner M, Regele H, Lirman DD, Rabold R, Hoffman SJ, Karp CL, Kleeberger SR, Wills-Karp M, Karron RA (2002) A Role for Immune Complexes in Enhanced Respiratory Syncytial Virus Disease. J Exp Med 196:859–865

Power UF, Nguyen TN, Rietveld E, De Swart RL, Groen J, Osterhaus AD, De Groot R, Corvaia N, Beck A, Bouveret-Le-Cam N, Bonnefoy JY (2001) Safety and immunogenicity of a novel recombinant subunit respiratory syncytial virus vaccine (BBG2Na) in healthy young adults. J Infect Dis 184:1456–1460

Pribul PK, Harker J, Wang B, Wang H, Tregoning JS, Schwarze J, Openshaw PJ (2008) Alveolar macrophages are a major determinant of early responses to viral lung infection but do not influence subsequent disease development. J Virol 82:4441–4448

Prince GA, Horswood RL, Berndt J, Suffin SC, Chanock RM (1979) Respiratory syncytial virus infection in inbred mice. Infect Immun 26:764–766

Sigurs N, Aljassim F, Kjellman B, Robinson PD, Sigurbergsson F, Bjarnason R, Gustafsson PM (2010) Asthma and allergy patterns over 18 years after severe RSV bronchiolitis in the first year of life. Thorax 65:1045–1052

Taylor G, Stott EJ, Bew M, Fernie BF, Cote PJ, Collins AP, Hughes M, Jebbett T (1984a) Monoclonal antibodies protect against respiratory syncytial virus infection in mice. Immunology 52:137–142

Taylor G, Stott EJ, Hughes M, Collins AP (1984b) Respiratory syncytial virus infection in mice. Infect Immun 43(2):649–655

Wang SZ, Xu H, Wraith A, Bowden JJ, Alpers JH, Forsyth KD (1998) Neutrophils induce damage to respiratory epithelial cells infected with respiratory syncytial virus. Eur Respir J 12:612–618

Human Airway Epithelial Cell Cultures for Modeling Respiratory Syncytial Virus Infection

Raymond J. Pickles

Abstract Respiratory syncytial virus (RSV) is an important human respiratory pathogen with narrow species tropism. Limited availability of human pathologic specimens during early RSV-induced lung disease and ethical restrictions for RSV challenge studies in the lower airways of human volunteers has slowed our understanding of how RSV causes airway disease and greatly limited the development of therapeutic strategies for reducing RSV disease burden. Our current knowledge of RSV infection and pathology is largely based on in vitro studies using nonpolarized epithelial cell-lines grown on plastic or in vivo studies using animal models semipermissive for RSV infection. Although these models have revealed important aspects of RSV infection, replication, and associated inflammatory responses, these models do not broadly recapitulate the early interactions and potential consequences of RSV infection of the human columnar airway epithelium in vivo. In this chapter, the *pro et contra* of in vitro models of human columnar airway epithelium and their usefulness in respiratory virus pathogenesis and vaccine development studies will be discussed. The use of such culture models to predict characteristics of RSV infection and the correlation of these findings to the human in vivo situation will likely accelerate our understanding of RSV pathogenesis potentially identifying novel strategies for limiting the severity of RSV-associated airway disease.

Contents

R. J. Pickles (✉)
Department of Microbiology and Immunology, UNC School of Medicine, 7021 Thurston Bowles, Campus Box 7248, Chapel Hill, NC 27599-7248, USA
e-mail: raymond_pickles@med.unc.edu

L. J. Anderson and B. S. Graham (eds.), *Challenges and Opportunities for Respiratory Syncytial Virus Vaccines*, Current Topics in Microbiology and Immunology 372,
DOI: 10.1007/978-3-642-38919-1_19,

1 Why Study RSV Infection of Human Columnar Airway Epithelium In Vitro?

The predominant site of RSV infection in the immunocompetent human respiratory tract is the columnar epithelium lining the conducting airways. Columnar airway epithelium serves as an important innate barrier in the respiratory tract and possesses highly specialized functions due to the distinct features of diverse columnar epithelial cell-types present throughout the airways. How RSV infection affects the integrity and innate defenses of the human airway epithelium resulting in airway disease is still not clearly defined. Whether outcomes of RSV infection of nonpolarized epithelial cell-lines or semipermissive animal models in vivo reflect those of RSV infection of human columnar airway epithelial cells is currently under investigation. Although in vivo human challenge protocols are available and are an excellent model system for assessing consequences of RSV infection and testing the efficacy of RSV vaccine candidates and anti-virals, these studies are limited to inoculation of the nasal epithelium and provide little information on RSV pathogenesis in lower airway regions. Human in vivo challenge studies are also cost-prohibitive to most in the RSV field; they require specialized facilities, personnel and appropriate review by regulatory agencies. Therefore, more cost effective and widely available models of human columnar airway epithelium have been sought to further understand how RSV infection affects the normal functions of the airway epithelium and how altering epithelium function influences RSV pathogenesis. Such in vitro models also hold great utility for preclinical testing of infection and growth phenotypes of live, attenuated RSV vaccine candidates and assessing the efficacy of RSV antiviral approaches.

Culture models of human columnar airway epithelium developed by several groups are gaining acceptance for studying particular aspects of respiratory virus infection and airway innate defenses (Pickles et al. 1998; Zhang et al. 2002; Sims et al. 2006; Zabner et al. 2000; Wright et al. 2005; Villenave et al. 2012; Ilyushina et al. 2012). Technical aspects associated with isolation and culture of human airway epithelial cell cultures will not be further discussed here and the reader is referred to a comprehensive description of these protocols (Fulcher et al. 2005). The availability of airway cell culture models relies on an abundant primary source of airway epithelial cells isolated from the lungs of human donors. In our laboratory, human airway epithelial cell cultures (HAE) are generated from epithelial cells isolated from human nasal or tracheobronchial epithelium obtained from cadaver airways or excess airway tissues after lung transplantation or elective surgery. An alternative source of human airway cells is from volunteers willing to undergo harvest of epithelial cells by scraping or brushing of nasal or tracheobronchial epithelium. Regardless of the method of procurement, isolated epithelial cells can then be grown using appropriate culture conditions to generate differentiated cultures of columnar airway epithelium. In our experience and that of others, HAE cultures exhibit morphologic and physiologic properties similar to those of the human columnar airway epithelium in vivo.

2 What is the Cellular Composition of HAE Cultures and What Airway Regions do they Represent?

To appreciate the usefulness of human columnar airway cultures it is important to describe the diversity of airway epithelial cells distributed in the human airways in vivo and how HAE reproduce this cell diversity in vitro.

Respiratory columnar epithelium lines the nasopharyngeal (*upper conducting airways*) and the tracheal, bronchial, and bronchiolar airways (*lower conducting airways*). The respiratory epithelium also lines the alveolar lung regions but does not possess morphologic features of a columnar epithelium. Although ciliated airway epithelial cells are predominant throughout the conducting airways of all species, both ciliated cell density and the phenotype of nonciliated cells are airway region and species dependent (Harkema et al. 1994). For humans, the cartilaginous conducting airway epithelium (nasal, tracheobronchial, bronchial) is a pseudostratified columnar epithelium with predominant ciliated cells interspersed with variable numbers of mucin-secreting cells. In these airway regions, columnar cells overlie basal epithelial cells directly attached to the underlying basement membrane. Ciliated cells are also prominent in the human small, noncartilaginous bronchiolar airways but their numbers decrease as the bronchioles approach the airspaces of the alveolar regions. In the cartilaginous airways, nonciliated columnar cells largely represent the morphology and properties of classic mucin-secreting cells (Goblet cells) whereas, the nonciliated columnar cells in the bronchiolar regions are most often identified as Clara cells. Distal to the conducting airways, the lung epithelium consists of Type I and Type II pneumocytes.

For human ciliated airway epithelium models, three anatomical airway regions are routinely used to obtain primary epithelial cells; the nasal epithelium, the nasopharyngeal epithelium (adenoids), and the tracheobronchial epithelium. HAE cultures derived from cells procured from these regions result in the generation of a pseudostratified mucociliary columnar epithelium in which ciliated cells and mucin-secreting (Goblet) cells make up the vast majority of columnar cells. Nonciliated columnar cells with morphological features of Clara cells are usually scarce in these culture models likely reflecting the low numbers of Clara cells present in the proximal airway tissues used for harvesting the epithelial cells.

3 RSV Infects Human Ciliated Airway Epithelium In Vivo and In Vitro

Histopathologic specimens of RSV-infected lung tissues have immunolocalized RSV antigen to columnar airway epithelium of the large conducting airways and in ciliated cells and nonciliated cells (Clara cells) in the bronchiolar regions (Johnson et al. 2007; Welliver et al. 2008; Neilson and Yunis 1990). Although RSV antigen can be detected by immunohistologic studies in alveolar regions, data supporting RSV infection of alveolar epithelial cells in immunocompetent patients are less robust than for RSV infection of the bronchiolar regions (Neilson and Yunis 1990). Although Type II pneumocytes can be infected by RSV, especially in immunocompromised patients, the frequency and timing of infection of bronchiolar versus alveolar epithelium and its relationship to outcomes of disease remain unclear.

Human airway epithelial cell cultures derived from cartilaginous airway regions have been used by us and by others to show RSV robustly infects ciliated cells (Fig. 1 Zhang et al. 2002; Villenave et al. 2012). However, ciliated cell tropism in these models is not unique to RSV since parainfluenza viruses (human PIV1-5, Sendai virus), human/avian influenza viruses, and coronaviruses exhibit preferential infection of ciliated cells in these models (Zhang et al. 2005, 2012; Scull et al. 2009; Sims et al. 2005; Villenave et al. 2012; Bartlett et al. 2008a; Schaap-Nutt et al. 2010). Indeed, by comparing infection of HAE with different respiratory viruses we and others have found human influenza viruses are the only respiratory viruses so far tested that infect both ciliated and nonciliated columnar cells of HAE (Scull et al. 2009; Matrosovich et al. 2004). The significance of expanded virus tropism for both ciliated and nonciliated cells with regards to outcomes of influenza virus infection is unknown but we speculate influenza virus evolution in human hosts towards nonciliated columnar cells may provide an advantage to the virus against innate host defenses.

For RSV infection of HAE, ciliated cells are the exclusive target for infection (Zhang et al. 2002). An estimated 60–90 % of lumenal surface cells in HAE are ciliated cells (Zhang et al. 2009), with the remainder of columnar cells resembling mucin-secreting Goblet cells found in human airway epithelium in vivo (Fig. 2). However, other less common "secretory columnar cell-types" are occasionally identified in HAE although these columnar cell-types are not commonly found in human airways in vivo (Fig. 2). Thus, for the most part, the columnar cell composition of HAE closely mimics columnar cell density and phenotype of the

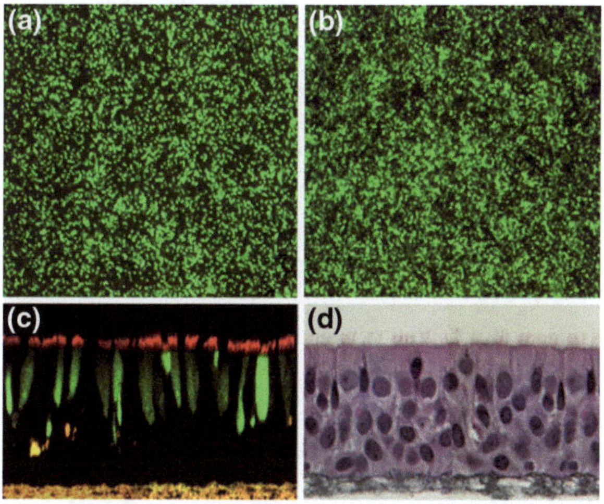

Fig. 1 Human airway epithelial cell cultures infected by RSV and PIV3 expressing GFP. (**a** + **b**) *En face* XY views of HAE infected with RSV (**a**) or PIV3 (**b**) and infected cells detected by expression of GFP (*green*). **c** Confocal XZ microscopy of HAE infected with PIV3 shows GFP (*green*) is localized to columnar cells positive for β-tubulin IV immunoreactivity (*red*) indicating PIV3 infects ciliated cells. **d** Histological cross-section of HAE demonstrating the morphologic characteristics of a pseudostratified mucociliary airway epithelium

columnar epithelium of the human nasal and tracheobronchial airway regions in vivo. In our experience, we have not routinely identified Clara cells in HAE; an expected finding given the isolation of cells from large airway regions where Clara cells are scarce. Isolation of airway epithelial cells from human bronchiolar airways where Clara cell density is high has not yet successfully resulted in differentiated cultures with columnar epithelial cells displaying morphologic characteristics of Clara cells in vivo. Primary cultures for human alveolar Type II pneumocytes are available and currently await studies with RSV infection (Bove et al. 2011).

4 Is RSV Infection of Human Ciliated Cells Different from Infection of NonPolarized Epithelial Cell-Lines?

Early studies of RSV infection of HAE revealed that although ciliated cells were robustly infected by RSV these cells were less susceptible to infection than human nonpolarized epithelial cell-lines, e.g., HEp-2 and A549 cells (*unpublished observations* and (Wright et al. 2005). One explanation for these observations is the greater abundance of extracellular glycocalyx present on the lumenal surface of ciliated cells when compared to that on the surfaces of nonpolarized epithelial cell-lines (Kesimer et al. 2009, 2012). It is expected RSV and other viruses that

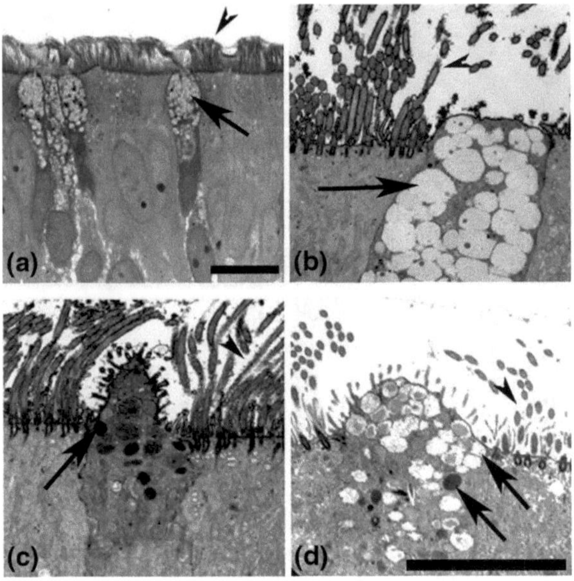

Fig. 2 Morphological features of columnar epithelial cells in HAE. **a** Histological section of HAE showing ciliated and nonciliated columnar epithelial cells with distribution similar to human tracheobronchial airway epithelium in vivo. **b–d** Transmission electron micrographs detailing morphological characteristics of nonciliated columnar epithelial cells in HAE. **b** A frequently identified nonciliated columnar cell containing clear mucin-rich granules and resembling a Goblet cell. **c** An infrequently identified nonciliated columnar epithelial cell with electron dense granules and resembling a Serous cell. **d** An infrequently identified nonciliated columnar epithelial cell with mixed phenotypes of mucin-secreting and serous-like cell features. *Arrowheads* indicate cilia of ciliated columnar cells and *long arrows* indicate clear and/or dense granules in nonciliated columnar epithelial cells. Scale bars represent 20 μm

infect ciliated cells will have to negotiate the more complex glycocalyx structure and these viruses have likely evolved strategies to enable this. Indeed, we have shown columnar cell glycocalyx in HAE poses a significant diffusion barrier to respiratory viruses reaching the apical membranes of columnar cells (Stonebraker et al. 2004; Kesimer et al. 2012). The robust glycocalyx of the columnar airway epithelium is largely due to the abundance of several families of carbohydrate-rich molecules, including proteoglycans, glycolipids, and glycoproteins most notably, the highly O-glycosylated 'tethered' mucins MUC1, 4, and 16. These tethered mucins constitute a significant portion of the glycocalyx structure in human and murine airway epithelia (Kesimer et al. 2012; Stonebraker et al. 2004; Button et al. 2012).

Despite the barrier properties of the glycocalyx on HAE some respiratory viruses, including RSV, are relatively efficient at negotiating this carbohydrate-rich network. We have speculated the RSV G glycoprotein may serve to facilitate the passage of RSV into the glycocalyx layer. This idea was based on data showing recombinant RSV deleted the G protein (RSVΔG) infected nonpolarized epithelial cell-lines just as efficiently as wild-type virus suggesting the G protein was not required for infection (Karron et al. 1997). In contrast to these previous findings in nonpolarized epithelial cells, we recently showed RSV infection of HAE is significantly increased when the virus expresses a full-length G protein (Kwilas et al. 2009). While RSV efficiently infects and spreads in HAE, RSVΔG poorly infects ciliated cells and displays attenuated growth characteristics in HAE (Kwilas et al. 2009). Since RSV G has been previously described as a "mucin-like" glycoprotein with the potential for multiple O-glycosylation sites (Collins 1990), we speculate one function of G may be to facilitate the negotiation of RSV through the robust glycocalyx layer present on the lumenal surface of polarized airway epithelial cells. These studies highlight just one example of the utility of HAE to better model in vivo virus–host airway cell interactions when compared to nonpolarized epithelial cell-lines.

The importance of the RSV G protein for efficient infection of human ciliated cells is further supported by experiments showing RSV propagated in HEp-2 cells efficiently infects HAE whereas RSV propagated in VERO cells poorly infects HAE (Kwilas et al. 2009). However, in contrast to RSVΔG growth in HAE, only the initial inoculum of VERO cell-derived RSV was attenuated in HAE and subsequent rounds of infection within HAE cultures proceeded similarly to those after inoculation with RSV propagated in HEp-2 cells (Fig. 3). Subsequently, it was discovered the G glycoprotein of RSV propagated in HEp-2 cells had a molecular mass of 90 kD while the G from VERO cell-derived RSV was only 55 kD suggesting VERO-derived G was deficient is some component important for efficient infection of ciliated cells. Although the numbers of ciliated cells infected by VERO cell-derived RSV were lower than for HEp-2 derived RSV, progeny virions budding from ciliated cells infected by VERO-derived RSV quickly regained normal infection kinetics suggesting deficiencies in G function for RSV propagated in VERO cells were possibly due to altered post-transcriptional modification of G in VERO cells and not due to genetic alterations in the virus genome. The significance of the different molecular masses for G protein derived from different producer cell-lines remains to be determined but only became apparent using HAE as the target cells for infection. Precisely, how the RSV G protein may be modified in producer cell-lines to account for these differences in mass and infectivity for ciliated cells is under investigation.

These findings associated with RSV G may also have technical implications for RSV vaccine development. Since VERO cells remain the only FDA-approved cell-line for propagating RSV vaccines for clinical use, the inoculating dose of vaccine candidates may exhibit reduced infection efficiency of ciliated cells of the human respiratory tract. For live, attenuated RSV vaccines that replicate similarly to wild-type RSV in the upper airways such an infection deficiency may not be a significant concern as further rounds of replication and infection would rapidly generate RSV expressing a fully functional G protein (see chapter by J.S. McLellan et al., this volume).

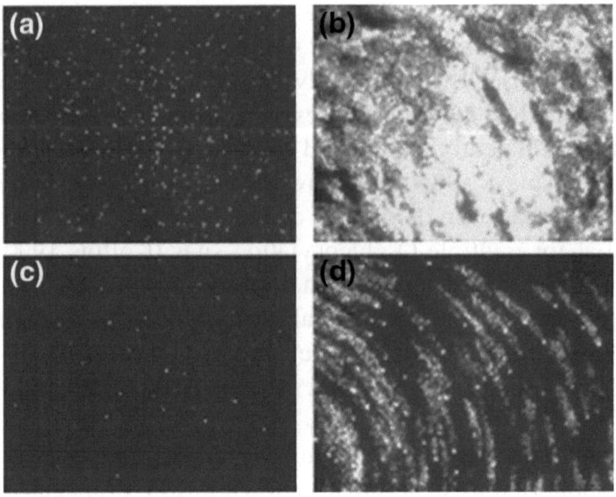

Fig. 3 Decreased initial infection of HAE by RSV propagated in VERO cells compared to HEp-2 cells. HEp-2 or VERO cell derived RSV titered on cell-lines was inoculated at equal titer (10^6 PFU) onto HAE and GFP expression (*white signal*) assessed at 1 (**a**, **c**) and 2 (**b**, **d**) days pi. *En face* views of: **a** RSV (HEp2-derived) infected HAE at 1 day pi; **b** RSV (HEp-2-derived) infected HAE at 2 days pi; **c** RSV (VERO-derived) infected HAE at 1 day pi; and **d** RSV (VERO-derived) infected HAE at 2 days pi. One day after inoculation RSV (HEp-2) infects far more ciliated cells than RSV (VERO) and by day 2 pi, RSV (HEp-2) has spread throughout the culture while RSV (VERO) exhibits delayed spread. *Note* Spread of RSV infection in (**d**) demonstrates patterns of 'comet-like' spread of infection accounted for by the directionality of cilia beat in HAE

5 Variability of Virus Infection in HAE Derived from Different Donors

In contrast to HAE, epithelial cell-lines offer unlimited and continuous use of genetically homogenous cell populations. HAE cultures are derived from individual donors with a strong possibility of genetic and environmental variability between donors. When using HAE there are two practical considerations when designing experiments. First, the total number of primary cells and hence, HAE cultures, derived from an individual donor will always be limiting. Human primary airway epithelial cells grown on plastic can be passaged several times to increase cell availability, but the ability to reproducibly differentiate these cells into ciliated columnar epithelium wanes in later passages. Attempts to generate 'human primary epithelial cell-lines' have successfully generated cultures of pseudostratified columnar airway epithelium however, the extent of ciliated cell differentiation in these cultures does not approach the ciliated cell densities achievable with freshly isolated primary epithelial cells (Fulcher et al. 2009). The second practical concern when using HAE relates to the potential for variability due to differences in host genetics and/or environmental exposures between individual donors. Currently, the only feasible method to determine variability in experimental outcomes between HAE derived from different donors is empirical testing of the specific experimental outcome, e.g., virus growth kinetics, across cultures derived from several different donors (i.e., 4–7 different donors).

In our experience, the most satisfactory approach to these experimental considerations is to generate sufficient numbers of HAE cultures (24–48 wells) from a single donor and plan experiments to include the necessary replicates of test and controls conditions within this cohort. HAE cultures derived from different donors can then be used in identical experiments and variability assessed. If variability is low, data can be aggregated across all HAE used. For example, PIV1 infection and growth kinetics in human ciliated cells of HAE was highly reproducible both between cultures derived from an individual donor and HAE derived from 5 different donors (Bartlett et al. 2008a, b). Similar data were obtained for influenza viruses (Scull et al. 2009) and PIV2 (Schaap-Nutt et al. 2012). However, in our experience, not all experimental outcomes are as consistent between donors. For example, although PIV2 infection kinetics between donors were identical, significant differences in inflammatory mediator production was determined between donors (Schaap-Nutt et al. 2012); thus, highlighting the requirement to confirm reproducibility of experimental outcomes both between cultures and donor sources.

6 Using HAE to Predict Respiratory Virus Growth Kinetics In Vivo

We have recently found striking differences between the growth kinetics of PIV serotypes 1–3 in HAE (Schaap-Nutt et al. 2012). In contrast, PIV1-3 grows with identical kinetics in nonpolarized epithelial cell-lines (HEp-2, LLC-MK2 and VERO cells). Why HAE reveal differences in PIV growth kinetics when epithelial cell-lines do not remains to be determined but requires detailed experimentation in ciliated cells to identify the underlying cause of this phenomenon.

The ability of HAE to discriminate between respiratory virus growth kinetics has also been shown for live, attenuated PIV, RSV, and influenza virus vaccine candidates (Schaap-Nutt et al. 2010; Bartlett et al. 2008a; Wright et al. 2005; Ilyushina et al. 2012). Generally, live, attenuated vaccine candidates exhibit a greater degree of growth attenuation in HAE models than in nonpolarized epithelial cell-lines and the degree of attenuation in HAE more closely mimics that measured in vivo. A human adenoid airway epithelium culture model was used to demonstrate that live, attenuated RSV vaccine growth kinetics in differentiated cultures, but not that in cell-lines, closely resembled the kinetics of vaccine growth in airways of human and nonhuman primates in vivo (Wright et al. 2005). Figure 4 illustrates similar experiments with a live, attenuated PIV2 vaccine that showed decreased virus growth in HAE compared to PIV2 wild-type at 32 °C (the

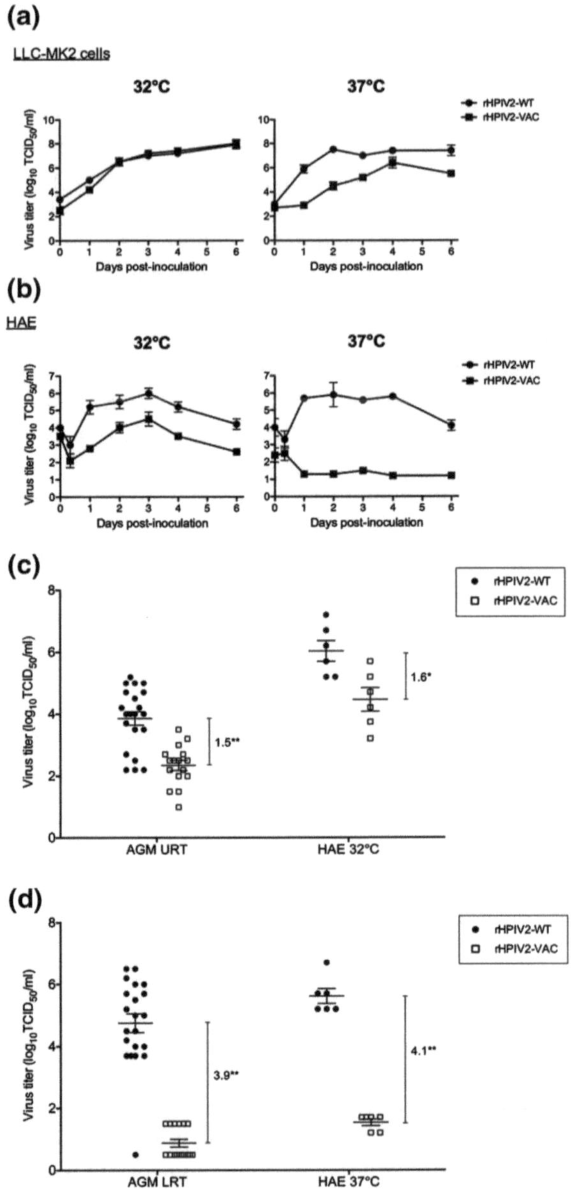

Fig. 4 PIV2 vaccine candidate growth kinetics in HAE but not in epithelial cell-lines correlate with growth kinetics in the upper and lower airways of nonhuman primates in vivo. **a** Growth kinetics of rHPIV2-WT (*closed circles*) and rHPIV2-VAC (*closed squares*) in LLC-MK2 cells at 32 and 37 °C. **b** Single cycle growth curves in HAE inoculated with rHPIV2-WT (*closed circles*) or rHPIV2-VAC (*closed squares*) at either 32 or 37 °C. Virus titers were determined in the apical compartments at the indicated times pi. **c** Titers of rHPIV2-WT and rHPIV2-VAC in the upper respiratory tract (URT) of NHPs and in the apical wash of HAE incubated at 32 °C. **d** Virus titers in the lower respiratory tract (LRT) of NHPs and in the apical wash of HAE incubated at 37 °C. In **c** and **d**, peak virus titer in NHPs or HAEs is indicated by a *closed circle* (rHPIV2-WT) or *open square* (rHPIV2-VAC). * indicates statistical significance with P value <0.01, while ** indicates a P value <0.001. *Adapted with permission from* Schaap-Nutt et al. (2010)

temperature of the nasal cavity) whereas, cell-lines showed no differences in growth kinetics between the PIV2 vaccine and PIV2 wild-type (Schaap-Nutt et al. 2010). The discrepancy between differences in attenuation of the PIV2 vaccine in HAE versus cell-lines was more dramatic when experiments were performed at 37 °C (the temperature of the lower airways). While PIV2 wild-type grew well in

HAE at 37 °C, the vaccine candidate was severely attenuated essentially showing zero growth. In contrast to these data in HAE, the vaccine candidate was only modestly attenuated compared to PIV2 wild-type in epithelial cell-lines at 37 °C. The relevance of these in vitro findings was demonstrated when PIV2 wild-type and vaccine candidate growth kinetics were determined in the upper and lower airways of nonhuman primates in vivo where the degree of growth attenuation for the vaccine candidate versus wild-type virus closely mimicked that in HAE but not in epithelial cell-lines. Recently, similar conclusions were reached for influenza virus vaccines where differentiated airway epithelium culture models far better predicted in vivo growth kinetics of vaccine candidates than epithelial cell-lines (Ilyushina et al. 2012).

Together, these published data with RSV, PIV, and influenza virus vaccine candidates strongly support the use of human columnar airway epithelium cultures for predicting growth attenuation of live, attenuated respiratory virus vaccine candidates. The capacity to predict in vivo growth attenuation in an appropriate human in vitro culture model will greatly benefit respiratory virus vaccine programs and reduce the number of animals required for initial testing of such vaccines.

7 The Epithelium is Only One Section of the Orchestra Playing the Immunologic Symphony in the Airways

The airway epithelium plays a central role in identifying, responding to, and resolving respiratory virus infections. The cellular cross-talk between epithelial cells, macrophages, dendritic cells, and infiltrating cells of the immune system is highly complex and ideally results in an appropriate, but not exaggerated, inflammatory response sufficient to clear virus infection. A significant drawback of human airway epithelium culture models for respiratory virus vaccine research is the lack of these other immune-related cell-types in this culture model.

Several groups have begun to develop the next generation of human airway cultures with co-cultures of other cellular components of the lung. Co-cultures of HAE (grown on the upper surface of the semipermeable membrane support) and dendritic cells (grown on the other side of the support) have been used to investigate cellular cross-talk between epithelial and dendritic cells (Ilona Jaspers, *personal communication*). Similar co-culture HAE models could be developed to include macrophages, fibroblasts, and immune cells normally recruited to the airways in response to infections.

Cultures containing only airway epithelial cells can be useful for assessing the inflammatory profile of a virus infection by measuring inflammatory mediators generated before and after virus infection. Global ‘whole culture’ analysis of secreted inflammatory mediators in culture washes can be performed at the protein level by ELISA or antibody/bead-based technologies. For example, the magnitude and duration of two secreted pro-inflammatory mediators, IL-8 and CXCL10 were measured by Luminex bead-based technology in the serosal media of RSV-infected HAE (Fig. 5). Alternatively, whole culture mRNA gene expression analysis using qRT-PCR provides a cost-effective approach for determining inflammatory responses to virus infections. These commonly used assays for measuring inflammatory mediator profiles measure protein secretion or changes in mRNA expression from whole HAE cultures in which differences in cell-types present, some infected, others not, can complicate interpretation of the data. Unfortunately, in our hands, attempts to isolate ciliated cell populations or RSV-infected ciliated cell populations from HAE by proteolytic dispersion or laser-capture dissection, significantly reduced mRNA integrity and invalidated conclusions drawn from these assays.

8 How Well are Ciliated Airway Epithelial Cells from Different Species Infected by Human RSV?

The development of differentiated columnar airway epithelial cell models from species other than human has allowed studies on species tropism of respiratory viruses. In addition to human cultures, we have also generated tracheal airway ciliated columnar epithelial cell cultures from bat, mouse, hamster, ferret, dog, cow, pig, and nonhuman primate. Although culture conditions need to be tailored for each species we have successfully obtained sufficient ciliated cells from all these species for comparative virus infection studies. Comparative in vitro versus in vivo morphologic studies indicate, depending on the species, that the proportion of ciliated versus nonciliated cells in airway cultures reflects the cellular distribution in the tracheal epithelium in vivo.

The use of differentiated cultures from different species also allows confirmation of the predicted tropism of respiratory viruses for columnar cells (Fig. 6). Using these models, we find human RSV robustly infects human ciliated cells and to a lesser extent, ciliated cells from nonhuman primates. In our hands, RSV poorly infects ciliated cells from hamsters and mice corroborating our in vivo studies in these semipermissive animal models (*unpublished observations*). Human PIV3 (and PIV1) shows a broader species tropism in vitro with robust infection of human, nonhuman primate, and hamster ciliated cells. These observations match the expected tropism of human PIV3 for humans, NHPs, and hamsters in vivo. While human PIVs only modestly infect mouse ciliated cells, Sendai virus (mouse PIV1) shows broad species tropism for ciliated cells robustly infecting cells in mouse, hamster, NHP, and human cultures.

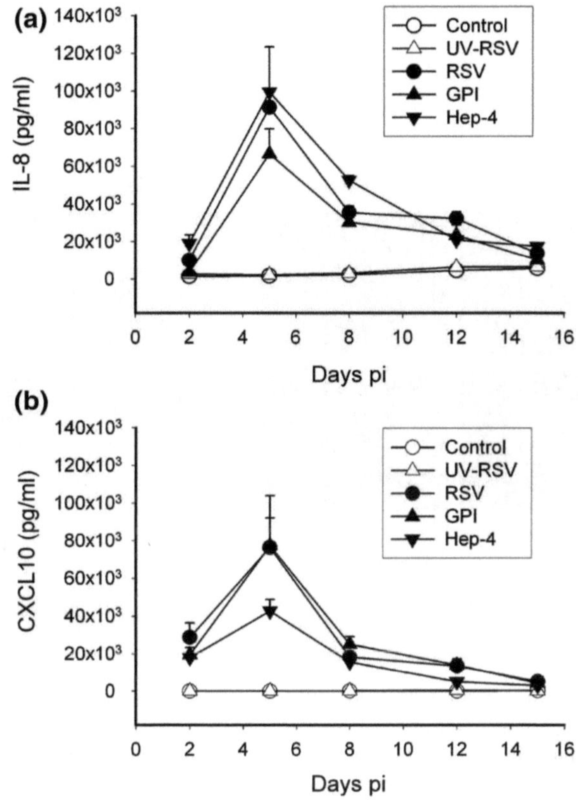

Fig. 5 Inflammatory mediator secretion from HAE after infection with RSV. HAE cultures were inoculated with recombinant RSV A2 expressing GFP (RSV, *solid circles*), recombinant RSV A2 (GPI, solid upright *triangles*), biologically derived RSV A2 (Hep-4, *solid inverted triangles*), ultraviolet light-inactivated RSV expressing GFP (UV-RSV, *open upright triangles*) or mock-inoculate (Control, *open circles*) and over time samples of serosal media harvested. Luminex bead-based technology was used to detect and measure 23 different pro-inflammatory analytes in the samples. (**a**) and (**b**) show concentrations of the pro-inflammatory mediators IL-8 and CXCL10 harvested in the serosal media, respectively. Of the 23 analytes probed only IL-8, CXCL10, IL-6, and RANTES were increased above baseline levels by RSV infection. Analytes probed for but not increased by RSV infection, were: IL-1α/β, IL-2, IL-3, IL-4, IL-5, IL-7, IL-9, IL-10, IL-12p40, IL-13, IL-15, Eotaxin, GM-CSF, IFNγ, IFNα2, TNFα, MCP-1, MIP-1α, and EGF. Data represent $n = 5$ replicate cultures derived from one donor. NOTE: Extent of inflammatory mediator secretion paralleled RSV titers. At maximal secretion (day 5) despite viral titers being equal there are differences in the extent of IL-8 and CXCL10 induced by the different RSV inoculates. For example, the GPI virus does not induce as much IL-8 as the recombinant expressing GFP or Hep-4. In contrast, for CXCL10 secretion, Hep-4 does not induce as much CXCL10 secretion as the other two RSV inoculates. Since cultures are derived from the same donor, host genetic variability is unlikely to account for these differences. Rather, these differences may be influenced by how host cells respond to subtle differences the virus genome sequences

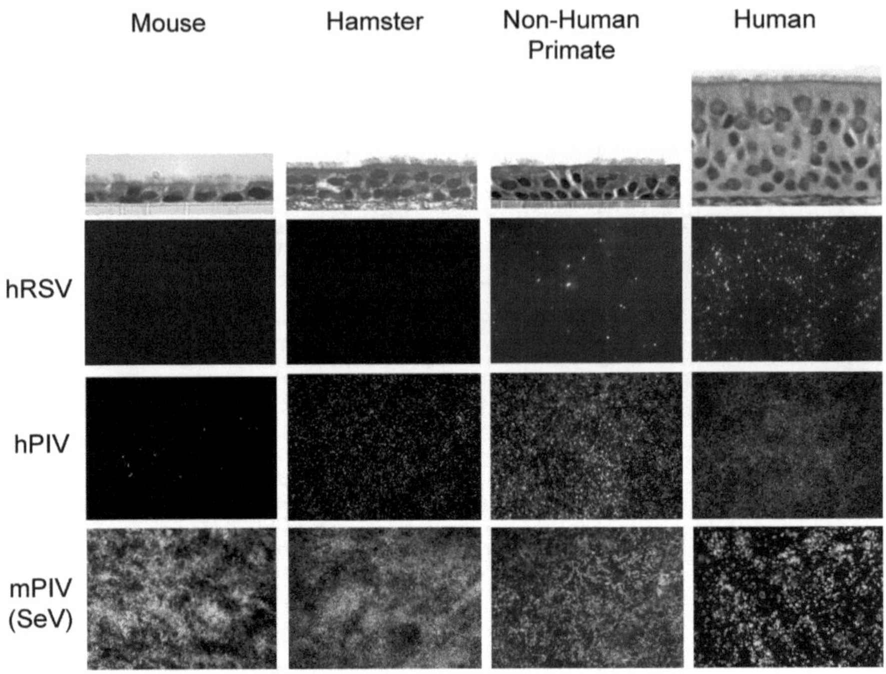

Fig. 6 In vitro models of ciliated tracheobronchial airway epithelium derived from multiple species for investigating respiratory virus tropism. (*Top row*) Histological sections of cultures of ciliated airway epithelium derived from cells isolated from the trachea of mouse, hamster, nonhuman primates, and humans. Ciliated and nonciliated cells are present in cultures from all species tested. (*Lower panels*) *En face* views of cultures of ciliated airway epithelium for each species inoculated with recombinant, GFP expressing, human RSV (hRSV), human PIV3 (hPIV), or mouse PIV1/Sendai virus (mPIV1/SeV). Infected cells were identified 3 days pi by monitoring GFP expression (*white signal*). NOTE: RSV infects human ciliated cells well and nonhuman primate cells modestly but does not infect hamster or mouse ciliated cells. In contrast, Sendai virus (SeV) robustly infects ciliated cells from all species tested. Human PIV3 infects human, nonhuman primate, and hamster ciliated cells well but poorly infects mouse ciliated cells. These in vitro data correlate with the known in vivo tropism of SeV for airways of all these species and for hPIV3 for human, nonhuman primates, and hamster airways

9 Conclusions and Future Directions

Human ciliated airway epithelium cultures closely resemble the morphologic and physiologic characteristics of the human ciliated airway epithelium in vivo; the primary infection site of many respiratory viruses including RSV. Existing evidence already indicates differentiated airway epithelium culture models more accurately predict in vivo virus-host airway cell interactions, virus growth kinetics and consequences of infection than nonpolarized epithelial cell-lines. The ability to compare virus infection in ciliated airway epithelium in vitro versus in vivo, e.g., for PIV infection of hamster airways, will strengthen the concept that airway cultures can be used as a precursor and/or surrogate model for in vivo studies. For RSV, direct comparison between infection of HAE in vitro versus in vivo is a significant challenge due to the inability to perform detailed experiments on humans especially in lower airway regions. Therefore, more emphasis should be placed on correlating the consequences of RSV infection of human airway epithelium culture models to histopathologic findings in lower airways of patients naturally infected by RSV.

Acknowledgments The author thanks Drs. Peter Collins and Mark Peeples for many years of enjoyable and fruitful collaboration and for teaching a cell biologist about the wonders of respiratory viruses. Thanks also to members of the laboratory who contributed to the thoughts and experiments discussed here, in particular, Drs. Liqun Zhang, Meg Hennessey, Racheal Liesman. I would also like to thank the Directors and teams of the UNC Cystic Fibrosis Center Tissue Culture Core, the UNC Morphology and Morphometry Core, and the Michael Hooker Microscopy Facility for supplying reagents and technical expertise. Portions of these studies were funded by grants from the UNC University Research Council, the Cystic Fibrosis Foundation, and the National Institutes of Health (R01HL103940, R01 HL77844, P50HL084934).

References

Bartlett EJ, Cruz AM, Esker J, Castano A, Schomacker H, Surman SR, Hennessey M, Boonyaratanakornkit J, Pickles RJ, Collins PL, Murphy BR, Schmidt AC (2008a) Human parainfluenza virus type 1 C proteins are nonessential proteins that inhibit the host interferon and apoptotic responses and are required for efficient replication in nonhuman primates. J Virol 82(18):8965–8977

Bartlett EJ, Hennessey M, Skiadopoulos MH, Schmidt AC, Collins PL, Murphy BR, Pickles RJ (2008b) Role of interferon in the replication of human parainfluenza virus type 1 wild type and mutant viruses in human ciliated airway epithelium. J Virol 82(16):8059–8070

Bove PF, Grubb BR, Okada SF, Ribeiro CM, Rogers TD, Randell SH, O'Neal WK, Boucher RC (2011) Human alveolar type II cells secrete and absorb liquid in response to local nucleotide signaling. J Biol Chem 285(45):34939–34949

Button B, Cai LH, Ehre C, Kesimer M, Hill DB, Sheehan JK, Boucher RC, Rubinstein M (2012) A periciliary brush promotes the lung health by separating the mucus layer from airway epithelia. Science 337(6097):937–941

Collins PL (1990) O glycosylation of glycoprotein G of human respiratory syncytial virus is specified within the divergent ectodomain. J Virol 64(8):4007–4012

Fulcher ML, Gabriel S, Burns KA, Yankaskas JR, Randell SH (2005) Well-differentiated human airway epithelial cell cultures. Methods Mol Med 107:183–206

Fulcher ML, Gabriel SE, Olsen JC, Tatreau JR, Gentzsch M, Livanos E, Saavedra MT, Salmon P, Randell SH (2009) Novel human bronchial epithelial cell lines for cystic fibrosis research. Am J Physiol Lung Cell Mol Physiol 296(1):L82–L91

Harkema J, Mariassy A, St. George J, Hyde DM, Plopper CG (1994) Epithelial cells of the conducting airways: a species comparison. In: Farmer SG, Hay DWP (eds) The airway epithelium, vol 55. Lung Biology in Health and Disease. Marcel Dekker, New York, pp 3–39

Ilyushina NA, Ikizler MR, Kawaoka Y, Rudenko LG, Treanor JJ, Subbarao K, Wright PF (2012) Comparative study of influenza virus replication in MDCK cells and in primary cells derived from adenoids and airway epithelium. J Virol

Johnson JE, Gonzales RA, Olson SJ, Wright PF, Graham BS (2007) The histopathology of fatal untreated human respiratory syncytial virus infection. Mod Pathol 20(1):108–119

Karron RA, Buonagurio DA, Georgiu AF, Whitehead SS, Adamus JE, Clements-Mann ML, Harris DO, Randolph VB, Udem SA, Murphy BR, Sidhu MS (1997) Respiratory syncytial virus (RSV) SH and G proteins are not essential for viral replication in vitro: clinical evaluation and molecular characterization of a cold-passaged, attenuated RSV subgroup B mutant. Proc Natl Acad Sci USA 94(25):13961–13966

Kesimer M, Ehre C, Burns KA, Davis CW, Sheehan JK, Pickles RJ (2012) Molecular organization of the mucins and glycocalyx underlying mucus transport over mucosal surfaces of the airways. Mucosal Immunol

Kesimer M, Kirkham S, Pickles RJ, Henderson AG, Alexis NE, Demaria G, Knight D, Thornton DJ, Sheehan JK (2009) Tracheobronchial air-liquid interface cell culture: a model for innate mucosal defense of the upper airways? Am J Physiol Lung Cell Mol Physiol 296(1):L92–L100

Kwilas S, Liesman RM, Zhang L, Walsh E, Pickles RJ, Peeples ME (2009) Respiratory syncytial virus grown in Vero cells contains a truncated attachment protein that alters its infectivity and dependence on glycosaminoglycans. J Virol 83(20):10710–10718

Matrosovich MN, Matrosovich TY, Gray T, Roberts NA, Klenk HD (2004) Human and avian influenza viruses target different cell types in cultures of human airway epithelium. Proc Natl Acad Sci USA 101(13):4620–4624

Neilson KA, Yunis EJ (1990) Demonstration of respiratory syncytial virus in an autopsy series. Pediatr Pathol 10(4):491–502

Pickles RJ, McCarty D, Matsui H, Hart PJ, Randell SH, Boucher RC (1998) Limited entry of adenovirus vectors into well-differentiated airway epithelium is responsible for inefficient gene transfer. J Virol 72(7):6014–6023

Schaap-Nutt A, Scull MA, Schmidt AC, Murphy BR, Pickles RJ (2010) Growth restriction of an experimental live attenuated human parainfluenza virus type 2 vaccine in human ciliated airway epithelium in vitro parallels attenuation in African green monkeys. Vaccine 28(15):2788–2798

Schaap-Nutt A, Liesman R, Bartlett EJ, Scull MA, Collins PL, Pickles RJ, Schmidt AC (2012) Human parainfluenza virus serotypes differ in their kinetics of replication and cytokine secretion in human tracheobronchial airway epithelium. Virology 433(2):320–328

Scull MA, Gillim-Ross L, Santos C, Roberts KL, Bordonali E, Subbarao K, Barclay WS, Pickles RJ (2009) Avian Influenza virus glycoproteins restrict virus replication and spread through human airway epithelium at temperatures of the proximal airways. PLoS Pathog 5(5):e1000424

Sims AC, Baric RS, Yount B, Burkett SE, Collins PL, Pickles RJ (2005) Severe acute respiratory syndrome coronavirus infection of human ciliated airway epithelia: role of ciliated cells in viral spread in the conducting airways of the lungs. J Virol 79(24):15511–15524

Sims AC, Yount B, Burkett SE, Baric RS, Pickles RJ (2006) SARS CoV replication and pathogenesis in human airway epithelial cultures. Adv Exp Med Biol 581:535–538

Stonebraker JR, Wagner D, Lefensty RW, Burns K, Gendler SJ, Bergelson JM, Boucher RC, O'Neal WK, Pickles RJ (2004) Glycocalyx restricts adenoviral vector access to apical receptors expressed on respiratory epithelium in vitro and in vivo: role for tethered mucins as barriers to lumenal infection. J Virol 78(24):13755–13768

Villenave R, Thavagnanam S, Sarlang S, Parker J, Douglas I, Skibinski G, Heaney LG, McKaigue JP, Coyle PV, Shields MD, Power UF (2012) In vitro modeling of respiratory syncytial virus infection of pediatric bronchial epithelium, the primary target of infection in vivo. Proc Natl Acad Sci USA 109(13):5040–5045

Welliver TP, Reed JL, Welliver RC Sr (2008) Respiratory syncytial virus and influenza virus infections: observations from tissues of fatal infant cases. Pediatr Infect Dis J 27(10 Suppl):S92–S96

Wright PF, Ikizler MR, Gonzales RA, Carroll KN, Johnson JE, Werkhaven JA (2005) Growth of respiratory syncytial virus in primary epithelial cells from the human respiratory tract. J Virol 79(13):8651–8654

Zabner J, Seiler M, Walters R, Kotin RM, Fulgeras W, Davidson BL, Chiorini JA (2000) Adeno-associated virus type 5 (AAV5) but not AAV2 binds to the apical surfaces of airway epithelia and facilitates gene transfer. J Virol 74(8):3852–3858

Zhang L, Bukreyev A, Thompson CI, Watson B, Peeples ME, Collins PL, Pickles RJ (2005) Infection of ciliated cells by human parainfluenza virus type 3 in an in vitro model of human airway epithelium. J Virol 79(2):1113–1124

Zhang L, Button B, Gabriel SE, Burkett S, Yan Y, Skiadopoulos MH, Dang YL, Vogel LN, McKay T, Mengos A, Boucher RC, Collins PL, Pickles RJ (2009) CFTR delivery to 25 % of surface epithelial cells restores normal rates of mucus transport to human cystic fibrosis airway epithelium. PLoS Biol 7(7):e1000155

Zhang L, Collins PL, Lamb RA, Pickles RJ (2012) Comparison of differing cytopathic effects in human airway epithelium of parainfluenza virus 5 (W3A), parainfluenza virus type 3, and respiratory syncytial virus. Virology 421(1):67–77

Zhang L, Peeples ME, Boucher RC, Collins PL, Pickles RJ (2002) Respiratory syncytial virus infection of human airway epithelial cells is polarized, specific to ciliated cells, and without obvious cytopathology. J Virol 76(11):5654–5666

Part V
Conclusions

Challenges and Opportunities for Respiratory Syncytial Virus Vaccines

Barney S. Graham and Larry J. Anderson

Abstract Respiratory syncytial virus (RSV) causes a significant proportion of the global burden of respiratory disease. Here we summarize the conclusions of a series of chapters written by investigators describing and interpreting what is known about the virology, clinical manifestations, immunity, pathogenesis, and epidemiology of RSV relevant to vaccine development. Several technological and conceptual advances have recently occurred that make RSV vaccine development more feasible, and this collected knowledge is intended to help inform and organize the future contributions of funding agencies, scientists, regulatory agencies, and policy makers that will be needed to achieve the goal of a safe, effective, and accessible vaccine to prevent RSV-associated disease.

Contents

B. S. Graham (✉)
Vaccine Research Center, NIAID, NIH, Bethesda, MD, USA
e-mail: bgraham@nih.gov

L. J. Anderson
Emory University Department of Pediatrics and Children's Healthcare of Atlanta, Atlanta, GA, USA

L. J. Anderson and B. S. Graham (eds.), *Challenges and Opportunities for Respiratory Syncytial Virus Vaccines*, Current Topics in Microbiology and Immunology 372,
DOI: 10.1007/978-3-642-38919-1_20,

1 Introduction

In this collection of brief reviews, we have attempted to consolidate much of the existing data relevant to the development of effective RSV vaccines. In addition, we sought to capture current views and opinions of the leaders in their respective disciplines to help frame the major issues important for developing new candidate vaccines and for navigating the regulatory pathways into clinical trials. RSV has been with us for a long time and continues to fill our pediatric hospital wards during each wintertime epidemic. The pathology and clinical syndrome of epidemic RSV bronchiolitis were probably first described in 1941 by Adams (1941), the virus was first discovered as chimpanzee coryza agent in 1956 (Blount et al. 1956), and RSV was identified as the major cause of bronchiolitis in infants in 1957 by Robert Chanock (Chanock et al. 1957). RSV is a global pathogen, causing yearly wintertime epidemics in temperate climates and more unpredictable and continuous outbreaks in tropical climates generally associating with rainy seasons (see chapter by C.B. Hall et al., this volume). Despite the global disease burden and extended time since its discovery, we still have not developed an effective vaccine for RSV. The reason for assembling these interpretive reviews at this time is based on a confluence of scientific events that have created the opportunity for an effective RSV vaccine to finally be realized.

2 Opportunities for Success

There has been much recent work on the clinical and molecular epidemiology of RSV on a global level including data from developing countries. These studies have confirmed the magnitude of the RSV-associated disease burden and the scope and dynamics of RSV genetic diversity. Second, the combined efforts of many groups over the last 3 decades have resulted in a better understanding of the vaccine-enhanced disease that occurred when children were immunized with formalin-inactivated alum-precipitated whole RSV vaccine in the 1960s. These studies, largely conducted in animal models provide immunological parameters and biomarkers that can help estimate the likely safety or potential risk of a candidate vaccine. Third, advances in RSV virology, particularly the development of reverse genetics and understanding of virus assembly and architecture, have provided more precise understanding of the specific roles of individual RSV proteins in the virus life cycle and immune evasion, and have provided the basis for multiple potential vaccine approaches. Fourth, new technologies used to rapidly isolate new human antibodies and breakthroughs in the structure of the RSV F glycoprotein have provided a blueprint for designing better vaccine antigens. Fifth, advances in immunology have suggested new vaccine formulation strategies for achieving protective immunity in the settings of immature and senescent immune responses. Understanding the immunological limitations of the very young and very old is especially critical for RSV because these groups experience the greatest disease severity. Sixth, technological advances in gene delivery and the ability to construct and manufacture a variety of gene-based vaccine vectors allows more selective control over the specificity and pattern of vaccine-induced immune responses than more traditional vaccine approaches based on whole virus.

2.1 RSV is a Global Disease

Physicians, epidemiologists, and virologists have always recognized that RSV was a ubiquitous pathogen and caused annual global epidemics. Recently, because of efforts of a few investigators, the availability of multiplex PCR and other rapid diagnostics, and improved surveillance for respiratory viruses in general due to heightened awareness and investment fueled by outbreaks of SARS, avian influenza, and pandemic influenza, there are more data documenting the importance of RSV as a respiratory pathogen in developing countries (see chapter by C. B. Hall et al., this volume). These advances have been punctuated by publications estimating the global disease burden (Nair et al. 2010) and mortality (Lozano et al. 2012) caused by RSV. The work on RSV in developing countries has demonstrated that the seasonality and age of peak illness severity may vary based on geographic and climatic parameters and should be taken into consideration in the planning of trials and development of target product profiles.

2.2 FI-RSV Vaccine-Enhanced Illness

Data from animal models and human immunology and pathology have provided guidance for minimizing the likelihood of inducing RSV vaccine-enhanced disease in future trials (see chapters by P.L. Collins et al., S.M. Varga and T.J. Braciale, G. Taylor, M.S. Boukhvalova and J.C.G. Blanco, and by P.J. Openshaw, this volume). These findings have been supported by more recent work on human genetic polymorphisms and transcriptional profiles associated with severe RSV disease (see chapters by E.H. Choi et al. and by R.A. Tripp et al., this volume). In aggregate, the data suggest that two major immunological goals should be achieved before advancing a candidate RSV vaccine into sero-negative infants.

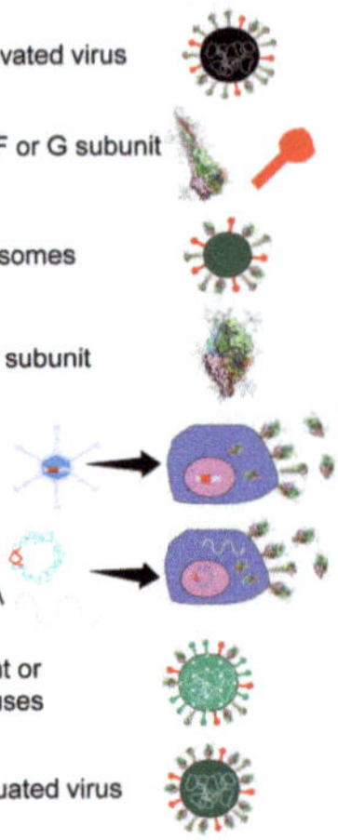

	Neutralizing epitopes	MHC pathway	CD8 T cell induction	IL-4	Delivery route	Immune modulation	Replication competence
Whole-inactivated virus	+/-	II	-	++	IM	+/-	-
Post-fusion F or G subunit	+	II	-	+/-	IM	-/+	-
VLPs or virosomes	+	II +/- I	+/-	-/+	IM	-/+	-
Pre-fusion F subunit	+++	II	-	+/-	IM	-	-
Vectors	++	I & II	++	-	IM or nasal	-	- or +
Naked DNA or RNA	++	I & II	+	-	IM	-	-
Recombinant or chimeric viruses	++	I & II	+	-	nasal	+/-	+
WT or attenuated virus	++	I & II	+	-	nasal or IM	+	+

Fig. 1 Biological profiles of candidate RSV vaccines. When new vaccine candidates emerge they will be compared to the FI-RSV vaccine associated with enhanced disease. Instead of categorizing vaccines as "killed" or "live" there should be a more precise biological profile described. There will be nuances in each product that could distinguish it from the FI-RSV. With new antigen designs that display targets for potent neutralizing antibody, modern adjuvants with established safety databases, and new vaccine antigen delivery approaches, there should be acceptable and rational avenues for moving several new RSV vaccine approaches into sero-negative infants where the need for protective immunity is the greatest. The chart depicts F being expressed by gene-based vectors. It is representative of other vaccine antigens that could be chosen, just as the recombinant adenovirus vector is representative of other potential gene delivery vectors. The categories shown are not exhaustive, but illustrate some of the properties that can determine the safety and immunogenicity of a candidate vaccine. "Neutralizing epitopes" refer to the likelihood the vaccine approach will induce antibodies against all or some of the known neutralizing epitopes. "MHC pathway" indicates the major antigen processing and presentation pathway engaged by the vaccine. "CD8 T cell induction" is the relative potency for the vaccine platform to generate this response. "IL-4" is a representative term for the potential for inducing Th2-type cytokines following RSV challenge after vaccination. "Immune modulation" indicates the potential for the representative vaccine to contain elements that alter or avoid RSV-induced immune responses based on in vitro or animal model data. "Delivery route" and "replication competence" are self-explanatory

First, potent neutralizing antibody should be induced so that virus spread and antigen load are controlled, and binding antibody that does not neutralize virus should be minimized to avoid immune complex deposition in airways. Secondly, Th2-biased CD4 T cell responses should be avoided to diminish the potential for allergic inflammation associated with airway hyperreactivity. With the advent of well-characterized vaccine antigens that can induce potent neutralizing activity and vaccine delivery systems that induce CD8 T-cells and a balanced Th1/Th2 response, these goals should be achievable. The immune responses induced by new candidate vaccines will need to be carefully evaluated and found to be distinct from those induced by FI-RSV to be tested in RSV naïve children (Fig. 1).

2.3 RSV Virology

The molecular mechanisms of RSV immune evasion and modulation are better described each year, and may eventually explain why durable immunity against infection is never achieved. These advances have been driven largely by the development of reverse genetics and ability to construct molecular clones of virus that can be selectively engineered to answer questions about viral pathogenesis (see chapter by P.L. Collins et al., this volume). Viral proteins like NS1, NS2, and G have multiple effects on cells responsible for initiating immune responses (see chapters by S. Mukherjee and N.W. Lukacs, and by S. Barik, this volume), and suggest that vaccines that elicit responses that block or avoid the immunomodulation associated with wild-type RSV infection without enhancing disease could provide more potent and durable immunity than natural infection. Improving on immunity induced by natural infection has the potential to affect transmission efficiency, and could have a profound effect on the ecology of RSV which seems to depend largely on the niche of very young infants with immature immune systems (see chapter by A.M.W Malloy et al., this volume).

2.4 RSV Structure

Structural biology is changing the approach to antigen design for many vaccine development programs (Kwong et al. 2012). RSV vaccine development may particularly benefit from defining the atomic structure of antigenic sites on the fusion (F) glycoprotein (see chapter by J.S. McLellan et al., this volume). The structures for three of the four known antigenic sites on the F glycoprotein associated with neutralization have now been solved by x-ray crystallography. Most recently the structure of the pre-fusion F trimer has been solved and the atomic structure revealed "antigenic site ∅" which is at the apex of the native trimer and the binding site for a new group of extremely potent monoclonal antibodies (McLellan et al. 2013). Antibodies to antigenic site ∅ may explain the observation that most neutralizing activity in human serum cannot be absorbed by the post-fusion F protein (Magro et al. 2012). Live-attenuated virus vaccines and gene-based vectors expressing the full-length F will express this vulnerable antigenic site, but prior subunit protein products have exclusively been in the post-fusion form of F. It is not known whether the FI-RSV vaccine product expressed antigenic site ∅, but that question and many others can now be addressed with the new reagents coming from this work.

2.5 Human Immunology

Advances in immunology have impacted several aspects of RSV vaccine development. Studies to define the antibody repertoire in young infants responding to either rotavirus or RSV have shown that infants do not have significant somatic mutation until 4–5 months of age, which limits affinity maturation (see chapters by S.M. Varga and T.J. Braciale and by A.M.W. Malloy et al., this volume). For infants infected with RSV early in life this would result in expanding the precursor frequency of B cells with relatively low affinity for RSV antigens. This may compromise the potency of vaccine-induced neutralizing antibodies, and raises the question of whether immunizing children beginning at 6 months of age would be a better strategy for achieving durable immunity. The improved understanding of Toll-like receptors (TLRs) and other receptors for pathogen-associated molecular patterns (PAMPS) has led to a better understanding of how RSV disables the innate immune system (see chapters by S. Mukherjee and N.W. Lukacs and by S. Barik, this volume). It has also led to new adjuvants approaches that may help in overcoming the immunological immaturity of the neonate and senescence of the elderly, which have been impediments to effective immunization in these target populations. In addition, we now have the ability to quantify and characterize the immune response patterns of human T cells using a variety of flow cytometric techniques to better recognize vaccine responses likely to be safe and those that may present risk.

2.6 Gene-Based Vaccine Vectors

Largely driven by the investment in HIV, malaria, and Tb vaccine development, a variety of approaches are now available for delivery of genes encoding vaccine antigens (see chapter by R.J. Loomis and P.R. Johnson, this volume). Although originally designed for inducing T cell-mediated immunity, many vaccine vectors or combinations can elicit substantial antibody responses. The great advantage of this type of vaccination is that it mimics live virus infection in that vaccine antigens are produced and presented by host cells, an immunization approach known to be safe in sero-negative infants. An added advantage to gene-based vaccination over live virus infection is that virus proteins important for inducing protective immunity can be selectively included, while excluding proteins associated with immunomodulation or other undesirable properties. Thus, vaccine vectors could potentially be designed to improve upon natural immunity induced by RSV infection. An illustration of this occurred in experiments performed over 20 years ago in which chimpanzees were immunized with recombinant vaccinia vectors expressing F or G prior to RSV challenge. Although the vaccinia recombinant did not induce a potent primary immune response and did not prevent infection, the neutralizing antibody responses post-RSV challenges were among the highest ever recorded (Collins et al. 1990). This observation suggests that if a properly designed RSV vaccine antigen could be appropriately delivered prior to the first RSV infection, a much more robust and durable immunity may be achievable.

3 Remaining Challenges

There are other issues relevant to RSV that need more attention to fill our gaps in knowledge and to provide the experimental tools and intellectual framework needed to complete the job of RSV vaccine development. These include the need for improved animal models, better understanding of mucosal immunity, more definitive clinical endpoints to use in efficacy trials, alternate vaccination strategies to protect the young infant (e.g., vaccinating pregnant women) and other high risk populations for whom vaccination may have limited effectiveness, and remedies for liability concerns.

3.1 Animal Models

Animal models have been critical for hypothesis generation related to the pathogenesis of RSV disease, FI-RSV vaccine-enhanced illness, and virus-induced airway hyperreactivity. Rodent models were also instrumental in the development and licensing of passive antibody prophylaxis for children at high risk of severe disease (see chapters by M.S. Boukhvalova and J.C.G. Blanco, and by P.J. Openshaw, this volume). However, all reported animal models of human RSV infection are semi-permissive except for chimpanzees. Chimps have been especially useful for gauging the attenuation of cold-adapted and temperature-sensitive strains and selection of live-attenuated virus vaccine candidates for clinical trials. African green monkeys are perhaps the next most permissive of the nonhuman primate but still require a large virus inoculum to establish a significant infection. Baboons have recently been reported to be susceptible to RSV but also require relative large inocula to establish modest levels of infection (Papin et al. 2013). The rodent models (mice and cotton rats) can be informative about the patterns of immune response and pathology following immunization and challenge, but neither system recapitulates the sequence of upper airway infection and spread to the lower airway that occurs over about 3 days. The large amount of virus and large volume of the inoculum required to infect the lower airway in current animal models essentially bypasses the first 3 days of natural infection. Using pneumoviruses matched to their natural host (e.g., PVM in mice or bovine RSV in cattle or sheep) may be a more authentic model of natural infection, but still not necessarily a reliable model for evaluating safety and efficacy of human RSV vaccines (see chapter by G. Taylor, this volume). Therefore, animal model data cannot guarantee the safety, immunogenicity, or efficacy of a candidate RSV vaccine in humans. Clinical trials in the target population, at this time, are the only way to determine which vaccines should be advanced to licensure. Since animal models will still be used for rank ordering and for determining "no-go" decisions when an inappropriate immune response is encountered, there are some important factors that should be remembered in interpreting results. First, the preparation of viral stocks is critical. The cell line used for production, the level of fetal bovine serum and other factors in the final media, the quality of the stock in terms of defective interfering particles, and minimizing the amount of cytokines and other biologically active substances in the final preparation can affect immune response patterns and pathology. Use of purified or semi-purified challenge virus should minimize the above noted factors that might confound results. Second, the possibility of enhanced disease after vaccination cannot be adequately assessed unless challenge is associated with some virus replication. The titer of RSV at the peak of the replication post challenge in untreated animals should be at least 10e5 pfu/gram lung, or the vaccine effect will be difficult to assess. Finally, it is important to consider sample type. For example, the expected type of inflammatory infiltrate is different for bronchoalveolar lavage specimens (primarily includes cells present in the larger airways) or from homogenized lung tissue (includes cells from intra-epithelial and lower airways of the lung).

3.2 Mucosal Immunology

One of the key decisions a vaccine developer has to make is whether induction of vaccine-elicited immune responses will occur systemically or mucosally. The vaccines that have advanced into sero-negative infants are live-attenuated viruses delivered intranasally. Some of the other vaccine approaches in development can be delivered mucosally, but most subunit proteins or vaccine vectors will be delivered intramuscularly (IM). Palivizumab provides the proof-of-concept that serum antibody with a sufficient level of neutralizing activity can prevent severe disease in infants at high risk. However, palivizumab does not prevent infection of the upper airways and it is not known if it alters transmission. A vaccine that decreases transmission has the potential to be given in contacts of high risk persons to decrease their risk of infection. There is also, very little known about the role of mucins in viral clearance or in interactions with secreted antibodies. Muc5ac is known to be induced in airways of animals infected with RSV, and is associated with airway hyperreactivity (see chapter by M.T. Lotz et al., this volume). The RSV G glycoprotein is heavily O-glycosylated and is rich in proline, serine, and threonine, so its chemical composition more resembles Muc proteins than a typical viral glycoprotein. It seems likely that a better understanding of how Muc proteins interact with RSV and with RSV-specific antibodies and a more complete understanding of the role of the RSV G glycoprotein in the airway may help guide vaccine development.

3.3 Target Populations

The major goals for an RSV vaccine are to prevent severe disease in young infants, to establish durable protection against reinfection, and to reduce excess mortality in the elderly. There are several potential approaches to achieve these goals without focusing directly on the neonate as the primary target population for vaccination. In some respects, immunizing neonates may be counter-productive if ineffective immune response patterns are established that are perpetuated throughout life. More work is needed to determine whether vaccination of infants beginning at 6 months of age, when about 70% of infants are still uninfected, could be a more effective strategy for achieving immunity in the general population. That would allow the vaccine to be the first RSV antigen exposure in the majority of infants at a time when effective somatic hypermutation was occurring, the Th2-biased tendencies of the infant are waning, and antigen presenting cells have a more adult-like phenotype. It is possible that establishing effective immunity in the majority of children or immunizing the parents and siblings of young infants could reduce the exposure of newborns to RSV resulting in a progressively larger number of uninfected children reaching the 6 month vaccination time point. More data on the epidemiology of transmission and more sophisticated mathematical modeling of transmission are needed, especially in developing country settings. Understanding transmission and mathematically modeling were a critical part of the successful immunization campaign to eliminate rinderpest, another paramyxovirus (Mariner et al. 2012). Immunizing pregnant women is another approach to protect infants by passive transfer of antibody from mother to children transplacentally or through breast milk. Vaccines for the various target populations will require different considerations and potentially different vaccine formulations as outlined in Table 1.

3.4 Clinical Trial Designs

There is a rich pipeline of candidate RSV vaccines and other prevention approaches in various stages of development (see chapters by H.Y. Chu and J.A. Englund , R.A. Karron et al., T.G. Morrison and E.E. Walsh, and by R.J. Loomis and P.R. Johnson, this volume, and Fig. 1). Since animal models are limited, and there are many potential target populations, the clinical development plans and clinical trial designs will be critical for achieving success. There are several issues to consider such as the seasonality of RSV which dictates when vaccine trials should be timed and makes North–South clinical collaborations appealing. In tropical regions where there is not distinct seasonality, the timing of trials will require more consideration. Having clear diagnostic and disease severity endpoints to judge efficacy are essential to determine the size and expense of the studies and ultimately the safety and efficacy of the vaccine. For example, strict diagnostic criteria and establishing a composite index of illness severity will be needed for studies with multiple sites to account for site differences in managing patients. It would be helpful to also use similar criteria and composite indices across studies. Endpoints will need to be tailored to the target population. For example, for the sero-negative children, hospitalization has been used for passive antibody and vaccine trials. For

Table 1 Target populations for candidate RSV vaccines

Target populations	Subgroups	Challenges		Preferred approaches[a]
		Group-specific	Common	
Neonates and infants (<6 month)		• Immature dendritic cells, T-cells, and B cell capacity for somatic hypermutation • Maternal antibody • Breast-feeding • Idiosyncratic adverse respiratory events (e.g., apnea and airway hyperreactivity) • Small airway size predisposes to inflammation-induced obstruction • Th2-biased immune responses	Concern about vaccine-enhanced disease	• Live attenuated virus • Gene-based vector with WT or pre-fusion F • Pre-fusion F protein with adjuvant that avoids Th2 bias • Combine with vaccines for other childhood respiratory diseases like influenza, metapneumovirus, and parainfluenza viruses
Infants and children (>6 month)	Sero-negative			• Live attenuated virus • Gene-based vector with WT or pre-fusion F • Pre-fusion F protein with adjuvant that avoids Th2 bias • Pre-fusion F protein with potent adjuvant • Gene-based vector with WT F or pre-fusion F

(continued)

Table 1 (continued)

Target populations	Subgroups	Challenges		Preferred approaches[a]
		Group-specific	Common	
Infants and children (>6 month)	Sero-positive	Unknown infection status	-Repeated infections despite natural immunity	• Pre-fusion F protein with balanced or Th2-biasing adjuvant
Siblings and parents of neonates Young adult women	Child-bearing age Pregnant women	Ring vaccination may not work Responses may not be maintained adequately by the time pregnancy occurs Pregnancy-related toxicity	-Difficult to boost pre-existing antibody	
Elderly (>65 years)		Senescent immune responses Difficult efficacy endpoint		• Pre-fusion F protein with potent adjuvant • Gene-based vector with WT F or pre-fusion F • Combined gene-based vector boosted with pre-fusion F protein • Combine with influenza vaccine

[a] Gene-based vector could be replication-defective (e.g. recombinant adenovirus or alphavirus or adeno-associated virus vector). Live-attenuated virus could be RSV or chimeric parainfluenza or Newcastle disease virus. Pre-fusion F protein refers to either the soluble purified protein or the protein presented on a particle (e.g. ferritin or VLP). Pre-fusion F is listed as the "preferred" and simplest antigen choice since it includes antigenic site Ø in addition to neutralizing determinants on the post-fusion F. However, this should not be interpreted as excluding the potential value of post-fusion forms of F, WT F expressed in a gene-based vector, or additional vaccine antigens that may have value. Importantly, there is at least one additional target for broadly neutralizing antibody in G and in animal models antibodies to G reduce immunopathology. In addition, some vaccine approaches may benefit by the addition of genes for internal structural and regulatory proteins as a source of additional T cell epitopes or to add constructs designed to stabilize glycoprotein structure. It is reasonable to include these additional vaccine antigens if care is taken to avoid proteins that interfere with induction or maintenance of immunity or epitopes that may elicit immunopathological responses

older children, medically attended lower respiratory tract illness may be used. In adults, diagnosing RSV infection is more difficult and the presence of underlying disease makes assigning causality for acute respiratory illnesses less precise. It is possible that experimental human challenge studies may help understand vaccine efficacy in adults, especially the effect of vaccination on virus shedding and mild illness.

3.5 *Liability Concerns*

Because of the legacy of FI-RSV vaccine-enhanced illness and the fact that many of the key target populations for an RSV vaccine are especially vulnerable (young infants, pregnant women, and the frail elderly), a concern about liability has been a significant part of the risk:benefit analysis for companies contemplating RSV vaccine development. If concerns present a road block to vaccine development, it seems reasonable to explore legislative solutions for diminishing the financial risks to developers, investigators, and study participants by providing assurance of indemnification for unanticipated adverse events.

4 Conclusions

At this point in time, the opportunities for success far outweigh the remaining challenges for providing safe and effective vaccine-induced immunity to prevent RSV infection and to reduce the substantial disease burden that it still causes. To complete that task will require the sustained commitment of funding agencies, ongoing combined and organized efforts of government, university, and industry scientists across multiple disciplines, thoughtful guidance from regulatory agencies, and facilitation by prescient law and policy makers.

Acknowledgments We thank the authors of these collected chapters for their outstanding scholarship and commitment to finding vaccine approaches for preventing RSV-associated disease. We dedicate this book to the memory of Dr. Caroline Hall whose career was committed to the care of children and prevention of viral diseases. Her life embodied the mission of this book and her fingerprints will be evident in the final solution for controlling RSV. This work was supported by intramural funding from the National Institute of Allergy and Infectious Diseases.

References

Adams JM (1941) Primary virus pneumonitis with cytoplasmic inclusion bodies: study of an epidemic involving thirty-two infants, with nine deaths. J Am Med Assoc 116(10):925–933

Blount RE Jr, Morris JA, Savage RE (1956) Recovery of cytopathogenic agent from chimpanzees with coryza. Proc Soc Exp Biol Med 92(3):544–549

Chanock R, Finberg L (1957) Recovery from infants with respiratory illness of a virus related to chimpanzee coryza agent (CCA). II. Epidemiologic aspects of infection in infants and young children. Am j hyg 66(3):291–300

Chanock R, Roizman B, Myers R (1957) Recovery from infants with respiratory illness of a virus related to chimpanzee coryza agent (CCA). I. Isolation, properties and characterization. Am j hyg 66(3):281–290

Collins PL, Purcell RH, London WT, Lawrence LA, Chanock RM, Murphy BR (1990) Evaluation in chimpanzees of vaccinia virus recombinants that express the surface glycoproteins of human respiratory syncytial virus. Vaccine 8(2):164–168

Kwong PD, Mascola JR, Nabel GJ (2012) The changing face of HIV vaccine research. J Intl AIDS Soc 15(2):17407. doi:10.7448/IAS.15.2.17407

Lozano R, Naghavi M, Foreman K, Lim S, Shibuya K, Aboyans V, Abraham J, Adair T, Aggarwal R, Ahn SY, Alvarado M, Anderson HR, Anderson LM, Andrews KG, Atkinson C, Baddour LM, Barker-Collo S, Bartels DH, Bell ML, Benjamin EJ, Bennett D, Bhalla K, Bikbov B, Bin Abdulhak A, Birbeck G, Blyth F, Bolliger I, Boufous S, Bucello C, Burch M, Burney P, Carapetis J, Chen H, Chou D, Chugh SS, Coffeng LE, Colan SD, Colquhoun S, Colson KE, Condon J, Connor MD, Cooper LT, Corriere M, Cortinovis M, de Vaccaro KC, Couser W, Cowie BC, Criqui MH, Cross M, Dabhadkar KC, Dahodwala N, De Leo D, Degenhardt L, Delossantos A, Denenberg J, Des Jarlais DC, Dharmaratne SD, Dorsey ER, Driscoll T, Duber H, Ebel B, Erwin PJ, Espindola P, Ezzati M, Feigin V, Flaxman AD, Forouzanfar MH, Fowkes FG, Franklin R, Fransen M, Freeman MK, Gabriel SE, Gakidou E, Gaspari F, Gillum RF, Gonzalez-Medina D, Halasa YA, Haring D, Harrison JE, Havmoeller R, Hay RJ, Hoen B, Hotez PJ, Hoy D, Jacobsen KH, James SL, Jasrasaria R, Jayaraman S, Johns N, Karthikeyan G, Kassebaum N, Keren A, Khoo JP, Knowlton LM, Kobusingye O, Koranteng A, Krishnamurthi R, Lipnick M, Lipshultz SE, Ohno SL, Mabweijano J, MacIntyre MF, Mallinger L, March L, Marks GB, Marks R, Matsumori A, Matzopoulos R, Mayosi BM, McAnulty JH, McDermott MM, McGrath J, Mensah GA, Merriman TR, Michaud C, Miller M, Miller TR, Mock C, Mocumbi AO, Mokdad AA, Moran A, Mulholland K, Nair MN, Naldi L, Narayan KM, Nasseri K, Norman P, O'Donnell M, Omer SB, Ortblad K, Osborne R, Ozgediz D, Pahari B, Pandian JD, Rivero AP, Padilla RP, Perez-Ruiz F, Perico N, Phillips D, Pierce K, Pope CA, 3rd, Porrini E, Pourmalek F, Raju M, Ranganathan D, Rehm JT, Rein DB, Remuzzi G, Rivara FP, Roberts T, De Leon FR, Rosenfeld LC, Rushton L, Sacco RL, Salomon JA, Sampson U, Sanman E, Schwebel DC, Segui-Gomez M, Shepard DS, Singh D, Singleton J, Sliwa K, Smith E, Steer A, Taylor JA, Thomas B, Tleyjeh IM, Towbin JA, Truelsen T, Undurraga EA, Venketasubramanian N, Vijayakumar L, Vos T, Wagner GR, Wang M, Wang W, Watt K, Weinstock MA, Weintraub R, Wilkinson JD, Woolf AD, Wulf S, Yeh PH, Yip P, Zabetian A, Zheng ZJ, Lopez AD, Murray CJ, AlMazroa MA, Memish ZA (2012) Global and regional mortality from 235 causes of death for 20 age groups in 1990 and 2010: a systematic analysis for the global burden of disease study 2010. Lancet 380(9859):2095–2128. doi:10.1016/S0140-6736(12)61728-0

Magro M, Mas V, Chappell K, Vazquez M, Cano O, Luque D, Terron MC, Melero JA, Palomo C (2012) Neutralizing antibodies against the preactive form of respiratory syncytial virus fusion protein offer unique possibilities for clinical intervention. Proc Nat Acad Sci U.S.A 109(8):3089–3094. doi:10.1073/pnas.1115941109

Mariner JC, House JA, Mebus CA, Sollod AE, Chibeu D, Jones BA, Roeder PL, Admassu B, van 't Klooster GG (2012) Rinderpest eradication: appropriate technology and social innovations. Science 337(6100):1309–1312. doi:10.1126/science.1223805

McLellan JS, Chen M, Leung S, Graepel KW, Du X, Yang Y, Zhou T, Baxa U, Yasuda E, Beaumont T, Kumar A, Modjarrad K, Zheng Z, Zhao M, Xia N, Kwong PD, Graham BS (2013) Structure of RSV fusion glycoprotein trimer bound to a prefusion-specific neutralizing antibody. Science. doi:10.1126/science.1234914

Nair H, Nokes DJ, Gessner BD, Dherani M, Madhi SA, Singleton RJ, O'Brien KL, Roca A, Wright PF, Bruce N, Chandran A, Theodoratou E, Sutanto A, Sedyaningsih ER, Ngama M, Munywoki PK, Kartasasmita C, Simoes EA, Rudan I, Weber MW, Campbell H (2010) Global burden of acute lower respiratory infections due to respiratory syncytial virus in young children: a systematic review and meta-analysis. Lancet 375(9725):1545–1555. doi:10.1016/S0140-6736(10)60206-1

Papin JF, Wolf RF, Kosanke SD, Jenkins JD, Moore SN, Anderson MP, Welliver RC Sr (2013) Infant baboons infected with respiratory syncytial virus develop clinical and pathologic Changes that parallel those of human infants. Am J Physiol Lung Cell Mol Physiol. doi:10.1152/ajplung.00173.2012

Index

L. J. Anderson and B. S. Graham (eds.), *Challenges and Opportunities for Respiratory Syncytial Virus Vaccines*, Current Topics in Microbiology and Immunology 372, DOI: 10.1007/978-3-642-38919-1, © Springer-Verlag Berlin Heidelberg 2013

MIX
Papier aus verantwortungsvollen Quellen
Paper from responsible sources
FSC® C105338

If you have any concerns about our products,
you can contact us on
ProductSafety@springernature.com

In case Publisher is established outside the EU,
the EU authorized representative is:
Springer Nature Customer Service Center GmbH
Europaplatz 3, 69115 Heidelberg, Germany

Printed by Libri Plureos GmbH
in Hamburg, Germany